Life is pain.
anyone who says
different is
selling something...

Okay

Engineering Psychology and Human Performance

SECOND EDITION

CHRISTOPHER D. WICKENS

University of Illinois at Champaign-Urbana

HarperCollins*Publishers*

Sponsoring Editor: Laura Pearson
Project Editor: Claire M. Caterer
Design Supervisor: Heather A. Ziegler
Cover Design: Wanda Lubelska Design
Production Administrator: Jeffrey Taub
Compositor: Digitype, Inc.
Printer and Binder: R. R. Donnelley & Sons Company
Cover Printer: New England Book Components, Inc.

Engineering Psychology and Human Performance, Second Edition
Copyright © 1992 by HarperCollins Publishers Inc.

Library of Congress Cataloging-in-Publication Data

Wickens, Christopher D.
    Engineering psychology and human performance / Christopher D.
Wickens. — 2nd ed.
        p.   cm.
    Includes bibliographical references and indexes.
    ISBN 0-673-46161-0
    1. Human engineering.   2. Man-machine systems. 3. Psychology,
Industrial.   I. Title.
TA166.W53   1992   91-23579
620.8′2 — dc20   CIP

        92 93 94   9 8 7 6 5 4 3 2

Brief Contents

Contents

done

done

Peoria
museum
w/center
of solar
system

Preface

I wrote this book, both the first and second editions, because I saw a need to bridge the gap between the problems of system design and much of the excellent theoretical research in cognitive experimental psychology and human performance. Many human-machine systems do not work as well as they could because they impose requirements on the human user that are incompatible with the way a person attends, perceives, thinks, remembers, decides, and responds, that is, the way in which a person processes information. Over the past five decades, tremendous gains have been made in understanding and modeling human information processing. My goal is to show how these theoretical advances have been, or might be, applied to improving human-machine interactions.

Although engineers encountering system design problems may find some answers or guidelines either implicitly or explicitly stated in this book, it is not a handbook of human factors or engineering psychology. References in the text provide a more encyclopedic tabulation of such guidelines. Instead, I have organized the book directly from the perspective of human information processing. The chapters generally correspond to the flow of information as it is processed by a human being and are less clearly organized from the perspective of different system components or engineering concerns, such as displays, illumination, controls, computers, and keyboards. Furthermore, although the following pages contain recommendations for certain system design principles, many of these are based only on laboratory research and theory; they have not been tested in real-world systems.

It is my firm belief that a solid grasp of theory will provide a strong base from which the specific principles of good human factors can be more readily derived. My intended audience, therefore, is (1) the student in psychology, who will begin to recognize the relevance to many areas in the real-world applications of the theoretical principles of psychology that he or she may have encountered in other courses; (2) the engineering student, who while learning to design and build systems with which humans interact, will come to appreciate not only the nature of human limitations — the essence of human factors — but also the theoretical principles of human performance and information processing underlying them; and (3) the actual practitioner in engineering psychology, human performance, and human factors engineering, who should understand the close cooperation that should exist between principles and theories of psychology and issues in system design.

The 12 chapters of the book span a wide range of human performance components. Following the introduction in Chapter 1, in which engineering psychology is put into the broader framework of human factors and system design, Chapters 2 through 7 deal with perception, attention, memory, learning, and decision making, emphasizing the potential applications of these areas of cognitive psychology. Chapters 8 through 11 cover reaction time, action, time-sharing, error, stress, and manual control, thereby addressing areas that are more traditionally associated with the engineering field. Finally, Chapter 12 is systems-oriented, discussing process control and automation. This chapter shows how many of the principles explained in earlier chapters are pertinent to one specific application of rapidly growing importance.

Although the 12 chapters are interrelated (just as are the components of human information processing), I have constructed them in such a way that any chapter may be deleted from a course syllabus and still leave a coherent body. Thus, a course on applied cognitive psychology might include Chapters 1 through 7, and another emphasizing strictly engineering applications might include Chapters 1, 2, 4, 5, 8, 9, 10, 11, and 12.

I have made a number of changes in the second edition that set it apart from the first. First, my increased involvement in applied aviation research since the publication of the first edition has made me increasingly aware of the need for psychologists to articulate more clearly the *implications* of their theories to system design. Hence, I have shifted the nature of the coverage somewhat toward an applications orientation, and I have deleted some of the more theoretical material whose practical implications in the first edition were less clear. Thus, I hope the book will serve engineering-oriented audiences better than did the first edition. In particular, Chapter 1 expands on the engineering orientation to the field.

Second, with one exception (Chapter 11), all chapter supplements have been deleted and the relevant material from the supplements integrated into the chapter text. The supplement to Chapter 11 remains because of its fairly technical nature.

Third, I have made substantial additions in a number of areas. Chapter 10, discussing stress and human error, is new, and half of Chapter 6 is now devoted to learning and training, an area that was neglected in the first edition. Chapter 3 (attention and display perception), Chapter 4 (spatial perception and cognition), and Chapter 12 (process control and automation) have been nearly totally rewritten. Chapter 3 contains major sections on eye movements and color coding; Chapter 4, substantial material on three-dimensional displays; and Chapter 12, a reorganized treatment of process control and an added section on expert systems. In addition, substantial expansions are included in Chapter 5 (dealing with communications), Chapter 7 (risk perception), Chapter 8 (compatibility), and Chapter 11 (control devices).

In any project of this kind, one is indebted to numerous people for assistance. In my case, the list includes my family, especially my wife, Linda, as well as several colleagues who have read and commented on various chapters, have provided feedback on the first edition, or have stimulated my thinking. In addition to all acknowledgments in the first edition (the text of which, of course, remains very much at the core of the current book), I would like to acknowledge the contributions of Art Kramer, Gavan Lintern, Pamela Tsang, Kim Vincente, Dennis Beringer, Jim Buck, Bruce Coury, and countless students who, in one form or another, have offered feedback regarding either bad or good sections of the first edition. In particular, I would like to single out the extensive and helpful feedback on the second edition offered by Neville Moray, Sallie Gordon, and Mike Venturino. Much credit for any improvements in the clarity, organization, or accuracy of coverage must surely go to these people.

Four specific individuals have contributed to the development of my interest in engineering psychology: My father, Delos Wickens, stimulated my early interest in experimental psychology; Dick Pew introduced me to academic research in engineering psychology and human performance; Stan Roscoe pointed out the importance of good research applications to system design; and Emanuel Donchin continues to emphasize the importance of solid theoretical and empirical research. Finally, it is impossible to do justice in crediting Karen Ayers's and Mary Welborn's contributions to this book. Without their hours of dedication at the word processor of a sometimes hostile computer, the project never would have succeeded.

Christopher D. Wickens

Chapter

1

Introduction to Engineering Psychology and Human Performance

At 4:00 A.M. on March 28, 1979, near Harrisburg, Pennsylvania, a temporary clog in the feedwater lines of Turbine No. 1 at the Three Mile Island nuclear power plant caused a rapid automatic shutdown of the feedwater pump and the turbine. A fraction of a second later, the redundant safeguards that are built into such systems functioned normally to supply an alternate source of feedwater. Immediately, four critical errors converged, demonstrating as never before how vulnerable is the human link in the performance of complex systems.

The first error had been committed before the clog developed. The pipe for the alternate feedwater supply had been blocked off during maintenance by personnel who had since gone off duty. As a result, the radioactive core, no longer receiving a continuous supply of cold water to remove its heat, began to increase in temperature, turning the surrounding coolant to vapor, and the pressure increased rapidly.

Automatic safeguards, however, continued to function properly. Cobalt rods descended into the core to slow down the process, and a pressure relief valve opened to "bleed off" some of the high-pressure overheated vapor from the primary cooling loop. Once pressure was reduced below the critical level, the automatic relief valve received a signal to close in the same manner that a furnace receives a signal from the thermostat to shut down when it has heated the room sufficiently. At this point the second error occurred. Because of a malfunction, the valve did not close.

1

Within a minute after the original shutdown, the supervisory crew at Three Mile Island was attempting to understand what was going on from a myriad of alarms, lights, and signals on massive display panels. Though their training allowed them to get a fairly accurate picture, they were led astray by one signal. The display for the pressure relief valve was designed to indicate what the valve was commanded to do rather than what it actually did, and the display indicated that the pressure relief valve had closed. This was the third critical error in the sequence.

Meanwhile, the pattern of redundant automated safeguards that are a mainstay of such complex systems continued to operate, and an emergency pump switched on to supply the system with a now badly needed source of coolant. At this point, the supervisors made a decision that turned what might have been a relatively minor incident into a major catastrophe. With instruments showing that pressure was already high and that the relief valve had closed, the operators decided to override the controls manually and shut off the emergency pump. This was the fourth error in the sequence: a decision based on their inference that coolant level was excessive rather than too low. The core was thereby deprived of the vital cooling it needed, and the incident soon accelerated to the point of no return.

This description of the Three Mile Island incident highlights only certain of the primary contributing factors. Rubinstein and Mason (1979) provide a fascinating, detailed, event-by-event account of the disaster (see also Adams, 1989). In the exhaustive inquiries, investigations, and hearings that followed the crisis, three things became clear. First, no single fault, mistake, event, or malfunction *caused* Three Mile Island. Rather, responsibility was distributed across a number of sources. Second, human error was involved at several levels, from the incorrect decision to shut down the emergency coolant to the human decisions in the design of a relief valve system that tells operators what the valve was commanded to do, not what it did. Third, and most important, the overwhelming complexity of information presented to the human operators and the confusing format in which it was displayed was probably sufficient to guarantee that somewhere along the line, the intrinsic limitations of human abilities to attend, perceive, remember, decide, and act—that is, to process information—would be overloaded (Electrical Power Research Institute [EPRI], 1977; Sheridan, 1981). Thus, in Three Mile Island, as in so many other accidents involving complex systems (Woods, O'Brian, & Hanes, 1987), although human error may be attributed as a cause (Reason, 1990), the human operators themselves are not at fault (Hurst & Hurst, 1985).

A major thesis of this book is that blame must be placed instead on the design of systems that overload human information-processing capabilities. Our own experiences offer many other, less dramatic examples of system-induced human error. One common irritation is mistakenly turning on a car's headlights instead of its windshield wipers because the controls are identically shaped and placed side by side. Making a mistake in operating a household appliance because of poorly worded instructions or looking up a ten-digit area code and phone number and then not remembering it long enough to finish dialing also are examples of problems arising from system designs that exceed human capacities.

One major purpose of this book is to examine human capabilities and limitations in the specific area of information processing. The second purpose is to demonstrate how knowledge of these limitations can be applied in the design of complex systems with which humans interact.

tracks 4

ENGINEERING PSYCHOLOGY AND HUMAN FACTORS

Designing machines that accommodate the limits of the human user is the concern of a field referred to as *human factors*. Meister (1989) defines human factors as "the study of how humans accomplish work-related tasks in the context of human-machine system operation, and how behavioral and nonbehavioral variables affect that accomplishment" (p. 2). The field is a very broad one—broader than this book, which is focused specifically on designing systems that accommodate the information-processing capabilities of the brain. Many of the principles that are important for designing machines that humans use, and that therefore belong in the province of human factors, are not related to this particular facet of human capacity. For example, designing an automobile's dashboard in such a way that all controls can be reached easily and manipulated without muscle fatigue, and all displays are visible without straining the neck, is a human factors concern. This design problem, however, must respond to the physical properties and constraints of the driver's body, not the brain's information-processing capabilities. These problems are real and legitimate concerns, but they are not within the purview of this text.

A focus on the information-processing capacities of the human brain, then, is a key characteristic delimiting the scope and contents of this book. Study of the processes of the brain falls within the realm of psychology. It is the discipline of engineering psychology, specifically, that applies a psychological perspective to the problems of system design.

Among the notable features of engineering psychology as it has emerged as a discipline in the last four decades are its solid theoretical basis (Howell and Goldstein, 1971) and its close relation to the discipline of experimental psychology. Although engineering psychology is similar to both human factors and experimental psychology, its goals are nevertheless unique. The goal of experimental psychology is to uncover the laws of behavior through experiments. However, the design of these experiments is unconstrained by a requirement to apply the laws. That is, it is not required that experiments generate immediately useful information. The goal of human factors, on the other hand, is to apply knowledge in designing systems that work, accommodating the limits of human performance *and* exploiting the advantages of the human operator in the process. Engineering psychology arises from the intersection of these two domains. "The aim of engineering psychology is not simply to compare two possible designs for a piece of equipment [which is the role of human factors], but to specify the capacities and limitations of the human [generate an experimental data base] from which the choice of a better design should be directly deducible" (Poulton, 1966, p. 178). That is, although research topics in engineering psychology are selected because of applied needs, the research goes beyond specific one-time applications and has the broader objective of providing a usable theory of human performance.

It should be reemphasized that the decision to exclude certain areas of human factors does not reflect a view that they are less important. In fact, much of the research in these areas is potentially more informative about actual design modifications and decisions than is research in human performance. The reader is referred to textbooks by Adams (1989), Bailey (1989), Boff and Lincoln (1988), Kantowitz and Sorkin (1983), Meister (1986), Salvendy (1987), and Sanders and McCormick (1987) for more detailed treatments of these areas. This book is intended to complement, rather than supplement or replace, these treatments. However, the following section provides a brief overview of the nature of these broader concerns.

The chapter will then conclude with a general model of the human as an information processor who attends to, perceives, translates, stores, and responds to events in the environment. This model provides a schematic framework for the chapters that follow, most of which examine component processes within the model in detail and consider their implications for system design.

A Brief History

Impetus for the development of human factors and engineering psychology as disciplines has arisen from three general sources: practical needs, technological advancements, and linguistic developments. Before the birth of human factors, or ergonomics, in World War II, emphasis was placed on "designing the human to fit the machine." That is, the emphasis was on training. Experience in World War II, however, revealed a number of instances in which systems, even with well-trained operators, simply weren't working. Airplanes were flying into the ground with no apparent mechanical failures; enemy contacts were missed on radar by highly motivated monitors. As a consequence, experimental psychologists from both sides of the Atlantic were brought in to analyze the operator-machine interface, to diagnose what was wrong, and to recommend the solution (Fitts & Jones, 1947; Mackworth, 1948). This represented the practical need underlying the origin of human factors engineering.

A second motivation has come from evolutionary trends in technology. With increased technological development in this century, systems have become increasingly complex, with more and more interrelated elements, forcing the designer to consider the distribution of tasks between human and machine. This problem has led system designers to consider what functions to allocate to humans and what to machines, which in turn necessitates a close analysis of human performance in different kinds of tasks. At the same time, with increased technology, the physical parameters of all systems have grown geometrically. For example, consider the increases in the maximum velocity of vehicles, progressing from the oxcart to the spacecraft; in the temperature range of energy systems, from fires to nuclear reactors; and in the physical size of vehicles, from wagons to supertankers and wide-bodied aircraft such as the Boeing 747. Particularly with regard to speed, this increase forces psychologists to analyze quite closely the operator's temporal limits of processing information. To the oxcart driver, a fraction of a second delay in responding to an environmental event will be of little consequence. But to the pilot of a supersonic aircraft, a delay of the same absolute magnitude may be critical in causing a collision.

Finally, a linguistic influence for the growth of human factors has come from the new language of information theory and cybernetics that after World War II began to replace the stimulus-response language of behavioral psychology as a description of human activities. Terms such as *feedback, channel capacity,* and *bandwidth* began to enter the descriptive language of human behavior. This new language enabled some aspects of human performance to be described in the same terms as the mechanical or electronic systems with which the operator was interacting. This shared terminology in turn facilitated the integration of humans and machines in system design and analysis.

The Human Factor in the Design Process

As we have noted, the study of engineering psychology is an integral part of the wider profession of human factors engineering. Human factors, in turn, is only one component (and sometimes a regrettably small one) of the total system design process. To understand how human factors and engineering psychology fit into the overall design process, it is important to consider first how this process is carried out. Whether in the construction of a revolutionary aircraft, in the development of an electronic banking system to handle customers' transactions, or in a nearly infinite variety of other modern systems with which a human operator must interact, there are certain common characteristics to the process of designing systems for the human user. In the following section, we will describe these characteristics as we examine a large map of the whole design process, focusing in more and more detail on the brain as an information-processing system—the focus of engineering psychology.

Mission and Function Analysis The initial phase of this design process is sometimes referred to as *mission analysis*. It describes the overall goals and objectives of the total system and the constraints under which it must operate. For example, a semiautomated electronic banking system for tellers' use might be proposed. Included in the analysis would be the range of transactions the system could handle, the maximum allowable delay, the amount of training users would need to have for operating it, and an upper limit on the cost per unit.

From here, a more detailed analysis, sometimes referred to as *function analysis*, identifies the specific sequence of functions (events, decisions, or information exchanges) that must be accomplished correctly for the mission to be successfully achieved. For example, to carry out the banking transaction properly, critical pieces of information (the account number and the nature of the transaction) have to be transmitted from the customer to the computer. This step may or may not require the intermediary of the teller. Information must be accessed from the bank's computer, authentication may be required, minimum time constraints are required for the information exchange, verification may be involved, privacy of the bank's response to the customer may need to be ensured, and various sorts of paperwork may need to change hands physically.

This information exchange flow is sometimes described by an *operational sequence diagram*, or OSD, an example of which is shown in Figure 1.1 for an airport's ground transportation information system (Bateman, 1979). The OSD essentially represents the information exchanges and decisions necessary to meet the system goals and requirements. These exchanges, or "transactions," are represented by specific symbols at the top of the figure. It is important to realize that the creation of an OSD does not by itself indicate what functions should be carried out by humans and what should be carried out by machines. It does not say, for example, that a bank transaction is better carried out by a human than by an automated teller. During the first three decades of the human factors profession, there were in fact strongly worded guidelines defining the kinds of tasks better performed by humans and those better performed by machines (Wickens, 1984). These guidelines were used to aid the designer in the allocation of functions. However, with the rapid growth in computer technology—allowing computers to become increasingly more proficient at a variety of tasks—such guidelines can no longer be offered with confidence (Kantowitz & Sorkin, 1987). For example, one standard guideline was that tasks involving perception and recognition of complex patterns should be performed by humans rather than machines. But the recent growth in computer vision research has now allowed computers to do a superior job in many sorts of pattern recognition tasks, particularly when compared with a human observer, who may suffer visual fatigue and periodic lapses of attention.

At present, the issue of what and how much to automate must in fact be considered from the standpoint of a number of other concerns besides simply who does what best. These include such factors as system cost, operator training time, or the level of trust in computer automation by the operator and customer. These factors will be considered in depth in Chapter 12, where we discuss the costs and benefits of automation.

Following completion of the OSD and an initial decision of what functions are to be carried out by the human operator, the next step in the design process is to ensure that the system is configured so that necessary information is available to human or machine at decision points in the system, at the times that information is needed. Indeed many complex systems fail because the flow of information is not managed in a proper manner. A traffic sign indicating a right turn may not be visible from a left-hand lane until it is too late to switch lanes. An airline's scheduling system may not have timely information on expected traffic delays, and so will schedule passengers for "impossible" connections. The critical importance of the OSD in this aspect of design lies in the fact that it

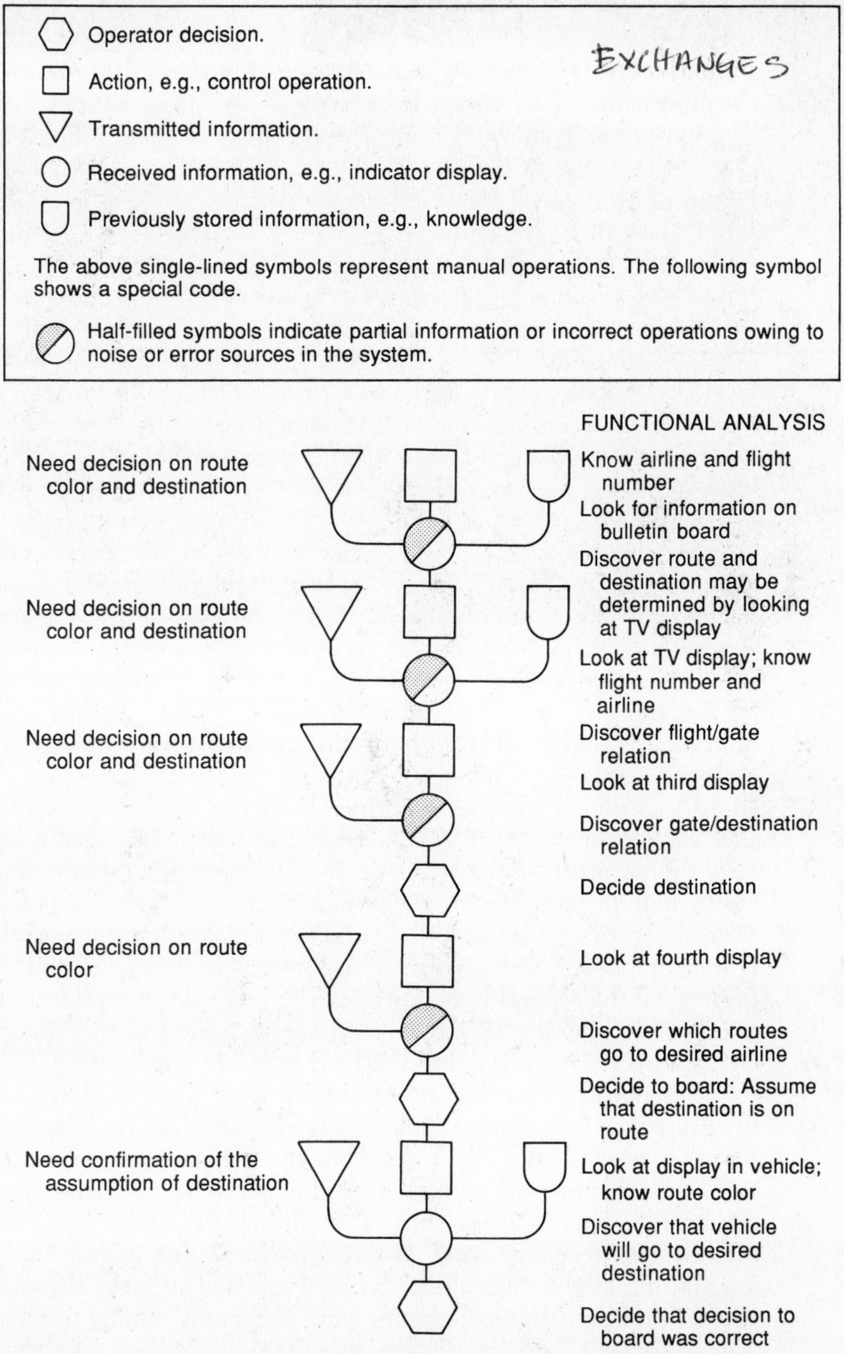

Figure 1.1 Operational sequence diagram for airport ground transportation information system. (*Source:* R. P. Bateman, "Design Evaluation of an Interface Between Passengers and an Automated Ground Transportation System." Reprinted with permission from *Proceedings of the Human Factors Society Annual Meeting* (1979). Copyright 1979 by The Human Factors Society, Inc. All rights reserved.)

allows for a structured physical reminder that forces the designer to consider the information flow. However, as we will see, good information flow is necessary but not sufficient to ensure the presence of good human factors.

Anthropometry and Biomechanics   Let us now focus on the human operator, sitting or standing in front of the "system" (for example, a computer terminal, a mechanical device, another human, or some combination of all three) and "exchanging information" with it to carry out the mission's goals. Even if careful attention to the OSD has ensured that necessary information is presented on the displays at the appropriate times and that controls (buttons, joysticks, and voice) are available for the exchange, there is no guarantee that this exchange will easily take place. Suppose the height of an aircraft seat is so low that pilots of short stature must strain their necks to see the approaching runway during a landing (Hawkins, 1987)? Suppose the operator of the machine tool, wishing to adjust a dial setting, must reach across (and therefore visually obscure) the dial indicator in order to turn the control knob. Or suppose a control knob is placed on the automobile dashboard so far away that 20 percent of the population must substantially shift the seating position to reach it?

These issues define the human factors field of *anthropometry*, the concern for designing systems that are compatible with the *physical constraints* of the human body — the line of sight of the eyes and the positioning and reach of the limbs. Closely related concerns, the *forces* imposed on the limbs by the need to control and manipulate, are addressed in the field of *biomechanics* (Chaffin, 1987; Kroemer, 1987). This field considers the complex interplay between the forces required to operate a piece of equipment and those that can be exerted by the muscles: Can the force necessary to rotate a valve or pull a lever located at an awkward spot under the seat be exerted by 95 percent of the population? Is the component that an aircraft maintenance technician must pull out of the fuselage of the aircraft located at such an unreasonable position that the risk of a fall is substantial? As you might imagine, many of the concerns of biomechanics are closely related to the field of safety engineering (Salvendy, 1987).

Although the consideration of biomechanics and anthropometry is critical to ensure that information exchange can be safely carried out, here too we find that this consideration is not sufficient to ensure good human factors. The human factors engineer must also be concerned with whether the displayed information can be adequately registered by the operator's sensory receptors (usually the eyes and ears). For example, the automobile dashboard display may be directly positioned in the driver's line of sight through the steering wheel (good anthropometry), but when the sun is shining through the window, it imposes such a reflective glare that the numbers and meter positions cannot be easily seen. Or in the telephone directory, the pages may be easy to turn, but the print is so small that even those with perfect eyesight have a hard time reading the names and numbers in anything less than perfect lighting. For many systems, similar consideration should also be given to the sensory characteristics of the ear. Can the machine supervisor hear the critical warning alarm over the whirring whine of the turbines? Does the static on the radio channel so distort the sound of the message that it cannot be understood?

Although we are now moving closer to the brain—the target of engineering psychology—these critical sensory limits are less related to how the brain processes information than to the operations of the eye and the ear themselves. Fortunately, a good deal is known about the limits and characteristics of both of these systems, and handbooks provide the human factors expert with such information as how much illumination is necessary to read print of a given size or how much noise of a certain level and description will be likely to mask the warning tone or the spoken word (Boff & Lincoln, 1988; Deatheridge, 1972; McCormick & Sanders, 1987; Salvendy, 1987). Such guidelines should be heeded, but in practice they are often violated. When they are violated, the information necessary for task performance will not be perceived, and naturally, the task will not be performed.

Human Performance   Given that concern for information flow, anthropometry, and sensory display characteristics have been satisfactorily addressed, our overview now concerns the final level of detail that represents the focus of this text: how the brain perceives and decides what to do with the sensory messages that are relayed from the eyes, ears, and other senses (perception); how the operator decides what to do with this information (decision); and how the actions that are decided on are carried out (action). The manner in which these three general stages are broken down into more detail (human performance) and their implications for the design of tasks, displays, and controls (engineering psychology) are the focus of this book and the subject of a further overview at the end of this chapter. However, the full range of topics related to human factors and engineering psychology must include mention of two other disciplines, both of crucial importance. These are the issues of training, which will be addressed to some extent in Chapter 6, and of personnel selection based on individual differences. Both of these, in turn, can be addressed either from the perspective of the engineering psychologist or from the perspective of the human factors engineer.

Training and Selection   The psychological perspective to training and selection examines the relevance of research on teaching the operator the skills that he or she must acquire to operate a system safely and efficiently and on selecting those operators whose abilities will match the skills required by the job. This perspective addresses such questions as the following: What is the importance of conceptual training and deep understanding versus rote memorization in learning the procedures to operate a complex piece of equipment? Is a complex task like driving a standard-shift automobile better taught by training in its separate components first or by practicing the whole task together from the beginning? What causes skills to be forgotten when they are not used? How stable are individual differences in verbal and spatial abilities, and should these differences be used as a basis for teaching people with words rather than with pictures and graphs?

The human factors engineering perspective, in contrast, often requires a broader view, considering the rich interplay of variables among training, individual differences, and system design (Booher, 1990; Rouse, 1985). To illustrate, consider the following complex decision: A conventional system is currently in use to make widgets in a factory. It is a simple system, and it is fairly easy to train workers of average ability to use it. However, it is not terribly reliable, and it imposes a heavy workload on the user because of its poor anthropometry and display design. Furthermore, it often produces shoddy widgets. Management is considering an investment into one of two alternative systems. One is a highly automated computer-based system that will require only a small number of highly skilled operators to supervise. There are great expenses in manufacturing this system, and the requirement for responsible, highly intelligent supervisors greatly restricts the criteria for selecting operators relative to the old version. The second alternative system is a more modest, less automated system—but one that requires the operator to do a lot more (than the other new system) and to do it in a substantially different way (than the old system). Hence, the cost of training the operators to use this second new system and the expense of designing training packages and simulators will be far greater than either of the other systems.

So what is the best system? Table 1.1 points out the comparative ratings of the three systems on each of six different dimensions, the first three describing human factors concerns. A plus (+) indicates a favorable rating, minus (−) indicates an unfavorable one, and zero (0) indicates an intermediate rating. Each dimension can probably ultimately be expressed by costs and savings. For example, poor human factors may translate to rapid employee turnover and high accident rates; these cost money. Higher operator selection criteria will translate to higher wages per operator. To make an informed decision about which system to select, there should be quantitative estimates of all of these dimensions, preferably in hard numbers that can be translated into dollar values, so that the trade-offs among dimensions can be considered. The U.S. Army, through its MANPRINT program, has implemented such a system, allowing for explicit consideration of all of these human factors considerations (Booher, 1990).

Attractive as this approach is to the human factors engineer because it

Table 1.1 MANUFACTURING SYSTEM

	Old	New 1 (Automation)		New 2 (Complex)	
Training costs	Low +	Modest	0	High	−
Operator selection criteria	None+	High	−	Modest	0
Operator workload (poor human factors)	High −	Low	+	Modest	0
Rate of widget production	Low −	High	+	Low	−
Product quality (company sales)	Low −	Moderate	0	Moderate	0
System cost (includes maintenance cost)	Low +	High	−	Moderate	0

accounts for the three human-centered factors (operator work load and performance, training, and selection), implementation is often thwarted by other considerations that may not involve human factors. These are *political considerations* (the selection of the better system might mean closing down a factory, or selection of the worse system might create jobs in the district of an influential member of Congress); *financial considerations*, shown at the bottom of Table 1.1 (the new system is just too expensive to build, and therefore, in the case of consumer products, will be too expensive to sell); *marketing considerations* (the carefully human-factored dashboard design is better, but it doesn't look "high tech" enough to sell); *matters of personal taste* (the top admiral who makes the decision doesn't like the new system because it just doesn't work the same way things did when he was sailing in the fleet); *the appeal of new technology* (computer voice alerts are built into the automobile, even though drivers find them disconcerting and often masked by the radio sound).

These facts of life in the design process are often frustrating for the human factors engineer (and to the system user as well), whose efforts to produce and design the "best" system are driven by the need to put top priority on the safety and comfort of the human operator and on the quality of system performance. (In this regard, the reader should note that three recent issues of the *Ergonomics Journal*, nos. 3, 4, 5, 1990, were devoted to the topic of "Marketing Ergonomics.") But although some or all of these potentially hostile forces to human factors will always be present, there are at least two directions the human factors engineer can take to increase the chances that human factors inputs will be heeded. These relate to the need for *early* input to the design process and the need for that input to be *quantitative* and *model driven* (Card, Moran, & Newell, 1986; Elkind, Card, Hochberg, & Huey, 1990). Because both of these needs relate directly to an understanding of the design cycle of system development, we will preface our discussion of them by first describing the nature of this cycle.

BORING

The Design Cycle

Figure 1.2 presents an example of the phases that are often seen as a system moves from conception to development to production and finally to its use in society. The concept for the new system often originates because of technological developments. For example, the development of high-speed, lightweight computer chips led to the availability of portable computers, which therefore became available to nontechnical users. Alternatively, the new concept may develop out of dissatisfaction with current systems designed to meet the original goals; for example, dissatisfaction with highly symbolic command strings in computer use gave rise to the concept of natural, icon-based manipulations, as implemented in the Xerox Star and Apple Macintosh computers.

The system concept is typically developed from paper plans into a prototype that can be tested to identify its flaws and shortcomings before full-scale production begins. After these tests are completed, the system finally goes into production and is manufactured for use in the "real world." This whole design

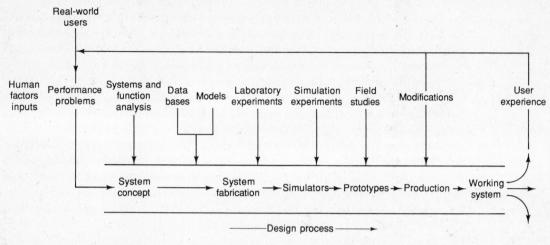

Figure 1.2 Inputs of human factors data into the design process.

cycle may vary tremendously in duration—from a few weeks for software products to several years for complex aerospace systems like the space shuttle or space station.

Data and Models   As shown across the top of Figure 1.2, there is a potential for human factors input at all points along the design cycle. First, the observation that a new system was necessary may be based heavily on the analysis of problems in human performance with the old system. Was the mental work load imposed by the old system unacceptably high? What kinds of errors did humans make? What caused those errors, and how can a better design alleviate them?

Next in the design cycle, the early concept development can benefit both from the structured systems and functions analysis just described and from the human factors data bases that already exist, regarding such areas as anthropometry (How high should the seat be built? Where should the controls and displays be placed?); vision (How big should the characters be? How much lighting is needed?); and many aspects of engineering psychology, as will be discussed later (e.g., What direction should knobs be rotated to increase a value? Should indicators be analog or digital?). Much of these data are available from handbooks (e.g., Boff & Lincoln, 1988; McCormick & Sanders, 1987; Salvendy, 1987). A second source of potential input closely related to the data bases is that information that comes from mathematical models of human performance and biomechanics (e.g., Card, Moran, & Newell, 1986; McMillan et al., 1989). Here a set of equations to represent some aspect of human performance can be calculated or "run" (often on a computer) as the model simulation receives input that would characterize the system under development. The computer-simulated "performance" of the model-simulated human operator and system together can be examined, and if it is unsatisfactory, changes in the design concept can be implemented.

For example, early in America's space program the issue was raised about whether the astronauts, riding the manned booster rocket in the early phases of launch, should be given the opportunity to "jump into the control loop" and steer the rocket on its desired course if the automated guidance system should fail. Although the concept of a human backup was promising, its success rested on whether human response time was fast enough to react to deviations in the rocket's course and put corrections of the appropriate direction and size into the steering system (see Chapter 11). As it turned out, the same answer to this question was reached in two very different ways. One answer was obtained by NASA engineers, who developed a very expensive simulation of the rocket dynamics and actually brought in the NASA astronauts to try to control it. The second answer was obtained by a company called Systems Technology Incorporated, who ran a computer model of the astronaut tracker, flying a computer model of the booster dynamics, in a way that will be discussed more fully in Chapter 11. The answer obtained by the two approaches was identical — the pilot could not fly the booster — but there was a tremendous difference in the cost of obtaining the answer when the expense of the simulator experiment was compared with the economy of the model run.

These results do not imply that model running will always produce answers that are cheaper and just as accurate as the answers produced from data. Models are often less reliable than experimental results, and until a model is extensively validated, its output should be treated with caution.

Experiments   Often during the earlier phases of development, neither data nor models are available to answer the human factors questions for the design process. Then an alternative source of input becomes the standard psychological experiment, familiar to most readers with training in experimental psychology. Should the designer employ voice or keyboard to enter data for the automated banking system? An experiment can be conducted that compares the two, using the kinds of transactions that the teller would be expected to carry out, under the levels of background noise and conversation that might typify the bank environment. A recommendation can be based on the magnitude of any substantial and statistically significant differences between the two response modes.

In performing experiments, of course, it is important for psychological theory to guide the design of the experiments and interpretation of the results. In this way, the results will gain in their generality and their usefulness for assisting in the design of other systems.

Simulation   As the development cycle continues, physical simulations of the system are produced, and it becomes possible to ask and answer human factors questions based on the results of simulator tests. This procedure is frequently used in the design of new aircraft, in which nonflyable simulations of the cockpit and aircraft dynamics are produced. The engine and flight characteristics are programmed in computer software, and pilots can "fly" the simulator in different conditions. The human factors expert studies the pilot's performance and obtains his or her opinions as the simulation is put through various benign and hostile conditions. Based on these objective and subjective measures, further recommendations for design changes can be made. Since the answers are obtained in an environment that is more like the final system than is the model or the laboratory experiment, they are oftentimes more credible to the design engineer.

Field Tests and User Evaluation   Still later, prototypes of the product are actually developed and become available for field tests, often involving the same kinds of tests as those used in simulation studies but now performed in conditions that are even closer to those of the final target environment. The design and certification of a new aircraft, for example, include many hours of flight tests on prototypes. Here again, aspects of the test that prove to be highly unsatisfactory from the human factors standpoint can be used to justify the recommendation of design changes. Finally, a working system is produced, marketed, and put into use in the real world. Of course, performance of the system is very much in the public eye, and here again, failures or difficulties with the system may indicate severe human factors design deficiencies that can provide inputs for modification. A classic example relates to the severe deficiencies in the human factors design of the nuclear reactor system at Three Mile Island. These deficiencies led eventually to modification in the way reactor control rooms are built (Woods, O'Brian, & Hanes, 1987).

Problems and Solutions

The previous section has described a number of potential influences that the human factors practitioner can have on the system design process. But these inputs are not always effective in practice because of two roadblocks or problems: One related to the timing of the input and the other to its form. The first problem is that despite the number of potential channels of human factors input, in practice only a few of them are usually effective. In particular, the farther along a system is in the design process—the farther to the right in Figure 1.2—the less likely it will be for human factors input to have an effect, even if the input can be well justified on the basis of human performance problems with the current system. This increasing inflexibility of systems to modification results because of the high cost of modifications once engineering plans are established and fabrication of the system has begun. From this point onward the cost of changing the system design will grow geometrically because almost any change has all sorts of implications that impose the need for still other changes. The expense of the additional changes usually reaches a point at which it is not cost effective to make them, particularly if the human factors engineer cannot really make a strong case that the system will be substantially improved by the modification. This brings us to the next roadblock to the input process.

The second problem relates to the form of advice and guidance that the human factors engineer can offer, which often comes as guidelines or recommendations: "A mouse rather than a keyboard should be employed here"; "The pressure gauge should be a moving needle display, not a digital one"; or "Our experiment reveals that display layout A gives significantly faster response time than display layout B." The trouble is, such answers do not tell the design engineer anything about how much better one format is than another, in terms of *operational performance*. Hence these recommendations typically cannot be balanced against the hard numbers that represent the added dollar cost of changing to the recommended design or the hard numbers that the engineer can often provide regarding mechanical system properties. The design engineer may rightfully complain; "You've told me that based on your experimental results one display takes one-tenth of a second less time to interpret than the other; but you haven't convinced me that this difference will create substantially fewer errors or less work load as people use the display in the fully developed system"; or "You've told me to use a mouse rather than push buttons, but you can't say anything about how much better the mouse will be to justify the extra cost of its inclusion and the added space requirements of a mouse surface."

These two problems can both be addressed by a common solution — the development and validation of quantitative human performance models, those that can provide hard numbers as guidelines and outputs rather than simply statements of "better" or "worse." These numbers can then often be translated into values that characterize human system performance; for example, "How many days of training will be saved by this particular training package?" "What is the probability of inputting an error, using this keyboard design?" The advantage of models are twofold. First, because their outputs are numbers, they provide more solid ammunition that can be balanced against the numbers that describe the financial cost of implementing human factors input. Second, because they are models, as shown in Figure 1.2, it is more feasible for them to be employed early in the design sequence, before the design decisions are "hardened" in production, when they will have a greater chance to influence the design process (Elkind, Card, Hochberg, & Huey, 1990; McMillan et al., 1989).

One example of the advantage of model application was already given — for the booster rocket. A second example relates to the size of a crew on certain military and civilian aircraft. Can a given complement adequately accomplish all functions that the aircraft must perform in its various missions? Or is an extra person required? If a quantitative model of pilot performance can be used to determine early in the design process that the extra person is needed, full engineering production can be devoted solely to the larger design. If there is no model to predict the needed complement and a smaller complement design is built and then proven impossible or dangerous, tremendous costs may be imposed. This is particularly true in high-speed military aircraft because the entire aircraft, including its handling and engine characteristics, would have to be redesigned to accommodate the extra size of, weight of, and environmental support for the additional person.

Human performance theory does indeed possess a number of quantitative models at various stages of development and various levels of complexity, many of which will be encountered in the later chapters of this book. Some of these models are relatively focused, describing only a small segment of human performance by a simple equation or even a single number. Examples would be the model describing the capacity of working memory (a fixed number of 7 ± 2 chunks, as discussed in Chapter 6), an equation for the time required to find a data entry as a function of its position on a linear list or menu (see Chapter 3), or an equation describing the human pilot that will predict how precisely an aircraft will be flown on a landing approach (see Chapter 11).

In contrast, some models are designed to describe more complex, global aspects of human performance, in which several processes and mental operations are working together (Baron, Kruser, & Huey, 1990; Elkind, Card, Hochberg, & Huey, 1990). Models of this category, because of their greater generality, sometimes sacrifice detailed accuracy of the component processes for the sake of avoiding unwieldy complexity. This sacrifice is compensated for by the model's applicability to a much broader range of activities. Some of these models will be described in Chapter 9.

Another class of models are those that attempt to describe how performance and cognitive processes are carried out, even though they may not reveal an estimate of the quantitative level of that performance. Instead, these models offer descriptions of the procedures and mechanisms that best describe the mental operations in complex performance. Examples are various models of decision making put forth in Chapter 7 and the model of signal detection described in Chapter 2, which contains both quantitative and qualitative aspects. Such models are often insightful because they highlight the different strategies that people use in carrying out tasks. Do people make decisions to maximize their gains or to minimize their losses? Or just to reach a satisfactory conclusion? The models are also useful because they may often reveal systematic ways in which actual performance departs from the best (optimal) performance possible.

A Model of Human Information Processing

As noted, models of human performance can be either specific or general, as well as quantitative or qualitative. Figure 1.3 presents a general qualitative model of human performance that provides a framework for organizing the contents of this book. The model describes the critical stages of information processing involved in human performance and is a composite of those presented by a number of previous investigators (e.g., Broadbent, 1958; Smith, 1968; Sternberg, 1969; Welford, 1976). It assumes that each stage of processing performs some transformation of the data and demands some time for its operation.

Although this conceptualization is not to be taken literally, it nevertheless aids the investigation of the components of human performance and so provides a useful technique for examining potential limitations in performance. These processing stages will be considered extensively in subsequent chapters. In the

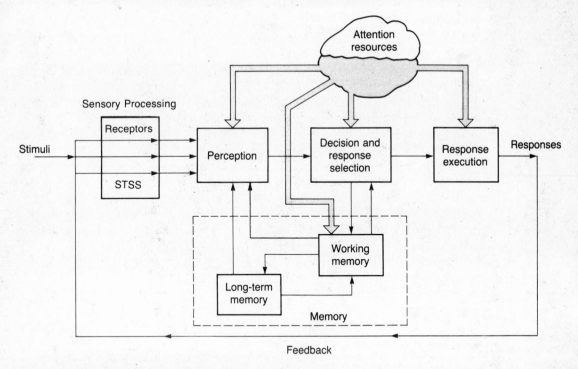

Figure 1.3 A model of human information processing.

following pages they are described briefly in the left-to-right order shown in Figure 1.3.

Sensory Processing   As we have noted, of primary importance in engineering psychology are the visual and auditory senses of the eyes and ears and the proprioceptive or kinesthetic senses of body and limb position. Unique limitations on each sensory system influence the quality and quantity of information that may be initially registered and so, potentially, all processes that follow. The characteristics of rod and cone receptors in the retina of the eye, for example, influence the conditions under which color can be used to display information. Characteristics of receptors and neural connections in the ear affect the ability of one sound to mask another in a noisy environment (Deatheridge, 1972). These properties are not considered extensively in the present text, however.

Short-Term Sensory Store   Each sensory system, or *modality*, appears to be equipped with a central mechanism that prolongs a representation of the physical stimulus for a short period of time after the stimulus has physically terminated. When attention is diverted elsewhere, the short-term sensory store (STSS) permits environmental information to be preserved temporarily and dealt with later. Three general properties are characteristic of STSS: (1) It is *preattentive*; that is, no conscious attention is required to prolong the image during the natural "time constant" of the store. (2) It is relatively *veridical*, preserving most of the physical details of the stimulus. (3) It is *rapidly decaying*. The particular time constant of the decay varies somewhat, normally being less than a second for short-term visual store (STVS, or iconic memory; Neisser, 1967; Sperling, 1960) and a bit longer for both short-term auditory store (STAS, or echoic memory; Neisser, 1967) and short-term kinesthetic store (Posner & Konick, 1966). Estimates for the last two range between 2 and 8 seconds, depending on the techniques employed in measurement (Lachman, Lachman, & Butterfield, 1979). The experience of iconic memory may be likened to the rapidly decaying image on a photographic print when a light is suddenly turned on in a darkroom or to a Polaroid picture in reverse time. Echoic memory possesses the characteristic of the decaying "internal echo" of a voice after it has ceased.

Perceptual Encoding   STSS preserves the details of the stimulus image only briefly and without attention. The information is then processed by progressively higher centers of the nervous system. This information is assumed to make contact with a unique neural code that was previously *learned* and *stored* in the brain. At this point the stimulus is said to be perceived or recognized. The result is a *perceptual decision* in which the physical stimulus is assigned to a single perceptual category. The dependence of perception on the previously learned neural code is represented in Figure 1.3 by the association between perception and *long-term memory*, the store of relatively permanent information.

The perceptual process is a many-to-one mapping; that is, a large number of different physical stimuli may often be assigned to a single perceptual category. For example, in perception of the letter *a*, different type styles (A, a, or *a*) all generate the same categorical perception, as does the sound "a" spoken by any voice. At the same time that there is a common response to all these forms of the letter *a*, their *differences* also are preserved at higher levels of processing. Mechanisms of focal attention allow us, if we choose, to attend to other dimensions of the stimulus as well, such as whether it is spoken by a male or female voice or written in upper or lower case. Even with this flexibility, however, the common response to the various physical forms of *a* will remain.

In our analysis of perception, it will be important to distinguish between a number of levels of complexity in the categorization task imposed by both the nature of the stimulus and the task confronting the operator. At a basic level, the task may call for the simple judgment of whether a stimulus is present. Only two perceptual categories are used, yes and no, and the operation is one of *detection* (Chapter 2). Thus the monitor of a radar scope is called on initially merely to detect the existence of a "blip" on the scope. At levels of greater complexity, the task may call for not just detection but also *recognition, identification* or *categorization* of the stimulus into one of several possible groups.

The complexity of these tasks is affected by whether one physical dimension alone is to be used in perceptual categorization of the stimulus, or whether instead several dimensions are to be considered in concert. As one example of the former case — the *absolute judgment task* — an operator may be asked to judge the loudness of a tone, the size of a crowd, or the smoothness of a plate of steel and assign the stimulus to a particular categorical level (Chapter 2).

In more complex tasks, at least two dimensions must be considered to match a particular stimulus category. Such multidimensional considerations typify the task of *pattern recognition* (Chapter 5). Each pattern is uniquely specified by a combination of levels (called features) along the several physical dimensions. For instance, recognition of a particular malfunction of a complex system will occur when the operator can associate a unique combination (pattern) of dial readings (features) on several different instruments (dimensions) to the perceptual category associated with a particular malfunctioning state. A particular collection of thermometer and barometer readings, satellite cloud-cover photographs, and wind speeds may signal "storm coming" to a trained meteorologist.

Finally, some aspects of perception may not involve any categorization into discrete states but rather judgment of continuous levels that have direct implications for action. For example, as we drive an automobile down the center of the lane, the amount of correction we apply to the steering wheel is directly proportional to the perceived size of the error and growth of error in our heading (Donges, 1978). We do not consciously assign error into "large" and "small" categories. This is called analog perception, in contrast to categorical perception, and describes much of our perceptual experience with space, the focus of Chapter 4.

Decision Making   Once a stimulus has been perceptually categorized, the operator must decide what to do with it. Sometimes the decision is a careful, thoughtful one, as when we decide where to attend college or whether to spend the evening studying or not (Chapter 7). At other times the decision may be a rapid, nearly automatic one. For example, recognizing that a traffic light has turned yellow (an absolute judgment task), the driver may decide to accelerate or to brake. In this case, the decision directly involves the selection of a response (Chapter 8). Alternatively, as indicated in Figure 1.3, a decision may be made to store the perceived information in memory before taking an action. If the decision is made to commit the information to memory, the information flow in Figure 1.3 follows the path leading to the memory box (Chapter 6). At this point, a second decision may be made to retain the information for a short time by actively rehearsing it in working memory or attempt to store it permanently (learn it) and so enter the information in long-term memory (Chapter 6). It is apparent that the point of decision making and response selection is a critical junction in the sequence of information processing. A large degree of choice is involved, and heavy potential costs and benefits depend on the correctness of the decision (Chapter 7).

Response Execution   If a decision is made to generate a response, an added series of steps is required to call up, and release with the appropriate timing and force, the necessary muscle commands to carry out the action (Chapters 8 and 11). The decision to initiate the response is logically separate from its execution. The baseball batter, for example, may decide to swing or not swing at a pitch (response selection) and if the swing is made, may execute one of a variety of successful or unsuccessful swings of the bat (response execution). One way to verify the complexity of the response execution process is to consider the extremely complex programming efforts required to enable robots to perform simple reaching, grabbing, and turning manipulations.

Feedback and Information Flow   It is evident that we typically monitor the consequences of our actions, forming the closed-loop feedback structure depicted in Figure 1.3 (see Chapter 11). Although feedback most often is considered in terms of a visual feedback loop (i.e., we see the consequences of our responses), feedback through the auditory, proprioceptive, and tactile (skin senses) modalities may be at least as important under some circumstances. For example, when changing gears in an automobile, the proprioceptive feedback from the hand is of considerable importance in evaluating when the gears have been engaged successfully. So too is the auditory "click" that tells us when we have depressed a button hard enough to be registered.

It is important to realize that information flow need not start with the stimulus. For example, sometimes our decisions or responses are internally triggered by "thoughts" in working memory. Furthermore, the flow need not be strictly left to right. We will see, for example, how immediate experience, represented in working memory, can affect our perceptions.

Attention   Much of the processing that occurs following STSS appears to require attention to function efficiently. In this context, we may model attention both as a searchlight that selects information sources to process and as a commodity or resource of limited availability. This pool of resources is shown at the top of Figure 1.3. If some processes require more of this resource, less is available for other processes, whose performance therefore will deteriorate. We need only note how one stops conversing in a car (perception, response selection, and execution) if there is a sudden need to scan a crowded freeway for a critical road sign (perception). Yet learning and practice decrease the demand for the limited supply of resources. We can walk while talking because walking, a well-practiced skill, requires few resources. The selective aspects of attention relating closely to phenomena in perception are discussed in Chapter 3. The aspects of attention that will be addressed in Chapter 9 concern how the allocation of attentional resources improves performance of tasks and how well humans can distribute attentional resources as they are required to various activities.

Two chapters are not directly incorporated within the framework of Figure 1.3. Chapter 10 deals with stress and human error. We will see that many stressors, like noise, sleep loss, or anxiety, have effects that are distributed across all processing components on Figure 1.3. Correspondingly, human error, much of which is attributable to stress, may result from breakdowns of any of the processing components.

Across all of the next ten chapters (2 through 11), we will illustrate the relevance of the different human performance components to the design of different systems, from computers to aircraft to medicine to graphs. However, Chapter 12 focuses directly on a single system—process control—because the task of the process control operation, as that of the operators on duty in the nuclear power plant at Three Mile Island, integrates issues addressed in all of the rest of the chapters in a way that cannot easily be disentangled. Finally, we conclude Chapter 12 with a discussion of automation, a conclusion that reintegrates our analysis of human performance into the broader context of human factors.

REFERENCES

Adams, J. A. (1989). *Human factors engineering.* New York: Macmillan.

Bailey, R. W. (1989). *Human performance engineering* (2nd ed.). Englewood Cliffs, NJ: Prentice Hall.

Baron, S., Kruser, D. S., & Huey, B. M. (eds.). (1990). *Quantitative modeling of human performance in complex, dynamic systems.* Washington, DC: National Academy Press.

Bateman, R. P. (1979). Design evaluation of an interface between passengers and an automated ground transportation system. *Proceedings of the 23rd annual meeting of the Human Factors Society* (pp. 120–124). Santa Monica, CA: Human Factors Society.

Boff, K., & Lincoln, J. (1988). *Engineering data compendium.* Wright Patterson AFB, OH: Harry Armstrong Aerospace Medical Research Laboratory.

Booher, H. (1990). *Manprint: An approach to systems integration.* New York: van Nostrand, Reinhold.

Broadbent, D. E. (1958). *Perception and communications.* New York: Pergamon Press.

Card, S., Moran, T., & Newell, A. (1986). "The model human processor" in K. Boff, L. Kaufman, & J. Thomas (eds.), *Handbook of perception and human performance* (vol. 2). New York: Wiley.

Chaffin, D. (1987). Biomechanical aspects of workplace design. In G. Salvendy (ed.), *Handbook of human factors.* New York: Wiley.

Deatheridge, B. H. (1972). Auditory and other sensory forms of information presentation. In H. P. Van Cott & R. G. Kinkade (eds.), *Human engineering guide to system design.* Washington, DC: U.S. Government Printing Office.

Donges, E. (1978). A two level model of driver steering behavior. *Human Factors, 20,* 691–708.

Electrical Power Research Institute (EPRI). (1977, March). *Human factors review of nuclear power plant design: Final report* (Project 501, NP–309–SY). Sunnyvale, CA: Lockheed Missiles and Space Co.

Elkind, J., Card, S., Hochberg, J., & Huey, T. S. (1990). *Human performance models for computer-aided engineering.* Orlando, FL: Academic Press.

Fitts, P. M., & Jones, R. E. (1947, July). *Analysis of factors contributing to 460 "pilot error" experiences in operating aircraft controls* (Memorandum Report TSEA

4–694–12, Aero Medical Laboratory). Wright Patterson AFB, OH: Harry Armstrong Aerospace Medical Research Laboratory. Reprinted in H. W. Sinaiko (ed.), *Selected papers on human factors in the design and use of control systems.* New York: Dover, 1961.

Hawkins, F. H. (1987). *Human factors in flight.* Brookfield, VT: Gower Technical Press.

Howell, W. C., & Goldstein, I. L. (eds.). (1971). *Engineering psychology: Current perspectives in research.* New York: Appleton-Century-Crofts.

Hurst, R., & Hurst, L. (1985). *Pilot error.* London: Aronson.

Kantowitz, B. H., & Sorkin, R. D. (1983). *Human factors: Understanding people-system relationships.* New York: Wiley.

Kantowitz, B., & Sorkin, R. (1987). Allocation of function. In G. Salvendy (ed.), *Handbook of human factors.* New York: Wiley.

Kroemer, K. (1987). Biomechanics of the human body. In G. Salvendy (ed.), *Handbook of human factors.* New York: Wiley.

Lachman, R., Lachman, J. L., & Butterfield, E. C. (1979). *Cognitive psychology and information processing.* Hillsdale, NJ: Erlbaum.

McCormick, E. J., & Sanders, M. S. (1987). *Human factors engineering and design.* New York: McGraw-Hill.

McMillan, G., Beevis, D., Salas, E., Strub, M., Sutton, R., & Van Breda, L. (1989). *Applications of human performance models to system design.* New York: Plenum Press.

Mackworth, N. H. (1948). The breakdown of vigilance during prolonged visual search. *Quarterly Journal of Experimental Psychology 1,* 5–61.

Meister, D. (1986). *Behavioral foundations of systems development.* London: Wiley.

Meister, D. (1989). *Conceptual aspects of human factors.* Baltimore, MD: Johns Hopkins Press.

Neisser, U. (1967). *Cognitive psychology.* New York: Appleton-Century-Crofts.

Posner, M. I., & Konick, A. F. (1966). Short-term retention of visual and kinesthetic information. *Organizational Behavior and Human Performance, 1,* 71–88.

Poulton, E. C. (1966). Engineering psychology. *Annual Review of Psychology, 17,* 177–200.

Reason, J. (1990). *Human error.* New York: Cambridge University Press.

Rouse, W. (1985). Optimal allocation of system development resources to reduce and/ or tolerate human error. *IEEE Transactions on Systems, Man, and Cybernetics, SMC–15,* 620–630.

Rubinstein, T., & Mason, A. F. (1979, November). The accident that shouldn't have happened: An analysis of Three Mile Island. *IEEE Spectrum,* pp. 33–57.

Salvendy, G. (1987). *Handbook of human factors.* New York: Wiley.

Sanders, M. S., & McCormick, E. J. (1987). *Human factors in engineering and design* (6th ed.). New York: McGraw-Hill.

Sheridan, T. (1981). Understanding human error and aiding human diagnostic behavior in nuclear power plants. In J. Rasmussen & W. B. Rouse (eds.), *Human detection and diagnoses of system failures.* New York: Plenum Press.

Smith, E. (1968). Choice reaction time: An analysis of the major theoretical positions. *Psychological Bulletin, 69,* 77–110.

Sperling, G. (1960). The information available in brief visual presentations. *Psychological Monographs, 74,* 498.

Sternberg, S. (1969). The discovery of processing stages: Extension of Donders' method. *Acta Psychologica, 30,* 276–315.

Welford, A. T. (1976). *Skilled performance.* Glenview, IL: Scott, Foresman.

Wickens, C. D. (1984). *Engineering psychology and human performance.* New York: Harper & Row.

Woods, D., O'Brian, J. F., & Hanes, L. F. (1987). Human factors challenge in process control: The case of nuclear power plants. In G. Salvendy (ed.), *Handbook of human factors.* New York: Wiley.

Chapter
2

Signal Detection, Information Theory, and Absolute Judgment

OVERVIEW

Information processing in most systems begins with the detection of some environmental event. In some cases, such as the Three Mile Island incident, the event is so noticeable that immediate detection is assured. The information-processing problems in these circumstances are those of recognition and diagnosis. However, there are many other circumstances in which detection itself represents a source of uncertainty or a potential bottleneck in performance because it is necessary to detect events that are near the threshold of perception. Will the security guard monitoring a bank of television pictures detect the abnormal movement on one of them? Will the radiologist detect the abnormal x-ray as it is scanned? Will the industrial inspector detect the flaw in the product?

This chapter will first deal with the critical question of detection in which perceptual categorization is made into one of two states: A signal is present or it is absent. The detection process will be modeled within the framework of signal detection theory, and we will show how the model can assist engineering psychologists in understanding the complexities of the detection process, in diagnosing what goes wrong when detection fails, and in recommending corrective solutions.

The process of detection may in fact involve more than two states of categorization. It may, for example, require the human operator to choose between three or four levels of uncertainty about the presence of a signal or to detect more than one kind of signal. At this point our consideration moves into the realm of identification and recognition. This situation requires the introduction of information theory to describe human performance in recognition tasks. Then the final pages of the chapter will deal with the simplest form of multilevel categorization, the absolute judgment task.

SIGNAL DETECTION THEORY

The Signal Detection Paradigm

Signal detection theory is applicable in any situation in which there are two discrete *states of the world* (call these signal and noise) that cannot easily be discriminated. Signals must be detected by the human operator, and in the process two response categories are produced: Yes (I detect a signal) and no (I do not). This situation may describe activities such as the detection of a concealed weapon by an airport security guard, the detection of a contact on a radar scope (N. H. Mackworth, 1948), a malignant tumor on an X-ray plate by a radiologist (Parasuraman, 1985; Swets & Pickett, 1982), a malfunction of an abnormal system by a nuclear plant supervisor (Less & Sayers, 1976), a critical event in air traffic control (Bisseret, 1981), typesetting errors by a proofreader (Anderson & Revelle, 1982), an untruthful statement from a polygraph (Szucko & Kleinmuntz, 1981), a crack on the body of an aircraft, or a communications signal from intelligent life in the bombardment of electromagnetic radiation from outer space (Blake & Baird, 1980).

The combination of two states of the world and two response categories produces the 2×2 matrix shown in Figure 2.1, generating four classes of joint events, labeled hits, misses, false alarms, and correct rejections. It is apparent that perfect performance is that in which no misses or false alarms occur.

State of the world

		Signal	Noise
Response	Yes	Hit	False alarm
	No	Miss	Correct rejection

Figure 2.1 The four outcomes of signal detection theory.

However, since the signals are not very intense in the typical signal detection task, misses and false alarms do occur, and so there are normally data in all four cells. In signal detection theory (SDT) these values are typically expressed as probabilities, by dividing the number of occurrences in a cell by the total number of occurrences in a column. Thus if 20 signals were presented, and there were 5 hits and 15 misses, we would write $P(\text{hit}) = 5/20 = 0.25$.

The SDT model (Green & Swets, 1988) assumes that there are two stages of information processing in the task of detection: (1) Sensory evidence is aggregated concerning the presence or absence of the signal, and (2) a decision is made about whether this evidence constitutes a signal or not. According to the theory, external stimuli generate neural activity in the brain. Therefore, on the average there will be more sensory or neural evidence in the brain when a signal is present than when it is absent. This neural evidence, X, may be conceived as the rate of firing of neurons at a hypothetical "detection center." The rate increases in magnitude with stimulus intensity. We refer to the quantity X as the *evidence variable*. Therefore, if there is enough neural activity, X exceeds a critical threshold X_c, and the operator decides "yes." If there is too little, the operator decides "no."

Because the amount of energy in the signal is typically low, the average amount of X generated by signals in the environment is not much greater than the average generated when no signals are present (noise). Furthermore, the quantity of X varies continuously even in the absence of a signal because of random variations in the environment and in the operator's own "baseline" level of neural firing (e.g., the neural "noise" in the sensory channels and the brain). This variation is shown in Figure 2.2. Therefore, even when no signal is present, X will sometimes exceed the criterion X_c as a result of random variations alone, and the subject will say "yes" (generating a false alarm at point A of Figure 2.2). Correspondingly, even with a signal present, the random level of activity may be low, causing X to be less than the criterion, and the subject will say "no" (generating a miss at point B of Figure 2.2). The smaller the difference in intensity between signals and noise, the greater these error probabilities become because the amount of variation in X resulting from randomness increases relative to the amount of energy in the signal. In Figure 2.2 the average level of X is increased slightly in the presence of a weak signal and greatly when a strong signal is presented.

For example, consider the monitor of a noisy radar screen. Somewhere in the midst of the random variations in stimulus intensity caused by reflections from clouds and rain, there is an extra increase in intensity that represents the presence of the stimulus — an aircraft. The amount of noise will not be constant over time but will fluctuate; sometimes it will be high, completely masking the stimulus, and sometimes low, allowing the plane to stand out. In this example, "noise" varies in the environment. Suppose instead you were standing watch on a ship, searching the horizon on a dark night for a faint light. It becomes difficult to distinguish the flashes that might be real lights from those that are just "visual noise" in your own sensory system. In this case, the random noise is internal. Thus "noise" in signal detection theory is a combination of noise from external and internal sources.

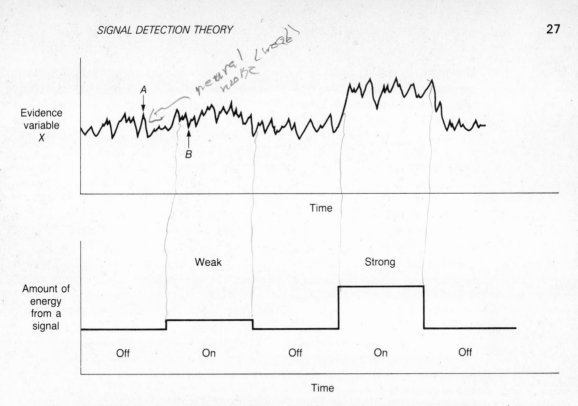

Figure 2.2 The change in the evidence variable x caused by a weak and a strong signal. Notice that with the weak signal there can sometimes be less evidence when the signal is present (point B) than when the signal is absent (point A).

The relations between the presence or absence of the signal, random variability of X, and X_c can be seen in Figure 2.3. The figure plots the probability of observing a specific value of X, given that a noise trial (left curve) or signal trial (right curve) in fact occurred. These data might have been tabulated (from the graph of the evidence variable at the top of Figure 2.2) by counting the relative frequency of different X values during the intervals when the signal was off, creating the probability curve on the left of Figure 2.3, and making a separate count of the probability of different X values while the weak signal was on, generating the curve on the right of Figure 2.3. As the value of X increases, it is relatively more likely to have been generated while a signal was present than while it was absent.

The value of X at which the absolute probability that X was produced by the signal equals the probability that it was produced by only noise is the point where the two curves intersect. The criterion value X_c chosen by the operator is shown by the vertical line. All X values to the right ($X > X_c$) will cause the operator to respond "yes." All to the left generate "no" responses. The different shaded areas represent the occurrences of hits, misses, false alarms, and correct rejections. Since the total area within each curve is one, the two shaded regions within each curve must also add to one. That is, $P(H) + P(M) = 1$ and $P(FA) + P(CR) = 1$.

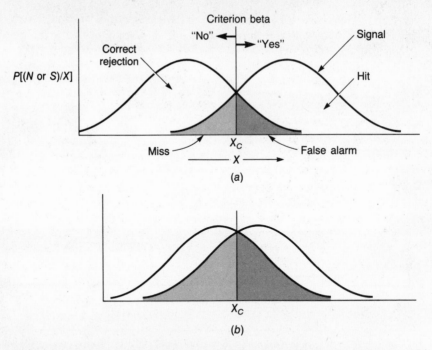

Figure 2.3 Hypothetical distributions underlying signal detection theory: (*a*) high sensitivity; (*b*) low sensitivity.

Setting the Response Criterion: Optimality in SDT

In any signal detection task, whether an experimental one in the laboratory or a real-world situation, operators may be described in terms of their *response bias*. They may be prone to say yes, thereby detecting most of the signals that occur but producing many false alarms ("risky responders"), or they may be "conservative," saying no most of the time and making few false alarms but missing many of the signals.

Different circumstances may dictate whether a conservative or a risky strategy is best. For example, when the radiologist scans the X ray of a patient who has been referred because of other symptoms of illness, it is probably appropriate to be more biased to say yes (detecting a tumor) than when examining the X ray of a totally healthy patient, for whom there is no reason to suspect any malignancy (Swets & Pickett, 1982). Consider, on the other hand, the monitor of the power-generating station who has been cautioned repeatedly by the supervisor not to make any unnecessary shutdowns of a turbine because of the resulting loss of revenue to the company. The operator will probably become conservative in monitoring the dials and meters for an indication of a malfunction and may be prone to miss (or delay responding to) a malfunction when it does occur.

As can be seen in Figure 2.3, an operator's conservative or risky behavior is determined by placing the decision criterion X_c. If X_c is placed to the right, much evidence is required for it to be exceeded, and most responses will be "no" (conservative responding). If it is placed to the left, little evidence is required, most responses will be "yes," and the strategy is risky. An important variable that is positively correlated with the X_c is the quantity *beta*, which is the ratio measure of neural activity:

$$\beta = \frac{P(X|S)}{P(X|N)} \qquad (2.1)$$

This is the ratio of the ordinate of the two curves of Figure 2.3, at a given level of X_c. Intuitively we see that higher values of beta (and X_c) will generate fewer yes responses and therefore fewer hits, but also fewer false alarms. Lower beta settings generate more yes responses, more hits, and more false alarms. Thus the beta parameter belongs to the category of *bias* measures, which have costs and benefits at both ends of the scale. Sometimes it is better to make fewer misses, sometimes fewer false alarms. Both beta and X_c define the *response bias* or *response criterion*.

More formally, the actual probability values appearing in the matrix of Figure 2.1 would be calculated from obtained data. These data would describe the areas under the two probability distribution functions of unit area shown in Figure 2.3, to the left and right of the criterion. Thus, for example, the probability of a hit, with the criterion shown, is the relative area under the "signal" curve (a signal was presented) to the right of the criterion (the subject said yes). One can determine by inspection that if the two distributions are of equal size and shape, then the setting of beta = 1 occurs where the two curves intersect and will provide data in which $P(H) = P(CR)$ and $P(M) = P(FA)$, that is, a truly "neutral" criterion setting.

Signal detection theory is able to prescribe exactly where the optimum beta should fall, given environmental conditions related to (1) the likelihood of observing a signal and (2) the costs and benefits of the four possible outcomes (Green & Swets, 1988; Swets & Pickett, 1982). We shall consider this optimal setting first as it is indicated by signal probability alone and then as the costs and benefits of different outcomes are considered.

1. *Equal probabilities.* Consider the situation in which signals occur just as often as they don't, and furthermore, there is neither a different cost to the two bad outcomes nor a different benefit to the two good outcomes of Figure 2.1. In this case *optimal* performance minimizes the number of errors (misses and false alarms). It can be shown that the particular symmetrical geometry of Figure 2.3 dictates that optimal performance will occur when X_c is placed at the intersection of the two curves, that is, when beta = 1. Any other placement, in the long run, would reduce the probability of being correct.

2. *Effect of probabilities.* It is intuitively reasonable that if a signal is more likely, the criterion should be lowered. For example, if the radiologist has other information to suggest that a patient is likely to have a malignant tumor, or the physician has received the patient on referral, the physician should be more likely to categorize a possible abnormality on the X ray as a tumor than to

ignore it as mere noise in the X-ray process. On the other hand, if signal probability is reduced, beta should be adjusted conservatively. For example, suppose a quality control inspector searching for defects in computer microchips is told that the batch under inspection has a very low estimated fault frequency because the manufacturing equipment has just received maintenance. In this case, the inspector should be more conservative in searching for defects. Formally, this adjustment of the optimal beta in response to changes in signal and noise probability is represented by the prescription

$$\beta_{\text{opt}} = \frac{P(N)}{P(S)} \tag{2.2}$$

This quantity will be reduced (made riskier) as $P(S)$ increases, thereby moving the value of X_c that will produce the optimal beta to the left of Figure 2.3. If this setting is adhered to, performance will still maximize the number of correct responses (hits and correct rejections). It is important to emphasize that the setting of optimal beta will not produce perfect performance. There will still be false alarms and misses as long as the two curves overlap. However, optimal beta is the best that can be expected for a given signal strength and a given level of sensitivity.

The reader should note that the formula for beta (2.1), $\beta = \dfrac{P(X|S)}{P(X|N)}$, and the formula for optimum beta shown in Equation 2.2, $\beta_{\text{opt}} = \dfrac{P(N)}{P(S)}$, are sometimes confused. β_{opt}, defining where beta *should* be set, is determined by the ratio of the *absolute* probability with which noise and signals occur in the environment. In contrast, where beta *is* set by an observer is determined by the ratio of probabilities of X, given signal and noise. These latter probabilities cannot be observed but must be inferred.

3. *Effect of payoffs.* The optimal setting of beta is also influenced by *payoffs*. In this case, *optimal* is no longer defined in terms of minimizing errors but is now maximizing the total expected financial gains (or minimizing expected losses). Again, intuitively, if it were important for signals never to be missed, the operator might be given high rewards for hits and high penalties for misses, leading to a low setting of beta. This payoff would be in effect for a power station monitor who, unlike his or her colleague described above, is admonished by the supervisor that severe costs in company profits (and the monitor's own paycheck) will result if faulty microchips pass through the inspection station. The monitor should therefore be more likely to discard good chips (a false alarm), even if he or she catches all the faulty ones. Conversely, in different circumstances, if false alarms are to be avoided, they might be heavily penalized. These costs and benefits can be translated into a prescription for the optimum setting of beta by expanding Equation 2.2 to

$$\beta_{\text{opt}} = \frac{P(N)}{P(S)} \times \frac{V(CR) + C(FA)}{V(H) + C(M)} \tag{2.3}$$

where V is the value of the specified desirable event (hit, H, or correct rejection, CR), and C is the cost of the specified undesirable event (miss, M, or false alarm, *FA*). In Equation 2.3 an increase in any quantity in the denominator will decrease the optimal beta and should lead to risky responding. Conversely, an increase in numerator values should lead to conservative responding. Notice also that the value and probability portions of the function theoretically combine independently. Thus, although such an event as the malfunction of a turbine may occur only very rarely, thereby raising the optimum beta, the consequence of a miss in detecting it might be so severe (cost of miss) as to reduce the optimal beta to a relatively low value.

Human Performance in Setting Beta   Since the actual value of beta that an operator uses can be computed on the basis of the known probabilities of hits and false alarms obtained from a series of detection trials, we may ask how well humans actually perform in setting their criteria in response to changes in payoffs and probabilities. In fact, humans do adjust beta as dictated by changes in these quantities. However, laboratory studies suggest that beta is not adjusted as much as it should be. That is, subjects demonstrate a "sluggish beta," a relationship shown in Figure 2.4. They are less risky than they should be if the ideal beta is high. As shown in Figure 2.4, this sluggishness is found to be

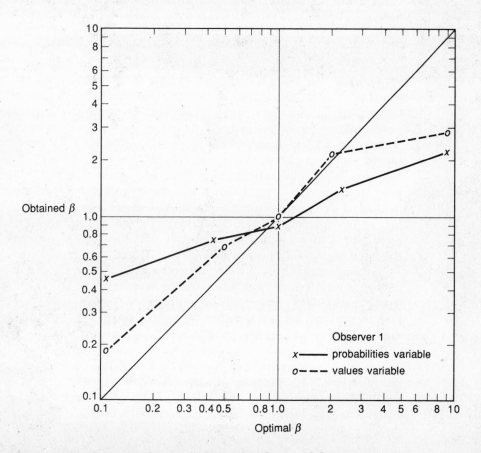

Figure 2.4 Relationship between obtained and optimal decision criteria.

more pronounced when beta is manipulated by probabilities than by payoffs (Green & Swets, 1988).

A number of explanations have been proposed to account for why beta is sluggish in response to probability manipulations. It is possible that it is simply a reflection of the operator's need to respond "creatively," by introducing the rare response more often than is optimal, since extreme values of beta optimally dictate long strings of either yes (low beta) or no (high beta) responses. Another explanation may be that the operator misperceives probabilistic data. There is some convincing evidence that people tend to overestimate the probability of very rare events and underestimate that of very frequent events (Peterson & Beach, 1967; Sheridan & Ferrell, 1974). This behavior, which will be discussed in more detail in Chapter 7, would produce the observed shift of beta toward unity.

The sluggish beta phenomenon can be demonstrated most clearly in the laboratory, where precise values of probabilities and values can be specified to the subjects. In real-world environments there is evidence that operators do indeed adjust beta with changes in probabilities although not necessarily as far as optimally prescribed. For example, quality control inspectors examining sheet metal for defects will adjust their response criterion according to the estimated defect rate of a batch (Drury & Addison, 1973). However, Harris and Chaney (1969), describing performance of inspectors in a Kodak plant, report that as the defect rate falls below about 5 percent, inspectors fail to lower beta accordingly, very clearly demonstrating a sluggish beta. Bisseret (1981) has applied signal detection theory to the air traffic controller's task of judging whether an impending conflict between two aircraft will (signal) or will not (noise) require a course correction. He finds that controllers will adjust beta to lower values (be more willing to detect a conflict and therefore command a correction) as the difficulty of the problem and therefore the uncertainty of the future increases. Bisseret also compared the performance of experts and trainees and found that the former group as a whole was lower in the setting of beta, being more willing to call for a correction. Bisseret attributes this finding to the fact that trainees are more uncertain about how to carry out the correction and therefore more reluctant to take the action that would require implementation. Bisseret argues that a portion of their training should be devoted to the issue of criterion placement.

In medicine, Lusted (1976) reports evidence that doctors do adjust the response criterion in diagnosis according to how frequently the disease occurs in the population — essentially an estimate of P(signal) — but they adjust less than the optimal amount specified by the changing probabilities. However, the difficulty of specifying precisely the costs or benefits of all four of the joint events in medical diagnosis, as in air traffic control, makes it quite difficult to determine the exact level of optimal beta (as opposed to the direction of its change). How, for example, can the physician specify the precise cost of an undetected malignancy (Lusted, 1976), or the air traffic controller specify the costs of an undetected conflict that produced a midair collision? How can the power plant operator specify the cost of a turbine that will be off the line for an unpredictable amount of time, with ill-defined damages, before the shutdown has actually occurred? The problems associated with specifying costs define the limits of applying signal detection theory to determine optimal response criteria.

Sensitivity

One of the most important contributions of signal detection theory is that it has made a conceptual and analytical distinction between the response bias parameters described above and the measure of the operator's *sensitivity*, the keenness or resolution of the detection mechanisms. We have seen that the operator may fail to detect a signal if his or her response bias is conservative. Correspondingly, the signal may be missed simply because the resolution of the detection process is low in discriminating signals from noise, even if the response bias is neutral or risky.

In terms of the theoretical representation of Figure 2.3, sensitivity refers to the separation of the internal noise and signal-plus-noise distributions along the X axis. If the separation is great, sensitivity is great and a given value of X is quite likely to be generated by either S or N but not by both. If it is low, sensitivity is low. Since the curves are assumed to be internal, their separation could be reduced either by physical properties of the signal (e.g., a reduction in its intensity or salience) or by properties of the subject (e.g., a loss of hearing for an auditory detection task or a lack of training of a medical student for the task of detecting complex tumor patterns on an X ray). Figures 2.2 and 2.3 present an example of two signals, one for which sensitivity is low, one for which it is high.

In the formal theory of signal detection, the sensitivity measure is called d' and corresponds to the separation of the means of two distributions in Figure 2.3, expressed in units of their standard deviations. This is a value that in most applications of signal detection theory ranges roughly between 0.5 and 2.0. Appendix A presents values of d' generated by different hit and false alarm rates.

Like bias, sensitivity also has an optimal value (which is not perfect). The computation of this optimal is more complex and is based on an ability to characterize precisely the statistical properties of the physical energy in signal and no-signal trials. Although this can be done in carefully controlled laboratory studies with acoustic signals and white-noise background, it is difficult to do in more complex environments. Nevertheless, data from auditory signal detection investigations showing the ways in which human sensitivity departs from optimal sensitivity (Green & Swets, 1988) have extremely important practical implications. These data suggest that the major cause for the departure results from the operator's lack of *memory* of the precise physical characteristics of the signal. When memory aides are provided that continuously remind the operator of what the signal sounds like or looks like, d' approaches the optimal levels. This point will be important when we consider the nature of vigilance tasks later in the chapter.

THE ROC CURVE

Theoretical Representation

It should be apparent that all detection performance that has the same sensitivity is in some sense equivalent, no matter what its level of bias. A graphic method of representation known as the receiver operating characteristic (ROC) curve is used to portray this equivalence of sensitivity across changing levels of bias. The ROC curve is useful for obtaining an overall understanding of the joint effects of sensitivity and response bias on the data from a signal detection analysis. In this section we will describe the ROC curve by emphasizing its relation both to the typical signal detection data matrix in Figure 2.1 and to the theoretical signal and noise curves in Figure 2.3.

Figure 2.1 presents the raw data that might be obtained from a signal detection theory (SDT) experiment. Of the four values, only two are critical. These are normally $P(H)$ and $P(FA)$ since $P(M)$ and $P(CR)$ are then completely specified as $1 - P(H)$ and $1 - P(FA)$, respectively. Figure 2.3 shows the theoretical representation of the neural mechanism within the brain that generated the matrix of Figure 2.1. As the criterion is set at different locations along the X axis of Figure 2.3, a different set of values will be generated in the matrix of Figure 2.1. Figure 2.5 is a more detailed picture of the third way of represent-

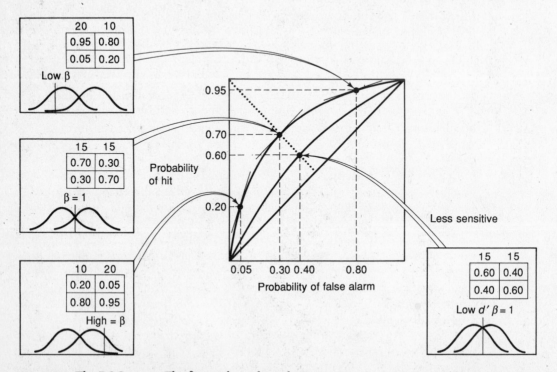

Figure 2.5 The ROC curve. The figure shows how three points on an ROC curve of high sensitivity relate to the stimulus-response matrix data and the underlying signal and noise curves. At the right, the figure also shows one point of lower sensitivity.

ing signal detection data: the ROC curve. The ROC curve plots on a single graph the joint value of P(hit) and P(FA) as each is obtained at a number of different settings of the response criterion.

Each signal detection condition generates one point on the ROC. If the signal strength and the observer's sensitivity remain constant, then by decreasing beta (either through changing payoffs or increasing signal probability) from one block of signal detection trials to another, a series of points are produced that move from conservative responding in the lower left of Figure 2.5 to risky responding in the upper right. When connected, these points make the ROC curve. Figure 2.5 shows the relationship among the raw data, the ROC curve, and the assumed underlying hypothetical distributions, collected at three different beta sets. One can see that sweeping the criterion placement from left (low beta or "risky" responding) to right (high beta or "conservative" responding), along the evidence variable produces progressively more "no" responses and moves points on the ROC curve from upper right to lower left.

It is often time-consuming to carry out the same signal detection experiment several times, each time changing only the response criterion by a different payoff or signal probability. A more efficient means of collecting data from several criterion sets is to have the subject provide a rating of confidence that a signal was present (Green & Swets, 1988). If three confidence levels are employed (e.g., "1" = confident that no signal was present, "2" = uncertain, and "3" = confident that a signal was present), as shown in Table 2.1, the data may be analyzed twice in different ways. During the first analysis, categories 1 and 2 would be considered a "no" response and 3 a "yes" response. This would produce data corresponding to a conservative beta setting since roughly two-thirds of the responses would be called "no." The second time, two responses (2 and 3) would be assigned instead to the "yes" category, representing the data matrix of a risky beta. Thus, two beta settings are available from only one set of detection trials. This economy of data collection is realized because the subject is asked to convey more information on each trial.

Table 2.1   ANALYSIS OF CONFIDENCE RATING IN SIGNAL DETECTION TASKS[a]

Subject's Response	Stimulus Presented		How Responses Are Judged	
	Noise	Signal		
"1" = "No Signal"	4	2	NO	NO
"2" = "Uncertain"	3	2	NO	YES
"3" = "Signal"	1	4	YES	YES
TOTAL NO. OF TRIALS	8	8	↓	↓
			Conservative Criterion P(FA) = 1/8 P(HIT) = 4/8	Risky Criterion P(FA) = 4/8 P(HIT) = 6/8

[a] The table shows how data with three levels of confidence can be collapsed to derive two points on the ROC curve. Entries within the table indicate the number of times the subject gave the response on the left to the stimulus (signal or noise) presented.

It is important at this point to scrutinize in more detail some of the general characteristics of the ROC curve and the way in which they relate to detection performance. Notice in Figure 2.5 that the ROC curve for a more sensitive observer is more bowed, being located closer to the upper left. Formally, the value of beta (the ratio of ordinate heights in Figure 2.3) at any given point along the ROC curve is equal to the slope of a tangent drawn to the curve at that point. This slope (and therefore beta) will theoretically always be equal to 1.0 at points that fall along the negative diagonal (shown by the dotted line). If the hit and false-alarm values of these points are determined, we will find that $P(H) = 1 - P(FA) = P(CR)$, as can be seen at the two points on the negative diagonal of Figure 2.5. Performance here is equivalent to performance at the point of intersection of the two distributions in Figure 2.3. Note also that points on the *positive* diagonal of Figure 2.5, running from lower left to upper right, represent chance performance: No matter how the criterion is set, $P(H)$ always equals $P(FA)$, and the signal cannot be discriminated at all from the noise. A representation of Figure 2.3 that gives rise to chance performance and corresponds to the points on the positive diagonal would be one in which the signal and noise distributions were perfectly superimposed. Finally, points in the lower right region of the ROC space represent "worse than chance" performance, when the subject says "signal" when no signal is perceived and vice versa.

Although Figure 2.5, plotted with a linear probability scale, shows a typically bowed curve, an alternative way of plotting the curve is on probability paper (Figure 2.6). This representation has the advantage that the bowed lines of Figure 2.5 now become straight lines parallel to the chance diagonal. Constant units of distance along each axis represent constant numbers of standard scores of the normal distribution. These are sometimes known as Z scores. For a given point, d' is then equal to $Z(H) - Z(FA)$, reflecting the number of standardized scores that the point lies to the upper left of the chance diagonal. A measure of response bias that correlates very closely with beta, and is easy to derive from Figure 2.6, is simply the z score of the false-alarm probability for a particular point (Swets & Pickett, 1982). It is also possible to look up values of d' and beta in published tables in which these parameters are listed as functions of $P(FA)$ and $P(H)$ (Swets, 1964; see Appendix A).

Empirical Data

It is important to realize the distinction between the theoretical, idealized curves in Figures 2.3, 2.5, and 2.6 and the actual empirical data collected in a signal detection experiment or a field investigation of detection performance. The most obvious contrast is that the representations in Figures 2.5 and 2.6 are continuous, smooth curves, whereas the data that would be collected would of course consist of a set of discrete points. More important, empirical results in which data are collected from a subject as the criterion is varied often provide points that do not fall precisely along a line of constant bowedness (Figure 2.5) or a 45-degree slope (Figure 2.6). More often the slope is slightly shallower. This situation arises because the distributions of noise and signal-plus-noise

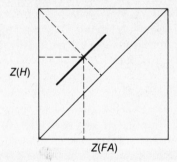

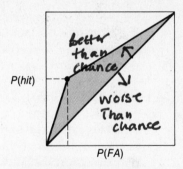

Figure 2.6 The ROC curve on probability paper.

Figure 2.7 Example of the sensitivity measure $P(A)$, the area under the ROC curve, derived from one point.

energy are not in fact precisely normal and of equal variance, as the idealized curves of Figure 2.3 portray, particularly if there is variability in the signal itself. This tilting of the ROC curve away from the ideal presents some difficulties for the use of d' as a measure of sensitivity. If d' is measured as the distance of the ROC curve of Figure 2.6 from the chance axis, and this distance varies as a function of the criterion setting, what is the appropriate setting for measuring d'? One approach is to measure the distance at unit beta arbitrarily (i.e., where the ROC curve intersects the negative diagonal). This measure is referred to as d_a and may be employed if data at two or more different beta settings are available so that a straight-line ROC can be constructed on the probability plot of Figure 2.6 (Green & Swets, 1988).

Although it is therefore desirable to generate two or more points on the ROC curve, there are some circumstances in which it may be impossible to do so, particularly when evaluating detection data in many real-world contexts. In such cases, the experimenter often has neither the luxury nor the feasibility of manipulating beta or using rating scales and so must use the data available from only a single stimulus-response matrix.

Under these circumstances, the measure $P(A)$ or the area under the ROC curve is an alternative measure of sensitivity (Calderia, 1980; Craig, 1979; Green & Swets, 1988). Such a measure represents the area to the right and below the line segments connecting the lower left and upper right corners of the ROC space to the measured data point (Figure 2.7). Craig (1979) and Calderia (1980) have argued that the advantage of this measure is that it is "parameter free." That is, its value does not depend on any assumptions concerning the shape or form of the underlying signal and noise distributions. For this reason, it is a measure that may be usefully employed even if two or more points in the ROC space are available but do not fall along a 45-degree line. (This suggests that the data do not meet the equal variance assumptions.) The measure $P(A)$ may be calculated from the formula

$$P(A) = \frac{P(H) + [1 - P(FA)]}{2} \tag{2.4}$$

Values of A' are represented in Appendix B. The reader is referred to Calderia (1980), Craig (1979), Swets and Pickett (1982), and Green and Swets (1988) for discussions of the relative merits of different sensitivity and bias measures.

APPLICATIONS OF SIGNAL DETECTION THEORY

Signal detection theory has had a large impact on experimental psychology, and its concepts are highly applicable to many problems of human factors as well. Its benefits can be divided into two general categories: (1) It provides the ability to compare sensitivity and therefore the quality of performance between conditions or between operators that may differ in response bias. (2) By partitioning performance and therefore performance change into bias and sensitivity components, it provides a diagnostic tool that recommends different corrective actions depending on whether a deterioration of performance results from a loss of sensitivity or a shift in response bias.

The implications of the first category are clear. The performance of two operators or the hit rate obtained from two different pieces of inspection equipment are compared. If A has a higher hit rate but also a higher false-alarm rate than B, which is superior? Unless the explicit mechanism for trading off hits for correct rejections and false alarms for misses is available, this comparison is impossible. Signal detection theory provides the mechanism.

The importance of the second point—the diagnostic value of signal detection theory—will be evident as we consider some actual examples of applications of signal detection theory to real-world tasks. In the myriad of possible environments in which the operator must detect an event and does so imperfectly, generating either misses or false alarms, the existence of these errors presents a challenge for the engineering psychologist: Why do they occur, and what corrective actions can prevent them? Two areas of application, medical diagnosis and eyewitness testimony, will be considered, leading to a more extensive discussion of vigilance.

Medical Diagnosis

The realm of medical diagnosis is a fruitful environment for the application of signal detection theory (Lusted, 1971, 1976; Parasuraman, 1985). Abnormalities (diseases, tumors) are either present in the patient or they are not, and the physician's initial task is often to make a yes or no decision. The strength of the signal (and therefore the sensitivity of the human operator) is related to factors such as the salience of the abnormality or the number of converging symptoms, as well as the training of the physician to focus on relevant cues. Response bias, meanwhile, can be influenced by both signal probability and payoffs. In the former category, influences include the disease prevalence rate and whether the patient is examined in initial screening (probability of disease low, beta high) or referral (probability higher, beta lower). Lusted (1976) has argued that physicians' detections generally tend to be less responsive to variation in the disease prevalence rate than optimal. Parasuraman (1985) found that radiologist residents were not responsive enough to differences between screening and referral in changing beta. Both of these results illustrate the sluggish beta phenomenon.

Factors that influence payoffs include the difficult to quantify consequences of hits (e.g., a detected malignancy that will lead to its surgical removal with associated hospital costs and possible consequences), false alarms (an unnecessary operation), and misses. Placing values on these events based on financial costs of surgery, malpractice suits, and intangible costs of human life and suffering is clearly difficult. Yet there is little doubt that they do have an important influence on a physician's detection rate. Lusted (1976) and Swets and Pickett (1982) have shown how diagnostic performance can be quantified with an ROC curve. In a very thorough treatment, Swets and Pickett go into considerable detail describing the appropriate methodology that should be employed when using signal detection theory to examine performance in medical diagnosis.

Several investigations have examined the more restricted domain of tumor diagnosis by radiologists, in which performance, at least theoretically, is far from optimal. Rhea, Potsdaid, and DeLuca (1979) have estimated the rate of omission in the detection of abnormalities to run between 20 and 40 percent. Swennsen, Hessel, and Herman (1977) have examined the effect on detection of directing the radiologist's attention to a particular area of an X-ray plate in which an abnormality is likely to occur. They find that such focusing of attention will indeed increase the likelihood of the tumor's detection, but will do so by reducing beta rather than increasing sensitivity. Parasuraman (1985), comparing the detection performance of staff radiologists and residents, found major differences between the two populations in terms of sensitivity (favoring the staff radiologists) and bias (radiologists showing a more conservative criterion in general).

Recognition Memory and Eyewitness Testimony

The domain of recognition memory represents a somewhat different application of signal detection theory. Here we are not assessing the correctness of a decision concerning whether or not a physical signal is present but rather a decision concerning whether or not a physical stimulus (the person or name to be recognized) "matches" a trace in memory. One important application of signal detection theory to memory is found in the study of eyewitness testimony (Buckhout, 1974; Ellison & Buckhout, 1981; Wells, Lindsay, & Ferguson, 1979). The witness to a crime may be called on to recognize or identify a suspect as the perpetrator. The four kinds of joint events in Figure 2.1 can readily be specified. The suspect examined by the witness either is (signal) or is not (noise) the same individual actually perceived at the scene of the crime. The witness in turn can either say, "That's the one" (Y) or "No, it's not" (N).

In this case, the joint interests of criminal justice and protection of society are served by maintaining a high level of sensitivity while keeping beta neither too high (many misses, with criminals more likely to go free) nor too low (a high rate of false alarms, with an increased likelihood that innocent individuals will be prosecuted). Signal detection theory has been most directly applied to a witness's identification of suspects in police lineups (Buckhout, 1974; Ellison & Buckhout, 1981). In this case, the witness is shown a lineup of five or so individuals, one of whom is the suspect detained by the police, and the others are "foils." The decision is now a two-stage process: Is the suspect in the lineup, and if so, which one is it?

Ellison and Buckhout (1981) have expressed concern that witnesses generally have a low response criterion in lineup identifications and will therefore often say yes to the first question. This bias would present no difficulty if their recognition memory was also accurate, enabling them to identify the suspect accurately. However, considerable research on staged crimes shows that visual recognition of brief events is notoriously poor (Buckhout, 1974; Loftus, 1979; Wells, Lindsay, & Ferguson, 1979). Poor recognition memory coupled with the risky response bias in turn allows those conducting the lineup to use techniques that will capitalize on witness bias to ensure a positive identification. These techniques include such things as ensuring that the suspect is differently dressed, is seen in handcuffs by the witness before the lineup, or is quite different in appearance from the foils. In short, techniques are used that would lead even a person who had *not seen* the crime to select the suspect from the foils or would lead the witness to make a positive identification from a lineup that *did not contain the suspect* (Ellison & Buckhout, 1981). This process is not testing the sensitivity of recognition memory. It is emphasizing response bias.

In applying signal detection theory to this procedure, investigators are interested in characteristics of the lineup process that might alter the magnitude of response bias toward the optimum. Three procedures have been suggested: (1) Buckhout (1974) has found that those who express greater confidence in their positive identifications ("I'm sure that's the one") are actually less sensitive observers than those who are less confident. (2) Ellison and Buckhout (1981) suggest the simple procedure of informing the witness that the suspect may not be in the lineup. As in reducing the apparent probability of a signal, the procedure will drive beta upward toward a more optimal setting. (3) Ellison and Buckhout argue that individuals in the lineup should all be equally similar to one another (such that a nonwitness would have an equal chance of picking any of them). Although such greater similarity will reduce the hit rate slightly, it will reduce the false-alarm rate considerably more. The result will be a net increase in sensitivity.

VIGILANCE

The vigilance paradigm is one of the most common applications of signal detection theory. In this paradigm the operator is required to detect signals over a relatively long period of time (commonly referred to as the watch), and the signals are intermittent, unpredictable, and infrequent. Examples include the radar monitor, who must observe infrequent contacts; the airport security inspector, who examines x-rayed carry-on luggage; the supervisory monitor of complex systems, who must detect the infrequent malfunctions of system components; and the quality control inspector, who examines a stream of products (sheet metal, circuit boards, microchips, fruits) to detect and remove defective or flawed items.

Two very general conclusions emerge from the analysis of operators' performance in these situations: (1) Operators are far from optimal, in particular, often showing a higher miss rate (or detection at longer latencies) than desirable. (2) In some situations, operators' detection performance as measured by P(hit) declines during the first half hour or so of the watch. This phenomenon was initially noted in radar monitors during World War II (N. H. Mackworth, 1948) and observed in industrial inspectors (Harris & Chaney, 1969; Parasuraman, 1986). Referred to as the vigilance *decrement* (loss in performance over time), it is distinguished from the vigilance *level*, which is the steady-state level of vigilance performance.

Vigilance Phenomena

It is important to distinguish two classes of vigilance situations, or paradigms. The *free-response* paradigm, such as that confronting the power plant monitor, is one in which a target event may occur at any time and nonevents are not defined. Event frequency in this case is defined by the number of targets per unit of time. The *inspection* paradigm, the quality control inspector's task, is one in which events occur at fairly regular intervals. A few of these are targets (defects), but most are nontargets (normal items). *Event frequency* is therefore an ambiguous term in the inspection paradigm because it may be defined either by the number of targets per unit of time or by the ratio of targets to total events (targets and nontargets). The latter measure, of course, may be held constant as the number of targets per unit of time is increased simply by increasing the total event rate, by speeding up a conveyor belt, for example. In typical industrial inspection tasks, target event frequency may be fairly high — around 5 to 25 percent. However, in other tasks, such as that of the airport security inspector, it will be much lower.

A large number of investigations of the vigilance decrement have been conducted over the last four decades with a myriad of experimental variables in various paradigms. Books by Broadbent (1971), Davies and Parasuraman (1980), Mackie (1977), J. F. Mackworth (1970), and Warm (1984), and chapters by Parasuraman (1986) and Parasuraman, Warm, and Dember (1987) provide a comprehensive compilation and summary of these studies. Some of the most salient general results are as follows:

1. The level of vigilance performance *decreases*, and the size of the vigilance decrement *increases*, as a target signal is made less *salient* by reducing its intensity or its duration or by increasing its similarity with background nontarget events.

2. The vigilance level will decrease and the decrement will increase to the extent that there is *time* or *location uncertainty* regarding when and where the target signal will appear. This uncertainty is particularly great if there are long intervals between signals.
3. For inspection tasks, which have background as well as target events, the vigilance level will be reduced and the decrement will increase when the background event rate is increased. The change will hold true, no matter whether the target/background ratio is held constant so both targets and backgrounds are more frequent (e.g., speeding up a conveyer belt for inspectors) or whether background events are more densely spaced between targets.
4. Successful performance on vigilance tasks requires sustained attention, which is particularly demanding of mental resources (Deaton & Parasuraman, 1988).

Specifying vigilance performance in terms of signals detected and missed has important implications for the design of systems. However, it should be apparent by now that defining vigilance performance only by P(hit) is ambiguous since in terms of signal detection theory $P(H)$ may decline either through a loss of sensitivity or through an increase in beta with no change in sensitivity (i.e., the subject makes fewer false alarms but also makes fewer hits).

In inspection tasks, when nontarget events are clearly defined, the application of the analysis of signal detection theory is straightforward since false alarms and the false-alarm rate may be easily computed. In the free-response paradigm, however, when there is no "nontarget event," further assumptions must be made in computing a false-alarm rate so that the opportunity to make a false alarm can be specified numerically. In the laboratory this is typically accomplished by defining an appropriate response interval after each signal within which a subject's response will be designated a hit. The remaining time during a watch is partitioned into a number of "false-alarm intervals" equal in duration to the response intervals. $P(FA)$ is simply the number of false alarms divided by the number of false-alarm intervals (Parasuraman, 1986; Watson & Nichols, 1976; Wickens & Kessel, 1979). However, there are certain cautions that need to be considered when signal detection theory is applied to vigilance, particularly when the false-alarm rates are quite low. Articles by Craig (1977); Jerison, Pickett, and Stenson (1965); and Long and Waag (1981) discuss some of these cautions.

Theories of Vigilance

An exhaustive listing of all of the experimental results of vigilance studies is far beyond the scope of this chapter. In any case, such a list would also be less than fully informative because methodological and procedural differences between laboratories have often generated seemingly contradictory results. Instead, it is more instructive to present three major theories of vigilance (in particular, the vigilance decrement) that predict when a loss in vigilance performance is apt to occur. No single theory is likely to be entirely correct, to the exclusion of others. Rather, under some circumstances the influence of factors underlying one theory will be stronger than others. Also, it is valuable to realize that although the theories have been offered to account for a change in vigilance performance *over time* (i.e., the decrement), their underlying constructs may account for vigilance differences between conditions that differ as a result of any other factor. These might include, for example, differences between displays, between tasks, or between environmental stressors. The advantage of such theories is that they provide somewhat parsimonious ways of accounting for vigilance loss and thereby suggest techniques to improve vigilance performance. After outlining three theories of vigilance, we will then show how they suggest corrective improvements.

The reduction of P(hit) over the watch has been observed in different laboratories to be a consequence of either a conservative shift in beta, a loss in sensitivity, or both. Of the three theories that are considered, the first will concern the cause of sensitivity loss, and the last two will account for criterion shifts.

Sensitivity Loss: Fatigue and Memory Load The earliest laboratory demonstrations of the vigilance decrement were reported by N. H. Mackworth (1948) in the classical "clock task." In this task, the subject monitored a clock hand which would tick at periodic intervals (events) and occasionally undergo a "double tick" (signal), moving twice the angle of the normal events. This task was chosen as the laboratory analog of a radar monitor's task, in which each sweep of the scope defines an event and an occasional contact on the scope defines the target. In another version, the clock hand would move continuously and occasionally pause. The former version is an inspection task, the latter a free-response task. In this paradigm, as well as in subsequent ones that have primarily employed visual signals, a loss in sensitivity usually occurs over time. Broadbent (1971) has argued that the sustained attention necessary to fixate the clock hand or other visual signals continuously extracts a toll in *fatigue*. Because of this fatigue, the subject looks away more often as the watch progresses, and therefore signals are missed.

More recently, investigators have concluded that any vigilance task that imposes a sustained load on working memory, to recall what the target signal looks or sounds like, will demand the continuous supply of processing resources (Parasuraman, 1979). This demand is every bit as fatiguing as the sustained demand to keep one's eyes open and fixated, and here too the eventual toil of fatigue will lead to a loss in sensitivity. This loss will be greatest when the event rate is high (Parasuraman, 1979; Parasuraman, Warm, & Dember, 1987). A further implication of the resource-demanding nature of vigilance tasks with working memory load is their susceptibility to interference from concurrent tasks, as will be discussed in Chapter 9.

Criterion Shifts: Arousal Theory In most vigilance tasks in which the hit rate declines, the decline is paralleled by a reduction in false alarms. On some occasions the shift in the false-alarm rate is sufficient to suggest that there is no loss in sensitivity at all but only a conservative adjustment of the response criterion (Broadbent & Gregory, 1965). *Arousal theory*, proposed by Welford (1968), postulates that in a prolonged low-event environment, the "evidence variable" X (see Figure 2.3) shrinks while the criterion stays constant. This change is shown in Figure 2.8. The shrinking results from an overall loss in total activity (both signal and noise) in the nervous system with decreasing arousal. An examination of Figure 2.8 reveals that such an effect will reduce both the hit rate and the false-alarm rate (a change in beta) while keeping the separation of the two distributions, as expressed in standard scores, at a constant level (a constant d').

The arousal view has been supported by several findings: Physiological manifestations of arousal correlate over the watch with the criterion shift (e.g., Dardano, 1962); drugs that increase arousal will reduce the vigilance decrement (N. H. Mackworth, 1950); and background noise (which increases arousal) will decrease the decrement, as will any extraneous interruption like a phone call (e.g., McGrath, 1963; see Welford, 1968, for a summary of this evidence).

Despite the success of arousal theory in accounting for some aspects of vigilance performance, there is good evidence that this theory is insufficient to explain all occurrences of criterion shifts. This evidence is provided by instances in which a manipulated variable that would be expected to influence

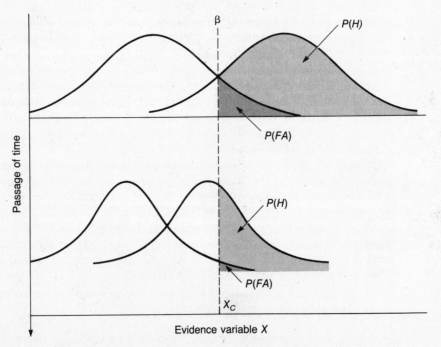

Figure 2.8 Welford's arousal theory of the vigilance decrement.

the arousal level does not produce the expected effect on the criterion. For example, as we have seen, increasing the total event rate, while keeping the absolute target frequency constant (thereby decreasing the target/background event ratio), will increase the decrement (Baddeley & Colquhoun, 1969; Jerison, Pickett, & Stenson, 1965). Yet arousal theory should predict the opposite effect since the more frequent nontarget events should increase arousal, not decrease it. Hence, another theory is needed. Expectancy theory, therefore, has been proposed to account for these and other nonarousal effects on the criterion shift.

Criterion Shifts: Expectancy Theory The expectancy theory proposed by Baker (1961) attributes the vigilance decrement to an upward adjustment of the response criterion in response to a reduction in the perceived frequency (and therefore expectancy) of target events. Part of the cause of the decrement, expectancy, often reflects an artifact of the preexperimental training sessions. During these sessions subjects typically hear (or see) a series of targets at a fairly high rate to give them several examples of what to listen (look) for. Then the experiment begins and the target rate is reduced to the lower frequency of the vigilance condition. In the early part of the experiment, beta then "drifts" upward to a new steady-state value, actually approaching the optimum value provided by Equation 2.1 (Craig, 1978). Colquhoun and Baddeley (1967) have shown that if the pretraining is provided at a very low signal rate, the decrement is reduced.

Beyond the pretraining artifact, vigilance decrements do occur, both in experiments when the pretraining signal rate is low and in situations outside of the laboratory (Parasuraman, Warm, & Dember, 1987). In these circumstances, expectancy theory explains a criterion shift in the following fashion. It is assumed that the subject sets beta on the basis of a subjective perception of signal frequency, $P_s(S)$. Then if a signal is missed for any reason, subjective probability $P_s(S)$ is reduced because the subject believes that one less signal occurs. This reduction in turn causes an upward adjustment of beta, which further increases the likelihood of a miss. The consequent increase in the miss rate decreases $P_s(S)$ further, and so on. Broadbent (1971) has labeled this phenomenon the "vicious circle" hypothesis, which would lead to an upward spiraling of beta and a downward spiraling of $P(H)$. Although in theory this behavior could lead to an infinite beta and a negligible hit rate, in practice other factors will operate to level off the criterion at a stable but higher value. Since the vicious circle depends on signals being missed in the first place, it stands to reason that the kinds of variables that reduce sensitivity (short, low-intensity signals) should also increase the expectancy effect in vigilance. This, in fact, is the case (Broadbent, 1971).

Techniques to Combat the Loss of Vigilance

It is probably true that no single theory of the vigilance decrement is totally right and the others are wrong. Instead, some aspects of all theories appear to be operating on performance at any given time, and total vigilance performance reflects some combination of the various effects described. Very often, for example, both d' and beta shifts are observed in a single vigil. It is also important to reemphasize that the theoretical mechanisms proposed to account for the vigilance decrement also account for differences between conditions or tasks that are not related to time. Thus, although Teichner (1974) has pointed out that the vigilance decrement is typically small (around 10 percent) in most laboratory studies, and as we have seen, in some real-world inspection tasks these miss rates can be substantial (Craig, 1989), the point remains that corrective techniques that can reduce the decrement will also improve the absolute level of performance. Like the theories of vigilance, these corrective techniques may be categorized into those that affect the response criterion and those that enhance sensitivity.

Shift in Response Criterion   Often, an unsatisfactory vigilance or inspection performance occurs simply because the operator's perceptions of the probability of signals or the costs of errors do not agree with reality. For example, in quality control, an inspector may believe that detecting defects is much more important than falsely rejecting good parts, when in fact both kinds of events are equally important. Sometimes simple instructions in industrial or company policy can adjust beta to an appropriate level. Thus, for example, in airlines security inspection, increased stress on the seriousness of misses (failing to detect a weapon smuggled through the inspection line) caused a substantial decrease in the number of misses from 1987 to 1988.

Less direct means can also adjust the response criterion to a more optimal level. For example, where possible, *knowledge of results* should be provided to allow an accurate estimation of the true $P(S)$ (N. H. Mackworth, 1950). Where this is not possible, Baker (1961) and Wilkinson (1964) have argued that introducing *false signals* will keep beta lower than it might be. This introduction will raise the subjective $P_s(S)$ and might raise the arousal level as well. Furthermore, if the false signals are physically similar to the real signals, by refreshing the standard of memory this procedure should lower the burden on working memory. As discussed earlier, it will reduce one source of departure of d' from its optimal. For example, as applied to the quality control inspector, a certain number of predefined defectives might be placed on the inspection line. These would be "tagged," so that if missed by the inspector they would still be removed. Their presence in the inspection stream should guarantee a higher $P_s(S)$ and therefore a lower beta than would be otherwise observed. There is, however, a considerable danger in employing this technique if the actions that the operator would take after detection should have undesirable consequences for an otherwise stable system. An extreme example would occur if false warnings were introduced into a chemical process control plant and these led the operator to shut down the plant unnecessarily.

Other techniques to combat the decrement have focused more directly on arousal and fatigue. Welford (1968) has argued persuasively that any event (such as a phone call, drugs, or noise) that will sustain or increase arousal should reduce the decrement. Using biofeedback techniques, Beatty, Greenberg, Deibler, and O'Hanlon (1974) have shown that operators trained to suppress theta waves (brain waves at 3–7 Hz, indicating low arousal) will also reduce the decrement.

Increasing Sensitivity   A logical outgrowth of Parasuraman's (1979) theory of vigilance loss is that any technique that will enhance the subject's memory of signal characteristics should reduce sensitivity decrements and preserve a higher overall level of sensitivity. Furthermore, as noted, any technique that combats the loss of sensitivity will also help reduce the decrements caused by expectancy. Indeed, in the previous section it was noted that the introduction of false signals could improve sensitivity by refreshing memory. It would seem also that the availability of a "standard" representation of the target should help. For example, Kelly (1955) reported a large increase in detection performance when quality control operators could look at television pictures of idealized target stimuli.

A study by Childs (1976), which found that subjects perform better when monitoring for only one target than when monitoring for one of several, is also consistent with the importance of memory aids in vigilance. Childs also observed an improvement in performance when subjects were told specifically what the target stimuli were rather than what they were not. His recommendation is that inspectors should have access to visual representation of possible defectives rather than simply the representation of those that are normal. Parasuraman's (1979) theory also suggests that high-event rates can produce larger losses in vigilance performance. As Saito (1972) showed in a study of bottle inspectors, a reduction of the inspection rate from 300 to under 200 bottles per minute markedly improved inspection efficiency.

Various artificial techniques of *signal enhancement* are closely related to the reduction in memory load. A trivial example of such a technique is, of course, simply to amplify the energy characterizing the signal events. But this approach may magnify the noise as much as the signal and therefore will do nothing to change the overall signal-to-noise ratio. More ingenious solutions capitalize on procedures that will differentially influence signal and nonsignal stimuli. Luzzo and Drury (1980) developed a signal-enhancement technique known as "blinking" that also reduces memory load. When successive nontarget events are physically similar to one another (as in the inspection of wired circuit boards), detection of miswired boards can be facilitated by a display technique that rapidly and alternately projects, at a single location, an image of two successive items: a known good prototype and the item to be inspected. If the latter is "normal," the image will be identical, fused, and continuous. If the item contains a malfunction (e.g., a gap in wiring), the gap location will appear as a stimulus that will turn on and off in a highly salient fashion as the displays are alternated.

A related signal enhancement technique is visual monitoring tasks is to induce coherent motion into the targets but not into the noise, thereby taking advantage of the human being's high sensitivity to target motion. For example, an operator scanning a radar display for a target blip among many similar noise blips encounters a very difficult detection problem. Scanlan (1975) has demonstrated how system design can capitalize on the fact that the radar target undergoes a coherent but slow motion whereas the noise properties are random. If the successive radar frames are stored, a number of recent frames can be replayed in fast time, forward and backward. Under these conditions the target's coherent motion suddenly stands out as a highly salient and easily detectable feature, greatly improving detection performance.

An alternative novel approach is to transcribe the events to the alternate sensory modality. This technique takes advantage of the *redundancy gain* that occurs when the signal is presented in two modalities at once. Employing this technique, Colquhoun (1975) and Lewandowski and Kobus (1989) found that sonar monitors detected targets more accurately when the target image was simultaneously displayed on a visual scope and as a transcribed auditory pattern than when either mode was employed by itself.

A technique closely related to the enhancement of signals through display manipulations is one that emphasizes operator training. Fisk and Schneider (1981) demonstrated that the magnitude of a vigilance decrement could be greatly reduced by training subjects to respond repeatedly to the target elements. This technique of developing *automatic processing* of the stimulus, which will be described in Chapter 5, tends to make the target stimulus "jump out" of the train of events just as one's own name is heard in a series of words. The conclusion of their study is quite consistent with Parasuraman's theory since a characteristic of automatic processing is its limited dependence on working memory (Schneider & Shiffrin, 1977).

Conclusions

Despite the plethora of vigilance experiments and the wealth of experimental data the application of research results to real-world vigilance phenomena has not yet been extensive. This is somewhat surprising in light of the clear shortcomings in many inspection tasks, with miss rates sometimes as high as 30 to 40 percent (Craig, 1984; Parasuraman, Warm, & Dember, 1987), and certain relatively salient findings from vigilance research. One possible reason why the results of laboratory studies have not been more fully applied relates to the discrepancy between the fairly simple stimuli, with known location and form, employed in many laboratory tasks and the more complex stimuli existing in the real world. The monitor of the nuclear power plant, for example, does not know precisely what configuration of warning indicators will signal the onset of an abnormal condition, but it is unlikely that it will be the appearance of a single near-threshold light in foveal vision. Some laboratory investigators have examined the effects of signal complexity and uncertainty (e.g., Adams, Humes, & Stenson, 1962; Childs, 1976; Howell, Johnston, & Goldstein, 1966). These studies are consistent in concluding that increased complexity or signal uncertainty will lower the absolute vigilance level. However, their conclusions concerning its influence on other vigilance effects (e.g., the size of the vigilance decrement), and therefore the generalizability of these effects to complex signal environments, have not been consistent.

A second possible reason why laboratory results have not been fully exploited relates to the differences in motivation and signal frequency between laboratory data and real vigilance phenomena. In the laboratory, signal rates may range from one an hour to as high as three or four per minute — low enough to show decrements; lower than fault frequencies found in many industrial inspection tasks; but far higher than rates observed in the performance of reliable aircraft, chemical plants, or automated systems, in which defects may occur at intervals of weeks or months. This difference in signal frequency may well interact with differences in motivational factors between the subject in the laboratory, performing a well-defined task and responsible only for its performance, and the real-time system operator confronted with a number of other competing activities and a level of motivation potentially influenced by large costs and benefits. This motivation level may be either lower or far higher than those of the laboratory subjects, but it will probably not be the same.

A study by Robert Earing, for example, found that there was little effect of expectancy on operators' detection of system failures as long as the failures were relatively salient. But if they were subtle, differences in signal expectancy, which might be anticipated between laboratory and real-world environments, exerted a strong effect on the response (Wickens & Kessel, 1981).

These differences do not mean that the laboratory data should be discounted. The basic variables causing vigilance performance to improve or deteriorate that have been uncovered in the laboratory will still probably affect detection performance in the real world, although the effect may be attenuated or enhanced. Rather, more data must be collected in real or highly simulated environments — such as those employed by Lees and Sayers (1976) and Crowe, Beare, Kozinsky, and Hass (1983) in the process control task, by Schmidke (1976) in maritime ship navigation monitoring, or by Ruffle-Smith (1979) in aviation — to verify the generalizability of the conclusions.

INFORMATION THEORY

The Quantification of Information

The discussion of signal detection theory was our first direct encounter with the human operator as a transmitter of information: An event (signal) occurs in the environment; the human perceives it and transmits this information to a response. Indeed, a considerable portion of human performance theory revolves around the concept of transmitting information. In fact, in any situation in which the human operator either perceives changing environmental events or is responding specifically to events that have been perceived, the operator is encoding or transmitting information. The aircraft pilot, for example, must process a multitude of visual signals bearing on the status of the aircraft while listening to auditory messages from air traffic control concerning flight plans and the status of other aircraft. A fundamental issue in engineering psychology is how to quantify this flow of information in terms that allow the diverse tasks confronting the human operator to be compared. Information theory accomplishes this goal. Given that we can associate processing efficiency with the amount of information an operator can process per unit of time and given that task difficulty can be associated with the rate at which information is presented, information theory provides a metric with which these measures can be compared across a wide number of different tasks.

Information is potentially available in a stimulus any time there is some uncertainty about what the stimulus will be. How much information is delivered by a stimulus depends in part on the number of possible stimuli that could occur in that context. If the same stimulus occurs on every trial, its occurrence conveys no information. If four events are equally likely, the amount of information conveyed by one of them when it occurs, expressed in *bits*, is simply equal to the base 2 logarithm of this number, for example, $\log_2 4 = 2$ bits. If there were only two alternatives, the information conveyed by the occurrence of one of them is $\log_2 2 = 1$ bit.

In human performance we are less interested in the amount of information in a stimulus than in the amount *transmitted* by the human operator to the response, a quantity designated H_T. The formal technique for computing information transmission will be described in the following pages, but an intuitive description will be given here. Obviously, if the operator responds correctly to all of the 2-bit stimuli, 2 bits of information are transmitted. If the operator ignores the stimuli and responds randomly, 0 bits are transmitted. If the operator makes some errors, performance ranges between these two limits. Thus the number of alternative stimuli, and therefore the amount of information in the input, places an upper bound on the maximum amount of information that an operator can transmit: $H_T \leq H_S$. With this intuitive description of information theory in mind, the reader may now skip to the heading "Absolute Judgment" at the end of the chapter to see its application in the absolute judgment task, rather than read the more thorough treatment of information theory that follows.

Formally, information is defined as the *reduction of uncertainty* (Shannon & Weaver, 1949). If before the occurrence of an event, you are less sure of the state of the world (you possess more uncertainty) than after, that event conveyed information to you. The statement "The sun rose this morning" conveys little information because before the occurrence of the event, you could anticipate it. On the other hand, the statement "Nepal is at war with Liechtenstein" conveys quite a bit of information. Your knowledge and understanding of the world are probably quite different after hearing the statement than they were before. Information theory formally quantifies the amount of information conveyed by a statement. This quantification is influenced by three variables: the number of possible events that could occur, N; the probabilities of those events, P; and their sequential constraints, or the context in which they occur. We will now describe how each of these three variables quantitatively influences the amount of information conveyed by an event.

Number of Events Before an event (which conveys information), a person has a state of knowledge that is characterized by uncertainty about some aspect of the world. After the event, that uncertainty is normally less. The amount of uncertainty reduced by the event occurrence in *bits* (binomial digits) is formally defined to be the average *minimum* number of true-false questions that would have to be asked to reduce the uncertainty. For example, the information conveyed by the statement "Bush won" after the 1988 election is 1 bit because the answer to one true-false question (e.g., "Did Bush win?" — "True" or "Did Dukakis win?" — "False") is sufficient to reduce the previous uncertainty. If, on the other hand, there were four candidates, all running for office, two questions would have to be asked and answered to eliminate uncertainty. In this case one question, Q1, might be "Was the winner from the liberal (or conservative) pair?" After this question was answered, Q2 would be "Was the winner the more conservative (or liberal) member of the pair?" Thus, if you were simply told the winner, that statement would formally convey 2 bits of information. This question-asking procedure assumes that all alternatives are equally likely to occur. Formally, then, when all alternatives are equally likely, the information conveyed by a stimulus H_S, in bits, can be expressed by the formula

$$H_S = \log_2 N \tag{2.5}$$

where N is the number of equally likely alternatives.

Because information theory is based on the minimum number of questions and therefore arrives at a solution in a minimum time, it has a quality of optimal performance. It is this optimal aspect that makes the theory attractive in its applications to human performance.

Probability In fact, events in the world do not always occur with equal frequency or likelihood. If you lived in the Arizona desert, much more information would be conveyed by the statement "It is raining today" than the statement "It is sunny." Your certainty of the state of the world is changed very little by knowing that it is sunny, but it is changed quite a bit (uncertainty is reduced) by hearing of the low-probability event of rain. In the example of the four election candidates, less information would be gained by learning that the favored candidate won than by learning that the candidate of the Socialist Workers or Libertarian party won. The probabilistic element of information is quantified by making rare events convey more bits. This in turn is accomplished by revising Equation 2.5 for the information conveyed by stimulus event i to be

$$H_S = \log_2 \left(\frac{1}{P_i} \right) \tag{2.6}$$

where P_i is the probability of occurrence of event i. This formula will increase H for low-probability events. Note that if N events are equally likely, each event will occur with probability $1/N$. In this case, Equations 2.5 and 2.6 are equivalent.

As noted, information theory is based on a prescription of optimal behavior. This optimum can be prescribed in terms of the order in which the true-false questions should be asked. If some events are more common or expected than others, we should ask the question about the common event first. In our four-candidate example we will do the best (ask the minimum number of questions on the average) by first asking, "Is the winner Bush?" or "Is the winner Dukakis?" assuming that Bush and Dukakis have the highest probability of winning. If instead the initial question was "Is the winner the Libertarian candidate?" or "Is the winner from one of the minor parties?" we have clearly "wasted" a question since the answer is likely to be no, and our uncertainty will probably be reduced by only a small amount.

The information conveyed by a single event of known probability is given by Equation 2.6. However, psychologists are often more interested in measuring the *average* information conveyed by a series of events with differing probabilities that occur over time — for example, a series of warning lights on a panel or a series of communication commands. In this case the average information conveyed is computed as

$$H_{ave} = \sum_{i=1}^{n} P_i \left[\log_2 \left(\frac{1}{P_i} \right) \right] \tag{2.7}$$

In this formula, the quantity within the outer brackets is the information per event as given in Equation 2.6. This value is now weighted by the probability of that event, and these weighted information values are summed across all events. Accordingly, frequent low-information events will contribute heavily to this average, whereas rare high-information events will not. If the events are equally likely, this formula will reduce to Equation 2.5.

An important characteristic of Equation 2.6 is that if the events are not equally likely, H_{ave} will always be less than its value if the same events are equally probable. For example, consider four events, A, B, C, and D, with probabilities of 0.5, 0.25, 0.25, and 0.125. The computation of the average information conveyed by each event in a series of such events would proceed as follows:

Event P_i	A 0.5	B 0.25	C 0.125	D 0.125
$\frac{1}{P_i}$	2	4	8	8
$\log_2 \frac{1}{P_i}$	1	2	3	3
$\sum P_i \left(\log_2 \frac{1}{P_i} \right)$ =	0.5 +	0.5 +	0.375 +	0.375 = 1.75 bits

This value is less than $\log_2 4 = 2$ bits, which is the value derived from Equation 2.5 when the four events are equally likely. In short, although low-probability events convey more information because they occur infrequently, the fact that they do occur infrequently causes their high-information content to contribute less to the average.

Sequential Constraints and Context   In the preceding discussion, probability has been used to reflect the long-term frequencies, or *steady-state* expectancies, that stimuli will occur. However, there is also a third contributor to information that reflects the short-term sequences of stimuli, or their *transient* expectancies. A particular stimulus may occur rarely in terms of its absolute frequency. However, given a particular *context*, it may be highly expected, and therefore its occurrence conveys very little information in that context. In the example of rainfall in Arizona, we saw that the absolute probability of rain is low. But if we heard that there was a large front moving eastward from California, our expectance of rain, given this information, would be much higher. That is, information can be reduced by the context in which it appears. As another example, the letter *u* in the alphabet is not terribly common and therefore normally conveys quite a bit of information when it occurs; however, in the context of a preceding *q*, it is almost totally predictable and therefore its information content, given that context, is nearly 0 bits.

Contextual information is frequently provided by *sequential constraints* on a series of stimuli. In the series of stimuli ABABABABAB, for example, $P(A) = P(B) = 0.5$. Therefore, according to Equation 2.7, each stimulus conveys 1 bit of information. But clearly the next letter in this sequence is almost certainly an A. Therefore, the sequential constraints reduce the information content in the same manner as a change in stimulus probabilities reduces information from the equiprobable case. Formally, the information provided by an event, given a context, may be computed in the same manner as in Equation 2.6, except that the absolute probability of the event P_i is now replaced by a *contingent* probability $P_i|X$, which is ready, "The probability of event i given that X has occurred." The variable X refers to the context.

Redundancy   To recapitulate, we note that three variables influence the amount of information that a series of stimuli can convey. The number of possible events, N, sets an upper bound on the maximum number of bits if all stimuli are equally likely, and changes in stimulus probability away from the equiprobable case and increases in sequential constraints both serve to reduce information from this maximum. The term *redundancy* formally defines this potential loss in information. Thus, for example, the English language is highly redundant because of two factors: All letters are not equiprobable (*e*'s vs. *x*'s), and sequential constraints such as those found in common digraphs like *ed*, *th*, or *nt* reduce uncertainty.

Formally, the *percent redundancy* of a stimulus set is quantified by the formula

$$\% \text{ redundancy} = \left(1 - \frac{H_{\text{ave}}}{H_{max}}\right) \times 100 \tag{2.8}$$

where H_{ave} is the actual average information conveyed taking into account all three variables (approximately 1.5 bits per letters for the alphabet) and H_{max} is the maximum possible information that would be conveyed by the N alternatives if they were equally likely ($\log_2 26 = 4.7$ bits for the alphabet). Thus the redundancy of the English language is $100(1 - 1.5/4.7) = 68$ percent. Wh-t th-s sug-est- is t-at ma-y of t-e le-ter- ar- not ne-ess-ry fo- com-reh-nsi-n. However, to stress a point that will be emphasized in Chapter 5, this fact does not negate the value of redundancy in many circumstances. We have seen already in our discussion of vigilance that redundancy gain can improve performance when perceptual judgments are difficult. At the end of this chapter we will see its value in absolute judgment tasks.

Information Transmission of Discrete Signals

In much of human performance theory, investigators are concerned not only with how much information is *presented* to an operator but also with how much is *transmitted* from stimulus to response, the *channel capacity*, and how rapidly it is transmitted, the *bandwidth*. Using these concepts, the human being is sometimes represented as an information channel, an example of which is shown in Figure 2.9. Consider the typist. First, information is present in the stimulus, the printed page from which the typist is typing. This value of stimulus information, H_S, can be computed by the procedures described, taking into account probabilities of different letters and their sequential constraints. Second, each response on the keyboard is an event, and so we can also compute response information, H_R, in the same manner. Finally, we ask if each letter on the page was appropriately typed on the keyboard. That is, was the information faithfully transmitted, H_T. If it was not, there are two types of mistakes: First, information in the stimulus could be lost, H_L (known as *equivocation*), which would be the case if a certain letter was not typed. Second, letters may be typed that were not in the original text. This is referred to as noise. Figure 2.9*a* indicates the relationship among these five information measures. Notice that it is theoretically possible to have a high value of both H_S and H_R but to have H_T equal to zero. This result would occur if the typist were totally ignoring the printed text (equivocation), creating his or her own message. (Formally this would be considered noise; more charitably it might be called creativity.) A schematic example of this case is shown in Figure 2.9*b*.

We will now compute H_T in the context of a four-alternative stimulus-response reaction time task rather than the more complex typing tasks. In this

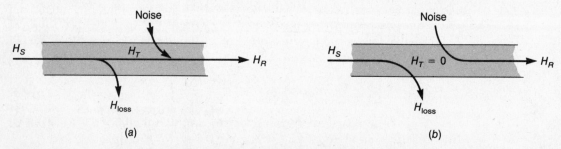

(a)   (b)

Figure 2.9 Information transmission and the channel concept: (*a*) information transmitted through the system; (*b*) no information transmitted.

task the subject is confronted by four possible stimuli, any of which may appear with equal probability, and must make a corresponding response for each.

When deriving a quantitative measure of H_T, it is important to realize that for an ideal information transmitter, $H_S = H_T = H_R$. In optimal performance of the reaction time task, for example, each stimulus (conveying 2 bits of information if equiprobable) should be processed ($H_T = 2$ bits) and should trigger the appropriate response ($H_R = 2$ bits). As we saw, in information-transmitting systems this ideal state is rarely obtained because of the occurrence of equivocation and noise.

The computation of H_T is performed by setting up a stimulus-response matrix, such as that shown in Figure 2.10, and converting the various numerical entries into three sets of probabilities: the probabilities of stimuli, shown along the bottom row; the probabilities of responses, shown along the right column; and the probabilities of a given stimulus-response pairing. These latter values are the probability that an entry will fall in each cell, where a cell is defined jointly by a particular stimulus and a particular response. In Figure 2.10a, there are four filled cells, with $P = 0.25$ for each entry. Each of these sets of probabilities can be independently converted into the information measures by Equation 2.7.

Once the quantities H_S, H_R, and H_{SR} are calculated, the formula

$$H_T = H_S + H_R - H_{SR} \tag{2.9}$$

allows us to compute the information transmitted. The rationale for using this formula is as follows: The variable H_S establishes the maximum possible transmission for a given set of stimuli and so contributes positively to the formula. Likewise, H_R contributes positively. However, to guard against situations such as that depicted in Figure 2.9b, in which stimuli are not coherently paired with

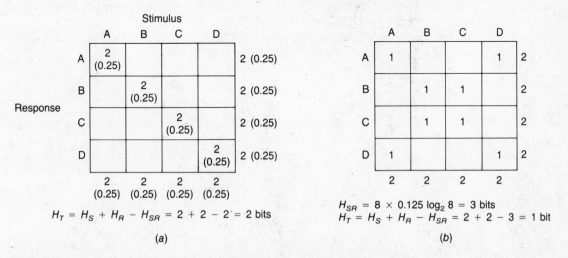

Figure 2.10 Two examples of the calculation of information transmission.

responses, H_{SR}, a measure of the dispersion or lack of organization within the matrix, is subtracted. If each stimulus generates consistently only one response (Figure 2.10*a*), the entries in the matrix should equal the entries in the rows and columns. In this case, $H_S = H_R = H_{SR}$, which means that by substituting the values in Equation 2.9, $H_S = H_T$. However, if there is greater dispersion within the matrix, there are more bits within H_{SR}. In Figure 2.10*b* this is shown by eight equally probable stimulus-response pairs, or 3 bits of information in H_{SR}. Therefore, $H_{SR} > H_S$ and $H_T < H_S$. In terms of Venn diagrams the relation between these quantities is shown in Figure 2.11. In this representation it is possible actually to add and subtract areas to obtain results similar to Equation 2.9.

Often the investigator may be interested in the information transmission *rate* expressed in bits/second rather than the quantity H_T expressed in bits. To find this rate, H_T is computed over a series of stimulus events, along with the average time for each transmission (i.e., the mean reaction time, *RT*). Then the ratio H_T/RT is taken to derive a measure of the *bandwidth* of the communication system in bits/second. This becomes a critical measure for much of human factors research because it expresses a measure of processing efficiency that accounts for both speed and accuracy and transcends a number of different tasks and measures. For example, a measure of typing or reaction time performance can be expressed in terms of bandwidth, and the same measure can also be used to express tracking performance. The following section will show how performance on continuous tasks such as tracking can also be described by information theory.

Information of Continuous Signals

Much of the information that is processed in complex systems (and in our daily lives) is continuous. Consider, for example, the curves in a road as we drive or the continuous deflections of a needle monitored by a pilot. Although these are

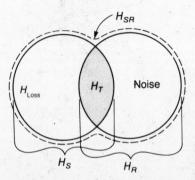

Figure 2.11 Information transmission represented in terms of Venn diagrams.

not discrete events, they clearly convey information to the operator in the same fashion as a warning light or a printed letter. When quantifying the bandwidth or stimulus information in continuous signals, it is first necessary to define the molecular event. This event is a reversal in the time series (a change in needle direction or the curve in the road). When expressed as a position function of time, the information conveyed by each reversal (in bits) is derived from the number of possible positions at which a reversal can take place. In Figure 2.12a each downward reversal conveys 1 bit since there are two locations at which the reversal could occur. The upward reversal is completely predictable in its location and so conveys no information. Thus there is, on the average, 0.5 bit/reversal across the continuous signal of Figure 2.12a.

When a signal is random, such as that shown in Figure 2.12b, it is necessary to make certain assumptions before the information content of each reversal can be assessed. If a reversal can occur at any level, the information, formally, is infinite. Therefore, it is necessary to assume the precision with which the position must be localized by the operator. If a meter is deviating between scale markers of 0 through 16 and the operator must read only the nearest scale marker, each deflection conveys $\log_2 16 = 4$ bits of information. From here it is a simple step to compute the information content of the continuous signal. This

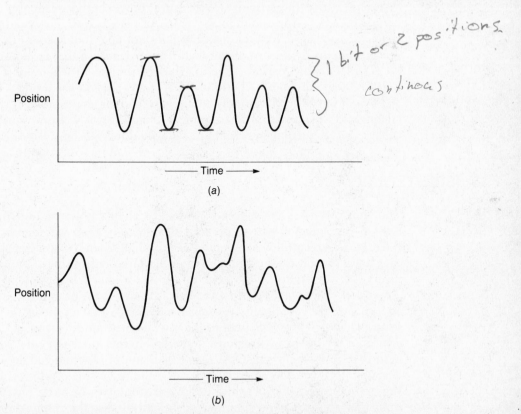

Figure 2.12 Continuous information: (a) one bit; (b) random.

simply becomes the information/reversal multiplied by the number of reversals/second, sometimes referred to as the *upper cutoff frequency* of the continuous signal. The product will be the *bandwidth* of the continuous signal expressed in bits per second. Continuous information transmission may be calculated by the same S-R matrix technique shown in Figure 2.12, once the continuous signal and response are divided into their discrete levels. There are, however, also more elegant continuous techniques that apply *Fourier analysis* (see Chapter 11) to compute an equivalent measure of *transinformation* (Baty, 1971). These techniques are beyond the scope of this treatment, however, and the reader is referred to Sheridan and Ferrell (1974) for a description of their use. In this text, the utility of continuous information theory will be emphasized in Chapter 5 on speech perception and Chapter 11 on continuous manual control.

Conclusion

In conclusion, it should be noted that information theory has its clear benefits —it provides a single combined measure of speed and accuracy that is generalizable across tasks—but it also has its limitations, discussed more fully by Wickens (1984). In particular, H_t measures only whether responses are *consistently* associated with stimuli in the matrix of Figure 2.10, not whether they are *correctly* associated, and the measure does not take into account the size of an error. Sometimes the size of the error is important, so that large errors should be penalized more than small ones. These errors occur when the stimulus and response scales lie along a continuum. Here we should use either a correlation coefficient or some measure of the integrated error across time, as discussed in Chapter 11. The relationship among H_t, measures of d' from signal detection theory, and percentage correct are described in Wickens (1984).

ABSOLUTE JUDGMENT

The human senses, although not perfect, particularly in the underload condition of the vigilance paradigm, are still relatively keen when contrasted with the detection resolution of machines. In this light then it is somewhat surprising that the limits of absolute judgment—the task confronting the operator who must assign a stimulus into one of three or more levels along a sensory dimension—are relatively severe. This is the task, for example, that confronts an inspector of wool quality who must categorize a given specimen into one of several quality levels; the police officer who must rapidly estimate the size of a crowd; or the operator who must immediately recognize the color of a warning light that appears on a display. Our discussion of absolute judgment will first describe performance when stimuli vary on only a single physical dimension. We will then consider absolute judgment along two or more physical dimensions that are perceived simultaneously and discuss the implications of these findings to principles of display coding.

Single Dimensions

Experimental Results In the typical laboratory experiment to investigate absolute judgment, a particular sensory stimulus continuum (e.g., tone pitch, light intensity, or texture roughness) and a number of discrete levels along that continuum (e.g., four tones of different intensities) are selected. These stimuli are then presented randomly to the subject one at a time, and the subject is asked to associate a different response to each one. From these data the experimenter can assess the extent to which each response matched the presented stimulus. In our discussion of a "typical" absolute judgment experiment, when four discriminable stimuli (2 bits) are presented, transmission (H_T) is usually perfect—at 2 bits. Then the stimulus set is enlarged, and additional data are also collected with five, six, seven, and more discrete stimulus levels, and H_T is computed each time by using the procedures described in the preceding section. Typically the results indicate that errors begin to be made when about five to six stimuli are used, and the error rate increases as the number of stimuli increase further. These results indicate that stimulus sets of this magnitude have somehow saturated the subject's capacity to transmit information about the magnitude of the stimulus. We say the subject has a maximum *channel capacity*. Bottle hecting

Graphically, these data can be represented in Figure 2.13, in which the actual information transmitted (H_T) is plotted as a function of the number of absolute judgment stimulus alternatives (expressed in informational terms as H_S). The 45-degree slope of the dashed line indicates perfect information transmission, and the "leveling" of the function takes place at the region in which errors began to occur (i.e., $H_T < H_S$). The level of the flat part or asymptote of the function indicates the channel capacity of the operator: somewhere between 2 and 3 bits. George Miller (1956), in a classic paper entitled "The Magical Number Seven Plus or Minus Two," noted the similarity of this level of asymptote across a number of different absolute judgment functions with different sensory continua. Miller concluded that the limits of absolute

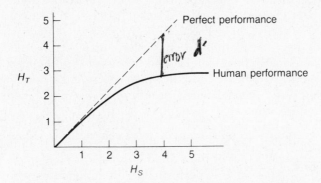

Figure 2.13 Typical human performance in absolute judgment tasks.

judgment at 7 ± 2 stimulus categories (2–3 bits) is a fairly general one that characterizes a number of different stimulus modalities. This limit does, however, vary somewhat from one sensory continuum to another; it is less than 2 bits for saltiness of taste and about 3.4 bits for judgments of position on a line.

The level of the asymptote does not appear to reflect a basic limit in sensory processing in the same manner that the signal detection parameter d' assesses a limit of sensory resolution. First, to associate the crude limits of absolute judgment with shortcomings of sensory processing seemingly is at odds with the fact that the senses are extremely keen in their ability to make *discriminations* between two stimuli (are they alike or different?). In fact, the number of adjacent stimulus pairs on a sensory continuum, such as tone pitch, that a human being can accurately discriminate is in the range of 1800 (Mowbray & Gebhard, 1961). Second, Pollack (1952) has observed that the limits of absolute judgment are little affected by whether the stimuli are closely spaced on the physical continuum, producing fewer possible discriminations, or are widely dispersed. Apparently, then, the limit is not sensory but is in the accuracy of the subject's *memory* for the representation of the four to eight different standards (Siegel & Siegel, 1972).

If, in fact, absolute judgment limitations are related to memory, there should be some association between this phenomenon and difference in learning or experience since differences in memory are closely related to those of learning.

It is noteworthy that sensory continua for which we demonstrate good absolute judgments are those for which such judgments in real-world experience occur relatively often. For example, judgments of position along a line (3.4 bits) are made in measurements on rulers, and judgments of angle (4.3 bits) are made in telling the time from analog clocks. High performance in absolute judgment also seems to be correlated with professional experience with a particular sensory continuum in industrial tasks (Welford, 1968) and is demonstrated by the noteworthy association of absolute pitch with skilled musicians (Carroll, 1975; Klein, Coles, & Donchin, 1984; Siegel & Siegel, 1972).

The question of pitch judgment and absolute pitch (the ability to identify correctly any heard note on the musical scale) has been of particular interest to psychologists. Naturally, the correlation between musical ability and absolute pitch does not necessarily imply that performance was learned since it is possible that superior abilities determined the musical proficiency rather than vice versa. The critical data are those in which controlled differences in experience or training can be associated with differences in pitch judgment. However, these data are somewhat ambiguous. Hartman (1954) has tried to train naive subjects to a high level of absolute pitch. He succeeded in improving performance somewhat, but only with relatively extensive training. The level obtained rarely reached that of the skilled musician or conductor. In contrast, Carroll (1975), comparing the performance of individuals who claimed to possess absolute pitch innately with that of one subject who taught himself absolute pitch, found that the latter performed equivalently in all respects to the former group.

A different source of evidence concerning the role of learning in absolute pitch is provided by the cross-cultural comparisons between pitch judgments of Javanese musicians and American musicians. In Javanese music there is no "standard" scale; musicians tune instruments differently on each occasion. In the absence of such a standard, performance of the Javanese musicians on an absolute pitch task was found to be inferior to that of Western musicians (Siegel & Siegel, 1972). In conclusion, the evidence clearly suggests that learning and experience play some role in the development of accurate absolute pitch judgment. Although differences in basic ability undoubtedly contribute as well, the exact proportion of this contribution is still uncertain.

Applications   The conclusions drawn from research in absolute judgment are highly pertinent to the performance of any task that requires operators to sort stimuli into levels along a physical continuum, particularly for industrial inspection tasks in which products must be sorted into various levels for pricing or marketing (e.g., fruit quality) or into categories for different uses (e.g., steel or glass quality). The data from the absolute judgment paradigm indicate the kind of performance limits that can be anticipated and suggest the potential role of training. Absolute judgment data are equally relevant to the issue of *coding*. In many cases some physical or conceptual continuum of importance in the performance of a task will be coded for display by variation along a displayed sensory continuum. For example, the size of socket wrenches may be coded by color so that they can be easily differentiated even when the digital size indicator cannot be read (Pond, 1979). A more conceptual dimension, such as the hierarchical level of a personnel unit in an organization or the degree of danger of a particular environment, may be coded into a number of different levels. It is, of course, possible to use letters or digits to identify the various levels, but in conditions of low visibility, high visual clutter, or high stress, these may not be read accurately. In this case basic data on the appropriate dimensions and the number of conceptual categories that can be employed without error are highly relevant to the development of such nonverbal display codes.

Moses, Maisano, and Bersh (1979) have cautioned that any conceptual continuum should not be arbitrarily assigned to a given physical dimension. They have argued that certain conceptual continua have a more "natural" association or compatibility with some physical display dimensions than with others. The designers of codes should be wary of the potential deficiencies (decreased accuracy, increased latency) imposed by an arbitrary assignment. For example, Moses, Maisano, and Bersh suggest that the representation of danger and unit size should be coded by the color and size of a displayed object, respectively, and not the reverse.

Multidimensional Judgment

If our limits of absolute judgment are so severe and seemingly can only be overcome by extensive training, how is it that we are able to recognize stimuli in the environment so readily? A major reason is that most of our recognition is based on the identification of some combination of two or more stimulus dimensions rather than levels along a single dimension. When a stimulus can vary on two (or more) dimensions at once, we make an important distinction between orthogonal and correlated dimensions. When dimensions of a stimulus are orthogonal, the level of the stimulus on one dimension can take on any value, independent of the other—for example, the weight and hair color of an individual. When dimensions are correlated, the level on one constrains the level on another—for example, height and weight, since tall people tend to weigh more than short ones.

Orthogonal Dimensions   The importance of multidimensional stimuli in increasing the total amount of information transmitted in absolute judgment has been repeatedly demonstrated (Garner, 1974). Egeth and Pachella (1969), for example, demonstrated that if 3.4 bits (10 levels) of information could be transmitted concerning position on a line, then when two lines were combined into a square, subjects could improve performance by transmitting 5.8 bits (57 levels) of information concerning the spatial position of a dot in the square. Note, however, that this improvement does not represent a perfect addition of channel capacity along the two dimensions. If processing along each dimension were independent and unaffected by the other, the predicted amount of information transmitted would be $3.4 + 3.4 = 6.8$ bits, or around 100 positions (10×10) in the square. Egeth and Pachella's results suggest that there is some loss of information along each dimension resulting from the requirement to transmit information along the other.

Going beyond the two-dimensional case, Pollack and Ficks (1954) combined six dimensions of an auditory stimulus orthogonally. As each successive dimension was added, subjects showed a continuous gain in total information transmitted but a loss of information transmitted per dimension. These relations are shown in Figure 2.14, with seven bits the maximum total capacity transmitted. In a related observation, Carroll (1975) suggested that the reason why people with absolute pitch are superior to those without does not lie in greater discrimination along a single continuum. Rather, those with absolute pitch make their judgments along two dimensions: the pitch of the octave and the value of a note within the octave. They have created a multidimensional stimulus from a stimulus that others treat as unidimensional.

Correlated Dimensions   The previous discussion and the data shown in Figure 2.14 suggest that combining absolute judgment dimensions orthogonally leads to a loss in information transmitted. As noted, however, there is another way of combining dimensions, which is to combine them in a *correlated* or *redundant* fashion. For example, the position and color of an illuminated traffic light are redundant dimensions. When the top light is illuminated, it is always red. In this case, H_s, the information in the stimulus, is no longer the sum of H_s across dimensions since this sum is reduced by the degree of correlation or redundancy between levels on the two dimensions. If the correlation is 1.0, as with the traffic light, total H_s is just the H_s on any single dimension (since all

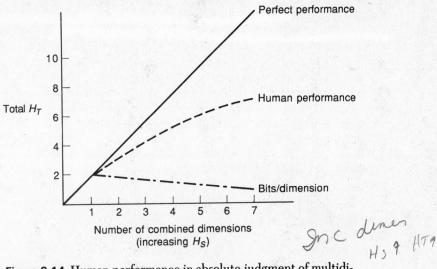

Figure 2.14 Human performance in absolute judgment of multidimensional auditory stimuli. As more dimensions are added, more total information is transmitted but less information is transmitted per dimension.

other dimensions are completely redundant). Clearly the maximum possible H_S for all dimensions in combination is now less than its value could be in the orthogonal case. However, Eriksen and Hake (1955) found that by combining dimensions redundantly, the information loss ($H_S - H_T$) is much less for a given value of H_S than it is when they are combined orthogonally, and the information transmitted (H_T) is greater than it would be along any single dimension.

In summary, we can see that orthogonal and correlated dimensions accomplish two different objectives in absolute judgment of multidimensional stimuli. Orthogonal dimensions maximize H_T, the efficiency of the channel; correlated dimensions minimize H_{loss}, that is, maximize the *security* of the channel.

Dimensional Relations: Integral, Separable, and Configurable   Orthogonal or correlated dimensions refer to properties of the *information* conveyed by a multidimensional stimulus, and not the physical form of the stimulus. However, it also turns out that combining dimensions in multidimensional absolute judgment tasks has considerably different effects depending on the nature of the *physical* relationship between the two dimensions. In particular, Garner (1974) has made the important distinction between an *integral* and a *separable* pair of physical dimensions, which behave quite differently when subjects sort stimuli according to their dimensional values. These pairs differ in the degree to which levels on the two dimensions in a pair can be independently specified. Separable dimensions occur when the levels along each of the two dimensions can be specified without requiring the specification of the level along the other. For example, the length of the horizontal and vertical lines radiating from the dot in

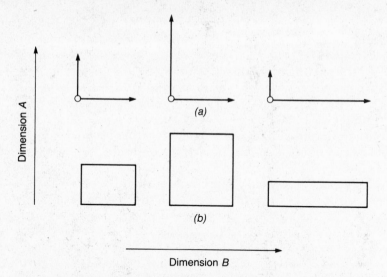

Figure 2.15 (*a*) Separable dimensions (height and width of a line segment), (*b*) integral dimensions (height and width of a rectangle). Dimension *A* is height; dimension *B* is width.

Figure 2.15*a* are two separable dimensions; one can be specified without specifying the other. For integral dimensions, this independence is impossible. The height and width of a single rectangle are integral because to display the height of a rectangle, its width must be given implicitly; otherwise, it would not be a rectangle (Figure 2.15*b*). Correspondingly, the color and brightness of an object are integral dimensions. Color cannot be physically represented without some level of brightness.

To reveal the differences between integral and separable dimensional pairs, experiments are performed in which subjects first sort or categorize along different levels of each dimension of a set of stimuli in which the other dimension is held constant. This is a *control condition*. In the stimulus example in Figure 2.15*b*, they might sort the rectangles by height while the width remains constant. In the *orthogonal condition*, they sort along each dimension while the other dimension takes on different and randomly changing values, which are to be ignored. Thus, as in Figure 2.15*b*, they might again sort by height, as width changes, even though the width is irrelevant to their task and hence should be ignored. Finally, in the *redundant sort condition*, the two dimensions are perfectly correlated. An example would be sorting rectangles whose height and width are positively correlated. Thus, the rectangles would all be of the same shape but would vary in size.

An experiment by Garner and Fefoldy (1970) revealed that when compared to sorting performance in the control condition, sorting with integral dimensions (e.g., rectangles), *helps* performance for the *redundant* condition, and *hurts* performance for the *orthogonal* condition. In contrast, when sorting with separable dimensions (the stimuli in Figure 2.15*a*), performance is little helped by redundancy and little hurt by the orthogonal variation of the irrelevant dimension. These differences between integral and separable dimensions are observed no matter whether performance is measured by accuracy (i.e., H_T), when several levels of each dimension are used, or speed, when only two levels of each dimension are used and accuracy is nearly perfect.

Exactly what makes dimensions integral or separable and how consistently the two symptoms of integrality appear together (redundancy gain and orthogonal cost) has been a matter of continued debate (e.g., Carswell & Wickens, 1990; Cheng & Pachella, 1984). It appears that there is really a continuum of integrality, and some dimensional pairs are quite integral (hue and brightness), others are somewhat so (height and width of a rectangle), others are fairly separable (color and shape of a geometric figure), and others are quite separable (the height of two parallel bar graphs). The fact that the concept of integrality is not absolute, with two categories (integral and separable), but is rather defined along a continuum does not diminish its importance. However, it is also important to consider two important qualifications to the general principles of separability and integrality.

First, dimensions that are normally separable *can* provide redundancy gain under degraded sensory conditions when differences in one dimension, the other, or both are quite hard to perceive (Garner, 1974). This aspect of redundancy gain was described in our discussion of vigilance. For example, auditory and visual signals are clearly separable but offer redundancy gain in vigilance.

The second qualification—really an elaboration—concerns correlated dimensions, such as the rectangles of varying shape shown in Figure 2.15. When dimensions are separable, like the color and shape of a symbol, it matters little which level of one dimension is paired with which level of the other (e.g., whether the red symbol is a square and the blue symbol is a circle, or vice versa). For integral dimensions, however, it often does make a difference. For example, when the height and width of rectangles are positively correlated, creating rectangles of constant shape and different size, performance is not as good as if the dimensions are negatively correlated, such as those shown in Figure 2.15, creating rectangles of different shapes (Lockhead & King, 1977; Weintraub, 1971). Pairs of dimensions for which the pairing of particular levels makes a difference are referred to as *configurable* (Carswell & Wickens, 1990; Pomerantz & Pristach, 1989). In the case of the rectangles in Figure 2.15, the dimensions *configure* in such a way as to make a new property, shape, which simply does not exist when the dimensions are separate. Pomerantz (1981) has referred to the *emergent properties* that can be produced when configurable dimensions are combined. These emergent properties, or emergent features, like the shape and size of a rectangle, will have important implications for object displays, as will be discussed in the next chapter.

Summary Before we discuss the practical implications of multidimensional absolute judgment, it is important to summarize briefly the concepts we have introduced. The information conveyed by a set of stimuli may vary independently on several dimensions at once. If the human operator is asked to classify all of these dimensions, more total information can be transmitted than if stimuli vary on only one dimension, but the information transmitted per dimension will be less (H_{loss} will increase). This loss will increase as more dimensions are varied, although it will not be so great if the dimensions are separable rather than integral.

The information conveyed by stimulus dimensions may also be correlated, which produces a redundancy gain in the speed and accuracy of information transmission; the gain will be greatest when the dimensions are integral but may also be found with separable dimensions when perceptual processing is degraded. Finally, stimulus dimensions may combine differently, sometimes configuring to make a new emergent property. If they do, the dimensions are said to be configurable, and performance will differ depending on which levels of one dimension are paired with which levels of the other.

Implications of Multidimensional Absolute Judgment As with single dimensions, the implications of multidimensional absolute judgment performance for engineering psychology may fall into two different classes, depending on the nature of the task. In the *sorting task*, the human operator sorts natural stimuli over whose physical form the human factors designer has little control. This may be the industrial inspector who is sorting a manufactured product (e.g., a piece of sheet metal) into categories along some dimension. In this task it is important to realize that the quality of sort will be less if the stimuli vary along other uncorrelated and integral dimensions in addition to the relevant one for sorting. This will be the case, for example, if the product sorted by hue varies not only in its hue but also in its shininess or texture.

In contrast, in the *symbolic categorization task*, the operator is confronted by an artificial symbol, created on a display by the system designer, which is to be interpreted as representing levels on two or more information dimensions. The meteorologist, for example, will view a weather station symbol on the display. The symbol might represent levels on several ordered dimensions: temperature, humidity, altitude, wind speed, and direction. How should these dimensions be physically represented? Of course, one choice is to represent them digitally. Such a display is good for precise checking of a single-station reading. But if the stations are close together, digital listing becomes cluttered and makes inefficient use of display space. In addition, it is not so easy to take a rapid glance at a digitally coded display and, for example, visualize all of the stations with a temperature reading of greater than 20 degrees to form a global understanding of the temperature gradient. The task would be much easier if temperature were coded by an analog variable like color. (See also Chapters 3 and 4.)

Given that a choice has been made to code displays by continuous dimensions (rather than digitally), the aspects of multidimensional absolute judgment become relevant. Is there a correlation between the displayed variables (e.g., higher altitudes would probably be associated with colder temperatures)? If so, it can lead to an improvement in performance on each dimension, particularly if integral dimensions are used. But if the correlation is less than perfect (i.e., less than complete redundancy), the ability to read one dimension and ignore differences on the other will be hurt by using integral dimensions.

Finally, if it is critical to avoid information loss on any particular variable, complete redundancy is the answer. The designer should use two physical dimensions to code changes in one of the represented quantities (e.g., use the height of a bar and its color to represent temperature). The redundancy gain will probably be greater if the dimensions are integral but will be realized even with separable dimensions. For example, Kopala (1979) found that Air Force pilots were better able to encode information concerning the level of threat of displayed targets when such information was presented redundantly by shape and color than when it was presented by either dimension alone. Stoplights are also a good example of redundant coding: The position of the light is perfectly correlated with the color. So are auditory warning systems, in which the loudness of the horn will be correlated with its distinct pitch to indicate the seriousness of an alarm condition. These dimensions are integral since a pitch must have a level of loudness.

TRANSITION

In this chapter, we have seen how people classify stimuli into two levels along one dimension, several levels along one dimension, and several levels along several dimensions. In our discussion of the first of these tasks, we saw that signal detection was characterized by the probabilistic element of decisions under uncertainty; this characteristic will be addressed in much more detail in Chapter 7 as we discuss more complex forms of decision making. In our treatment of multidimensional absolute judgment, we saw that more information could be transmitted as dimensions were combined, and indeed most complex patterns that we encounter and classify in the world are multidimensional. We discuss these elements of *pattern recognition* in Chapter 5. Finally, it is apparent that when human operators transmit information along all dimensions of a two-dimensional (or more) stimulus, they must *divide attention* between the dimensions. When they are asked to process one dimension and ignore changes on the others, they are *focusing attention*. These concepts of divided and focused attention will be considered in much more detail in the next chapter, when we will consider their broader relevance to the issue of attention to events and objects in the world as well as to dimensions.

REFERENCES

Adams, J. A., Humes, J. M., & Stenson, H. H. (1962). Monitoring of complex visual displays III: Effects of repeated sessions on human vigilance. *Human Factors, 4*, 149–158.

Anderson, K. J., & Revelle, W. (1982). Impulsivity, caffeine, and proofreading: A test of the Easterbrook hypothesis. *Journal of Experimental Psychology: Human Perception & Performance, 8,* 614–624.

Baddeley, A. D., & Colquhoun, W. P. (1969). Signal probability and vigilance: A reappraisal of the "signal rate" effect. *British Journal of Psychology, 60,* 169–178.

Baker, C. H. (1961). Maintaining the level of vigilance by means of knowledge of results about a secondary vigilance task. *Ergonomics, 4,* 311–316.

Baty, D. L. (1971). Human transinformation rates during one-to-four axis tracking. In *Proceedings of the seventh annual conference on manual control* (NASA SP–281). Los Angeles: University of Southern California.

Beatty, J., Greenberg, A., Deibler, W. P., & O'Hanlon, J. P. (1974). Operator control of occipital theta rhythm affects performance in a radar monitoring task. *Science, 183,* 871–873.

Bisseret, A. (1981). Application of signal detection theory to decision making in supervisory control. *Ergonomics, 24,* 81–94.

Blake, T., & Baird, J. C. (1980). Finding a needle in a haystack when you've never seen a needle: A human factors analysis of SET I. In G. Corrick, E. Hazeltine, & R. Durst (eds.), *Proceedings of the 24th annual meeting of the Human Factors Society.* Santa Monica, CA: Human Factors Society.

Broadbent, D. E. (1971). *Decision and stress.* New York: Academic Press.

Broadbent, D. E. & Gregory, M. (1965). Effects of noise and of signal rate upon vigilance as analyzed by means of decision theory. *Human Factors, 7,* 155–162.

Buckhout, R. (1974). Eyewitness testimony. *Scientific American, 231*(6), 23–31.

Calderia, J. D. (1980). Parametric assumptions of some "nonparametric" measures of sensory efficiency. *Human Factors, 22,* 119–130.

Carroll, J. B. (1975). Speed and accuracy of absolute pitch judgments: Some latter-day results. *The L. L. Thurston Psychometric Laboratory Research Bulletin.* Chapel Hill: University of North Carolina.

Carswell, C. M., & Wickens, C. D. (1990). The perceptual interaction of graphical attributes: Configurality, stimulus homogeneity, and object integration. *Perception & Psychophysics, 47,* 157–168.

Cheng, P. W., & Pachella, R. G. (1984). A psychophysical approach to dimensional separability. *Cognitive Psychology, 16,* 279–304.

Childs, J. M. (1976). Signal complexity, response complexity, and signal specification in vigilance. *Human Factors, 18,* 149–160.

Colquhoun, W. P. (1975). Evaluation of auditory, visual, and dual-mode displays for prolonged sonar monitoring in repeated sessions. *Human Factors, 17,* 425–437.

Colquhoun, W. P., & Baddeley, A. D. (1967). Influence of signal probability during pretraining on vigilance decrement. *Journal of Experimental Psychology, 73,* 153–155.

Craig, A. (1977). Broadbent and Gregory revisited: Vigilance and statistical decision. *Human Factors, 19,* 25–36.

Craig, A. (1978). Is the vigilance decrement simply a response adjustment towards probability matching? *Human Factors 20,* 447–451.

Craig, A. (1979). Nonparametric measures of sensory efficiency for sustained monitoring tasks. *Human Factors, 21,* 69–78.

Craig, A. (1984). Human engineering, the control of vigilance. In J. S. Warm (ed.), *Sustained attention in human performance* (pp. 247–291). Chichester, Eng.: Wiley.

Crowe, D. S., Beare, A. N., Kozinsky, E. J., & Hass, P. M. (1983). *Criteria for safety-related nuclear power plant operator action* (NUREG/CR–3123 ORNL/TM–8626). Oak Ridge, TN: Oak Ridge National Laborabory.

Dardano, J. F. (1962). Relationships of intermittent noise, intersignal interval, and skin conductance to vigilance behavior. *Journal of Applied Psychology, 46,* 106–114.

Davies, D. R., & Parasuraman, R. (1980). *The psychology of vigilance.* London: Academic Press.

Deaton, J. E., & Parasuraman, R. (1988). Effects of task demands and age on vigilance and subjective workload. *Proceedings of the 32nd annual meeting of the Human Factors Society* (pp. 1458–1462). Santa Monica, CA: Human Factors Society.

Drury, C. G., & Addison, S. L. (1973). An industrial study of the effects of feedback and fault density on inspection performance. *Ergonomics, 16,* 159–169.

Egeth, H., & Pachella, R. (1969). Multidimensional stimulus identification. *Perception & Psychophysics, 5,* 341–346.

Ellison, K. W., & Buckhout, R. (1981). *Psychology & criminal justice.* New York: Harper & Row.

Eriksen, C. W., & Hake, H. N. (1955). Absolute judgments as a function of stimulus range and number of stimulus and response categories. *Journal of Experimental Psychology, 49,* 323–332.

Fisk, A. D., & Schneider, W. (1981). Controlled and automatic processing during tasks requiring sustained attention. *Human Factors, 23,* 737–750.

Garner, W. R. (1974). *The processing of information and structure.* Hillsdale, NJ: Erlbaum.

Garner, W. R., & Fefoldy, G. L. (1970). Integrality of stimulus dimensions in various types of information processing. *Cognitive Psychology, 1,* 225–241.

Green, D. M., & Swets, J. A. (1988). *Signal detection theory and psychophysics.* New York: Wiley.

Harris, D. H., & Chaney, F. D. (1969). *Human factors in quality assurance.* New York: Wiley.

Hartman, E. G. (1954). The influence of practice and pitch distance between tones on the absolute identification of pitch. *American Journal of Psychology, 67,* 1–14.

Howell, W. C., Johnston, W. A., & Goldstein, I. L. (1966). Complex monitoring and its relation to the classical problem of vigilance. *Organizational Behavior & Human Performance, 1,* 129–150.

Jerison, M. L., Pickett, R. M., & Stenson, H. H. (1965). The elicited observing rate and decision process in vigilance. *Human Factors, 7,* 107–128.

Kelly, M. L. (1955). A study of industrial inspection by the method of paired comparisons. *Psychological Monographs, 69,* (394), 1–16.

Klein, M., Coles, M. G., & Donchin, E. (1984). People without absolute pitch process tones without producing a P300. *Science, 223,* 1306–1308.

Kopala, C. (1979). The use of color-coded symbols in a highly dense situation display. In C. Bensel (ed.), *Proceedings of the 23rd annual meeting of the Human Factors Society.* Santa Monica, CA: Human Factors Society.

Lees, F. P., & Sayers, B. (1976). The behavior of process monitors under emergency conditions. In T. Sheridan & G. Johannsen (eds.), *Monitoring behavior and supervisory control*. New York: Plenum Press.

Lewandowski, L. J., & Kobus, D. A. (1989). Bimodal information processing in sonor performance. *Human Performance, 2*(1), 73–84.

Lockhead, G. R., & King, M. C. (1977). Classifying integral stimuli. *Journal of Experimental Psychology: Human Perception & Performance, 3*, 436–443.

Loftus, E. F. (1979). *Eyewitness testimony*. Cambridge, MA: Harvard University Press.

Long, C. M., & Waag, W. L. (1981). Limitations and practical applicability of d′ and β as measures. *Human Factors, 23*, 283–290.

Lusted, L. B. (1971). Signal detectability and medical decision making. *Science, 171*, 1217–1219.

Lusted, L. B. (1976). Clinical decision making. In D. Dombal & J. Grevy (eds.), *Decision making and medical care*. Amsterdam: North Holland.

Luzzo, J., & Drury, C. G. (1980). An evaluation of blink inspection. *Human Factors, 22*, 201–210.

McGrath, J. J. (1963). Irrelevant stimulation and vigilance performance. In D. M. Buckner & J. J. McGrath (eds.), *Vigilance: A symposium*. New York: McGraw-Hill.

Mackie, R. R. (1977). *Vigilance: Relationships among theories, physiological correlates, and operational performance*. New York: Plenum Press.

Mackworth, J. F. (1970). *Vigilance and attention*. Baltimore: Penguin.

Mackworth, N. H. (1948). The breakdown of vigilance during prolonged visual search. *Quarterly Journal of Experimental Psychology, 1*, 5–61.

Mackworth, N. H. (1950). *Research in the measurement of human performance* (MRC Special Report Series No. 268). London: H.M. Stationery Office. Reprinted in W. Sinaiko (ed.), *Selected papers on human factors in the design and use of control systems*. New York: Dover, 1961.

Miller, G. A. (1956). The magical number seven plus or minus two: Some limits on our capacity for processing information. *Psychological Review, 63*, 81–97.

Moses, F. L., Maisano, R. E., & Bersh, P. (1979). Natural associations between symbols and military information. In C. Bensel (ed.), *Proceedings of the 23rd annual meeting of the Human Factors Society*. Santa Monica, CA: Human Factors Society.

Mowbray, G. H., & Gebhard, J. W. (1961). Man's senses vs. informational channels. In W. Sinaiko (ed.), *Selected papers on human factors in the design and use of control systems*. New York: Dover.

Parasuraman, R. (1979). Memory load and event rate control sensitivity decrements in sustained attention. *Science, 205*, 925–927.

Parasuraman, R. (1985). Detection and identification of abnormalities in chest x-rays: Effects of reader skill, disease prevalence, and reporting standards. In R. E. Eberts & C. G. Eberts (eds.), *Trends in ergonomics/human factors II* (pp. 59–66). Amsterdam: North-Holland.

Parasuraman, R. (1986). Vigilance, monitoring, and search. In K. Boff, L. Kaufman, & J. Thomas (eds.), *Handbook of perception and human performance. Vol. 2: Cognitive processes and performance* (pp. 43.1–43.39). New York: Wiley.

Parasuraman, R., Warm, J. S., & Dember, W. N. (1987). Vigilance: Taxonomy and utility. In L. S. Mark, J. S. Warm, & R. L. Huston (eds.), *Ergonomics and human factors* (pp. 11–31). New York: Springer-Verlag.

Peterson, C. R., & Beach, L. R. (1967). Man as an intuitive statistician. *Psychological Bulletin, 68,* 29–46.

Pollack, I. (1952). The information of elementary auditory displays. *Journal of the Acoustical Society of America, 24,* 745–749.

Pollack, I., & Ficks, L. (1954). The information of elementary multidimensional auditory displays. *Journal of the Acoustical Society of America, 26,* 155–158.

Pomerantz, J. R. (1981). Perceptual organization in information processing. In M. Kubovy & J. R. Pomerantz (eds.), *Perceptual organization* (pp. 141–180). Hillsdale, NJ: Erlbaum.

Pomerantz, J. R., & Pristach, E. A. (1989). Emergent features, attention and perceptual glue in visual form perception. *Journal of Experimental Psychology: Human Perception & Performance, 15,* 635–649.

Pond, D. J. (1979). Colors for sizes: An applied approach. In C. Bensel (ed.), *Proceedings of the 23rd annual meeting of the Human Factors Society.* Santa Monica, CA: Human Factors Society.

Rhea, J. T., Potsdaid, M. S., & DeLuca, S. A. (1979). Errors of interpretation as elicited by a quality audit of an emergency facility. *Radiology, 132,* 277–280.

Ruffle-Smith, H. P. (1979). *A simulator study of the interaction of pilot workload with errors, vigilance, and decision* (NASA Technical Memorandum 78482). Washington, DC: NASA Technical Information Office.

Saito, M. (1972). A study on bottle inspection speed-determination of appropriate work speed by means of electronystagmography. *Journal of Science of Labor, 48,* 395–400. (In Japanese, English summary.)

Scanlan, L. A. (1975). Visual time compression: Spatial and temporal cues. *Human Factors, 17,* 337–345.

Schmidke, I. I. (1976). Vigilance. In E. Simonson & P. C. Weiser (eds.), *Psychological and physiological correlates of work and fatigue.* Springfield, MA: Thomas.

Schneider, W., & Shiffrin, R. M. (1977). Controlled and automatic human information processing II: Perceptual learning, automatic attending, and a general theory. *Psychological Review, 84,* 127–190.

Shannon, C. E., & Weaver, W. (1949). *The mathematical theory of communications.* Urbana, IL: University of Illinois Press.

Sheridan, T. B., & Ferrell, W. A. (1974). *Man-machine systems: Information, control, and decision models of human performance.* Cambridge, MA: MIT Press.

Siegel, J. A., & Siegel, W. (1972). Absolute judgment and paired associate learning: Kissing cousins or identical twins? *Psychological Review, 79,* 300–316.

Swennsen, R. G., Hessel, S. J., & Herman, P. G. (1977). Omissions in radiology: Faulty search or stringent reporting criteria? *Radiology, 123,* 563–567.

Swets, J. A. (ed.). (1964). *Signal detection and recognition by human observers: Contemporary readings.* New York: Wiley.

Swets, J. A., & Pickett, R. M. (1982). *The evaluation of diagnostic systems.* New York: Academic Press.

Szucko, J. J., & Kleinmuntz, B. (1981). Statistical vs. clinical lie detection. *American Psychologist, 36,* 488–496.

Teichner, W. (1974). The detection of a simple visual signal as a function of time of watch. *Human Factors, 16,* 339–353.

Warm, J. S. (ed.). (1984). *Sustained attention in human performance.* Chichester: Wiley.

Watson, D. S., & Nichols, T. L. (1976). Detectability of auditory signals presented without defined observation intervals. *Journal of Acoustical Society of America,* 59, 655–668.

Weintraub, D. J. (1971). Rectangle discriminability: Perceptual relativity and the law of pragnanz. *Journal of Experimental Psychology, 88,* 1–11.

Welford, A. T. (1968). *Fundamentals of skill.* London: Methuen.

Wells, G. L., Lindsay, R. C., & Ferguson, T. I. (1979). Accuracy, confidence, and juror perceptions in eyewitness testimony. *Journal of Applied Psychology, 64,* 440–448.

Wickens, C. D. (1984). *Engineering psychology and human performance.* New York: HarperCollins.

Wickens, C. D., & Kessel, C. (1979). The effect of participatory mode and task workload on the detection of dynamic system failures. *IEEE Transactions on Systems, Man, and Cybernetics, SMC–13,* 24–34.

Wickens, C. D., & Kessel, C. (1981). The detection of dynamic system failures. In J. Rasmussen & W. Rouse (eds.), *Human detection and diagnosis of system failures.* New York: Plenum Press.

Wilkinson, R. T. (1964). Artificial "signals" as an aid to an inspection task. *Ergonomics, 7,* 63–72.

APPENDIX A

SOME VALUES OF d'

			P(false alarm)			
P(hit)	0.01	0.02	0.05	0.10	0.20	0.30
0.51	2.34	2.08	1.66	1.30	0.86	0.55
0.60	2.58	2.30	1.90	1.54	1.10	0.78
0.70	2.84	2.58	2.16	1.80	1.36	1.05
0.80	3.16	2.89	2.48	2.12	1.68	1.36
0.90	3.60	3.33	2.92	2.56	2.12	1.80
0.95	3.96	3.69	3.28	2.92	2.48	2.16
0.99	4.64	4.37	3.96	3.60	3.16	2.84

Selected values from Signal Detection and Recognition by Human Observers (Appendix 1, Table 1) by J. A. Swets, 1969, New York: Wiley. Copyright 1969 by John Wiley & Sons, Inc. Reproduced by permission.

APPENDIX B

VALUES COMPUTED FROM THE FORMULA

$$A' = 1 - \frac{1}{4}\left\{ \frac{P(FA.}{P(H)} + \frac{[1 - P(H)]}{[1 - P(FA)]} \right\}$$

				P(false alarm)			
P(hit)	0.00	0.05	0.10	0.15	0.20	0.25	0.30
0.05	0.762	0.500					
0.10	0.775	0.638	0.500				
0.15	0.787	0.693	0.597	0.500			
0.20	0.800	0.727	0.653	0.577	0.500		
0.25	0.812	0.752	0.692	0.629	0.566	0.500	
0.30	0.825	0.774	0.722	0.669	0.615	0.558	0.500
0.35	0.837	0.793	0.748	0.702	0.654	0.605	0.553
0.40	0.850	0.811	0.771	0.730	0.687	0.644	0.598
0.45	0.862	0.827	0.792	0.755	0.717	0.678	0.637
0.50	0.875	0.843	0.811	0.778	0.744	0.708	0.671
0.55	0.887	0.859	0.829	0.799	0.768	0.736	0.703
0.60	0.900	0.874	0.847	0.820	0.792	0.762	0.732
0.65	0.912	0.889	0.864	0.840	0.814	0.787	0.760
0.70	0.925	0.903	0.881	0.858	0.835	0.811	0.786
0.75	0.937	0.918	0.900	0.876	0.855	0.833	0.811
0.80	0.950	0.932	0.913	0.894	0.875	0.855	0.835
0.85	0.962	0.946	0.930	0.912	0.894	0.876	0.858
0.90	0.975	0.960	0.944	0.930	0.913	0.897	0.881
.095	0.987	0.974	0.960	0.946	0.932	0.917	0.903
1.00	1.00	0.987	0.975	0.962	0.950	0.937	0.925

Chapter
3

Attention in Perception and Display Space

OVERVIEW

The limitations of human attention represent one of the most formidable bottle-necks in human information processing. We can all easily relate instances when we failed to notice or attend to the words of a speaker because we were distracted or occasions when we had so many tasks to perform that some were neglected. These intuitive examples of failures of attention may be described more formally in terms of three categories.

1. *The limits of selective attention.* In some instances we select inappropriate aspects of the environment to process. For example, as we will discuss in Chapter 7, decision makers sometimes select salient rather than diagnostic cues. A dramatic example is provided by the behavior of the flight crew of the Eastern Airlines L-1011 flight that crashed in the Everglades in 1972. Because of their preoccupation with a malfunction elsewhere in the cockpit, none of the personnel on the flight deck attended to the critical altimeter reading and to subsequent warnings that the plane was gradually descending to ground level (Wiener, 1977; see also Chapter 12).

2. *The limits of focused attention.* Occasionally we are unable to concentrate on one source of information in the environment; that is, we have a tendency to be distracted. The clerical worker transcribing a tape in a room filled with extraneous conversation encounters such a problem. So also does the translator who must ignore the feedback provided by his or her own voice to concentrate solely on the incoming message. Another example is the process control room operator attempting to locate rapidly a critical item of information in the midst

of a "busy" display consisting of a multitude of other changing variables. The difference between failures of selective attention and failures of focused attention is that in the former case there is an intentional but unwise choice to process nonoptimal environmental sources. In the latter case this processing of nonoptimal sources is "driven" by external environmental information despite the operator's efforts to shut it out.

3. *The limits of divided attention*. On the other side of the coin are the limits of divided attention. When problems of focused attention are encountered, our attention is inadvertently directed to stimuli we do not wish to process. When problems of divided attention are encountered, we are unable to divide our attention between stimuli or tasks, all of which we wish to process. Here we may consider the automobile driver, who must scan the highway for road signs while maintaining control of the vehicle, or the example in Chapter 7 of the fault diagnostician, who must maintain several hypotheses in working memory while scanning the environment for diagnostic information and also entering this information into a recording device. Thus the limits of divided attention are often considered to be synonymous with our limited ability to *time-share* performance of two or more tasks.

Attention in perception may be described by the metaphor of a searchlight (Wachtel, 1967). The momentary direction of our attention is a searchlight. Its focus falls on that which is in momentary consciousness. Everything within the beam of light is processed whether wanted (successful focusing) or not (failure to focus). Within the context of this metaphor, two properties of the searchlight are relevant to phenomena of human experience: (1) the "breadth" of the beam and the distinction, if any, between the desired focus, or penumbra (that which we want to process), and the umbra (that which we must process but do not want to), representing the issues of divided and focused attention, respectively; (2) the direction of the searchlight — how it knows when, what, and where in the environment to illuminate — describing the properties of selective attention. Each of these will be considered in some detail as we consider examples of how operators search the complex stimulus world for critical pieces of information and how that information is processed once it has been found.

The searchlight metaphor conveniently describes the various characteristics of attention and perception, the topic of this chapter. Yet the concept of attention is relevant to a range of activities that are not only perceptual. We can speak of dividing attention between two tasks, no matter what stage of processing they require. The broader issue of divided attention as it relates to the time-sharing of activities will be the concern of Chapter 9, after we have discussed other stages of information processing. In this chapter we will present a basic overview of the experimental findings of selective, focused, and divided attention in perception and their relevance to display layout, addressing first those aspects of attention that are serial (highlighting visual scanning) before considering its parallel characteristics in first vision and then audition.

SELECTIVE ATTENTION

Visual Sampling

Our discussion of selective attention begins with the eye and with visual sampling, that is, when the operator seeks information and searches for targets. Although selective attention has many aspects that are more subtle than the direction of gaze, it is still the case that for much of the time, our gaze is driven by our need to attend; thus we can learn a lot about selective attention by studying visual sampling behavior, the perfect analog of the searchlight of attention (Fisher, Monty, & Senders, 1981; Moray, 1986; Senders, Fisher, & Monty, 1978).

Before we describe some of the human performance models and data that have been applied to visual sampling, it is important to understand a few basic characteristics of the eye fixation system. First, there is only a small region of the visual field in which we can perceive detail. This region, the *fovea*, is about 2 degrees of visual angle. To keep objects in foveal vision, to "look at" them, the eyeball itself exhibits two different kinds of movement: *Pursuit* movements occur when the eye follows a target moving at a constant speed across the visual field. As you follow the trajectory of a ball or a flying bird, your eyes will show constant velocity pursuit movements. *Saccadic* movements are a series of discrete, jerky movements that jump from one stationary point in the visual field to the next. They can sometimes be superimposed on pursuit movements. If the velocity of the moving ball or flying bird is too fast for a pursuit movement to follow, a saccade will be used to "catch up" and bring the target back into foveal vision (Young & Stark, 1963).

The saccadic behavior used in visual sampling, which will be the focus of the present discussion, is characterized by two critical aspects: the scanning process that characterizes the transition of visual fixations from one part of the visual field to another and the fixation itself. The latter is characterized by a *location* (the center of the fixation), a *useful field of view* (diameter around the central location from which information is extracted), and a *dwell time* (how long the eye remains at that location).

Visual sampling behavior has been studied in two somewhat different contexts. In what we shall refer to as the *supervisory/control* context, the operator scans the display of a complex system under supervision — an aircraft cockpit, for example — and allocates attention through visual fixations to various instruments, as these represent sources of information. In the *target search* context, the operator scans a region of the visual world, looking for something at an unknown location; it may be a failure in a circuit board examined by a quality control inspector (see Chapter 2), a search and rescue mission for a downed aircraft, or a receiver suddenly breaking into the open on the football field. The important difference between these two contexts is that in supervisory/control, the critical action is to read the value of several changing indicators whose location is known. In free field search, the critical action is to find the target whose location and existence is unknown (Liu & Wickens, 1989). We will discuss each of these situations in turn.

More on the Eye

The eye is both a servomechanism and a mécanisme de cerveau.
And sometimes it does its own thing and sometimes it goes
　　Where the brain wants it to go.

The eyes are the window to the mind and the mind's window
　　To the scene
So that one is never quite sure whether it's the world or
　　The mind that makes the eye shift to where it's going
　　From where it's been.
You can watch the eyes and catch the thought
While it's so hot that even the mind hasn't had it yet.

With a mind of its own the eye looks at the place best calculated
　　To let the mind's eye see what the mind wants to see;
And then all the world rushes in to be reduced
To common sense and percept before the next saccade is loosed.

　　　　　　　　　　　　　　　　　　　　—J. W. Senders

Source: J. W. Senders, *Visual scanning processes*. University of Tilburg, the Netherlands. Unpublished doctoral dissertation, 1980.

Supervisory Control Sampling

Optimality of Selective Attention In the aircraft cockpit or the process control console, there are many sources of information in the environment that must be sampled periodically. In these situations, engineering psychologists have studied how optimal the performance of subjects is when selecting the relevant stimuli to attend at the appropriate times. As in our discussions of signal detection theory, *optimal* is defined in terms of a behavior that will maximize an expected value or minimize an expected cost. For example, an aircraft pilot who continuously fixates (samples) the airspeed indicator but ignores the altimeter is clearly nonoptimal. One who samples both of these but never checks outside the aircraft is doing better but is still not optimal, for this pilot is incurring the expected costs of missing important stimuli (e.g., other aircraft) that can be seen only by looking out the window.

Engineering psychologists who have worked in this area divide the environment into *channels*, along which critical *events* may periodically occur. They have assumed that environmental sampling is guided by the expected cost that results when an event is missed. The *expected* cost of missing an event is equal to the true cost of missing it multiplied by the probability that the event will be missed. The probability of missing an event in turn is directly related to event frequency. Those events that occur often are more likely to be missed if the channels along which they occur are not sampled. Also in most real-world tasks, the probability of missing an event on a channel increases with the amount of time since the channel has last been sampled. For example, the probability of missing an aircraft outside the pilot's window increases with the time that has passed since the pilot has last looked out.

When optimum sampling is examined in the laboratory, the subject has typically been presented with two or more potential channels of stimulus information, along which events may arrive at semipredictable rates. For example, a channel might be an instrument dial, with an "event" defined as the needle moving into a danger zone, as in Figure 3.1 (e.g., Senders, 1964). The general conclusions of these studies, which are well summarized by Moray (1981, 1986), suggest the four conclusions discussed in the following paragraphs.

First, people are said to form a "mental model" of the statistical properties of events in the environment, and they use the model to guide their visual sampling. The mental model may be thought of as a set of expectancies about how frequently and when events will occur on each channel and about the correlation between events on pairs of channels. Because sampling strategies provide estimates of the operator's internal model of a system under supervision, the patterns of fixations will also be of help to the system designer in locating information displays more optimally. Dating from the pioneering work of Fitts, Jones, and Milton (1950), aviation psychologists have employed scan-

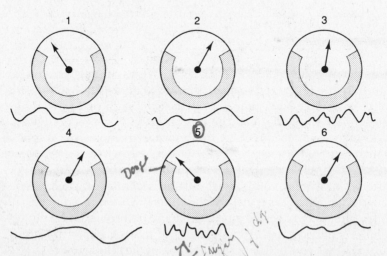

Figure 3.1 Display typical of those used for studying instrument scanning. Under each display is an example of the time-varying input the operator must sample to ensure that none of the needles move into the danger zones.

ning data to configure displays according to two principles: Frequently sampled displays should be placed centrally, and pairs of displays that are often sampled sequentially should be located close together (Elkind, Card, Hochberg, & Huey, 1990; Frost, 1972).

Second, people learn to sample channels with higher event rates more frequently and with lower rates less frequently. However, the sampling rate is not quite adjusted upward or downward with event frequency as much as it should be. This is similar to the sluggish beta phenomenon discussed in Chapter 2, describing the reluctance to adjust the response criterion in signal detection.

To elaborate on this second point, models developed by Carbonnell, Ward, and Senders (1968) and Sheridan (1972) propose that the time between the samplings of an instrument should be determined by the trade-off of two factors: the growth of uncertainty of the state of the unsampled channel (related to the event rate on that channel) and the cost of taking a sample. Since sampling, or switching visual attention, has some subjective cost, people will not scan too rapidly across all channels of a dynamic instrument panel. Nor is there any need for frequent sampling if the channels change their state slowly (channel 4 in Figure 3.1); hence, the operator's uncertainty about the state of the unsampled channel grows slowly. But eventually the operator's uncertainty will reach a high enough level so that it becomes worth the cost of a fixation to find out what is happening there (i.e., to "reset" uncertainty to zero). Modeling visual sampling in this way, Carbonnell, Ward, and Senders found that their model did a very accurate job of describing the fixation patterns of pilots making an instrument landing.

Third, human memory is imperfect and sampling reflects this fact. People therefore tend to sample information sources more often than they would need to if they had perfect memory about the status of an information source when it was last sampled. This fact explains the "oversampling" of channels with low event rates, described in the second conclusion. Also because of limits of memory, people may forget to sample a particular display source if there are several possible sources. This, for example, might well be the case for the monitor of a nuclear process control console. Such forgetfulness will become increasingly likely if the channels are not physically represented by a display location on a surface but are stored in a computer and must be accessed for inspection on a display screen by a manual or vocal request. These limitations in memory suggest the usefulness of a computer that presents "sampling reminders" (Moray, 1981).

Fourth, when people are given a *preview* of scheduled events that are likely to occur in the future, sampling and switching become somewhat more optimal. Now subjects' sampling can be guided by an "external model," the display of the previewed events. Thus the dispatcher or industrial scheduler can be helped by having a preview of anticipated demands on different resources (Sanderson, 1989; see also Chapter 12). The student is helped by having a preview of assignments in different courses. However, research by Tulga and Sheridan (1980) has found that as the number of channels increases, people fail to take advantage of preview information much beyond the immediate present, apparently because of the heavy load on working memory required to do so. This is one explanation for why predictive displays for industrial scheduling have not always been useful (Sanderson, 1989).

Processing Strategies The fact that scanning behavior reflects the operator's mental model of the environment, and therefore can indicate his or her information needs, has important implications for understanding operator strategies when interacting with information sources. In a study of a simulated process control plant, Moray and Rotenberg (1989), for example, used scanning analysis to suggest that operators were engaged in "cognitive tunneling" on a failed system, such as that described in the Three Mile Island incident in Chapter 1 (see also Chapters 7 and 12). When one of the systems under supervision "failed," operators stopped keeping track of the status of other systems (i.e., looking at them) as their diagnosis of the failed system was carried out. Moray and Rotenberg also used scanning measures to identify the problems associated with delayed feedback from control input settings. After making a control adjustment to one system, operators switch their visual attention to the indicator where feedback for that response is expected. Their fixation often appears to stay locked on to that indicator until it eventually reflects the control input. This can represent a substantial waste of visual attention if the delay is long.

Russo and Rosen (1975) applied scanning analysis to an information-seeking task that was a static version of supervisory/control sampling; *multiattribute decision making*, a concept that will be discussed in depth in Chapter 7. This is the task confronting a decision maker who must choose between several alternatives (e.g., cars to purchase) varying on several attributes (e.g., price and gas mileage). Russo and Rosen found the scan patterns useful in revealing the kinds of strategies subjects used when selecting the most desirable car to purchase out of a set of six cars, each characterized by three different attributes. The subjects tended to compare specific pairs of cars, those that were close together on a display. This latter feature of their data suggests how a desire to minimize the extent of eye movements can affect information-seeking strategies.

Eye Movements in Target Search

When the operator is looking for an object in the environment, such as a flaw in a piece of sheet metal or a piece of aircraft wreckage on the ground, the visual scan pattern tends to be far less structured and organized than in the supervisory/control task. As a consequence, scanning is considerably less amenable to optimal modeling. Nevertheless, a number of characteristics of visual search have emerged from research in this area.

Environmental Expectancies Like supervisory/control scanning, target search can also be described as driven in part by cognitive factors related to the expectancy of where in the visual field a target or the most useful information is likely to be found. These areas tend to be fixated first and most frequently. This characteristic of information-seeking and scanning behavior has been used to account for differences between novices and experts. In football, for example (Abernethy, 1988), the expert quarterback will know where to look for the open receiver with the highest probability. Kundel and La Follette (1972) have studied differences in the way that novice and expert radiologists scan X-ray plates in search of a tumor. The expert examines first and most closely those areas in which tumors are most likely to appear; the novice tends to search the whole plate evenly. Mourant and Rockwell (1972) have studied differences in scan patterns in automobile driving, noting that the experienced driver has a wider scan pattern, distributed farther ahead on the highway, and checks the rearview mirror more often. The novice driver, in contrast, tends to focus most scans on the road close ahead and to the right of the vehicle, as if aligning the vehicle with the curb. As we will see in Chapter 11, the expert's tendency to attend farther ahead on the highway allows information to be extracted from an aspect of the environment—automobile heading—that allows for smoother vehicle control with more anticipation.

The role of information in visual scanning has also been used to explain how we scan pictures (Gould, 1976; Yarbus, 1967). People have a tendency to fixate most on those areas that contain the most information (e.g., faces, contours, and other areas of high visual details). Furthermore, a scan path over the same picture will change, depending on what information the viewer is expected to extract (Yarbus, 1967).

Display Factors and Salience   The fact that much of visual search behavior is internally driven by cognitive factors means that there are no highly consistent *physical* patterns of display scanning (e.g., left-to-right or circular-clockwise) and no optimal scan pattern in search, beyond the fact that search should be guided by the expectancy of target location. Nevertheless, although they cannot be used to account for a great deal of systematic behavior, there are certain display factors that will guide the allocation of visual attention some of the time, although they are often overridden by the cognitive factors.

There is little doubt that visual attention will be drawn to items in a display that are large, bright, colorful, and changing (e.g., blinking), a characteristic that can be exploited when locating visual warnings (see Chapter 12) but that may bias decision making (see Chapter 7). There is some evidence also that search behavior is guided by other, more subtle factors related to the physical locations on the display. For example, Megaw and Richardson (1979) found that those subjects who did exhibit a systematic scan pattern when searching for targets tended to start at the upper left. This fact may reflect eye movement in reading. A search also tends to be most concentrated toward the center regions of the visual field, avoiding the edges of a display, a pattern that Parasuraman (1986) has described as an "edge effect." Also, as noted in the discussion of supervisory/control sampling, scans tend to be made most frequently between adjacent elements on a display.

These location-driven search tendencies are not strong and are usually dominated by internally driven scan strategies (Levy-Schoen, 1981). For example people sometimes search computer menus in a random or expectancy-driven fashion, rather than from top to bottom (Card, 1986). However, it seems reasonable that a knowledge of these tendencies should be employed in designing multielement displays to locate information of greatest importance, such as warning and hazard labels, in areas of greatest salience, an issue that we will return to in Chapter 7 in the discussion of the cues used for decision making.

Search Coverage and the Useful Field of View   How much visual area is covered in each visual fixation? This is not an easy question to answer. It is clear, on the one hand, that we can sometimes take in information from peripheral vision (see Chapter 4). On the other hand, resolution of fine visual detail requires the highest acuity region of the fovea, an angle of no more than about 2 degrees surrounding the center of fixation. Mackworth (1976) has addressed this uncertainty by defining the "useful field of view" (UFOV) as a circular area around the fixation point from which search information is extracted. Operators will search a region so that two adjacent UFOV may touch but not overlap. Thus, the size of the UFOV can be estimated from the minimum distance between successive fixations in a search task. The data collected by Mackworth and others suggest that the size of the UFOV may vary from around 1 degree to 4 degrees of visual angle. This size appears to be determined by the density of information and by the discriminability of the searched-for target from the background. Thus, looking for a dark flaw on a clear background in glass inspection will lead to a larger UFOV than scanning for a misaligned connection in a circuit board or microchip.

The size of the UFOV and the maximum rate with which different fixations can be made (2–4 per second) limit the amount of area that can be searched in a given time. However, even in the absence of time limits, it is apparent that humans do not tend to search in an exhaustive fashion, blanketing an entire area with UFOVs, and inevitably locating a target. Stagar and Angus (1978) studied airborne search and rescue experts who searched photographs for crash sites; the searches covered only 53 percent of the available terrain, a fact that led to less than perfect performance. To make matters worse, investigators have also noted the many occasions when targets may be fixated within a UFOV yet not detected (Abernethy, 1988; Kundel & Nodine, 1978; Stagar & Angus, 1978). As we describe below, this finding suggests that a signal detection stage with a decision criterion typically follows the search process.

Fixation Dwells   We have said little about how long the eye rests at a given fixation. Since the eye extracts information over time, one might think that long dwells should be associated with greater information pickups. It turns out that this is sometimes the case, but not always. In a study of pilot scanning, for example, Harris and Christhilf (1980) found that pilots fixated longer on those critical instruments from which information necessary to control the aircraft had to be extracted rather than on those requiring a mere check to assure that they were "in bounds." Indeed, the instrument that involved longer dwells—the attitude display indicator—was the same instrument established by Fitts, Jones, and Milton (1950) to be fixated on most frequently. The critical role of this instrument in flying will be discussed in more detail in Chapter 4. In target search, Kundel & Nodine (1978) distinguished between short *survey dwells*, used to establish those regions more likely to contain a target, and longer *examination dwells*, used to provide a detailed examination of the region for an embedded target.

In addition to scanning and sampling strategies, fixation dwells are also governed by the *difficulty* of information extraction. Thus, displays that are less legible or contain denser information will be fixated on longer (Mackworth, 1976). Williams and Harris (1985) report that digital displays in an aviation context will be fixated on for a longer time than analog displays. In normal reading, longer dwells are made on less familiar words and while reading more difficult text (McConkie, 1983; see Chapter 5). When examining pictures, people fixate longer on objects that are unusual and out of context (Friedman & Liebelt, 1981). As we saw in the last chapter, low familiarity, low frequency, and out of context translate directly to higher information content, suggesting that dwell time has some relation to the information content of a display.

However, there does not always appear to be a positive relationship between difficulty and fixation dwell. In their study of simulated process control, Moray and Rotenberg (1989) examined fixations on the same instrument before and after a failure of the system reflected by the instrument. They found that although the instrument was fixated on more frequently after the failure, dwell times were unchanged. Russo and Rosen (1975) found few differences in dwell time on decision options across different strategies. In visual search tasks, the quest for a systematic relationship between dwell time and search performance (or novice-expert differences) has not met with success. It may well be that experts can extract more information from a display fixation, but they also extract this information at a faster rate, thereby having a fixation dwell time roughly the same as the novice.

Peripheral Vision The fact that successive scans in target search or reading often cover a greater distance of visual angle than the size of the fovea itself suggest that the destination of the next saccade must be perceived in peripheral vision (McConkie, 1983). The visual appearance of this destination acts to guide the programming of scanning behavior. Yet it does not appear that peripheral vision plays much of a role in detecting objects in target search. For example, Johnston (1967) found no evidence that individual differences in peripheral vision beyond 4 degrees of visual angle had any influence on target-searching skills.

Conclusion The discussion of visual scanning behavior yields two general conclusions. First, scanning tells us a good deal about the internal expectancies and understandings that drive selective attention. Second, the greatest usefulness of scanning research to engineering psychology is probably in the area of diagnostics. Frequently watched instruments can be seen as those that are most important to an operator's task. This fact may lead to design decisions to place these instruments in prominent locations or close together (Elkind, Card, Hochberg, & Huey, 1990). Differences between novice and expert fixation patterns can indicate how the mental model or the search strategy of the novice departs from that of the expert, and display items that require long dwells may indicate nonoptimal formatting. Finally, before leaving the area of visual scanning we refer the reader to Chapter 5, where we deal with the issue of visual fixation in reading, a task that is neither search nor supervisory control but one of great importance in design.

Visual Search Models

Visual scanning is of course heavily involved in visual search. But there are other aspects of search that cannot be revealed by scanning, including such aspects as the uncertainty of target identification or differences in the physical makeup of targets (e.g., one-dimensional versus multidimensional). Furthermore, whereas scanning reveals details about the process of visual search, human factors engineers may often be interested in the *product* of that search: How long does it take to find a target? Or what is the probability that a target will be detected in a given period of time? Hence, engineering psychologists have been concerned with the development of visual search models that will allow these values to be predicted (e.g., Bloomfield, 1979; Bloomfield & Howarth, 1969).

One such model was developed by Drury (1975, 1982) to predict the time it would take an industrial quality control inspector to detect a flaw in an inspected product. Drury's particular interest was in the inspection of sheet metal. The model is a two-stage one. The first stage describes the target search and predicts that the probability of locating a target will increase if more search time is given. However, it will not increase linearly but at a diminishing rate, as shown in Figure 3.2. In light of our previous discussion of scanning, this is not surprising given that (1) a target may be fixated on more than once without being detected, and (2) search strategies don't usually cover the whole search field, even when adequate time is given. In the second stage of Drury's model, a decision stage, the operator uses the expectancy of flaws (the overall manufacturing quality) to set a decision criterion, in much the same way as in signal detection theory (Chapter 2). If the expectancy of a failure is high and the response criterion is low, the probability that a fixated-on element will be called a flaw will increase as the false-alarm rate increases as well.

The particular shape of the curve in Figure 3.2 has important implications for the designer of industrial inspection stations: There is an optimum amount of time that each product should be searched, given that one can specify a cost for inspection time (which increases linearly with longer time) and a cost for misses. To allow (or force) the operator to search for a greater time, in the effort to achieve a higher detection rate, will probably lead to diminishing gains in inspection accuracy (Drury, 1982). Drury (1975, 1982) discusses the manner in which this optimum time should be established, given such factors as the

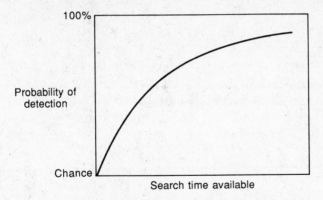

Figure 3.2 Probability of detection as a function of time available for search. (*Source:* Adapted from C. Drury, "Inspection of Sheet Metal: Model and Data." Reprinted with permission from *Human Factors* (vol. 17, 1975). Copyright 1975 by the Human Factors Society, Inc. All rights reserved.)

specification by the industrial manager of the desired rate at which products should be inspected, the probability of fault occurrence, and the desired overall level of inspection accuracy. Then industrial material to be inspected can be presented at a rate determined by this optimal time.

Another important approach to visual search has been to model the kinds of variables that affect the search speed when searching through a set of one- or multidimensional stimuli to locate a particular target. A vehicle dispatcher, for example, might need to scan a computerized city map to locate one of the vehicles that is coded as not in service and has a large carrying capacity. In these kinds of tasks, the operator might be asked to search through an array such as that shown schematically in Figure 3.3 and, for example, (1) report the location of the white *X*, (2) report the presence of a large *T*, or (3) search for a black target. Extensive research in this area reveals a number of general conclusions.

First, the dominant effect on search time is the number of elements to be searched (Drury & Clement, 1978; Mocharnuk, 1978). This is because the search is usually serial, as each item is inspected in turn. If there are more items, search time will be slowed. But if there are more examples of the target present, it is more likely that the target will be found early in the search.

Second, it matters relatively little if the elements are closely spaced, requiring little scanning, or are widely dispersed. The increased scanning that is required with wide dispersal does lengthen the search time slightly. However, the high density of nontarget elements when the items are closely spaced also has a small retarding influence on search. Thus the two factors, scanning distance and visual clutter, essentially trade off with one another as target dispersion is varied.

Third, exceptions to the first conclusion regarding serial search occur when the target is defined by one level along one salient dimension. For example,

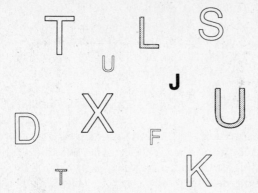

Figure 3.3 Stimuli for a typical experimental search task.

Third, exceptions to the first conclusion regarding serial search occur when the target is defined by one level along one salient dimension. For example, performance of task 3 above will be little affected by the number of items since the target in Figure 3.3 is defined by a single level (black) of one dimension (color).

Fourth, searching for any of *several* different target types is generally slower than searching for only one (Craig, 1981). An example in Figure 3.3 would be "Search for a *P* or a *Q*." The exception to this conclusion occurs when the set of two (or more) targets can be discriminated from all other nontargets by a *single common feature*. Varying levels of training may be necessary for the perceptual system to tune in to this critical discriminating feature. For example, in Figure 3.4, if the instructions were to "search for an *X* and a *K*," subjects might learn that given the particular set of nontarget stimuli used, *X* and *K* are the only letters that contain diagonal lines, and hence they will be able to search efficiently (Neissen, Novick, & Lazar, 1964). Thus, in an applied search task in industrial inspection, we can predict that there would be an advantage to training operators to focus on the common set of unique and defining features that would be common to all faults, and would therefore distinguish them from normal items.

Fifth, the role of extensive training in target search can sometimes bring performance to a level of *automaticity,* when search time is unaffected by the number of targets (Fisk, Oransky, & Shedsvold, 1988; Schneider, Dumais, & Shiffrin, 1984; Schneider & Shiffrin, 1977). Generally speaking, automaticity results only when over repeated trials the targets never appear as nontarget stimuli, a condition that Schneider and Shiffrin refer to as *consistent mapping.* This is contrasted with *varied mapping* search, when targets may sometimes appear as nontargets. We will discuss the concept of automaticity further in Chapter 5, in the context of reading; in Chapter 6, in the context of training; and again in Chapter 9, in the context of time-sharing.

Structured Search

The model proposed by Drury (1975) describes a search through a free field in which a target could be located anywhere and there is little organization to guide the search. Somewhat different is the process of structured search through a menu or array, such as when the computer user wishes to locate a particular item on a menu or the airline passenger is scanning a TV terminal for information concerning a particular flight. One feature of this search process that makes it more receptive to good models than does search in a free field is that the search is often linear, with each searched item adding a constant time to the search process (Neisser, Novick, & Lazar, 1964). In the letter-search task developed by Neisser, 1963), as shown in Figure 3.4, subjects scan a vertical column of random three- or five-letter sequences as rapidly as possible until they detect a particular target letter (or set of letters) that is specified prior to a trial. An example of two such search trials is shown at the top of Figure 3.4. The researchers observed a linear relationship between the serial position of the letter in the list and the time needed to detect the target. This function is shown at the bottom of Figure 3.4. The slope of the function reflects the average time

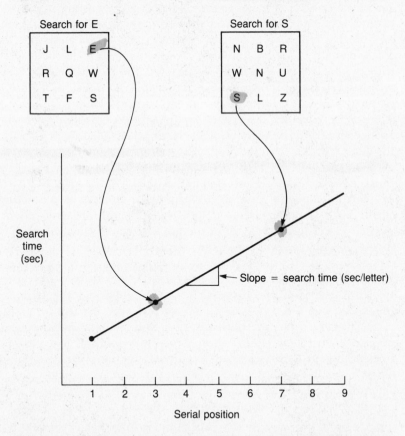

Figure 3.4 Neisser's letter-search paradigm. The top of the figure shows two lists with different targets. The bottom graph shows the search time for letters in each of the two lists as a function of their serial position on the list. Also presented are the data from other serial positions and the resulting linear slope. (*Source:* "Decision Time without Reaction Time" by U. Neisser, 1963, *American Journal of Psychology,* 76, p. 377. Copyright 1963 by the Board of Trustees of the University of Illinois. Reprinted by permission of the University of Illinois Press.)

required to decide that each given letter is not the target (or not in the target set if there are two or more targets).

One important application of structured search models is in the design of computer-based menu systems, a critical component in the study of human-computer interaction. In the typical menu task, the user must locate a target word, symbol, or command; must scan the list until the item is located; and then typically must press a key. Menus may of course be multilevel, in which case the target term may be reached only after a search through higher-level terms. Thus, a travel agent, searching for a flight from a particular city, may first access a menu of city names and, after selecting an option within that name level of the menu, scan an embedded menu of all flights departing from that city.

Menu designers would clearly like to structure a menu in such a way that target items are reached in the minimum average time, and the linear visual search model can serve as a useful guide. If menus are organized randomly, given the general tendency to search from the top downward (Somberg, 1987) and the linear search strategy, the target will be located after an average of $NT/2$ seconds, in a self-terminating search where N is the menu size and T is the time to read each item (Lee & MacGregor, 1985). Within each search, the time will be directly proportional to the distance of the item from the top of the menu.

It is possible for designers, capitalizing on this linear search strategy, to reduce the expected search time if they know that some items will be searched for more frequently than others. These items can be positioned toward the top of the menu in direct proportion to their frequency of use. Using these assumptions, Lee and MacGregor (1985) have developed quantitative models that predict the expected time needed to locate an item as a function of reading speed and computer response speed when there are embedded (multilevel) menus. Their model guidelines dictate that the optimal number of words per menu is between three and ten, depending on the reading speed and the computer response speed. Their data, consistent with others to be described in Chapter 8, argue against many embedded levels of short menus.

The question of how to locate items as a function of their frequency of use is important, but it is not simply answered because the use frequency of a given item will vary with different users and, within the same user, across different occasions. In the latter situation, a possible solution is a flexible case-based presentation order, in which an intelligent computer can make some inferences about a user's preference and position the item accordingly (see Chapter 12). The problem with this solution is that it leads to the violation of an important principle of display design—location consistency.

To evaluate the role of consistency, Somberg (1987) compared menu search time across four menu structures: random structure; frequency-based structure; consistent structure, in which items remained in the same position with every use of the menu; and alphabetic structure. The characteristic of the latter two systems is that operators can use internalized rules (knowledge of location and alphabetical order, respectively) to access the item directly, thereby avoiding the linear search requirements. Consistent with these predictions, Somberg found that retrieval time with the alphabetic menu was as fast as with the frequency-based menu and did not vary as a function of the serial position of the item on the menu. After sufficient practice, so that positions could be learned, search time with the constant position menu was the fastest of all. Such findings are not surprising. As in free-field search, knowledge and expectancy of environmental locations, acquired from experience, can be the most efficient guide to the allocation of visual attention.

In conclusion, we note that quantitative models of human visual search and scanning performance are fairly successful. Although they don't succeed in predicting exactly how an operator will accomplish a task or how long it will take for an item to be located, the answers they provide are usually at a more precise level than those offered by the engineer's intuition. Visual search is only a small component of human performance, but it offers a success story in the domain of performance models.

PARALLEL PROCESSING AND DIVIDED ATTENTION

The first part of this chapter has addressed those aspects of attention and perception that are distinctly serial. In the search or supervisory/control task, items are scanned one by one, and the dominant characteristic of all search models is the linear relationship between time and the number of items to be searched. Yet even in this discussion we have alluded to certain ways in which processing is parallel, rather than serial. In models of scanning, we have talked about a useful field of view, with the assumption that several items within that field might be processed together (in parallel). Thus in reading, there is good evidence that when we fixate on a short word, all letters within that word are processed in parallel (see Chapter 5). We have also discussed the potential role of peripheral vision in processing some information in parallel with the fovea. Finally, our discussion of search models has indicated that when a target is defined by one level on a salient feature or by an automatically processed stimulus, the search time to detect the target does not depend on how many nontargets are in the visual field. In other words, a whole set of nontarget items could be processed in parallel in the search for the target.

In the last half of this chapter then, we will focus on those aspects of perceptual processing that operate in parallel. We speak of *divided* rather than *selective* attention in this case. Although divided attention and parallel processing would seem to be a good thing for human performance—particularly in high-demand environments such as the air traffic control console or the busy office—this ability can be a two-edged sword if it becomes impossible to narrow the focus of attention when needed and shut out unwanted inputs. This failure occurs when divided attention becomes *mandatory* rather than optional. In this case we speak of a failure of *focused attention* as being the downside of successful divided attention. For example, psychologists have found that many of the same display principles that create successful divided attention are responsible for the failure to focus attention. In the previous chapter, we saw that integral dimensions help when operators can divide their attention between two redundant dimensions but hurt when they must focus attention on one while ignoring independent changes in the other. Because of the close relationship between divided and focused attention, our discussion will often treat the two topics in consort. We begin by considering parallel processing at its earliest phases of the visual information-processing sequence; then consider the role of space, objectness, and color in attention; and finally shift our discussion to parallel processing and focused and divided attention in the auditory modality.

Preattentive Processing and Perceptual Organization

Many psychologists have argued that the visual processing of a multiple-element world has two main phases: A *preattentive* phase is carried out automatically and organizes the visual world into objects and groups of objects; then *focal attention* selects certain objects of the preattentive array for further elaboration (Kahneman, 1973; Neisser, 1967). These two processes can be associated with short-term sensory store and perception, respectively, in the model of information processing presented in Figure 1.3. Thus, the perceptual judgment of which parts of the visual field are objects (figure) and which parts are in the background—known as *figure-ground* perception—is preattentive. So also in the grouping together of similar items on the display shown in Figure 3.5a. Psychologists of the *Gestalt* school (Boring, 1942) have made efforts to identify a number of basic principles that cause items to be preattentively grouped together on the display. An important characteristic of such displays is their *redundancy*, in information-processing terms (Garner, 1974). That is, knowledge of where one item is on the display will allow an accurate guess of the location of other display items in a way that is impossible with the less organized display shown in Figure 3.5b. Indeed, Tullis (1983) and Palmiter and Elkerton (1987) have developed a set of information-theory-based measures of display organization that can be used to quantify the organization of alphanumeric and analog displays, respectively. Because all items of an organized display must be processed together to reveal the organization, such parallel processing is sometimes called *global* or *holistic* processing, is contrast to the *local* processing of a single display within an array.

Two examples taken from relatively basic research paradigms illustrate the differences between global and local processing. Shown in Figure 3.6a is a stimulus presented to subjects by Navon (1977): a large *F* made up of a series of small *T*'s. When the subjects are asked to report the name of the large letter, there is a conflict. The small letters—perceived by local processing—lead one to respond *T*, whereas the large letters—global processing—lead one to respond *F*. This conflict is not present with the stimuli in Figure 3.6b. The phenomenon illustrated by the stimuli in *a* is known as *response conflict,* which we will discuss later in the chapter. From the perspective of global versus local processing, the important point is that when subjects are asked to report the

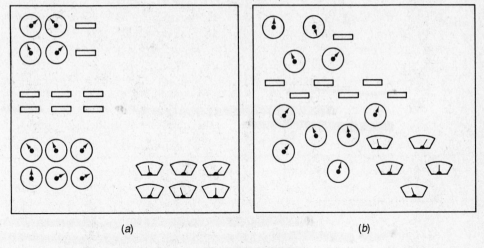

Figure 3.5 Gestalt principles of display organization.

small letters, they are disrupted when the large letter is incompatible. But this interference is asymmetric. When asked to report the large letters, there is little interference from the incompatibility of the small (Navon, 1977). Thus, the global aspects of the stimulus appear to be automatically processed in a way that makes them immune to the local aspects, for which a more focal mode of attention is required.

A second example is the texture segregation shown in Figure 3.7. At the top of the figure, the vertical *T*'s appear more different from the slanted *T*'s

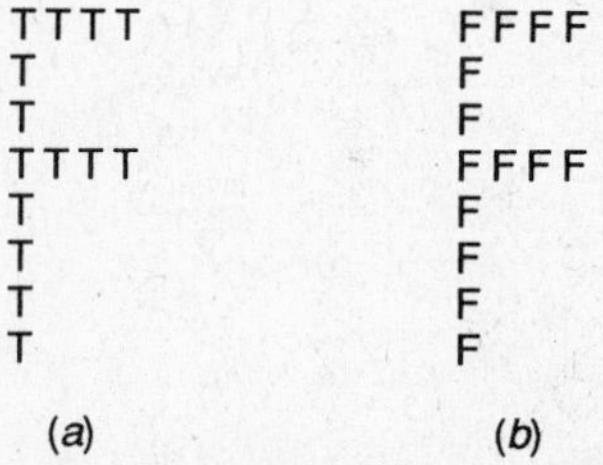

```
TTTT        FFFF
T           F
T           F
TTTT        FFFF
T           F
T           F
T           F
T           F
```

(a)   (b)

Figure 3.6 Global and local perception. (a) Global and local letters are incompatible; (b) global and local are compatible.

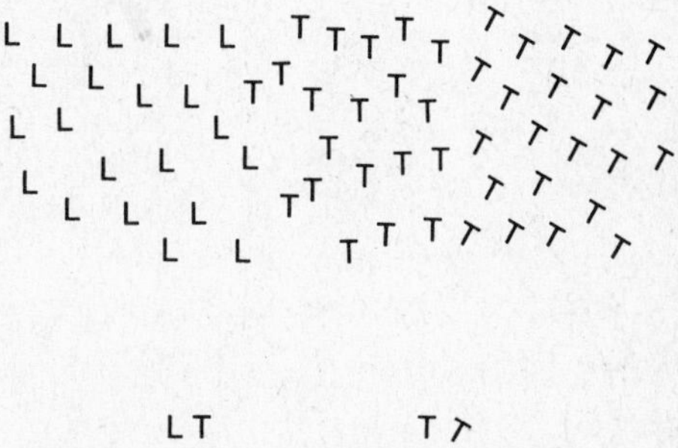

Figure 3.7 Global versus local perception. On the top (global perception), contrast the *L*'s (left) with the *T*'s (center) with the slanted *T*'s (right). The distinction between the *T*'s and slanted *T*'s is greater. However, in the bottom (local perception) the distinction between *L* and *T* is at least as great as between the two *T*'s.

than from the *L*'s on the left. This is a discrimination based on global perception (Olson & Attneave, 1970). At the bottom of the figure, however, illustrating local perception, the difference in discriminability between the two pairs is reduced, if not reversed.

The issue of global and local perception is closely related to the concept of *emergent features*, which was discussed briefly in Chapter 2. An emergent feature is a global property of the set of stimuli (or displays) that is not evident as each is seen in isolation. Consider the two sets of gauges shown in Figure 3.8, in which the normal setting of each gauge is vertical. However, it is apparent that the vertical alignment of the gauges on the top set allows more rapid detection of the divergent reading because of the emergent feature—the long vertical line—which is present in the top set but not in the bottom (Dashevsky, 1964).

Because global or holistic processing tends to be preattentive and automatic, one might say that it can reduce attention demands as an operator processes a multielement display. But this savings is only realized under two conditions: First, the Gestalt principles of proximity or the information principles of display organization, redundancy, and symmetry, must be used to produce groupings or emergent features in the first place. Second, the organization formed by the spatial proximity of different elements on the display panel must be *compatible* with the cognitive organization of the task. Thus, for example, in Figure 3.5*a*, the organization of the displays will be of little

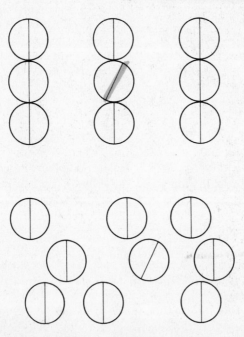

Figure 3.8 Global perception in the detection of misalignment.

assistance, and may even be of harm, if the task performed by the operator requires constant comparison of information presented in dials in the top left and bottom right group. We refer to this as a violation of *compatibility* between the display and task requirements. In a more serious example, some nuclear power consoles were designed with the panels for two reactors lying side by side, one the perfect mirror image of the other. This configuration provided wonderful symmetry, which at a global level provided organization, but it made it very difficult for the operator to go from one panel to the other.

Spatial Proximity

Overlapping Views: The Head-Up Display The previous discussion indicated that spatial organization is important for fostering preattentive processing in multielement displays. The next question we address is whether spatial proximity, or closeness in space, will help divided attention. Intuitively, one might thing that it would. For example, one cannot simultaneously look at the speedometer and look out the windshield at the road, and so it would seem that a display that could superimpose a view of the speedometer on a view of the road should facilitate divided attention or parallel processing between the two channels (Goesch, 1990).

The experimental data that exist in this area suggest that although spatial proximity will at least *allow* parallel processing, it certainly will not guarantee it. For example, a study by Wickens and Andre (1990), in which subjects had to divide their attention among three gauge indicators to determine the likelihood of a simulated aircraft stall, revealed that the spatial separation of the three indicators had no effect on performance. In an experiment by Neisser and Becklen (1975; Becklen & Cervone, 1983), subjects watched a display on which two video games were presented simultaneously, one superimposed over the other. One showed distant figures tossing a ball, the other two pairs of hands slapping one another. One game was designated as relevant, and critical elements were to be monitored and detected. Neisser and Becklen found that while monitoring one game, subjects failed to see events on the other game, even when these were unusual or novel (e.g., the ball tossers paused to shake hands). They also had a difficult time when detecting events in two games at once. These results suggest that separation may be defined not only in terms of differences in visual or retinal location but also in terms of the nature of the perceived activity or possibly the inferred distance from the observer.

Neisser and Becklen's (1975) finding has a direct counterpart in the automobile (Goesch, 1990) and in aviation, in which pilots may be presented with a *head-up display* (HUD) of critical instrument readings on the glass windscreen (Steenblik, 1989; Weintraub, 1987), shown in Figure 3.9. Although this procedure was meant to ensure that information inside and outside the cockpit could be processed simultaneously without visual scanning, the conclusions of Neisser and Becklen's study indicate that there is no guarantee that this will occur. A pilot may treat the two distances as different attentional channels and

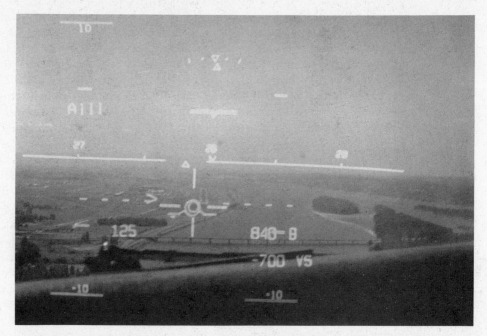

Figure 3.9 Head-up display used in aviation. HGS (Head-up Guidance System). (*Source: Courtesy of Flight Dynamics, Inc., Portland, OR.*)

become engrossed in processing instrument information on the HUD while ignoring critical cues from outside the aircraft, a phenomenon that was observed in experiments by Fischer, Haines, and Price (1980) and Larish and Wickens (1991). These data, however, should not be interpreted to mean that the HUDs are poor displays. The circumstances in which they do facilitate some parallel processing (or require less visual scanning) seem to outnumber those in which they do not (Larish & Wickens, 1991; Newman, 1987; Steenblik, 1989; Weintraub, 1987). The data suggest only that caution must be exercised before assuming that superimposing two items in the visual field guarantees parallel processing.

Visual Confusion, Conflict, and Focused Attention Although close proximity in space may sometimes allow more successful divided attention, it appears that it also imposes the potential for *confusion* between those items that represent the momentary desired focus of attention and those that are momentarily unwanted, that is, a failure to focus attention. Several lines of evidence may be cited to support this claim.

First, as we saw earlier in this chapter, in visual scanning the spatial density of the objects has little effect on visual search time. With a high-density field, any advantages that may be realized in terms of more items per fixation will be negated by the increased clutter. Second, in the study of divided attention across display indicators, Wickens and Andre (1990) found that the most critical variable in predicting performance is the degree of spatial separation of relevant from irrelevant items, not the spatial separation between the relevant items themselves. Third, a study by Holahan, Culler, and Wilcox (1978) found that the ability to locate and respond to a stop sign in a cluttered display is directly inhibited by the proximity of other irrelevant signs in the field of view.

The fourth piece of evidence is found in a study by Eriksen and Eriksen (1974), which will be discussed in more detail because it sets the stage for the discussion of object displays. In this experiment, subjects moved a lever to the right if the letter *H* appeared and to the left if the letter *F* appeared. Reaction time (RT) was measured in this *control condition*. In other conditions, the central target was closely flanked by two adjacent letters, which were irrelevant to the subjects' task and were therefore to be ignored. It turns out that almost any flanking letter will slow RT relative to the control condition. This is the result of *perceptual competition*, or *display clutter*, a failure of focused attention caused by the competition for processing resources between close objects in space. Only if the flanking letters are identical (i.e., an *H* surrounded by two other *H*'s) can RT sometimes be slightly faster than the control condition. This is another example of *redundancy gain*.

In the particular case in which the flanking letters are mapped to the opposite response (i.e., an *H* flanked by *F*'s), RT is slowed still further. There is now an added cost to processing, which Eriksen and Eriksen (1974) describe as *response conflict*, a concept introduced in the context of Navon's (1977) experiment. It illustrates more clearly the failure of focused attention. It is as if the navigator sitting next to the automobile driver were saying, "Turn left," while a passenger in the backseat, engaged in a different conversation, says, "Yeah, right."

Response conflict and redundancy gain are thus opposite sides of the same conceptual coin. If two perceptual channels are close together, they will both be processed, even if only one is desired. This processing will inevitably lead to some competition (intrusion or distraction) at a perceptual level. If they have common implications for action, the perceptual competition will be balanced by the fact that both channels activate the same response. If, however, their implications for action are incompatible, the amount of competition is magnified.

In Eriksen and Eriksen's (1974) study, the similarity between stimuli was defined by spatial location. When flanking letters were moved closer to the central letter, the similarity effects were enhanced. We might expect that the observed effects of response conflict and redundancy gain would be amplified even further if the different sources of information represented different attributes of a *single* stimulus object at one spatial location. Indeed several studies have shown that this is the case. Many of these investigations have employed some variation of the Stroop task (Stroop, 1935), in which the subject is asked to report the color of a series of colored stimuli as rapidly as possible. In a typical control condition (e.g., Hintzman et al., 1972; Keele, 1972), the stimuli consist of colored symbols—for example, a row of four *X*'s in the same color. In the critical *conflict* condition, the stimuli are color names that do not match the color of ink in which they are printed (e.g., the word *red* printed in blue ink). Here the results are dramatic. The reporting of ink color is painfully slow and fumbling when compared to the control condition, as the semantic attribute of the stimulus (*red*) activates a response that is fundamentally incompatible with information that the subject is instructed to process (the color blue). The mouth cannot articulate the words *red* and *blue* at the same time, yet both are called for by different attributes of the single stimulus.

In light of the strong Stroop effects resulting from response conflict, it is not surprising that *cooperation* through redundancy gain has also been observed when the color of the ink matches the semantic content of the word (Hintzman et al., 1972; Keele, 1972, 1973). In fact, several examples of both cooperation and competition resulting from the relationship between two attributes of a stimulus have been reported. Clark and Brownell (1975) observed that judgments of an arrow's direction (up or down) were influenced by the arrow's location within the display. "Up" judgments were made faster when the arrow was higher in the display. "Down" judgments were made faster when it was low. Similarly, Rogers (1979) found that the time it took to decide if a word was *left* or *right* was influenced by whether the word was to the right or left of the display.

Object Displays

The experimental results concerning the Stroop effect suggest that several dimensions belonging to a single object will guarantee their parallel processing; this characteristic that will help performance if parallel processing is required, but it will disrupt performance if one dimension is irrelevant and to be ignored, particularly if it triggers an automatic and incompatible response tendency. Since objects are more likely to define *integral dimensions*, this finding is consistent with the experimental results reviewed in Chapter 2: that integral dimensions produced a cost for the filtering task and a benefit for the integration of redundant dimensions.

It should therefore not be surprising to learn that designers of some systems have capitalized on the parallel processing of objects to create multidimensional object displays. Figure 3.10 illustrates four such examples. On the top is the attitude display indicator, a two-dimensional object display used in aircraft

Figure 3.10 Four examples of object displays discussed in the text: (*a*) aircraft attitude display indicator; (*b*) safety parameter display; (*c*) medical display of oxygen exchange; (*d*) decision aid display. (*Source:* (*c*) adapted from William G. Cole, "Medical Cognitive Graphics," *Proceedings of CHI '86 Human Factors in Computing Systems.* Copyright 1986 by the Association for Computing Machinery, Inc. Reprinted by permission. (*d*) adapted from B. J. Barnett and C. D. Wickens, "Display Proximity in Multicue Information Integration: The Benefits of Boxes." Reprinted with permission from *Human Factors* (vol. 30, 1988). Copyright 1988 by The Human Factors Society, Inc. All rights reserved.)

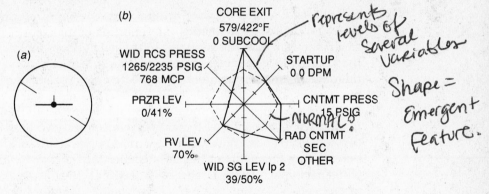

(a)

(b)

CORE EXIT
579/422°F
0 SUBCOOL

represents levels of several variables

Shape = Emergent Feature.

WID RCS PRESS
1265/2235 PSIG
768 MCP

STARTUP
0 0 DPM

PRZR LEV
0/41%

CNTMT PRESS
15 PSIG

NORMALS

RV LEV
70%

RAD CNTMT
SEC
OTHER

WID SG LEV lp 2
39/50%

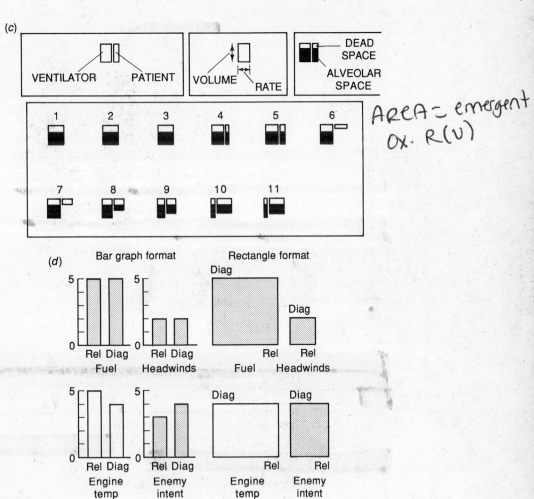

(c)

VENTILATOR   PATIENT

VOLUME   RATE

DEAD
SPACE

ALVEOLAR
SPACE

AREA = emergent ox. R(v)

1   2   3   4   5   6

7   8   9   10   11

(d)

Bar graph format

Rectangle format

Fuel

Rel Diag

Headwinds

Rel Diag

Diag

Fuel

Rel

Diag

Headwinds

Rel

Engine
temp

Rel Diag

Enemy
intent

Rel Diag

Diag

Engine
temp

Rel

Diag

Enemy
intent

Rel

control. The vertical location of the aircraft symbol relative to the horizon indicates aircraft *pitch* (nose up or nose down), and the angle between the aircraft and the horizon represents the *bank*, or roll. In addition to its objectness, this display is configured in a way that represents the aircraft, and is therefore *familiar* to the pilot. Next is the safety parameter display for nuclear power reactor operators designed by Westinghouse, in which the values of eight key parameters are indicated by the length of imaginary "spokes" extending from the center of the display and connected by line segments to form a polygon (Woods, Wise, & Hanes, 1981). In addition to its objectness, a potential advantage of this display is that certain failures, because of the nature of the parameter changes, cause a unique shape or configuration of the polygon to emerge, as seen on the right of the figure. Using a term presented in the previous chapter, in the context of configural dimensions, we say that the object display produces an *emergent feature*.

The display developed by Cole (1986) for medical applications in the third row of the figure illustrates another example of an emergent feature. The rectangular display represents the oxygen exchange between patient and respirator. The width represents the rate of breathing, and the height represents the depth of breathing. Hence, the *area* of the rectangle—an emergent feature—signals the total amount of oxygen exchanged, a critical variable to be monitored. This correspondence holds true because oxygen amount = rate × depth, and rectangle area = width × height.

One might ask, however, if the critical parameter is oxygen amount, why shouldn't that quantity be displayed directly rather than inferred from the area of the rectangle formed by the two contributing dimensions? The reason is that there may be times when it is necessary to *focus* attention on one of the variables contributing to the amount (rate or depth), rather than integrating across these dimensions. Then the question can be asked, Will the close proximity created by the single-object format disrupt the ability to focus attention on one of its dimensions, as when the operator must check one of the values being integrated? This question is addressed in the next section.

The Proximity Compatibility Principle

The issue of whether different tasks are served differently by more or less integrated displays is represented explicitly in the *proximity compatibility principle* (Barnett & Wickens, 1988; Carswell & Wickens, 1987; Wickens & Andre, 1990), which states that to the extent that information sources must be *integrated* (as in Cole's display), there will be a benefit to presenting those dimensions in an integrated (i.e., objectlike) format. That is, high *display proximity* helps in tasks with high mental proximity. To the extent that information must be *treated separately* (low mental proximity, as when one must focus on one variable and ignore values of the other), the benefit of the high-proximity object display will be reduced, if not sometimes reversed. The advantage of object displays for information integration results from two factors: (1) the parallel processing of object dimensions and (2) the fact that when these dimensions are all the same (e.g., all measures of extent, like the height and width of a rectangle), they will be more likely to produce an emergent feature (e.g., area or shape), which can directly serve the integration task requirements (Carswell, 1990). When focused attention on one dimension is required, the very emergent feature that helped integration by calling attention to itself can make the focused attention task more difficult.

The proximity compatibility principle has been addressed in a number of different contexts. The context illustrated at the bottom of Figure 3.10 is representative: an intelligent airborne decision aid that would make recommendations to the pilot about the advisability of continuing with a mission. Each recommendation is based on an *information source* (e.g., a weather advisory or an engine fuel check). The two dimensions of the display represent the two characteristics of each source: its *reliability* (how much it could be trusted) and its *diagnosticity* (relevance to the decision at hand). As we will see in Chapter 7, these two features combine to indicate the total information *worth of the source*. Barnett and Wickens (1988) found that tasks requiring the integration of the two dimensions to evaluate that total worth were served better by a rectangle display than by a bar graph display. As with Cole's (1986) experiment, the emergent feature of the area directly revealed the quantity to be inferred. But when attention had to be focused on one variable, to the exclusion of the other (e.g., a question might ask, "What was the reliability of weather information?"), the advantage for the rectangle over the bar graph disappeared. In Chapter 4, we will discuss some applications of the proximity compatibility principle to the design of graphs (Carswell & Wickens, 1988).

It is important to recognize that emergent features need not be created exclusively by single objects (Sanderson, Flach, Buttigieg, & Casey, 1989). Indeed we saw the emergent feature created by the vertical array of dials at the top of Figure 3.8. An array of parallel bar graphs representing, for example, four engine parameters, such as that in Figure 3.11, also creates an emergent feature—a horizontal line across the top—when all engines are running at the same level. In this case, display proximity is defined not by belonging to a

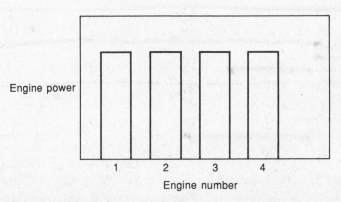

Figure 3.11 Emergent features in engine parameter display.

common object but by the identical form of representation (vertical extent) of all the indicators.

It should also be noted that the identical color of two objects on a display, like the integrality of the dimensions of an object, appears also to adhere to the principle of compatibility proximity (Wickens & Andre, 1990). That is, two items on a cluttered display will be more easily integrated or compared if they share the same color (different from the clutter), but the shared identity of color may disrupt the ability to focus attention on one while ignoring the other. A unique color code will help this focusing process, just as it disrupts the integration process.

The proximity compatibility principle also applies to spatial distance in a cluttered display. Two pieces of information that need to be integrated on a cluttered display should be placed in close spatial proximity, as long as this proximity does not also move them closer to irrelevant clutter (Wickens & Andre, 1990). For example, Sweller, Chandler, Tierney, and Cooper (1990) found that text materials lead to better learning if graphic material and relevant text (two information sources with close mental proximity) are adjacent to one another on the page. Bettman, Payne, and Staelin (1986) discuss the importance of spatial proximity between related items (costs and benefits of a product) in the design of product warning labels; and Milroy and Poulton (1978) point to the importance of close proximity between graphed lines and their labels. That is, labels should be set next to the lines, not in a legend below. An illustration of a violation of the principle is found in the design of a radar display, which may have been a contributing cause to the *Vincennes* incident. An Iranian passenger airplane was mistaken for an attacking fighter plane and inadvertently shot down by the American ship *Vincennes* in the Persian Gulf (U.S. Navy, 1988). In the radar display, the symbol, signifying the location of the aircraft itself, was in a separate location from the tag that labeled the "range gate," a critical piece of information for interpreting the actions of the approaching aircraft. It is likely that the lack of close spatial proximity prevented the operators from integrating the two pieces of information in a way that would lead to the correct classification of the aircraft: a climbing commercial air carrier, rather than a descending, attacking fighter.

In conclusion, three things can be accomplished by moving two (or more) displayed elements close together and/or integrating them as dimensions of a single object.

1. This close proximity will increase the possibility of parallel processing by moving both dimensions into foveal vision. Parallel processing will be most likely to occur if they are integrated as dimensions of a single object.

2. Both close spatial proximity and objectness can create useful emergent features if the two display dimensions are the same (e.g., the length of two lines can create an emergent feature; the length of a line and its color cannot). These emergent features can help information integration if they are mapped into key variables of the task. (The mapping calls for creativity and ingenuity by the display designer.) The emer-

gent features can hurt performance if they are not mapped into the task.

3. The close proximity, enhanced by objectness, can create response conflict. Both response conflict and emergent features will then be troublesome to the extent that the task calls for focused attention on one of the variables combined in the display.

Color Coding

Discussions of target search and visual attention must include a brief treatment of the specific effects of color coding in displays, although color coding relates to a number of other topics in this book, some already covered and some to be covered in future chapters.

For that 97 percent of the population who are not color-blind (roughly 7 percent of males cannot adequately discriminate certain wavelengths of light), differences in color are processed more or less automatically and in parallel with characteristics of shape and motion (Treisman, 1986). Although there are also costs, three distinct benefits result from the automatic characteristics of color processing.

1. Color stands out from a monochrome background. Therefore color-coded targets are rapidly and easily noticed. As suggested in our discussions of visual search, the speed of search for a target that is uniquely color coded in a cluttered field is independent of the size of the field. Therefore, color coding of targets or critical elements in a display is quite effective for rapid localization (Christ, 1975). Color, for example, can be extremely effectively as a means of highlighting an important item on a menu display (Fisher & Tan, 1989).

2. Color coding can act like a preattentive organizing structure to tie together multiple elements that may be spatially separated. As noted, this characteristic will be quite useful if the commonly colored items also need to be *integrated* as part of the task (Wickens & Andre,1990). Thus, for example, there will be an advantage to color coding different regions on a weather map according to temperature, thereby providing an indication of temperature patterns that can be integrated across space. We will discuss this aspect of color coding further in Chapter 12 in the context of process control displays.

3. The automaticity with which color is coded enhances its value as a *redundant* coding device, which will signal information otherwise signaled by shape, size, or position. As noted in Chapter 2, the traffic light is an example of redundant coding of color with location. Kopala (1979) demonstrated the importance of redundant coding of color with shape to indicate threat classifications in airborne tactical displays.

Because of its attractiveness and aesthetic appeal, color coding has become quite prevalent in many displays. However, there are a number of limitations, which are sometimes subtle yet may be critical for system design.

Broadment focus on one or the other.

Treisman Theory

switch them.

1. Like any other sensory continuum, as discussed in Chapter 2, color is subject to the limits of absolute judgment. To guarantee that the value (and therefore meaning) of a color will not be misidentified, the system designer should probably use no more than five or six colors in a display (Carter & Cahill, 1979). Furthermore, if colors are to be perceived under conditions of glare or changing or low illumination (e.g., the dashboard or cockpit), failures of absolute judgment may be even more prevalent; for example, red may be confused for brown (Stokes, Wickens, & Kite, 1990).

2. Color does not naturally define an ordered continuum. If people are asked, for example, to place five colors in an order from "least" to "most," there will be a great divergence of opinion about the appropriate ordering. Even the rainbow spectrum is far from universally recognized as a continuum. Since color ordering does not have a strong *population stereotype*, a concept that will be discussed in Chapter 8, it may be inefficient to use color coding to represent an ordered variable like temperature or speed.

3. Although color ordering does not have a strong population stereotype, such stereotypes are exhibited in the association of specific colors with specific meanings (e.g., red equals danger, hostility, the order to stop; green equals go and safety). These associations can be a strength if they are exploited in design or a weakness if a color-coding scheme associates a color with a conflicting meaning. For example, suppose a temperature-coding scheme is designed in which green represents low temperature, but in the system very low temperatures signal an unsafe operating condition. Hence, the natural association of green with "safe" or "go" is not the one that should be inferred by the operator.

4. Because of the automaticity with which it is processed, irrelevant color coding can be distracting. When different colors are used to highlight different areas or items, it is important that the distinction made by the colors is compatible with relevant cognitive distinctions that are intended to be interpreted by the viewer. This issue of display-cognitive compatibility was discussed in the context of spatial organization and will emerge again in later discussions of display motion (Chapter 4).

Finally, color is appealing and attractive but expensive. Operators usually express a preference for color over monochrome displays, but because of its greater expense and the potential shortcomings just outlined, considerable caution should be exercised before deciding on its implementation. As Shneiderman (1987) points out in his guidelines for display design, displays should be designed for monochrome first.

ATTENTION IN THE AUDITORY MODALITY

The auditory modality is different from the visual in two important respects that have implications for the study of attention. First, the auditory sense is *omnidirectional*. It can take in input from any direction and so, unlike vision, does not have an analogy to scanning as a valid index of selective attention. Second, most auditory input is transient. A word or a tone is heard and then it ends, in contrast to most visual input, which tends to be more continuously available. For example, the printed word usually remains on the page. Hence, the preattentive characteristics of auditory processing—those required to "hold on" to a stimulus before it is gone—are more critical in audition than in vision. As discussed briefly in Chapter 1, short-term auditory store is longer than short-term visual store.

There is a long history of research in auditory selective attention, which will not be discussed here (see Moray, 1969; Wickens, 1984). Much of this research is based on variations of a *dichotic listening task*, in which the listener hears two independent streams of sounds, words, or sentences, one in each ear. Sometimes both ears are to be attended to; sometimes only one is to be attended to and the other ignored. Interest has focused on the physical and semantic characteristics of messages that lead to successes and failures in these divided and focused attention tasks (Broadbent, 1958; Cherry, 1953; Moray, 1969; Treisman, 1969).

Filter theory — 2 messages / 2 ears

Auditory Divided Attention

Treiseman's Theory
concert — at hall
(dance — is playing)

A general model of auditory attention that is consistent with the data, and with theories proposed by Norman (1968) and Keele (1973), proposes that streams of auditory input that are not in the focus of attention remain in preattentive short-term auditory store for a period of three to six seconds (see Chapter 6). The contents of this store can be "examined" if a conscious switch of attention is made. Thus, if your attention wanders while someone is talking to you, it is possible to switch back suddenly and "hear" the last few words that the person has spoken, even if you were not attending to them at the moment they were uttered.

In the absence of a conscious attention switch, information in nonattended channels continues to make contact with long-term memory, in terms of the information-processing diagram shown in Figure 1.3. That is, nonattended words are not just meaningless "blobs of sound," but their meaning is analyzed at a preattentive level. Other functions of the brain meanwhile continue to monitor all channels for "pertinent" material. If the nonattended material is sufficiently pertinent, our adaptive mechanisms will often bring this material to the focus of attention (i.e., we will switch attention to the nonattended channel). For example, a loud sound will almost always grab our attention, as it may signal a sudden environmental change that must be dealt with. Our own name also has a continued pertinence, and so we will sometimes shift attention to it when it is spoken, even if we are listening to another speaker (Moray, 1959). So also does any material that is semantically related to the topic that is the current focus of attention (Treisman, 1964b).

And what of the fate of the words or sounds that never receive our attention—either because their pertinence is not high enough or because we do not voluntarily choose to listen to them? As Dr. Seuss says, "Oh, their future is dreary" (Seuss, 1971). There is little evidence that this material makes any permanent impact on long-term memory, beyond the brief, transient activation of the semantic unit. Hence the idea of "learning without awareness," whether in one's sleep or through techniques of "subliminal perception," has received little empirical validation (Swets & Druckman, 1988).

This discussion has focused more on the failures than on the successes of divided attention in the auditory modality. We also know, without the need of experimental data to convince us, how difficult it is to listen to two speakers at once. In our discussion of visual attention, however, we saw that close proximity, particularly as defined by objectness, was a key to producing a successful division of attention. We also saw that the same manipulations of proximity that allowed success in divided attention were responsible for the failure of focused attention. These manipulations and observations also have analogies in audition.

It is possible to think of an "auditory object" as a sound (or series of sounds) with several dimensions, which seem to enjoy the same benefits of parallel processing as do the dimensions of a visual object. Thus, Moore and Massaro (1973) found that subjects were able to judge the quality and pitch of a tone simultaneously as well as either dimension could be judged alone. We are all certainly aware of our ability to pay attention to both the words and melody of a song and to the meaning and voice inflections of a spoken sentence. Some effort has been made to capitalize on parallel processing in the design of auditory warning alerts, in which redundant dimensions like pitch, timbre, and interruption rate are combined to create the most noticeable sounds (Edworthy & Loxley, 1990; Sorkin, 1987).

Failure of Focused Auditory Attention

In vision, we saw that the use of close proximity to facilitate parallel processing was a two-edged sword because it disrupted the ability to focus attention. In the auditory modality, too, we find that focused attention on one channel is disrupted when there is increased similarity with competing messages that are to be ignored. The amount of disruption seems to depend on the different dimensions along which similarity is defined.

Spatial location is perhaps the most salient physical dimension of similarity. When attempting to attend to one auditory message and ignore the contents of another, our ability to do so declines as the two messages become closer in space (Egan, Carterette, & Thwing, 1954; Spieth, Curtis, & Webster, 1954; Treisman, 1964a). An extreme example of this difference in separation is provided by the distinction between monaural and dichotic listening. In monaural listening, two messages are presented with equal relative intensity to both ears, a stimulus that would be experienced when listening to two speakers at the same physical location. In dichotic listening, the two messages are delivered to opposite ears by headphones, so that no overlap is possible. Egan, Carterette, and Thwing (1954) found that there are large benefits of dichotic over monaural listening in terms of the operator's ability to filter out distraction from an unwanted channel.

We are, of course, also able to attend selectively to auditory messages even from highly similar locations. The classic "cocktail party effect" describes this ability to attend to a speaker at a noisy party and selectively filter out (with varying degrees of success) other conversations coming from the same spatial location. In this case, we must be able to use dimensions of similarity other than location to focus attention selectively. One such dimension for selection is defined by pitch. It is easier to process selectively a male or female voice in the presence of a second voice of the opposite sex (and thereby different pitch) than to attend selectively when confronted with two voices of the same sex (Treisman, 1964a). Interestingly, this pitch discrimination can be made even though there is a large overlap between the frequency content or spectrum of male and female voices.

Intensity may also serve as a dimension of selection. It is clearly easy to attend to a loud message and tune out a soft one. However, Egan, Carterette, and Thwing (1954) found that it is also possible to attend selectively to the softer channel and ignore a louder one as long as the differences in intensity are less than 10 decibels. When the difference is more, however, the advantage to the selective difference is more than outweighed by the fact that the louder channel physically masks the softer one.

We usually think about channels of selective attention defined in terms of physical properties of the stimulus. However, Treisman (1964b) has observed that *semantic content,* a cognitive dimension, can readily serve as a cue for selection. Bilingual subjects were asked to attend to an auditory message played to one ear as a second message to be ignored was played in the other. Treisman found that subjects' ability to process the attended-to message was disrupted if the message delivered to the opposite ear and spoken in a different language was of a similar semantic content. The fact that a different language was employed ensured that the failure of selection could not have resulted from any *physical* similarity between messages. Although these results show that similar semantic content impairs the ability to focus attention, they do not imply that differences in semantic content provide a perfect cue for selection. Two messages from identical locations and spoken with identical voices will still create considerable problems in selection, even if each is quite different in meaning.

It might be noted that the current discussion of semantically defined channels provides some insight into the findings of the visual attention experiment by Neisser and Becklen (1975), in which subjects attending to one video game were totally oblivious of events in a second game superimposed on the first. Although one was superimposed on another, the two games were defined by different and unrelated flows of events, which allowed them to define two different cognitive channels. This, in turn, allowed attention to be focused on one, with minimal processing of the concurrent channel.

To summarize, auditory messages differ from one another along a wide variety of different dimensions such as pitch, location, loudness, and semantic content. The greater the difference between two messages along a given dimension and the greater the number of dimensions of difference, the easier it will be to focus on one message and ignore the other. There is not, however, a great deal of evidence pertaining to the degree to which one dimension is more important than another.

Practical Implications

The characteristics of auditory attention have a number of practical implications for system design, some of which have been alluded to already. For example, we discussed the concept of the auditory object and the fact that system designers can capitalize on the parallel processing of several dimensions of an object to provide either more redundancy or a greater amount of information in a given auditory alert, an issue that will have further importance in the discussion in Chapter 12 of process control failure detection.

Another important issue in auditory display is the question of what features of an alert will grab attention, so that the alert will be processed (Sorkin, 1987). As described in Chapter 12, although loud tones have the advantage of usually calling attention to themselves, they also have disadvantages. They are annoying and startling, and sometimes their very intensity can create a stressful environment that leads to nonoptimal information processing (see Chapter 10). Hence it may be that designers can capitalize on the human operator's tendency to switch attention to "pertinent" material (that is not necessarily loud) to design less noxious alerts. A simple example would be personalized alerts that are prefaced with the operator's own name. These attention-grabbing but softer auditory warnings have been called *attensors* (Hawkins, 1987).

A third potential application concerns the steps that can be taken to avoid confusion between auditory messages and the disruption of focused attention. Because the auditory modality, unlike the visual, does not have a directional "ear ball" that can filter out unwanted information, relatively greater concern must be given to determining those auditory display features that can allow channels that may be heard concurrently to be distinguished and discriminated. For example, how can the automobile designer ensure that a dashboard auditory warning will not be confused with a radio channel, engine noise, or ongoing conversation? As with vision, the spatial dimension can be employed to help establish the discriminability of auditory channels. An experiment by Darwin, Turvey, and Crowder (1972) suggests that three "spatial" channels may be processed without distraction if one is presented to each ear and a third is presented with equal intensity to both ears, thereby appearing to originate from the midplane of the head.

In this manner, airplane pilots might have available three distinct audio channels: one for messages from the copilot, one for messages from air traffic control, and a third for messages from other aircraft or for synthesized voice warnings from their own aircraft. They could not process the three in parallel since all would call for common semantic analysis, which we saw was impossible, but they could at least focus on one without the unwanted intrusion of information from the others. The definition of channels in terms of the pitch dimension suggests that additional separation might be obtained by distinguishing the three spatial channels, redundantly, through variation in pitch quality. Thus, the center message that is most likely to be confused with the other two could be presented at a substantially different pitch (or with a different speaker's voice) than the others.

Cross-Modality Attention

The discussion up to this point has focused exclusively on attention within a modality. But we are often confronted with parallel inputs across modalities, as when we watch and listen to a TV program or when the pilot is monitoring the visual environment on takeoff or landing, while listening to critical auditory messages spoken by the copilot regarding key velocities. In Chapter 9, in the context of multiple-resource theory, we will discuss some of the experiments suggesting that dividing attention between modalities may be better than dividing attention within a modality. Here, however, we consider an important difference between vision and other modalities when they are put in conflict —the phenomenon of *visual dominance*.

The phenomenon of visual dominance opposes the instinctive tendency that humans have to switch attention to stimuli in the auditory and tactile modalities. These stimuli are intrusive and the peripheral receptors have no natural way to shut out auditory or tactile information. We cannot "close our earlids" in the same way that we close our eyelids. Nor can we "shift our ears" in the same manner that we shift our gaze. As a consequence, auditory devices are generally preferred to visual signals as warning indicators (Cooper, 1977; Simpson & Williams, 1980; Sorkin, 1987).

The superior ability of the auditory modality to alert was demonstrated by Posner, Nissen, and Klein (1976). These investigators found that an auditory warning stimulus will speed the response to both a subsequent auditory and visual stimulus; however, a visual warning will not speed the response to the auditory stimulus. For humans to counteract these automatic alerting tendencies of the auditory modality, they have a bias toward processing information in the visual channels—a bias that is intended to overcome this alerting handicap of the visual modality. This bias is known as visual dominance.

Examples of visual dominance over auditory or proprioceptive stimuli are abundant. For example, Colovita (1971) required subjects to respond as fast as possible to either a light (with one hand) or a tone (with the other hand). On very infrequent occasions, both stimuli were presented simultaneously. When this occurred, subjects invariably made *only* the response with the hand appropriate to the light. In fact, subjects were even unaware that a tone had been presented on these trials. Experiments by Jordan (1972) and by Klein and Posner (1974) supported the dominance of vision over proprioception. These investigators found that reaction time to a stimulus that was a compound of a light and a proprioceptive displacement of a limb was slower than reaction time to the proprioceptive stimulus alone. This result suggests that the light captured attention and slowed down the more rapid processing of the proprioceptive information. Different examples of visual dominance are observed in experiments in which the sense of vision and proprioception are placed in conflict through prismatically distorted lenses (Rock, 1975). Behavior in these situations suggests that the subject responds appropriately to the visual information and disregards that provided by other modalities.

Some time-sharing situations described in Chapter 8 also appear to manifest a form of visual dominance when an auditory task is performed concurrently with a visual one. In these circumstances, the auditory task tends to be hurt more by the division of attention than the visual task (Isreal, 1980; Massaro & Warner, 1977; Treisman & Davies, 1973).

Although visual dominance may be viewed as an adaptive and therefore useful adjustment of information processing made in response to anatomical differences between sensory channels, there are of course circumstances in which it can lead to nonadaptive behavior. Illusions of movement provide one such example. When the visual system gives ambiguous cues concerning the state of motion, the correct information provided by proprioceptive, vestibular, or "seat of the pants" cues is often misinterpreted and distorted. For example, while sitting in a train at a station with another train just outside the window, passengers may experience the illusion that their train is moving forward, while in fact their train is stationary and the adjacent train is moving backward. The passenger model of the world has discounted the proprioceptive evidence from the seat of the pants that informs them that no inertial forces are operating.

In summary, when an abrupt auditory stimulus may intrude on a background of ongoing visual activity, it will probably alert the operator and call attention to itself. However, if visual stimuli are appearing at the same frequency and providing information of the same general type or importance as auditory or proprioceptive stimuli, biases toward the visual source at the expense of the other two will be expected.

TRANSITION

In this chapter we have described attention as a filter to the environment. Sometimes the filter strains out irrelevant auditory or visual input, and sometimes the filter broadens to take in parallel streams of environmental information. The effective breadth of the filter is dictated by the limits of our senses (e.g., foveal vision), the differences and similarities between stimulus channels, and the strategies and understanding of the human operator. What happens, then, when material passes through the filter of attention? We saw in Chapter 2 that material may be provided with a simple yes-no classification (signal detection) or categorized into a level on a continuum (absolute judgment). But more often the material is given a more sophisticated and complex interpretation. It is perceived.

At this point it is convenient to distinguish between two different kinds of perceptual interpretations. The first kind are analog spatial interpretations, whose relevance to the perceiver is defined by the *continuous spatial* dimensions of how far away things are, where they are, how big they are, and how they are oriented. Many of the judgments that pilots make regarding the state of their aircraft, or drivers make regarding the state of their vehicle, are of this form. So also is the reading of a dial, a graph, or the mercury level in a thermometer. We say that stimuli have an *analog* correspondence to their relevance. The second class of interpretations is those that are *verbal and symbolic*. Some stimuli have meaning to us that is not directly embodied in their physical form (location, shape, and orientation). Rather this meaning is *interpreted* by *decoding* some symbolic representation, usually a written or spoken word. Hence, this form of perception is heavily language-based.

Attention to these stimulus sources, whether analog or symbolic, is necessary but not sufficient to guarantee a proper interpretation of the state of the world on which to base future actions. In Chapter 4 we will discuss the perception and interpretation of analog material. In Chapter 5 we will discuss symbolic verbal material. Finally, in Chapter 9 we revisit the concept of attention, in the context of dividing attention between tasks rather than between perceptual channels.

REFERENCES

Abernethy, B. (1988). Visual search in sport and ergonomics: Its relationship to selective attention and performer expertise. *Human Performance, 1*(4), 205–235.

Barnett, B. J., & Wickens, C. D. (1988). Display proximity in multicue information integration: The benefit of boxes. *Human Factors, 30,* 15–24.

Becklen, R., & Cervone, D. (1983). Selective looking and the noticing of unexpected events. *Memory & Cognition, 11*(6), 601–608.

Bettman, J. R., Payne, J. W., & Staelin, R. (1986). Cognitive considerations in designing effective labels for presenting risk information. *Journal of Marketing and Public Policy, 5,* 1–28.

Bloomfield, J. R. (1979). Visual search with embedded targets: Color and texture differences. *Human Factors, 21*(3), 317–330.

Bloomfield, J. R., & Howarth, C. I. (1969). Testing visual search theory. In H. W. Leibowitz (ed.), *Image evaluation* (pp. 203–214). Munich: NATO Advisory Group on Human Factors Symposium.

Boring, E. J. (1942). *Sensation and perception in the history of experimental psychology.* New York: Appleton-Century-Crofts.

Broadbent, D. E. (1958). *Perception and communications.* London: Pergamon Press.

Carbonnell, J. R., Ward, J. L., & Senders, J. W. (1968). A queueing model of visual sampling: Experimental validation. *IEEE Transactions on Man-Machine Systems, MMS-9,* 82–87.

Carswell, C. M. (1990). Graphical information processing: The effects of proximity compatibility. *Proceedings of the 34th annual meeting of the Human Factors Society* (pp. 1494–1498). Santa Monica, CA: Human Factors Society.

Carswell, C. M., & Wickens, C. D. (1987). Information integration and the object display: An interaction of task demands and display superiority. *Ergonomics, 30*(3), 511–527.

Carswell, C. M., & Wickens, C. D. (1988). *Comparative graphics: History and applications of perceptual integrality theory and the proximity compatibility hypothesis.* University of Illinois Technical Report (ARL–88-2-/AHEL–88-1; AHEL Technical Memorandum 8-88). Savoy, IL: Aviation Research Laboratory.

Carter, R. C., & Cahill, M. C. (1979). Regression models of search time for color-coded information displays. *Human Factors, 21,* 293–302.

Cherry, C. (1953). Some experiments on the reception of speech with one and with two ears. *Journal of the Acoustical Society of America, 25,* 975–979.

Christ, R. E. (1975). Review and analysis of color coding research for visual displays. *Human Factors, 17,* 542–570.

Clark, H. H., & Brownell, H. H. (1975). Judging up and down. *Journal of Experimental Psychology: Human Perception and Performance, 1,* 339–352.

Cole, W. G. (1986). Medical cognitive graphics. In *Proceedings of the ACM-SIGCHI: Human factors in computing systems* (pp. 91–95). New York: Association for Computing Machinery, Inc.

Colovita, F. B. (1971). Human sensory dominance. *Perception & Psychophysics, 16,* 409–412.

Cooper, G. E. (1977). *A summary of the status and philosophy relating to cockpit warning system* (NASA Contractor Report NAS 2–9117). Washington, DC: NASA.

Craig, A. (1981). Monitoring for one kind of signal in the presence of another. *Human Factors, 23,* 191–198.

Darwin, C., Turvey, M. T., & Crowder, R. G. (1972). An analog of the Sperling partial report procedure. *Cognitive Psychology, 3,* 255–267.

Dashevsky, S. G. (1964). Check-reading accuracy as a function of pointer alignment, patterning and viewing angle. *Journal of Applied Psychology, 48,* 344–347.

Drury, C. (1975). Inspection of sheet metal: Model and data. *Human Factors, 17,* 257–265.

Drury, C. (1982). Improving inspection performance. In G. Salvendy (ed.), *Handbook of industrial engineering.* New York: Wiley.

Drury, C. G., & Clement, M. R. (1978). The effect of area, density, and number of background characters on visual search. *Human Factors, 20,* 597–602.

Edworthy, J., & Loxley, S. (1990). Auditory warning design: The ergonomics of perceived urgency. In E. J. Lovesey (ed.), *Contemporary ergonomics 1990* (pp. 384–388). London: Francis and Taylor.

Egan, J., Carterette, E., & Thwing, E. (1954). Some factors affecting multichannel listening. *Journal of the Acoustical Society of America, 26,* 774–782.

Elkind, J. I., Card, S. K., Hochberg, J., & Huey, B. M. (eds.). (1990). *Human performance models for computer-aided engineering.* Orlando, FL: Academic Press.

Eriksen, B. A., & Eriksen, C. W. (1974). Effects of noise letters upon the identification of a target letter in a non-search task. *Perception & Psychophysics, 16,* 143–149.

Fischer, E., Haines, R., & Price, T. (1980, December). *Cognitive issues in head-up displays* (NASA Technical Paper 1711). Washington, DC: NASA.

Fisher, D. F., Monty, R. A., & Senders, J. W. (eds.). (1981). *Eye movements: Cognition and visual perception.* Hillsdale, NJ: Erlbaum.

Fisher, D. L., & Tan, K. C. (1989). Visual displays: The highlighting paradox. *Human Factors*, *31*(1), 17–30.

Fisk, A. D., Oransky, N. A., & Skedsvold, P. R. (1988). Examination of the role of "higher-order" consistency in skill development. *Human Factors*, *30*(6), 567–582.

Fitts, P., Jones, R. E., & Milton, E. (1950). Eye movements of aircraft pilots during instrument landing approaches. *Aeronautical Engineering Review*, *9*, 24–29.

Friedman, A., & Liebelt, L. S. (1981). On the time course of viewing pictures with a view towards remembering. In D. F. Fisher, R. A. Monty, & J. W. Senders (eds.), *Eye movements: Cognition and visual perception* (pp. 137–154). Hillsdale, NJ: Erlbaum.

Frost, G. (1972). Man-machine system dynamics. In H. P. Van Cott & R. G. Kinkade (eds.), *Human engineering guide to system design*. Washington, DC: U.S. Government Printing Office.

Garner, W. R. (1974). *The processing of information and structure*. Hillsdale, NJ: Erlbaum.

Goesch, T. (1990). Head up displays hit the road. *Information Display*, *7–8*, 10–13.

Gould, J. D. (1976). Looking at pictures. In R. A. Monty & K. W. Senders (eds.), *Eye movements and psychological processes* (pp. 323–345). Hillsdale, NJ: Erlbaum.

Harris, R. L., & Christhilf, D. M. (1980). What do pilots see in displays? In G. Corrick, E. Hazeltine, & R. Durst (eds.), *Proceedings of the 24th annual meeting of the Human Factors Society*. Santa Monica, CA: Human Factors Society.

Hawkins, F. H. (1987). *Human factors in flight*. Brookfield, VT: Gower Technical Press.

Hintzman, D. L., Carre, F. A., Eskridge, V. L., Ownes, A. M., Shaff, S. S., & Sparks, M. E. (1972). "Stroop" effect: Input or output phenomenon? *Journal of Experimental Psychology*, *95*, 458–459.

Holahan, C. J., Culler, R. E., & Wilcox, B. L. (1978). Effects of visual distraction on reaction time in a simulated traffic environment. *Human Factors*, *20*(4), 409–413.

Isreal, J. (1980). Structural interference in dual task performance: Behavioral and electrophysiological data. Unpublished Ph.D. dissertation, University of Illinois, Champaign.

Johnston, D. M. (1967). The relationship of near-vision peripheral acuity and far-vision search performance. *Human Factors*, *9*(4), 301–304.

Jordan, T. C. (1972). Characteristics of visual and proprioceptive response times in the learning of a motor skill. *Quarterly Journal of Experimental Psychology*, *24*, 536–543.

Kahneman, D. (1973). *Attention and effort*. Englewood Cliffs, NJ: Prentice Hall.

Keele, S. W. (1972). Attention demands of memory retrieval. *Journal of Experimental Psychology*, *93*, 245–248.

Keele, S. W. (1973). *Attention and human performance*. Pacific Palisades, CA: Goodyear.

Klein, R. M., & Posner, M. I. (1974). Attention to visual and kinesthetic components of skills. *Brain Research*, *71*, 401–411.

Kopala, C. (1979). The use of color-coded symbols in a highly dense situation display. In C. Bensel (ed.), *Proceedings of the 23rd annual meeting of the Human Factors Society*. Santa Monica, CA: Human Factors Society.

Kundel, H. L., & La Follette, P. S. (1972). Visual search patterns and experience with radiological images. *Radiology, 103,* 523–528.

Kundel, H. L., & Nodine, C. F. (1978). Studies of eye movements and visual search in radiology. In J. W. Senders, D. F. Fisher, & R. A. Monty (eds.), *Eye movements and the higher psychological functions* (pp. 317–328). Hillsdale, NJ: Erlbaum.

Larish, I., & Wickens, C. D. (1991). Attention and HUDs: Flying in the dark? *Proceedings of the Society of Information Display*. Playa del Rey, CA: Society of Information Display.

Lee, E., & MacGregor, J. (1985). Minimizing user search time in menu retrieval systems. *Human Factors, 27,* 157–162.

Levy-Schoen, A. (1981). Flexible and/or rigid control of oculomotor scanning behavior. In D. F. Fisher, R. A. Monty, & J. W. Senders (eds.), *Eye movements: Cognition and visual perception* (pp. 299–314). Hillsdale, NJ: Erlbaum.

Liu, Y., & Wickens, C. D. (1989). Visual scanning with or without spatial uncertainty and time-sharing performance. *Proceedings of the 33rd annual meeting of the Human Factors Society*. Santa Monica, CA: Human Factors Society.

McConkie, G. W. (1983). Eye movements and perception during resting. In K. Raynor (ed.), *Eye movements in reading*. New York: Academic Press.

Mackworth, N. H. (1976). Ways of recording line of sight. In R. A. Monty & J. W. Senders (eds.), *Eye movements and psychological processing* (pp. 173–178). Hillsdale, NJ: Erlbaum.

Massaro, D. W., & Warner, D. S. (1977). Dividing attention between auditory and visual perception. *Perception & Psychophysics, 21,* 569–574.

Megaw, E. D., & Richardson, J. (1979). Target uncertainty and visual scanning strategies. *Human Factors, 21*(3), 303–316.

Milroy, R., & Poulton, E. C. (1978). Labeling graphs for increasing reading speed. *Ergonomics, 21,* 55–61.

Mocharnuk, J. B. (1978). Visual target acquisition and ocular scanning performance. *Human Factors, 20,* 611–632.

Moore, J. J., & Massaro, D. W. (1973). Attention and processing capacity in auditory recognition. *Journal of Experimental Psychology, 99,* 49–54.

Moray, N. (1959). Attention in dichotic listening. *Quarterly Journal of Experimental Psychology, 11,* 56–60.

Moray, N. (1969). *Listening and attention*. Baltimore: Penguin.

Moray, N. (1981). The role of attention in the detection of errors and the diagnosis of errors in man-machine systems. In J. Rasmussen & W. Rouse (eds.), *Human detection and diagnosis of system failures*. New York: Plenum Press.

Moray, N. (1986). Monitoring behavior and supervising control. In K. R. Boff, L. Kaufman, & J. P. Thomas (eds.), *Handbook of perception and human performance*. New York: Wiley.

Moray, N., & Rotenberg, I. (1989). Fault management in process control: Eye movements and action. *Ergonomics, 32*(11), 1319–1342.

Mourant, R. R., & Rockwell, T. H. (1972). Strategies of visual search by novice and experienced drivers. *Human Factors*, *14*(4), 325–336.

Navon, D. (1977). Forest before the trees: The precedence of global features in visual processing. *Cognitive Psychology*, 9, 353–383.

Neisser, U. (1963). Decision time without reaction time. *American Journal of Psychology* 76.

Neisser, U. (1967). *Cognitive psychology*. New York: Appleton-Century-Crofts.

Neisser, U., & Becklen, R. (1975). Selective looking: Attention to visually specified events. *Cognitive Psychology* 7, 480–494.

Neisser, U., Novick, R., & Lazar, R. (1964). Searching for novel targets. *Perceptual and Motor Skills*, *19*, 427–432.

Newman, R. L. (1987). Response to Roscoe, The trouble with HUDs and MHDs. *Human Factors Society Bulletin*, *30*(10), 3–5.

Norman, D. (1968). Toward a theory of memory and attention. *Psychological Review*, *75*, 522–536.

Olson, R. K., & Attneave, F. (1970). What variables produce stimulus grouping. *American Psychologist 83*, 1–21.

Palmiter, S., & Elkerton, J. (1987). Evaluation metrics and a tool for control panel design. *Proceedings of the 31st annual meeting of the Human Factors Society*. Santa Monica, CA: Human Factors Society.

Parasuraman, R. (1986). Vigilance, monitoring and search. In K. R. Boff, L. Kaufman, & J. P. Thomas (eds.), *Handbook of perception and human performance*. New York: Wiley.

Posner, M. I., Nissen, J. M., & Klein, R. (1976). Visual dominance: An information processing account of its origins and significance. *Psychological Review*, *83*, 157–171.

Rock, I. (1975). *An introduction to perception*. New York: Macmillan.

Rogers, S. P. (1979). Stimulus-response incompatibility: Extra processing stages versus response competition. In C. Bensel (ed.), *Proceedings of the 23rd annual meeting of the Human Factors Society*. Santa Monica, CA: Human Factors Society.

Russo, J. E., & Rosen, L. D. (1975). An eye fixation analysis of multialternative choice. *Memory and Cognition*, *3*, 267–276.

Sanderson, P. M. (1989). The human planning and scheduling role in advanced manufacturing systems: An emerging human factors domain. *Human Factors*, *31*(6), 635–666.

Sanderson, P. M., Flach, J. M., Buttigieg, M. A., & Casey, E. J. (1989). Object displays do not always support better integrated task performance. *Human Factors*, *31*, 183–198.

Schneider, W., Dumais, S. T., & Shiffrin, R. M. (1984). Automatic and control processing and attention. In R. Parasuraman & D. R. Davies (eds.), *Varieties of attention* (pp. 1–27). Orlando, FL: Academic Press.

Schneider, W., & Shiffrin, R. (1977). Controlled and automatic human information processing I: Detection, search, and attention. *Psychological Review*, *84*, 1–66.

Senders, J. (1964). The human operator as a monitor and controller of multidegree of freedom systems. *IEEE Transactions on Human Factors in Electronics*, *HFE–5*, 2–6.

Senders, J. W. (1980). *Visual scanning processes*. Unpublished doctoral thesis, University of Tilburg, Netherlands.

Senders, J. W., Fisher, D. F., & Monty, R. A. (eds.) (1978). *Eye movements and the higher psychological functions*. Hillsdale, NJ: Erlbaum.

Seuss, Dr. (1971). *The lorax*. New York: Random House.

Sheridan, T. (1972). On how often the supervisor should sample. *IEEE Transactions on Systems, Sciences, and Cybernetics*, *SSC-6*, 140–145.

Shneiderman, B. (1987). *Designing the user interface: Strategies for effective human-computer interaction*. Reading, MA: Addison-Wesley.

Simpson, C., & Williams, D. H. (1980). Response time effects of alerting tone and semantic context for synthesized voice cockpit warnings. *Human Factors*, *22*, 319–330.

Somberg, B. L. (1987). A comparison of rule-based and positionally constant arrangements of computer menu items. *Proceedings of CHI & GI 87 Conference on Human Factors in Computing Systems*. New York: Association for Computing Machinery.

Sorkin, R. D. (1987). Design of auditory and tactile displays. In G. Salvendy (ed.), *Handbook of human factors* (pp. 549–576). New York: Wiley.

Spieth, W., Curtis, J., & Webster, J. (1954). Responding to one of two simultaneous messages. *Journal of the Acoustical Society of America*, *26*, 391–396.

Stagar, P., & Angus, R. (1978). Locating crash sites in simulated air-to-ground visual search. *Human Factors*, *20*, 453–466.

Stark, L., & Ellis, S. R. (1981). Scanpaths revisited: Cognitive models direct active looking. In D. F. Fisher, R. A. Monty, & J. W. Senders (eds.), *Eye movements: Cognition and visual perception*. Hillsdale, NJ: Erlbaum.

Steenblik, J. W. (1989, December). Alaska airlines' HGS. *Air Line Pilot*, pp. 10–14.

Stokes, A. F., Wickens, C. D., & Kite, K. (1990). *Display technology: Human factors concepts*. Warrendale, PA: Society of Automotive Engineers.

Stroop, J. R. (1935). Studies of interference in serial verbal reactions. *Journal of Experimental Psychology*, *18*, 643–662.

Sweller, O., Chandler, P., Tierney, P., & Cooper, M. (1990). Cognitive load as a factor in the structuring of technical material. *Journal of Experimental Psychology: General*, *119*, 176–192.

Swets, J., & Druckman, D. (1988). *Enhancing human performance*. Washington, DC: National Academy Press.

Treisman, A. (1964a). The effect of irrelevant material on the efficiency of selective listening. *American Journal of Psychology*, *77*, 533–546.

Treisman, A. (1964b). Verbal cues, language, and meaning in attention. *American Journal of Psychology*, *77*, 206–214.

Treisman, A. (1969). Strategies and models of selective attention. *Psychological Review*, *76*, 282–299.

Treisman, A. (1986). Properties, parts, and objects. In K. R. Boff, L. Kaufman, & J. P. Thomas (eds.), *Handbook of perception and human performance*. New York: Wiley.

Treisman, A., & Davies, A. (1973). Divided attention to eye and ear. In S. Kornblum (ed.), *Attention and performance IV*. New York: Academic Press.

Tulga, M. K., & Sheridan, T. B. (1980). Dynamic decisions and workload in multitask supervisory control. *IEEE Transactions on Systems, Man, and Cybernetics, SMC– 10*, 217–232.

Tullis, T. (1983). The formatting of alphanumeric displays. *Human Factors, 25*, 657–682.

U.S. Navy. (1988). *Investigating report: Formal investigation into the circumstances surrounding the downing of Iran air flight 655 on 3 July 1988*. Department of Defense Investigation Report.

Wachtel, P. L. (1967). Conceptions of broad and narrow attention. *Psychological Bulletin, 68*, 417–419.

Weintraub, D. J. (1987). HUDs, MHDs, and common sense: Polishing virtual images. *Human Factors Bulletin, 30*(10), 1–2.

Weintraub, D. J., Haines, R. F., & Randle, R. J. (1984). The utility of head-up displays: Eye focus vs. decision times. In *Proceedings of the 28th annual meeting of the Human Factors Society* (pp. 529–533). Santa Monica, CA: Human Factors Society.

Wickens, C. D. (1984). *Engineering psychology and human performance*. New York: HarperCollins.

Wickens, C. D., & Andre, A. D. (1990). Proximity compatibility and information display: Effects of color, space, and objectness of information integration. *Human Factors, 32*, 61–77.

Wiener, E. L. (1977). Controlled flight into terrain accidents: System-induced errors. *Human Factors, 19*, 171.

Williams, A. J., & Harris, R. L. (1985). *Factors affecting dwell times on digital displays* (NASA Technical Memorandum 86406). Hampton, VA: NASA Langley Research Center.

Woods, D., Wise, J., & Hanes, L. (1981). An evaluation of nuclear power plant safety parameter display systems. In R. C. Sugarman (ed.), *Proceedings of the 25th annual meeting of the Human Factors Society*. Santa Monica, CA: Human Factors Society.

Yarbus, A. L. (1967). *Eye movements and vision*. New York: Plenum Press.

Young, L. R., & Stark, L. (1963). Variable feedback experiments testing a sampled data model for eye tracking movements. *IEEE Transactions on Human Factors in Electronics, HFE–4*, 38–51.

Chapter
4

Spatial Perception and Cognition and the Display of Spatial Information

OVERVIEW

Space is a critical aspect of the environment in which many systems must operate. Vehicles must navigate through space; controls must often be pushed, pulled, or rotated across space; and displays are often intended to convey a sense of spatial distance, for example, the graph, in which large changes on the page are more significant than small ones. Our ability to perceive and understand spatial relations, the focus of this chapter, has several facets. The first relates to *analog* perception—perception in which large spatial or physical differences are more important or significant in interpretation than small ones, for example, when reading a graph or an analog meter but not when reading a digital meter or a word. In the digital meter, the spatial difference in the *physical* representation between, say, 79774 and 80000 is substantial; every digit is changed. But the difference in *meaning* between these two values is small. In contrast, the spatial difference between the words *altitude* and *attitude* is small, differing by only one letter, but its significance to a pilot is considerable. The first describes the height of the plane above the ground; the second describes its angle relative to the horizon.

A second important feature of spatial perception is the role of *motion*. Dials, meters, and vehicles move through space with velocity and direction, and the ability of humans to perceive and understand this motion is a critical feature in the design of such displays. Third, space is also *three-dimensional*, and although many displays are only two-dimensional, certain aspects of our perception of depth have implications for displays in two as well as in three dimensions. A fourth important feature concerns a person's understanding of the spatial relations in an environment through which he or she must navigate, that is, the understanding of *geography* and the interpretation of *maps*. Finally, the chapter concludes by discussing some efforts designers have made to capitalize on the role of space in systems that are not themselves inherently spatial, in particular, the digital computer.

ANALOG JUDGMENTS

Human performance often depends on accurate judgments of distance, extent, and depth. These are judgments in which *how far* and *how much* (analog quantities) are as important to us as *what*. The first area we shall consider is the perception and understanding of graphs. We will then address the role of motion as we consider the design of meters and dials.

Graphic Perception

A history of the graphic display of data dates back to the pioneering work of Playfair (1786), who first realized the power of allowing analog symbols, rather than words and digits, to represent quantitative data. The purposes of graphs are many, but using the example in Figure 4.1, we may divide them roughly into three categories (Carswell & Wickens, 1988): (1) to read a precise value (e.g., what was the level of factory A's production in 1986?), or (2) to compare two or more values (was there more production in 1986 or 1987?), or (3) to infer to deduce a more complex relationship (was the production growth in factory A qualitatively different from that in factory B?). As another example of the third task, the psychologist may examine a plot of the data from an experiment, trying to infer the relationship between dependent and independent variables.

Judgments 2 and 3 are typically analog judgments: Large differences between points or readings are more significant than small ones. Therefore, except for the instances when the precise judgments of task 1 are of primary importance, graphs, rather than tables of digits, are a preferred way to present data. However, even with graphic presentations, biases sometimes distort perception (Cleveland & McGill, 1984, 1985). Furthermore, not all graphs are equal, and some serve certain tasks better than others.

Perceptual Biases In understanding the human factors of graphs, the data on absolute judgment, described in Chapter 2, are relevant. We can't expect linear quantities to be read accurately without precise scale markings, as shown on the right but not the left axis of Figure 4.1. However, it also turns out that when making *relative* or *comparative* judgments between two quantities, our

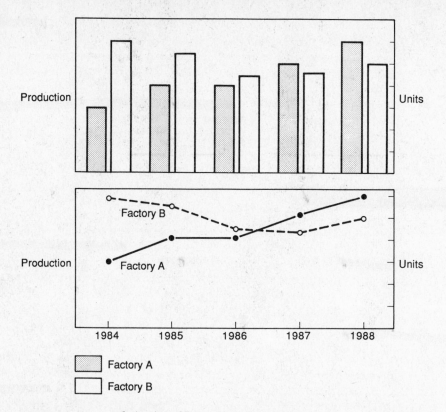

Figure 4.1 Bar graph (top) and line graph (bottom). Each graph represents the same data: the production of two factories, A and B, over five years. The graphs illustrate several points described in the text.

ability to do so depends very directly on the physical dimension used to represent the continuous quantity. Cleveland and McGill (1985) proposed that subjects' abilities to make such judgments fall off directly in the order of the graphic symbology arranged from the top to the bottom in Figure 4.2. The best comparative judgments are made with the evaluation of two linear scales, aligned to the same baseline. The poorest are when people are asked to judge differences in the volume of two cubes or the hue of two patches. Note that selecting the bigger quantity is relatively easy with all of the graphic formats. But judgments of *how much* they differ or whether the difference between two indicators is bigger than the difference between another two are not well served by symbols toward the bottom of Figure 4.2. Note also that the common line graph or bar graph (Figure 4.1) typically adheres to the optimum format for comparative judgments.

The judgments people make in extracting information from graphs may also be systematically *biased* by certain characteristics of the perceptual system. Some of these are related to fundamental optical illusions that distort our sense of perception (Coren & Girgus, 1978). For example, when viewing the *Poggen-*

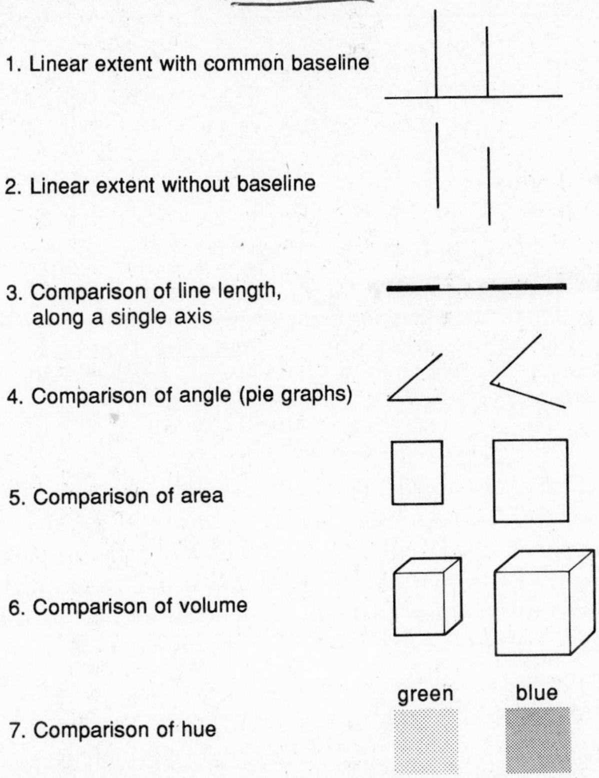

[handwritten: OUTWARD]

[handwritten: INWARD]

1. Linear extent with common baseline

[handwritten: Most ↓]

[handwritten: — Size]

2. Linear extent without baseline

[handwritten: — Slope estimation]

3. Comparison of line length, along a single axis

[handwritten: — Ratio estimation]

4. Comparison of angle (pie graphs)

[handwritten: — Proportion judgement of]

5. Comparison of area

[handwritten: — Change, J.O.]

6. Comparison of volume

7. Comparison of hue

green   blue

[handwritten: Least effective]

Figure 4.2 Examples of seven different graphic ways of presenting quantities to be compared. The graphs are arrayed from the most effective (top) to the least (bottom). (*Source:* W. S. Cleveland and R. McGill, "Graphical Perception and Graphical Methods for Analyzing Scientific Data." *Science, 229,* 828–833. Reprinted by permission of AT&T Bell Lab, 600 Mountain Avenue, Murray Hill, NJ 07974.)

dorf illusion, shown in Figure 4.3*a,* people tend to "flatten" the sloping lines away from the vertical, seeing the angle as less acute than it is. Poulton (1985) found that this same illusion was responsible for people's tendency to underestimate the value of points on sloping graphs, such as that shown in Figure 4.3*b.* Consistent with the form of the illusion, this bias shows that people "bend" their perception of the true location of the points downward, as indicated by the arrow. Poulton believes that the solution to this bias is to include calibration markers on both edges of the graph and to draw all four sides as shown in Figure 4.3*c.*

A second example of graphic bias is provided by Cleveland and McGill (1984) and is illustrated in Figure 4.4: the difficulty in comparing *differences* between two lines of different slope. For example, in the figure the difference

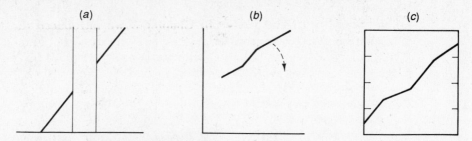

Figure 4.3 (*a*) The Poggendorf illusion: The two diagonal lines actually connect. (*b*) a line graph susceptible to "bending" from the Poggendorf illusion. (*c*) "Debiasing" of the Poggendorf illusion by marked edges on both sides. (*Source:* E. C. Poulton, "Geometric Illusions in Reading Graphs." *Perception & Psychophysics*, 37 (1985), 543. Reprinted with permission of Psychonomic Society, Inc.)

between the two curves is actually smaller in the left than the right regions. Yet perceptually the difference appears smaller in the right because judgments of differences along the Y axis are clearly biased by the *visual* separation (not the vertical separation) of the two curves, which is in fact less on the right side. Cleveland and McGill suggest that where the difference in differences (an interaction) is an important feature of a set of graphic data, it should be plotted directly, as shown at the bottom of the figure. Cleveland and McGill discuss several other biases of similar sorts.

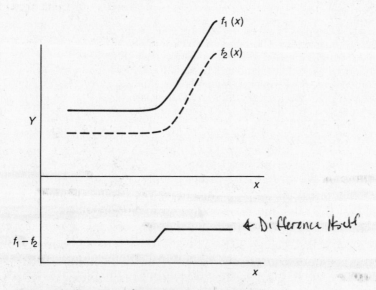

Figure 4.4 Illustrates biases in perceiving differences between pairs of lines $f_1(x)$ and $f_2(x)$ with changing slopes. The bottom curve plots the actual difference, $f_1(x) - f_2(x)$, which is larger on the right than on the left.

Compatibility of Proximity In Chapter 3 we introduced the notion of a *compatibility* between perception, as represented on the display, and cognition, as defined by the task requirements. We saw there that this display-cognitive compatibility could be defined in part by the *proximity compatibility principle* (Carswell & Wickens, 1987). Tasks requiring integration of information are better served by more integral, objectlike displays. A review of the literature on comparative graphics (studies in which different graphic formats are compared) by Carswell and Wickens (1988) reveals that the proximity compatibility principle applies to graphic displays as well. An objectlike graph (e.g., a line graph) was compared with a more separable format (e.g., a bar graph), as shown by the bottom and top half, respectively, of Figure 4.1. Each study was placed into one of four task categories, depending on whether the task required subjects to focus attention on one variable (e.g., report or locate a value), compare a subset of values within the display, synthesize information within the display, or make a judgment that involved integration across all values in the display. This ordering defines a continuum of task proximity, representing the extent to which the integration of all variables, rather than filtering out some, is necessary to carry out the task. Figure 4.5 shows the proportion of studies in each category that realized an "object display benefit" (i.e., better performance with the object than with the separate display) and those that showed the reverse effect. The figure clearly indicates the increasing benefit of objectlike graphs as tasks require more integration.

As a specific graphic example of this principle, using the data in Figure 4.1, consider this question: How is the rate of growth different between the two factories? The separate bar graph on the top, which contains eight objects, makes this integrative, complex comparison task more difficult than does the integrated line display on the bottom, which contains only two objects. In contrast, judgments requiring simple focused attention can be made as well or better with the bar graph than with the line graph, a conclusion supported by experimental evidence (Carswell, 1990).

An experiment by Goettl, Wickens, and Kramer (in press) is another example of the application of the proximity compatibility principle to graphic displays. The subjects were to think of themselves as scientists, examining the data from an experiment in which the speed and accuracy of performance were recorded. In one task, the subjects were asked to make inferences requiring an integration of speed and accuracy data from two hypothetical experimental conditions. Performance in this inference task was better when the data were presented as two points (two objects) in a two-dimensional speed-accuracy space, such as in Chapter 8 (Figure 8.3), than as four separate bar graphs (four objects) presenting speed and accuracy for each condition separately. In contrast, when subjects were asked to perform a task calling for precise judgment of one particular value, the bar graph display proved superior. The reader will note the similarity of these findings to the predictions made by Garner's theory of integral and separable dimensions, discussed in Chapter 2.

The idea that graphs should be compatible with the kinds of judgments people are expected to make goes beyond the concept of objectness. If an

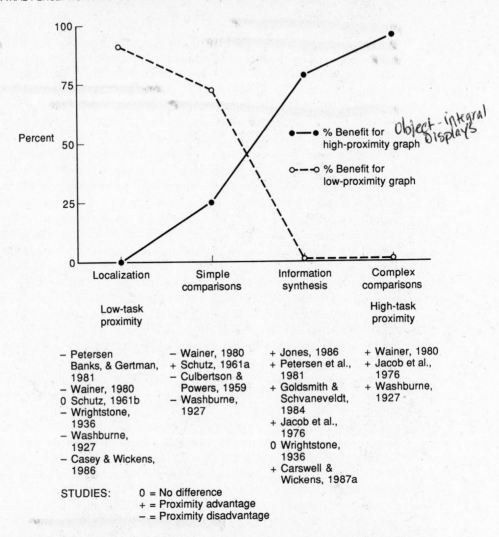

Figure 4.5 Proportion of studies showing an object-display advantage (solid line) or disadvantage (dashed line) as a function of task type (focused, left; integration, right). The figure illustrates the proximity compatibility principle. (*Source:* Carswell and Wickens, 1988.)

experiment examines three factorially crossed independent variables (e.g., a $2 \times 2 \times 2$ experimental design), how should their effect be displayed? If two panels are created, each presenting the effects of two variables (A and B) at a different level of the third (as in the upper panels of Figure 4.6), the graphic designer should be aware that differences and comparisons due to the third variable (variable C) will be hard to perceive because the two levels of this variable are spatially separated in two different panels. If those differences are critical to an understanding of the data's relationships, they should be included as one of the variables *within* each panel—ideally as the variables represented

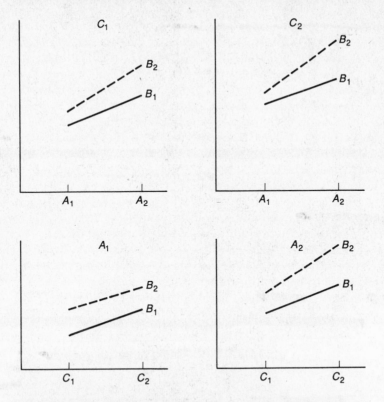

Figure 4.6 Hypothetical graph of three variables. At the top, the differences between the two levels of variable *C* will be hardest to perceive accurately because of their greater spatial separation. Between the top and the bottom, the consistency of representing the unchanged *B* variable has been maintained.

by the two points on each line (like *A*). In this way, the variable's effect is directly represented by an emergent feature — the slope — of the constructed graph. We discuss the bottom two panels of Figure 4.6 below.

A final example of the proximity compatibility principle in graphic data presentation is quite intuitive but sometimes neglected — the attachment of legends to points and lines. When legends refer to lines in a graph, they should, if possible, be placed close to the lines to which they refer (Milroy & Poulton, 1978; Sweller, Chandler, Tierney, & Cooper, 1990). Two pieces of information to be integrated (the legend and the line) should be close together in space (as in the line graph of Figure 4.1, but not in the bar graph, where the legend is placed below). If spatial proximity cannot be accomplished, the line(s) should be clearly distinguished from one another by texture or color (not fine detail differences in point shape), and this texture or color should be a prominent part of the legend.

In conclusion, we note that the graph's designer may have one of two general goals: to convey a specific message to the reader or to provide a way of looking at and exploring a data set. In the former case it is necessary to configure the data in a way that emphasizes the message to be conveyed, or at least in a way that is not hostile to the message. In the latter case, the guidelines are not as clear, and probably the flexibility to produce multiple representation should be the guiding principle. These representations should emphasize different comparisons and contrasts. However, it should be remembered that the goal of the graph in this case is to help people understand subtle relationships in the data that do, by definition, involve integration. Hence, the different formats should highlight different forms of visual proximity and integration (Kolata, 1982). The possible use of three-dimensional graphs in this regard will be addressed later in this chapter.

Multiple Graphs   The previous discussion has focused on the ideal, compatible properties of single graphs. An equally important issue lies in the presentation of multiple graphs, which may show related sets of data. Here the designer's concern must be focused on the relationship between successively viewed graphic formats, in addition to the optimization of each format by itself. Three specific concerns can be identified.

1. *Consistency*. When the same data are plotted in different ways, or different data are plotted as a function of the same variables, it is critical to maintain as much consistency as possible between graphs. For example, the variable coded by line tone (e.g., dashed versus dot) in one graph, should, where possible, be coded by the same physical distinction in all graphs; and of course the association of physical coding to meaning (solid line to one level, dashed line to another) should also be consistent. Thus, notice in the bottom two panels of Figure 4.6, where variables *A* and *C* have traded places from their positions in the top panels, to allow easier visualization of the effect of *C*, the coding on *B* is consistent with the coding in the top panels. If such consistency is needlessly violated, the reader will have cognitive difficulty in making a transition from one graph to the other. This issue defines poor *visual momentum*, a concept that we will consider at the end of this chapter.

2. *Highlighting differences*. Naturally such consistency cannot be absolute (otherwise the same graphs would be produced). But where related material is presented, it then becomes critical to highlight the changes or differences from graph to graph, either in the text or in the symbols themselves. For example, a series of graphs presenting different *Y* variables as a function of the same *X* variable should highlight the *Y* label. This system allows the same cognitive set to be transferred from graph to graph, while the single mental revision that is necessary is prominently displayed. The time- and effort-consuming visual search necessary to locate the changed element is eliminated.

3. *Distinctive legends*. Corresponding to point 2, legends of similar graphs must highlight the distinct features, not bury them as a single word that is nearly hidden in the last line of otherwise identical multiline legends.

This discussion leads to two further points that have more general relevance to the area of human-system interaction. First, some of the failures of consistency, or lack of highlighting (as well as other graphic shortcomings), arise not from short sightedness on the part of the graph's designer but from reliance on statistical graphics packages that may not have been designed with these perceptual and cognitive principles in mind. This point will be emphasized in Chapter 12. Too much reliance on automation can have its costs as well as its benefits.

Second, the concern with multiple graphs illustrates the point that there are different levels of analysis in system design. Each individual system (graph in this case) may be optimized, thereby producing what we might call *local* optimality. Yet the configuration of all systems together, with each one locally optimized, can produce *global* disharmony as certain emergent features (e.g., inconsistency) of the *set* of systems (controls or displays) are revealed that will disrupt their smooth perception, understanding, or use. Optimizing total systems design must often reflect a compromise between satisfying local compatibility and achieving global harmony (Wickens, Andre, & Haskell, 1990).

Dials, Meters, and Indicators: Display Compatibility

Graphs are generally static and are often paper representations. In contrast, many systems present analog information in the more dynamic form of dials, meters, or other changing elements, which represent the momentary state of some part of the system. As with graphs, it is equally important that dials and meters be compatible with the operator's information-processing needs, or *mental model* of the system. The mental model, a concept we will discuss further in Chapter 6, forms the basis for understanding the system, predicting its future behavior, and controlling its actions (Gentner & Stevens, 1983). As a consequence, there are three levels of representation that must be considered in designing display interfaces: (1) the physical system itself (which is analog); (2) the internal representation (which should be analog); and (3) the critical interface between these two, the display surface on which changes in the system are presented to update the operator's mental model and to form the basis for control action and decision. It is important to maintain a high degree of congruence, or *compatibility*, among these three representations. Compatibility between the real system and the mental model is clearly a matter of training (but may be influenced by ability differences as well). If an operator correctly understands the nature of the system dynamics, the mental model will be analog. If both the physical system and the mental representation are analog, then it is important that the third element—the display—be formatted in a manner that will be compatible with the other two. The following sections will discuss the issue of compatibility between display representation and mental representation. The related concept of compatibility between display and response, the issue of stimulus-response compatibility, will be discussed in Chapter 8.

ity: static aspects of orientation and dynamic properties of motion. As we will see, the two interact in terms of their implications for analog display design. However, in terms of processing mechanisms, there appears to be good evidence that there are separate perceptual channels to encode position and velocity information (Antsis, 1986; Sekular, 1974). For example, Lappin, Bell, Harm, and Kottas (1975) obtained experimental evidence that subjects could extract information about the velocity of a moving stimulus that was independent of the two dimensions, distance and time, that combine to create that velocity. The static and dynamic components of display compatibility will now be considered in detail.

The Static Component: Analog Compatibility of Orientation   In the context of aviation displays, Roscoe (1968) has dealt with the issue of display compatibility at some length as an example of what he refers to as "the principle of pictorial realism." The representation of aircraft altitude is a typical instance. Altitude physically is an analog quantity. Conceptually, it is also represented to the pilot in analog form, with large changes in altitude more important than small changes. Therefore, to achieve compatibility, a display of altitude (i.e., an altimeter) should also be of analog (moving needle) rather than digital format, a guideline echoing the earlier discussion of graphs. The transformation of symbolic digital information to analog conceptual representation imposes an extra processing step, which will lead to longer visual fixations, longer processing time, or a greater probability of error (Grether, 1949; Toal et al., 1982).

There are, of course, other factors that influence the choice of analog or digital representations of altitude or of other continuously varying quantities. For example, a requirement to read the absolute value of the indicator with high precision would favor the digital format. On the other hand, the need to perceive rate-of-change information, to estimate the magnitude of the variable when it is rapidly changing, or to estimate at a glance the distance of that variable from some limit, favors the analog format (Helander, 1987).

Although altitude represents an obvious example of an analog quantity with analog representation, numerous other variables may be identified whose internal representations are probably also analog (e.g., temperature, pressure, speed, power, or direction), as well as those conceptual dimensions for which the nature of internal representation is less well specified (degree of danger or readiness status).

Displaying information in analog format that is mentally represented in analog fashion is a necessary but not sufficient condition to ensure maximum display compatibility. Since analog continua necessarily are associated with some physical ordering, it is essential also that the orientation and ordering of the display be compatible with the ordering of the mental representation. This compatibility defines another aspect of Roscoe's (1968) principle of pictorial realism. Display compatibility may be violated in shapes, for example, if a circular altimeter (pointer or dial) represents the linear conception of altitude (Grether, 1949). (The main limitation of the linear display is that it must occupy a large physical extent of "real estate" to convey the same level of scale resolution as the more "compact" round dial display.)

Display compatibility may also be violated in direction. In the simple case of the altimeter, our mental model of altitude represents high altitude as up and low altitude as down. Therefore, the altimeter also should present high altitudes at the top of the scale and low ones at the bottom, rather than the reverse arrangement or a horizontal one. Similarly, high temperatures should be higher, and low temperatures lower. Displayed quantities, in short, should correspond with the operator's mental model of them. Finally, compatibility may be violated by dissecting and displaying in separate parts a variable that is unitary. Grether (1949), for example, reports that operators have a more difficult time extracting altitude information from three concentric pointers of a rotating display (indicating units of 100, 1,000, and 10,000 feet) than from a single pointer.

Compatibility of Display Movement   Roscoe (1968) and Roscoe, Corl, and Jensen (1981) propose the *principle of the moving part* — that the direction of movement of an indicator on a display should be compatible with the direction of movement of an operator's mental model of the variable whose change is indicated. In the case of the household thermometer, this principle is typically adhered to because a rise in the height of the mercury column indicates a rise in temperature. There are, however, circumstances in which the principle of the moving part and the compatibility of orientation operate in opposition to one another, and so one or the other must be violated. This occurs in so-called fixed-pointer/moving-scale indicators.

An example of this violation is shown in Figure 4.7, which could represent an altimeter. In the moving-pointer display (Figure 4.7*a*) both the principles of

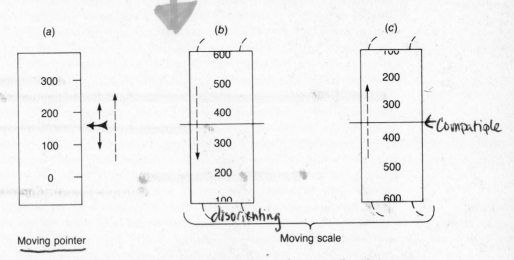

Moving pointer   Moving scale

Figure 4.7 Display movement: (*a*) moving-pointer altimeter; (*b* and *c*) moving-scale or fixed-pointer altimeters. The dashed arrows show the direction of display movement to indicate an increase in altitude.

the moving part and the compatibility of orientation are satisfied. High altitude is at the top, and an increase in altitude is indicated by an upward movement of the moving element on the display. Neither static nor motion compatibility is violated. However, in the moving-scale display, to print the numbers in such a way that high altitudes are at the top (Figure 4.7b), the scale must move *downward* to indicate an *increase* in altitude, in violation of the principle of the moving part. If the labeling is reversed to conform to the principle of the moving part (Figure 4.7c), this change will reverse the orientation and display high altitude at the bottom. A final disadvantage with both the moving-scale displays is that as in a digital display, scale values become difficult to read when the variable is changing rapidly since the digits themselves are moving.

Of course moving-scale displays can expose a narrower range of the scale at a higher precision than the fixed scale, in which the whole scale range must be displayed. A possible solution here is to employ a hybrid scale. The pointer moves as in Figure 4.7a, but only a small portion of the scale is exposed. When the pointer approaches the top or bottom of the window, the scale shifts more slowly in the opposite direction to bring the pointer back toward the center of the window. Thus the pointer moves at higher frequencies and the scale shifts at lower frequencies as needed. This concept of *frequency separation* has an important realization in aircraft displays, as we will now describe.

The Frequency-Separated Display

Roscoe (1968, 1981) was concerned with the best way to design the indicator of an aircraft's bank and pitch — the attitude display indicator — to satisfy both kinds of compatibility. Two conventional indicators of attitude are shown in Figure 4.8. In (a) the "outside-in" bird's-eye, or ground-referenced, display indicates the plane as it would appear to an outsider looking at the plane from behind. When the plane banks to the left, the display indicator also rotates left, and so the principle of the moving part is confirmed. On the other hand, the display intrinsically provides a frame of reference (a view from outside the aircraft) that is inconsistent with the pilot's actual frame of reference (inside the aircraft). The static picture that the display provides, a horizontal horizon and tilted airplane, is incompatible with what the pilot actually sees out of the cockpit window: a tilted horizon and an aircraft that from the pilot's frame of reference is horizontal. Thus it violates the principle of pictorial realism.

In contrast to the bird's-eye view, which violates compatibility of orientation, the congruent correspondence of frames of reference is given by the more conventional "inside-out," moving-horizon, or pilot's-eye display shown in Figure 4.8b. However, this display now violates the principle of the moving part. For example, banking the aircraft to the left, which will generate a leftward rotation of the pilot's mental model of the aircraft, produces a rightward rotation of the moving element on the display (the horizon) (Johnson & Roscoe, 1972).

To achieve a compromise between the static and dynamic aspects of the display, Fogel (1959) proposed the concept of *frequency-separated display*

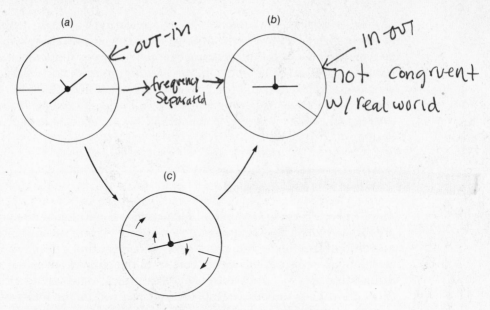

(a) — OUT-in

→ frequency → Separated

(b) — In-out

not congruent w/ real world

(c)

Figure 4.8 *(a)* Outside-in, *(b)* inside-out, and *(c)* frequency-separated display. All displays show an aircraft banking left. Low-frequency return to steady state is indicated by arrows.

(Johnson & Roscoe, 1972; Roscoe, 1968). In this display, lateral movement of the controls (thus changing the roll of the aircraft) will induce an outside-in (moving-aircraft) change on the display, as in Figure 4.8*b*, so that display and aircraft *motion* are compatible when conditions are dynamic. When, however, the pilot enters into a gradual turn, which requires the bank angle to be held constant for some period of time, and perception is thereby static, the horizon and the plane both rotate to an inside-out format, so that the frame of reference is now equated between the two. This progression is shown by the transition in Figure 4.8 from *a* to *c* to *b*. Thus, at high frequencies, when movement and motion perception are most dominant, the principle of the moving part is adhered to. At low frequencies of motion, to which people's velocity sensors are less sensitive, the static principle of compatibility of frame of reference is followed. An evaluation of this display by Roscoe and Williges (1975) indicate it to be more effective than either the inside-out or the outside-in display in terms of the accuracy of the pilot's control and the number of inadvertent control reversals that were made when flying. The advantages are shown even with skilled pilots who were used to the conventional moving-horizon display (Roscoe, 1981).

In summary, the principle of compatibility, which will be discussed again in the context of responses in Chapter 8, is probably one of the most important guidelines in the engineering psychology of display design. Compatible displays are read more rapidly and accurately than incompatible ones under normal conditions. More important, their advantages increase under conditions of stress. Yet the nature of the internal mental models of complex systems such as computers or chemical processes that guide compatibility decisions must be better understood (Carroll & Olson, 1987; Rouse & Morris, 1986). Until these are carefully specified and the extent of individual differences in the format of representation become better known, it will often be difficult to translate the generic principles of display compatibility to the specific instances of display design.

THREE DIMENSIONS AND DEPTH

Much of our previous discussion has focused on multidimensional displays that are intended to convey information regarding two dimensions at once. However, there are situations in which a third dimension on a display is represented as depth, or as the perceived distance from the observer along an axis perpendicular to the plane of the display. These displays are intended to represent three dimensions of Euclidian space, and they will be the focus of the current section. Such displays may be developed for one of two general purposes. First, the three dimensions may actually be designed to represent three Euclidian dimensions of space, as when a display is constructed to guide the pilot in a flight path. Second, the display may use the third (depth) dimension to represent another (nondistance) quantity. Examples of this usage are found in many three-dimensional graphics packages.

To understand the advantages and costs of three-dimensional displays, along with the causes of certain systematic distortions in our ability to use depth in the natural world, it is important to discuss briefly some of the fundamental characteristics of visual depth and distance judgment. The reader wishing more detail should refer to Boff, Kaufman, and Thomas (1986) or Wickens, Todd, and Seidler (1989).

Depth Judgments

The accurate perception of depth and distance is accomplished through the operation of several perceptual cues. Some of these are characteristics of the object or world we are perceiving, and others are properties of our own visual system. We refer to these as *object-centered* and *observer-centered* cues, respectively.

Monocular **Object-Centered Cues**   Object-centered cues are sometimes described as pictorial cues because they are the kinds of cues that an artist could build into a picture to convey a sense of depth. Figure 4.9 is an example of a three-dimensional scene that incorporates eight of the following cues:

1. *Linear perspective.* When we see two converging lines we assume that they are two parallel lines receding in depth (the road).

Figure 4.9 Illustrates several object-centered cues for depth as described in the text.

occlusion

2. *Interposition*. When the contours of one object obscure the contours of another, we assume that the obscured object is more distant (the buildings on the right).

3. *Height in the plane*. Because we normally view objects from above, we assume that objects higher in our visual field are farther away (the two trucks).

4. *Light and shadow*. When objects are lighted from one direction, they normally have shadows that offer some clues about the objects' orientation relative to us (Ramachandran, 1988) as well as its three-dimensional shape (the buildings and trucks).

5. *Relative size*. If objects are known to be the same true size, those subtending a smaller visual image on the retina (the retinal image) are assumed to be farther away (compare the two trucks).

6. *Textural gradients*. Most surfaces are textured, and when the plane of a texture is oriented toward the line of sight, the grain will grow finer at greater distance. This change in grain across the visual field is referred to as a textural gradient (the field on the left and the center line of the road).

7. *Proximity-luminance covariance.* Closely related to textural gradients is the fact that objects and lines are typically brighter as they are closer to us, and so continuous reductions in illumination and intensity are assumed to signal receding distance (the road lines) (Dosher, Sperling, & Wurst, 1986).

8. *Aerial perspective.* More distant objects often tend to be "hazier" and less clearly defined (the mountains).

9. *Relative motion gradient or parallax.* When we move relative to a three-dimensional scene, objects that are closer to us show greater relative motion than those that are more distant. Hence, our perceptual system assumes that distance from us is inversely related to the degree of motion.

Motion, like light and shadow, is also used as a cue to the three-dimensional *shape* of objects themselves, as well as their location. For example, the cloud of points in Figure 4.10 does not appear to be three-dimensional. Yet if these were points of light on a rotating three-dimensional cylinder like a rotating can, they would show a pattern of motion—slow near the edges, fastest at the center—that leads to an unambiguous interpretation of a rotating three-dimensional cylinder (Braunstein & Andersen, 1984). This property is referred to as *structure through motion* (Braunstein, 1990).

Binocular

Observer-Centered Cues Three sources of information about depth are functions of characteristics of the human visual system.

1. *Binocular disparity.* The images received by the two eyes, located at slightly different points in space, are disparate. The degree of disparity provides a basis for the judgment of distance, the principle employed in stereoscopes or, as we will see, in the emerging technology of three-dimensional displays.

2. *Convergence.* The "cross-eyed" pattern of the eyes, required to focus on objects as they are brought close to the observer, is necessary to

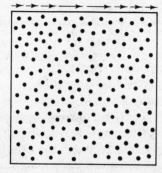

Figure 4.10 Potential stimulus for recovery of structure through motion. If the horizontal motion of the dots were proportional to the velocity vectors at the top of the figure, the flat surface would be perceived as a three-dimensional rotating cylinder.

bring the image onto the detail-sensitive retina of both eyes. Proprio-ceptive messages from the eye muscles to the perceptual centers of the brain inform the latter of the degree of convergence, and therefore of an object's distance.

3. *Accommodation.* Like convergence, accommodation is a cue provided to the brain by the eye muscles. In this case the muscles adjust the shape of the lens to bring the image into focus on the retina. Misac-commodation can be an important limitation in the quality of sensory information in many environments, such as the aircraft cockpit or vehicle cab, in which the eyes must often change their focus of atten-tion (and therefore depth) between a far domain and a near domain—the latter representing the instrument panel, maps, or checklists (Weintraub, Haines, & Randle, 1985). The relatively slow speed of the accommodation response from one domain to the other has led de-signers to consider systems in which the near-domain information is displayed at optically far distances, achieved by an optical process of *collimation.* This technique is often used to present information on head-up displays, as discussed in Chapter 3.

Perceptual Hypotheses and Ambiguity

In the description of object-centered cues, the phrase *we assume* was often used to describe the perceiver's interpretation of the world. These interpreta-tions in turn are used by the observer to make inferences about how close and how far away things are. Hence, we often speak of depth and distance percep-tion as governed by perceptual *hypotheses* about the way things are, based on our assumptions. In Figure 4.9, we hypothesize, for example, that the two trucks in the visual field are the same true size, and therefore the one with the smaller-sized retinal image is farther away (Gregory, 1977; Rock, 1983). These hypotheses and assumptions are relatively automatic and unconscious. Typi-cally as we observe the three-dimensional world, our hypotheses are supported by all depth cues working *redundantly* to provide the same information. How-ever, there are occasions when the hypotheses we assume do not correspond with reality because the cues are few in number, the assumptions we make about the world are incorrect, or the cues are ambiguous.

Consider, for example, a study by Eberts and MacMillan (1985) of the causes of rear-end collisions. The researchers noted that the frequency with which small cars are rear-ended is considerably greater than that for large cars. They reasoned that drivers' judgments about how far a car is in front, and therefore how soon they must apply the brakes, are based on the cue of relative size. Drivers assume an average size of vehicles and use it as the basis for mentally computing a perceived distance. Smaller-than-average cars therefore are perceived to be relatively farther away than they actually are (just as larger ones are perceived to be relatively closer). So when perceiving small cars, the braking process is initiated later than it should be, with the consequence of closures that are too fast. Eberts and MacMillan tested and confirmed this hypothesis in an experiment.

Similar faulty assumptions can occur in aviation. If a pilot is flying over unfamiliar terrain and encounters bushes shaped like trees, the pilot may assume that these are actually trees, believe the plane is farther from the ground than it is, and so become dangerously close to colliding with the earth (Hart, 1988). A related example of a false hypothesis is shown in Figure 4.11a.

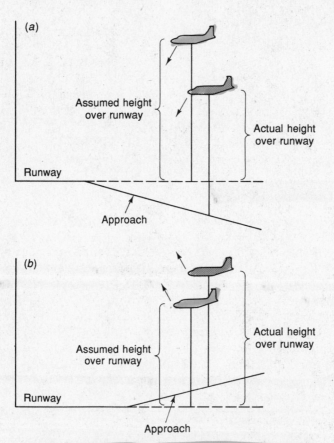

Figure 4.11 Misjudgment of the height over the runway as described in the text. The black plane is the position of the actual airplane. The white plane is the pilot's perceived position relative to the ground, based on the (faulty) assumption that the ground beneath is level and not upsloping (as in *a*) or downsloping (as in *b*). In each figure the direction of the inappropriate correction is shown by the arrow. (*Source:* F. H. Hawkins, *Human Factors in Flight*. Brookfield, VT: Gower Technical Press, 1987, p. 121. Reprinted by permission of the Flight Safety Foundation of Arlington, VA.)

Flying low over a flat but upward-sloping surface on approach to a runway, the pilot may assume that the surface is not only flat but level (the most probable assumption since most flat surfaces are also level). As a consequence of this assumption, the pilot will perceive the plane to be higher than it really is above the level of the runway. The pilot will then erroneously "correct" the altitude by descending, and therefore will be likely to fly an approach that is much closer to the upward-sloping terrain than is advisable, the possible consequence being a "short" landing. Correspondingly, the opposite assumption may be made while approaching over a flat but down-sloping terrain (Figure 4.11*b*) (Hawkins, 1987).

Gregory (1977) has pointed out that perceptual ambiguities of size and distance are particularly likely to occur when the three-dimensional world is represented on a two-dimensional display. Figure 4.12 shows a three-dimensional graphic plot of a set of data. In this case, the data happen to describe the proximity compatibility principle discussed in the previous chapter. The graph represents the height of a dependent variable (e.g., data interpretation accuracy) as a function of two independent variables (e.g., whether the task requires focused attention or integration and whether the display is an object or a bar graph). Figure 4.12 represents display type (bar graph vs. object) in the depth

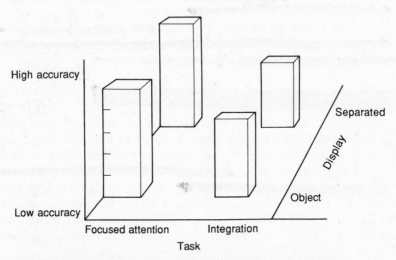

Figure 4.12 Perceptual distortions induced by three-dimensional graphics. The graph presents a hypothetical set of data that might be obtained to demonstrate the proximity compatibility principle, discussed in the previous chapter. On the left (focused attention), the two bars are actually the same height; but the sense of depth makes the more "distant" bar appear larger. On the right (integration task), performance with the separated display (rear bar) is actually lower than with the integrated display (close bar), but perspective makes them both appear to be the same. You may measure the bars to make these comparisons.

plane by use of perspective. The graph depicts the two displays of performance as being equivalent when used for focused attention tasks (the left side), but our perception is that performance with the separated display represented at the back is better because our hypothesis is that the more distant bar, of the same retinal image size, must be larger. When the integration task is considered (the right side), performance with the separate display is actually depicted as worse, but our perception is that the two displays provide equivalent accuracy because the bar graphs are of the same perceived size. However, the smaller displayed size of the more distant bar is not perceived because it is perceived to be farther away.

It is important to note that these misperceptions are relatively automatic. It is not easy to use our conscious awareness to de-bias the judgments of relative length. To compensate for these biases, different solutions have been suggested. For example, regarding the misjudgment of aircraft altitude shown in Figure 4.11, Kraft (1978) has argued that pilots must pay more attention to their flight instruments, even in good visual conditions. For the bias shown in Figure 4.12, one approach might be to capitalize on the perceiver's automatic tendency to compensate for the depth dimension, and therefore scale down the more "distant" bar graphs accordingly. However, the problem with this solution is that the amount by which display height should be reduced with perceived distance cannot easily be specified since it depends on how "compelling" the impression of distance is, an impression that may vary greatly from person to person and format to format. A better solution is to provide scale markings on the blocks themselves so that height can be read by counting the ticks, as shown on the left front bar of the graph.

Three-Dimensional Displays of Three-Dimensional Space

The problems that can arise when making precise, absolute distance judgments in three dimensions suggest some major constraints on using a graphic dimension of depth to represent a nondistance dimension. But there are also some very compelling reasons why three-dimensional displays *should* be used to represent three-dimensional worlds, such as the product designed at a computer-aided design (CAD) workstation, the contour map studied by the petroleum geologist, the map of magnetic forces around the human brain, the display of air traffic shown in Figure 4.13a, or the flight path display shown in Figure 4.14a.

In all of these cases, the display representation of a three-dimensional object is more compatible with the operator's mental model of the three-dimensional world represented by the display than is the two-dimensional counterpart, like that shown in Figure 4.13b and 4.14b. Even though the two-dimensional representations provide the necessary information to reconstruct the three-dimensional picture, these two-dimensional renderings typically require some mental gymnastics in order to integrate and reconstruct the picture. When this picture is used to maintain general spatial awareness or to control flight path guidance, rather than to make precise readings of altitude, orienta-

(a)

Perspective display

(b)

Plan-view display

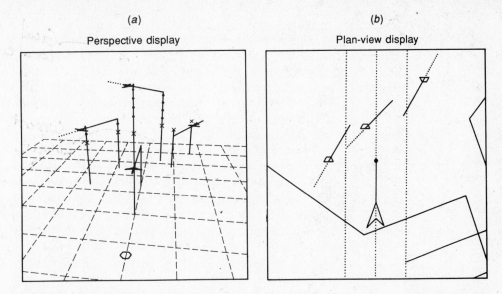

Figure 4.13 A modified air traffic display proposed for presentation in the cockpit to provide the pilot with greater situation awareness of surrounding air traffic: (*a*) three-dimensional representation; (*b*) two-dimensional representation. The symbols around each aircraft in (*b*) indicate whether they are above or below the altitude of the viewer's own aircraft. (*Source:* S. R. Ellis, M. W. McGreevy, and R. J. Hitchcock, "Perspective Traffic Display Format and Air Pilot Traffic Avoidance." Reprinted with permission from *Human Factors* (vol. 29, 1987). Copyright 1987 by The Human Factors Society, Inc. All rights reserved.)

tion, and distance, the advantages of the three-dimensional representation should be obvious (Wickens, Haskell, & Harte, 1989). Indeed, systematic comparisons of flight control, traffic control, and data interpretation have found better performance on these integration kinds of tasks (when information from the three dimensions must be integrated) with the single-frame, three-dimensional representation (Bemis et al., 1988; Ellis, McGreevy, & Hitchcock, 1984; Merwin & Wickens, 1991; Morello, Knox, & Steinmetz, 1977; Wickens & Todd, 1990, see Wickens, Todd, & Seidler, 1989, for a review).

Focused Attention and Artificial Frameworks The advantages of three-dimensional displays for flight control and situational awareness notwithstanding, such displays still have two specific limitations relative to their two-dimensional counterparts: focused attention and false hypotheses. These limitations must be overcome by giving careful concern to human factors design issues. First, as noted, three-dimensional representations in a two-dimensional display plane are inherently ambiguous in specifying absolute distances and depths if the true size of objects is not known in advance. Therefore, such displays will not be useful for the precise reading of values, such as answering the question relative to Figure 4.12: "How much better is performance with object than

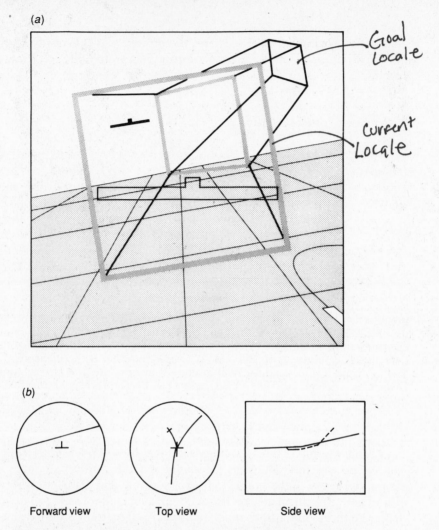

(a)

Goal Locale

Current Locale

(b)

Forward view   Top view   Side view

Figure 4.14 *(a)* Three-dimensional perspective flight path display with a future predicted position of the aircraft shown by the smaller black symbol. *(b)* The same information as it might be represented in two-dimensional plan views.

separate displays?" or to Figure 4.14a: "How far below the flight path is my aircraft predictor symbol?" This deficiency is another example of the proximity compatibility principle discussed in Chapter 3. Focused attention will be disrupted by the object integration of a three-dimensional display.

A potential solution to this limitation, however, may be offered by incorporating *artificial frameworks* into the display. The cockpit display of traffic information (CDTI), developed by Ellis, McGreevy, and Hitchcock (1984, 1987) and shown in Figure 4.13a is one example of such frameworks. Each aircraft is "attached" to a post that protrudes from the ground at its current geographical location, and its altitude is unambiguously specified by the markers on each post. Furthermore, the direction of the predicted flight path of each aircraft can also be unambiguously interpreted through these supports, creating a sort of a "wicket" for each aircraft. Wickens, Todd, and Seidler (1989) and McGreevy and Ellis (1986) also discuss other principles of designing three-imensional displays that can be used to minimize the biases and distortions we experience in resolving three-dimensional ambiguity.

Resolving Ambiguities   A second serious concern results when ambiguity can allow false hypotheses to be formed about depth and distance because the necessary cues for depth perception are not incorporated into the display. Consider the flight command path display shown in Figure 4.15*a*, in which the two boxes, like those in Figure 4.14*a*, represent segments of an imaginary tunnel to be "flown through." Cues of relative size and height in the plane help resolve the ambiguity of which box is closer. Both boxes are assumed to be the same true size, and the higher one is perceived as farther away. But consider the display in Figure 4.15*b*, in which the pilot is approaching the tunnel from beneath. Here height in the plane offers information contradictory to relative size, and the perceiver may note perceptual reversals in which the smaller box is sometimes perceived as closer, not more distant.

Such errors of display interpretation can be critical in spatial environments when objects must be approached, manipulated, and moved. These would include not only aviation but also the control of robots and teleoperators in space, under the sea, or in other hostile environments. The logical solution would be to incorporate additional redundant depth cues, such as the cues of interposition (hidden lines) and textural gradient shown in Figure 4.15*c*. A rough design guideline is to assume that each added cue increases the compellingness of depth experience and, therefore, the avoidance of ambiguities (Bruno & Cutting, 1988). But from the designer's point of view, the addition of cues, particularly in computer-driven dynamic displays, can be expensive and should be avoided if they are unnecessary. So which cues should be used? The answer to this question comes in part from studies of *cue dominance* in which ambiguous situations are set up and the "winning" cue is established as the one that governs the final perceptual interpretation (Dosher, Sperling, & Wurst, 1986). Thus, for example, the interpretation of Figure 4.15*b* that is perceived will establish whether relative size or height in the plane is more dominant. It is clear by looking at Figure 4.15*c* that interposition is dominant over height in the plane, as it removes ambiguity from the perception. A synthesis of a number of such studies reveals that three cues, in particular, *interposition*, *motion parallax*, and *binocular disparity* (created artificially in stereoscopic displays), are the most dominant (Wickens, Todd, & Seidler, 1989). Because of its special technological requirements, the third of these will be discussed in more detail.

Stereoscopic Displays   A relatively accurate sense of three-dimensionality can be created through stereopsis, presenting slightly different images to the two eyes. As we saw earlier, the amount of disparity can provide a direct

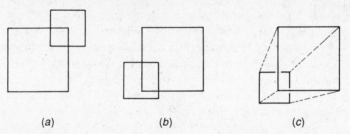

Figure 4.15 Perspective tunnel such as that shown in Figure 4.13a: (a) with height in the plane and relative size as redundant cues for depth; (b) with height in the plane and relative size conflicting, a reversal of perception may sometimes be experienced; (c) the conflict is resolved by including interposition and texture gradients on the connecting lines.

unambiguous cue for depth, and it dominates most other cues with which it is placed in competition. Emerging display technology has now also allowed fairly faithful production of stereoscopic cues in dynamic displays such as those in Figures 4.13 and 4.14 (Merritt & Fisher, 1990). Furthermore, comparative evaluations with perspective displays generally reveal that stereopsis enhances performance (Nataupsky & Crittendon, 1988; Way, 1988; Zenyuh, Reising, Walchli, & Biers, 1988).

In spite of these advantages, current computer-generated stereoscopic displays are expensive, require the viewer to wear special polarized glasses, are somewhat sensitive to disruption from vibration and poor viewing conditions, and are reduced in intensity and spatial resolution. Furthermore, not all people can accurately use stereoscopic cues. A display designer, therefore, may ultimately be asked to balance the added cost of the three-dimensional stereoscopic display against the performance benefits that it provides over its nonstereoscopic rivals. A study of three-dimensional tracking by Kim, Ellis, Tyler, Hannaford, and Stark (1987) found that the stereoscopic display provided benefits over a two-dimensional perspective display only as long as the cues of the latter were relatively sparse. When a richer set of pictorial cues was added (including texture to provide textural gradients), the advantages of stereopsis were eliminated. Similar results were found by Sollenberger and Milgram (1989), who examined stereo and motion parallax as tools for microscopic analysis in neurosurgery. Furthermore, in their review of the literature on three-dimensional displays, Wickens, Todd, and Seidler (1989) concluded that the enhancement of depth perception offered by stereoscopic displays was greatly diminished when the cue of relative motion or of motion parallax was available.

NAVIGATION AND SPATIAL COGNITION

An important use of spatial analog displays is to give the traveler through an environment a sense of location, locomotion, and direction: "Where am I and where am I headed?" Naturally such information does not need to be spatially displayed or represented. Displays can contain such directions as "pull up," "turn left," or "come to heading 045." These are often referred to as *command displays* (see Chapter 5). Furthermore, our mental representation of space may reside in categorical terms like "left of" or "further than" rather than in analog form. Indeed we will discuss some of the issues relating to the verbalization of spatial relations later in this chapter. However, since the goal of navigation is to achieve movement through space and because large movements are typically more significant than small ones, it is apparent that the task must be fundamentally anchored in our analog understanding of space.

We will begin by considering how people judge their movement through space and how displays may be designed to facilitate this judgment. We will next consider how people learn about a geographical environment: how it is mentally represented and what the biases are in this representation. Finally, we will consider how navigational performance can be supported through the design of maps and instructions.

Judgments of Egomotion

As we move through an environment, whether in a plane, in an automobile, in a boat, or on foot, our judgments of the direction and speed with which we are moving depend on information distributed across the visual field, not just in the area of foveal vision. Thus good drivers who primarily fixate far down the center of the highway are still making effective use of the flow of texture beside the highway as viewed in peripheral vision (Leibowitz, 1988; Leibowitz & Post, 1982). Correspondingly, performance on a variety of tracking tasks will be degraded to the extent that the amount of peripheral vision is restricted (Wickens, 1986).

As a consequence of this usefulness of peripheral vision, a number of engineering psychologists have recently argued that conventional aircraft navigation instruments, such as the attitude display indicator shown in Figure 4.8, are less than fully effective because perception of them is pretty much restricted to foveal vision. This restriction has led to the proposition that the pilot's perception of flight information can be augmented by peripheral displays. Although peripheral vision is not highly effective for recognizing objects, it is far more proficient at conveying information about motion and orientation (Leibowitz, 1988). One example is the Malcomb horizon, which extends, through a laser projection, a visible horizon all the way across the pilot's field of view (Stokes, Wickens, & Kite, 1990).

A second problem with the conventional instrument panel is that the information necessary for the pilot to obtain a good sense of location and motion is contained in at least four separate instruments, which must then be mentally integrated—the attitude display indicator, altimeter, compass, and airspeed indicator (Figure 4.16). In addition, the vertical speed indicator, depicting the rate of change of altitude, is often consulted. One solution to this integration problem is achieved through the development of integrated three-dimensional displays such as that shown in Figure 4.14*a*. A second solution lies in the design of ecological displays, which capitalize on the visual cues humans naturally use

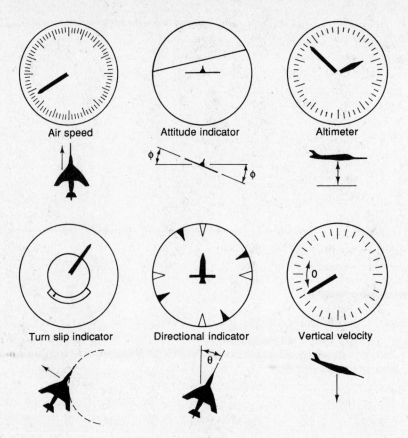

Figure 4.16 A typical aircraft instrument panel. (*Source:* C. D. Wickens, "The Effects of Control Dynamics on Performance," in K. Boff, L. Kaufman, and J. Thomas (eds.), *Handbook of Perception and Performance, Vol. II.* New York: Wiley, 1986. Reprinted by permission of John Wiley and Sons, Inc.)

to perceive their motion through the environment—the cues of direct perception that will support *egomotion* (Gibson, 1979; Larish & Flach, 1990; Owen & Warren, 1987; Weinstein, 1990). These cues have sometimes been referred to as *optical invariants* because they represent properties of the light rays that reach the eye (or any display surface) and have an invariant or unchanging relationship to the location and heading of the observer, whether walking, driving, or flying. Gibson (1979) has pointed out a number of such invariants, and four of these are as follows:

1. *Compression.* As we saw, textural gradient provides a cue to three-dimensionality. A change in the compression of a textured surface, such as that between the left and right of Figure 4.17, will accurately signal a change in altitude and viewing orientation.

2. *Splay.* Parallel receding lines, providing the depth cue of linear per-

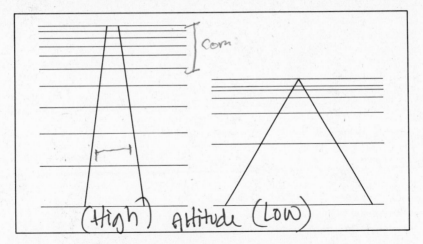

Figure 4.17 Illustrates perceptual cues of splay and compression: On the left, the perception is of being high above the field looking down. On the right, the observer is at low altitude looking forward.

spective, will also signal a change in altitude as given by the angle between them—the *splay*, which can also be seen by contrasting the two panels of Figure 4.17. Experimental evidence has established the value of both splay and compression in helping the pilot to control altitude (Johnson, Tsang, Bennett, & Phatek, 1987; Warren & Riccio, 1985). Furthermore, because these cues present altitude in a more natural, "ecological" fashion, there is some evidence that they are processed more automatically, leaving more attentional resources available for other tasks (Weinstein, 1990).

3. *Optical flow.* Whereas compression and splay are properties of a static environment, optical flow refers to the velocity of points traveling across the display surface (and therefore the retina) as we move through the world. This velocity is indicated by the arrows in Figure 4.18. The flow itself has certain critical features. The *expansion point* is that place where there is no flow but from which all flow radiates. It indicates the direction of momentary heading. For the pilot, the expansion point is critical because if it is below the horizon, its position forecasts a contact with the ground unless corrections are made. Furthermore, the relative rate of flow away from the expansion point, above, below, left, or right, gives a good cue regarding the *slant* of the surface relative to the path of motion. A flow that is of uniform rate on all sides indicates a heading straight into the surface, such as the landing of a lunar module or a helicopter descending straight down to the point of touchdown. In Figure 4.18, we see that the aircraft is angling into the surface because the optical flow is greater below than above the expansion point.

4. *Global optical flow.* The *rate* of flow of optical texture past the observer (Larish & Flach, 1990) is determined both by the observer's velocity over the ground and by the height above the ground. Thus, global optical flow will increase as we travel faster and also as we travel closer to the ground.

Figure 4.18 Optical flow. The arrows indicate the momentary velocity of texture across the visual field that the pilot would perceive on approach to landing.

A potential bias in human perception occurs because our subjective perception of speed is heavily determined by global optical flow. Thus, we feel as if we are traveling faster in a sports car than in a large sedan or bus, in part because the sports car is closer to the ground. When the Boeing 747 was first introduced, this bias was determined to be the cause of a problem encountered when pilots taxied the aircraft too fast and occasionally damaged the landing gear while turning on or off the runway. The reason for this error, in terms of global optical flow, was simple. The cockpit of the wide-bodied 747 is about twice as far above the runway as in the previous generation of narrow-bodied aircraft. Hence, for the same true taxiing speed, the global optical flow was half as fast. Pilots therefore perceived themselves to be traveling slower than necessary and would unsafely accelerate to obtain a global optical flow that matched their perception of the appropriate taxiing speed established through prior experience. As a result they achieved a true velocity that was unsafe (Owen & Warren, 1982).

Global optical flow (and hence perceived velocity) can be affected by altitude and velocity. It can also be affected by the density of texture over which a vehicle passes. As this density increases (texture is finer), the global optical flow increases and the traveler perceives a faster velocity. This characteristic of perceptual experience was exploited by Denton (1980) in an ingenious application of perceptual research to highway safety. Denton's concern was with automobile drivers in Great Britain who approached traffic circles at an excessive rate of speed. His solution was to decrease the spacing between road markers gradually and continuously as the distance to the stop point decreased. A driver who might be going too fast (not slowing down appropriately) would see the global optical flow as *increasing*. Believing the vehicle to be accelerating, the driver would compensate by imposing a more appropriate degree of braking or slowing. Denton's solution was imposed on the approach to a particularly dangerous traffic circle in Scotland. Not only was the average approach speed significantly slower following introduction of the markers, but also the rate of fatal accidents was greatly reduced.

Acquisition of Navigational Information

The human factors problems associated with maps and navigation should be self-evident to anyone who has ever encountered the following circumstances: (1) driving through a complicated series of intersections, heading in a generally southerly direction while navigating from a conventional north-up road map; (2) following a list of instructions on how to get somewhere (". . . go south two blocks and then turn right . . .") and missing a turn; and (3) having carefully studied the map of a strange place, suddenly finding oneself unable to locate one's position within the region depicted by the map because the surrounding landmarks are all unfamiliar. Examination of these problems requires that we first understand how the traveler learns geographical knowledge about an environment.

Thorndyke (1980) proposes that as we become increasingly familiar with a geographical environment, such as a city to which we have just moved, the nature of our knowledge of that environment undergoes qualitative as well as quantitative changes. The qualitative changes are characterized by a progression through three levels of knowledge. Initially our representation is characterized by *landmark knowledge*. We orient ourselves exclusively by highly salient visual landmarks (statues, buildings, etc.), and so our knowledge may be characterized in large part by direct visual images of those features. The prominence of landmark knowledge at early stages of navigational understanding of an environment suggests the importance of including salient landmarks at intermittent locations in the design and planning of any city or neighborhood. These provide the skeletal frame of reference around which to build the two subsequent phases of learning: route knowledge and survey knowledge (Anderson, 1979).

Landmark knowledge is followed by *route knowledge*. Here, understanding is characterized by the ability to *navigate* from one spot to another, utilizing landmarks or other visual features to trigger the decisions to turn left, turn right, or continue straight at a given intersection. If we are able to recall these features, we may offer verbal directions to someone else on how to navigate the route ("turn left when you get to the church"). If, however, we can only recognize them, we may navigate ourselves but cannot give directions to others ("I can take you there, but I can't explain how to get there"). In this case, the landmarks must be perceived directly to trigger the action decisions. Although route knowledge is based on recognition of the visual features and therefore possesses spatial elements, it does contain a verbal component in the somewhat categorical statements of action (turn left; turn right). Also, route knowledge is knowledge from a uniquely *ego-centered* frame of reference.

Sufficient navigational experience eventually provides us with what Thorndyke refers to as *survey knowledge*. Here, our knowledge resides in the form of an internalized "cognitive map" (Tolman, 1948), the analog to the true physical map. Our ability to describe the relative location of two landmarks in a city, even though we may never have traveled a route that connects them, offers convincing evidence that survey knowledge can be acquired from navigation. Survey knowledge, unlike route knowledge, is based on a *world-centered* frame of reference.

Route and survey knowledge support different kinds of tasks (Thorndyke, 1980). Route knowledge supports tasks that are framed from an ego-referenced perspective, such as pointing the direction to a landmark in the environment or judging how long it would take to travel a given route. Survey knowledge, in contrast, should support more world-referenced tasks, such as judging the absolute location of a landmark or the relative direction or the distance between two landmarks.

Although navigation through an environment will directly produce route knowledge and eventually create survey knowledge, it is also apparent that a shortcut to survey knowledge can be obtained through map study. To test this hypothesis, Thorndyke and Hayes-Roth (1978) compared the performance between two groups of subjects on orienting and distance-estimation tasks. One group had acquired knowledge of the geography of the Rand Corporation building through extensive navigation training (route knowledge). The other group had acquired knowledge by map study (survey knowledge). Thorndyke and Hayes-Roth's results generally confirmed the predictions. At moderate levels of training, map-learning subjects showed better estimates of euclidian distance than of route distance and better judgments of object localization than of orientation. Subjects trained by navigation showed the reverse effect in both cases.

Further training of both groups, however, revealed an asymmetry of transfer. With more extensive learning, the subjects trained by navigation developed survey knowledge and eventually performed as well as or better than the map-learning subjects on all tasks. This finding is consistent with the order of the learning progression proposed by Thorndyke (1980). The map-learning subjects, on the other hand, failed to improve on the ego-centered route-estimation and orientation tasks with more extensive map training.

These results have some important implications in training operators in tasks in which different kinds of spatial judgments must be made. Depending on whether the task requires relative object localization from a neutral reference or navigation and orientation from one's own reference, different kinds of training should be employed. In particular, the results suggest that extensive map study may not be terribly effective in preparation for a task in which one must navigate through a strange environment (e.g., the helicopter pilot prior to embarking on a rescue mission). A more effective training procedure would be provided by an inside-out experience of the environment in which the operator actually "navigates" through videotapes or even views a highly abstracted video that indicates the twists, turns, and landmarks to be encountered in navigation.

One relatively obvious advantage of survey knowledge concerns the consequence of errors. When route knowledge had guided our navigation and we are found off course as the result of a wrong turn or an intentional sidetrack, the information provided by route knowledge, perfect while on course, suddenly becomes nearly meaningless. Survey knowledge, on the other hand, may be used to guide us back either to the required route or to the desired goal by a different route. In a sense, the trade-off between the two forms of knowledge seems to be one between automaticity and cognitive simplicity, on the one hand, favoring route knowledge, and flexibility and generality, on the other, favoring survey knowledge. The flexibility provided by survey knowledge will assist the lost traveler. The automaticity of route knowledge will be useless. Where the likelihood or potential costs of these errors are greater, survey knowledge becomes preferable.

Distortions of Spatial Cognition

Survey knowledge of an environment, derived either from maps or from navigational experience, is clearly spatial or analog in form. However, this knowledge, like the representation of visual images, is not a veridical photographic reproduction of the map but contains some interesting systematic distortions that are manifest as we try to use the information on the map (Tversky & Schiano, 1989).

Distance Estimations and the Filled Distance Effect   Maps or the mental representation of imagined maps must often be scanned. Kosslyn, Ball, and Reiser (1978) demonstrated that the time taken to scan a mental image is proportional to the distance of the scan. Thorndyke (1980) pursued this line of reasoning with the corollary logic that our estimation of size (area or distance) of an image is influenced by the time taken to scan its extent. If this premise is accepted, anything that will slow down the scan time will increase the perceived size. Applying this reasoning to the estimation of map distances, Thorndyke performed the following experiment: Subjects first memorized simple schematic maps of roads and cities. Subsequently, when they were asked to estimate distances between target cities on the memorized maps (scanning their spatial image), those distances were systematically overestimated to the extent that more cities occurred between the target end points. The data were modeled by a counting-timer scheme, in which perceived distance is judged proportional to the elapsed time on a mental counter from initiation to end point. This counter is slowed by intervening cities as the scan becomes disrupted by the added clutter. When the subjects estimated distances on a map that was directly perceived instead of imagined, the filled distance effect was also observed but was of smaller magnitude.

The implications of this research to the development of computer-generated maps are demonstrated by considering the biases that may occur when unimportant information concerning cities or other geographical features is displayed. Such information might be made callable on request but should not necessarily be permanently displayed when rapid distance estimates are required.

We saw that the filled distance effect was enhanced when a map image was scanned in working memory rather than viewed directly. It turns out that when we recall geographical information from long-term memory, still more distortions are evident (Tversky & Schiano, 1989). For example, we tend to overestimate the distances between landmarks that are close to us, relative to those that are farther away. Distances between cities in our own state are estimated to be greater than equivalent distances between cities that are farther away (Holyoak & Mah, 1982). Another example, which we will discuss in some detail because of its prevalence, is *rectilinear normalization*.

Rectilinear Normalization   Our navigation and orientation judgments tend to be carried out in a normalized, right-angle-grid world, with orientation of most of our maps and cities emphasizing north, south, east, and west. (This is particularly true for those of us living in flat, midwestern cities.) Within this framework, there is some evidence that judgments of north and south (or up and down) are made with greater facility than those of east and west (Maki, Maki, & Marsh, 1977). Relative to other directions of the compass, however, all four of these primary directions exert a dominating influence. For example, Loftus (1978) found that both novice subjects and skilled pilots, when asked to indicate the direction of an auditorily presented compass heading (''230 degrees''), showed the most rapid reactions to headings that corresponded with the four major compass directions. Furthermore, subjects showed relatively faster responses to directions close to the compass headings (e.g., 85 degrees, 182 degrees) than to those in the middle of a quadrant (45 degrees, 150 degrees). These data suggest that subjects employed the four directions as major orienting landmarks. The heard direction is interpreted initially in terms of the nearest compass point and is then modified in terms of its degree of departure from that point. The closer the point is to the middle of the quadrant (and farther from the landmark), the longer will be the contribution of the second component to the estimation time. This result thus emphasizes both the categorical role of the N-S-E-W landmarks and the analog operations of subsequent orientation from these landmarks.

The strength of the predominant N-S-E-W grid we impose on our internal representation of geography is sufficiently great to induce some very systematic biases in our judgments of survey knowledge. Chase and Chi (1979) refer to these biases collectively as our tendency to engage in *rectilinear normalization*, demonstrated by an experiment of Stevens and Coupe (1978) in which subjects were asked to make judgments of the relative direction of one city to another—to decide, for example, the direction of Reno, Nevada, from San Diego, or of Santiago, Chile, from New York. In these examples, the tendency is to report that Reno, Nevada, is northeast of San Diego (it is, in fact, northwest) and that Santiago, Chile, is southwest of New York (it is southeast). The erroneous responses result from applying a grid or a set of propositional statements to reconstruct the geographical location in the following fashion: ''Nevada is east of California; San Diego is in southern California; Reno is in northern Nevada; therefore, Reno is northeast of San Diego." Similarly, "New York is in eastern North America; Santiago is in western South America; South America is south of North America; therefore, Santiago is southwest of New York." The logic performed in making these judgments is in fact closer to a discrete propositional logic than to one based on visual imagery and is another example of the close cooperative interaction between verbal and spatial processes.

In a related example, Milgram and Jodelet (1976) note that Parisians tend to "straighten" their mental representation of the flow of the Seine River through Paris, forcing it into more of an east-west linear flow than it actually has. This cognitive distortion into the rectilinear grid thereby induces consequent errors and distortions in their survey knowledge of the area. It is important to note, however, that distortions of this kind will not generally disrupt navigation based on route knowledge. Parisians can still easily navigate along familiar routes because the navigation does not require the precise analog judgments that are distorted in their survey knowledge. City navigation instead depends on simple categorical decisions (left-right-straight), which are perfectly compatible with the distortions imposed by rectilinear normalization.

In yet another demonstration of rectilinear normalization. Chase and Chi (1979) asked subjects to draw a map of the Carnegie-Mellon University campus from memory. In their reconstruction, subjects tended to "force" nonrectilinear intersections into a rectilinear N-S-E-W grid (see Figure 4.19). Interestingly enough, only the architects in the sample failed to impose this distortion. Finally, Howard and Kerst (1981) found that when subjects were asked to recall maps of a campus, they tended to force their reconstruction into clusters of buildings oriented to the four primary directions.

These examples illustrate the manner in which errors of misjudgment might occur when people attempt to navigate without maps in environments in which the normal N-S-E-W grid pattern is not imposed. One implication of this finding is that schematic maps, which themselves often systematically distort

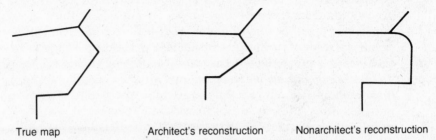

True map   Architect's reconstruction   Nonarchitect's reconstruction

Figure 4.19 Reconstruction of street map on Carnegie-Mellon University campus showing the bias due to rectilinear normalization. (*Source: Cognitive Skill: Implications for Spatial Skill in Large-Scale Environments* [Tech. Rep. No. 1] by W. Chase and M. Chi, 1979, Pittsburgh, PA: University of Pittsburgh Learning and Development Center.)

the analog environment and force it into the rectilinear grid, would serve to reinforce this tendency, with potentially unfortunate consequences when accurate survey knowledge is required.

Canonical Orientation   Our internal representations of spatial knowledge appear to have a preferred, or canonical, orientation (Tversky & Franklin, 1990). Thus, when we imagine an ego-referenced view of the world, this image is typically upright, not angled or inverted. When we mentally represent a geographical layout, our canonical axis appears to depend on how we have learned that knowledge. When we have learned by studying a map, which we typically do in a north-up orientation, the canonical orientation is north-up (Sholl, 1987). But when we learn by navigating through an environment, there is no longer a strong canonical orientation (Hintzman, O'Dell, & Arndt, 1981; Sholl, 1987), or the orientation may be that perspective from which we most often view the environment (i.e., the front window of our house or looking toward a city). Canonical orientations are important in designing navigational aids.

Maps and Navigational Aids

Maps versus Route Lists   Information presented to travelers on commercial transportation systems may be in the format of either a linear listing of the bus or subway stops (route list) or an actual map (sometimes highly schematized), showing the spatial relations between the stops. Bartram (1980) asked subjects to select the combination of bus routes they would need to take to navigate between two selected locations on the London bus system. The information was presented either in a route list (the sequential listing of all bus stops on a route, showing stops that were common between lines) or in the form of a map (either schematic or accurate). Bartram found that decision times were considerably more rapid in the map condition and were uninfluenced by decision complexity. In contrast, the time required to make slower route-list decisions increased with complexity. These results are consistent with the conclusions drawn in the preceding discussion since the judgment required in the task requires survey knowledge (e.g., the chosen route is constant and independent of the subject's frame of reference). In this context-free situation the data of Thorndyke and Hayes-Roth (1978) suggest that the map display should be superior.

A second contrast between the two forms of geographical representation is illustrated in an experiment performed by Wetherell (1979) in which the two means of providing navigational instructions were compared. Wetherell had subjects learn the directions to navigate between two points in a driving "maze" by using one of two techniques: either learning a linear list of turns (route knowledge) or studying a spatial map of how to get from A to B (map, or survey, knowledge). The subjects then actually drove the route (on an unfamiliar terrain). Those who were trained in the map condition made many more errors than those trained in the route-list condition. Through subsequent experimentation, Wetherell identified two contributing factors to the observed difference: (1) The spatial-processing demands of actually driving, seeing, and orienting interfered with the spatial demands of maintaining the mental map in working memory in the survey-knowledge condition. This kind of interference between two spatial tasks will be discussed in greater detail in Chapters 6 and 9. (2) Subjects had great difficulty reorienting or rotating the cognitive map, which was learned in a north-up direction, to their own subjective frame of reference as they approached any intersection heading east, south, or west. Corresponding findings were obtained by Streeter, Vitello, and Wonsiewicz (1985), who asked subjects to drive through neighborhoods using either a tape-recorded route list (series of verbal directions) or a customized map with the route highlighted. Both the speed and accuracy of following the route was best with the route list. Interestingly, many of the subjects using the map made errors in turning the wrong direction. This is the sort of confusion that can be avoided when instructions are phrased in ego-referenced terms.

These findings reiterate a point made when discussing the compatibility of frames of reference in visual displays. Route knowledge, because of its more verbal and ego-referenced characteristic, is less sensitive to an incompatibility of orientation. The analog survey knowledge, at least as possessed by novices, is more adversely affected by the reversed frame of reference that can occur when navigating.

Fixed versus Rotating Maps   When using paper maps, users have the option to rotate them to a canonical north-up orientation or in the direction they are heading; however, electronic maps are increasingly being introduced in both aircraft and automobiles, and the designer is confronted with the question of whether a fixed or rotating map (or both) is the desirable option. Navigating with a map that is rotated so that the direction that is up or forward on the map is aligned with the momentary direction of travel has a clear benefit because of the *congruence* between left-right distinctions on the map and those in the view of the world. In contrast, when a map is fixed in a north-up orientation (or as with the subjects in Wetherell's experiment, when one is navigating from a remembered map in the canonical north-up orientation), difficulties are imposed by the need to rotate mentally the image of the map so that "up" corresponds with the direction of travel—and therefore landmarks, roads, intersections, and geographical features on the imagined map are in the same relative orientation as they are in the forward field of view (Figure 4.20).

A series of studies have indeed identified this cost of mental rotation when a north-up, world-referenced map orientation is out of alignment with the ego-referenced field of view. These studies involve reading topographic maps (Eley, 1988), studying buildings (Warren, Rossano, & Wear, 1990), flying helicopter simulations (Aretz, 1991; Harwood & Wickens, 1991), or studying "you are here" (YAH) maps such as those in shopping malls (Figure 4.21). Levine (1982) conducted a survey of YAH maps in New York City and found that 75 percent of them were positioned so that they violated this principle of axis alignment, some by 180 degrees.

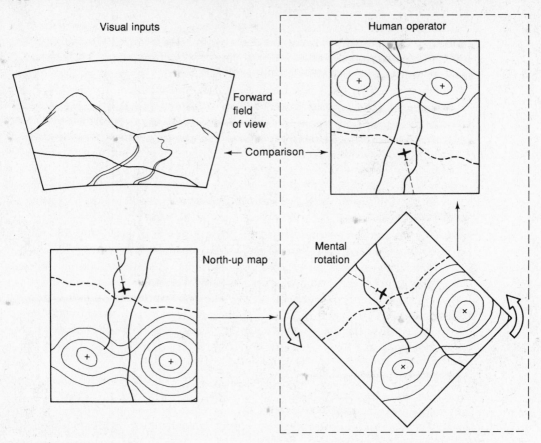

Figure 4.20 Illustrates the mental rotation required to compare the image seen in an ego-referenced forward field of view (top) with a world-referenced north-up map (below) when the aircraft is heading south. The map image is mentally rotated (right) to bring it into congruence with the forward field of view.

In light of the documented cost of mental rotation with fixed north-up maps, it might seem that the design of electronic maps should always be rotating. Yet there are at least three reasons why this is not necessarily the optimal solution. First, it is technologically quite expensive to update constantly the computer display of an electronic map as it rotates, particularly if the map contains a lot of printing and many symbols. Second, it is true that some navigators often do hold their maps in a north-up orientation, whether on the ground or in airborne navigation.

Third, comparative evaluations of the two map types have revealed that the advantages for one over the other are not consistent and appear to be task-dependent (Aretz, 1991; Baty, Wempe, & Huff, 1974; Hart & Wempe, 1979; Harwood & Wickens, 1991). Thus, the *consistent* location of landmarks relative to the frame on the north-up map will serve tasks that depend on locating landmarks on the map (Harwood & Wickens, 1991) and tasks that depend on

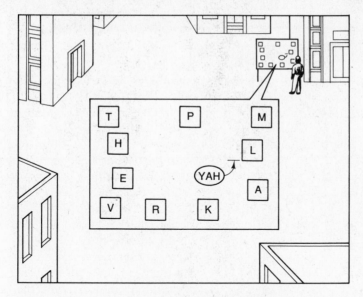

Figure 4.21 You-are-here (YAH) maps. Notice that the "up" orientation of the YAH map is congruent with the forward view beyond the map, and that the observer's perspective in viewing the map is shown on the map itself. (*Source:* M. Levine, "You-Are-Here Maps: Psychological Considerations," *Environment and Behavior, 14,* (1987), 221–237. Copyright © by M. Levine. Reprinted by permission of Sage Publications, Inc.)

learning the relative location of features in the environment (Aretz, 1991). Furthermore, a world-referenced, north-up map will support planning and communications with other users, in a different vehicle or at a different location, who may not share the same momentary ego frame of reference (e.g., another pilot or an air traffic controller).

In conclusion, two options would appear to be available in designing maps that satisfy both the needs for an ego-referenced and world-referenced representation. The obvious solution, of course, is to make both options available on a user-select basis (which still does not address the technological requirements of rotating complex maps and may introduce possibilities of user confusion). A second solution is to use a north-up map but within it provide a very salient representation of the forward field of view of the navigator, depicted so that the two frames of reference are presented in a common framework. This technique, which exploits the principle of *visual momentum* (to be described later in this chapter), was evaluated by Aretz (1991) in a north-up map for helicopter navigation and was found to provide performance that was as good or better than a track-up map. This equivalence was found even on tasks that required mental rotation to compare landmarks between the map and the forward view.

SPATIAL REPRESENTATIONS OF NONSPATIAL SYSTEMS

Direct Manipulation and Spatial Data Bases

We have described humans' perception and understanding of space and the development of displays to represent it. Within the last decade, designers of computer systems are beginning to capitalize on the fact that people have a lifetime's worth of experience in negotiating in a three-dimensional environment and manipulating three-dimensional objects (Hutchins, Hollan, & Norman, 1985). The spatial metaphor, therefore, is an important emerging concept in human-computer interaction, and several examples can be cited in which spatial arrays on a display are intended to mimic spatial characteristics of real-world scenes and objects. Thus, tasks can be performed directly on the basis of these arrays rather than by manipulating more abstract or arbitrary symbols (Shneiderman, 1987). Spatially arranged spreadsheet programs for accounts and record keeping, such as Lotus, are one example; so too are screen-oriented text editors in which a full page of text is displayed. These have nearly completely replaced the line-oriented editor. Indeed the omnipresent *cursor* and the mouse represent spatial devices that can be used to define areas of interest and operation within a spatial array—an analog to the hand-held pointer (see also Chapter 11).

The philosophy behind the introduction of these spatially guided interaction tools is that the computer should be manipulated in a way that is analogous to the way people directly manipulate objects in space; hence, these innovations have been described by the general label *direct manipulation interface* (DMI) (Hutchins, Hollan, & Norman, 1985; Jacob, 1989) and have been heavily utilized in personal computer models developed over the last decade, such as the Xerox Star (Smith, 1981) or the Apple Macintosh.

Shneiderman (1987) has noted that DMI contains the following three features, which are representative of the focus of this chapter on analog perception and spatial processing:

1. There is a continuous representation of objects and actions of interest.
2. Objects are manipulated by physical actions or directional keys, rather than complex syntax.
3. Operations are rapid and reversible.

To these may be added a fourth characteristic of many direct manipulation interfaces, the use of familiar *icons*—like wastebaskets, windows, or scissors—and the capitalization on *analogies* to real-world actions. If these icons are well chosen, this feature makes the actions of the DMI easy for the novice to learn and difficult to forget (see also Chapter 5).

A close cousin of the DMI is the spatially organized data base retrieval system. Cole (1982) made a careful survey of user needs for the electronic office of the future. The results established the extent to which office workers depended on their spatial memory to know where information was stored and provided a strong rationale for building the spatial metaphor into data base storage-retrieval systems. The important role of space in data base retrieval was also demonstrated in an experiment by Vicente, Hayes, and Williges (1987) in which subjects of varying levels of verbal and spatial ability were required to answer queries about a data base. The authors found that spatial ability was the strongest predictor of performance, and in particular, those of low spatial ability were far more likely to get lost in the data base. Some designers have thereby capitalized on the use of space to represent data bases. For example, a display might contain several distinct spatial locations into which the user can put various documents. The advent of computer-based "windows" allows the user to look at several data structures simultaneously on the display. Henderson and Card (1986) proposed the use of "doors" and "rooms" to give the user a sense of spatially entering and exiting various files. As another example, Knepp, Barrett, and Sheridan (1982) examined a spatial map for searching an inventory of products that might vary along three or four dimensions (e.g., screws that might vary in their length, width, thread shape, and material). Subjects could then spatially move through the three- or four-dimensional data base, their position represented as a point in a displayed cube defined by three of the current dimensions.

Even those who have advocated the spatial metaphor and DMI are cautious about their shortcomings and limitations. Shneiderman (1987), for example, notes that users must learn the meaning of graphic representations that may not always be clearly defined. He notes also that graphic representations may be misleading and that they may take up excessive display space. He also raises the concern that for the expert typist, moving a spatially guided mouse may be slower than typing.

Equally important, Hutchins, Hollan, and Norman (1985) note that the very feature that provides strength to the DMI in some circumstances — its adherence to the physical constraints of three-dimensional geometry — can limit its usefulness in powerful symbolic operations when the user must go beyond the constraints of three-dimensional analog operations. As the authors note, "If we restrict ourselves to only building an interface that allows us to do things we can already do and to think in ways we already think, we will miss the most exciting potential of new technology: to provide new ways to think of and to interact with a domain. Providing these new ways and creating conditions that will make them feel direct and natural is an important challenge to the interface designer" (p. 336).

What is the validated success of the DMI? On the one hand, much of its success lies in the intuitive value of visual spatial metaphor, the popularity of computer systems like the Macintosh, and the fact that spatial concepts and spatial relations are important features in the thinking of many nonexpert computer users (Cole, 1982; Shneiderman, 1987). Yet there is not an extensive empirical data base that has compared performance on spatially oriented versus symbolic displays.

Some of the most informative evaluations have examined different formats of data base retrieval systems. For example, Thomas and Gould (1975) found that the spatially oriented "query by example" data retrieval system offered better performance than one based on the more symbolic, verbal, key-word search. A study by Peters, Yastrop, and Boehm-Davis (1988), however, offers some qualifications to assertions regarding the superiority of spatially organized data bases. The authors compared graphic and tabular formats of a data base representing airline reservations information and found that the best format was determined by the kind of question asked. Questions emphasizing the spatial relations, such as "How do you get from X to Y?" were best answered with the spatial format; questions emphasizing the tabular information, such as "How many flights arrive at 9:00 A.M.?" were better answered with the tabular format. Hence, we see yet another form of compatibility — the compatibility of a display with the questions that need to be asked.

Along the same lines, Durding, Becker, and Gould (1977) emphasize that the spatial organization of a data base must be compatible with the user's mental model of the data, and that such models can be categorized into different organizational types. For example, the screws stored in the study by Knepp et al. (1982) fit into an orthogonal, four-dimensional matrix. Yet other data have hierarchical, network, or list relations between their elements, as illustrated in the different examples in Figure 4.22. These categories were used in an experiment by Durding, Becker, and Gould (1977), who asked subjects to memorize lists of the 16 words in each category. Although all the lists were presented in a linear array, subjects' ability to recall them was greatly improved when they were given a skeletal structure compatible with the underlying format of organization of the data base. As a guideline to designers of data base retrieval systems, the researchers suggest the importance of spatial maps that outline the structure of data bases — structures that should be compatible with the user's mental model.

Computer-Aided Design and Visualization

Closely related to the concept of direct manipulation are two developments in human-computer interaction: computer-aided design (CAD) and scientific visualization. Developments in both of these areas have been greatly facilitated by three-dimensional computer displays. In the CAD environment (Barfield et al., 1987), the designer's task is considerably aided by a three-dimensional representation in which objects can be rotated and examined and design options can be tried and erased. A major potential use of the CAD workstation is in the domain of anthropometry and biomechanics (Kroemer, 1987; McMillan et al., 1989; Chapter 1), in which computers can portray dynamic three-dimensional models of human operators engaging in the analog operations of lifting, reaching, and pulling. These actions produce forces on the human body that can be directly and spatially revealed by vectors within the display rather than derived from tables (Kroemer, 1987).

In *scientific visualization* (Alexander & Winarsky, 1989; Robbins & Fisher, 1989; Zorpette, 1989) the designers of powerful computer-driven display processors have used three-dimensional graphics, augmented with color, to

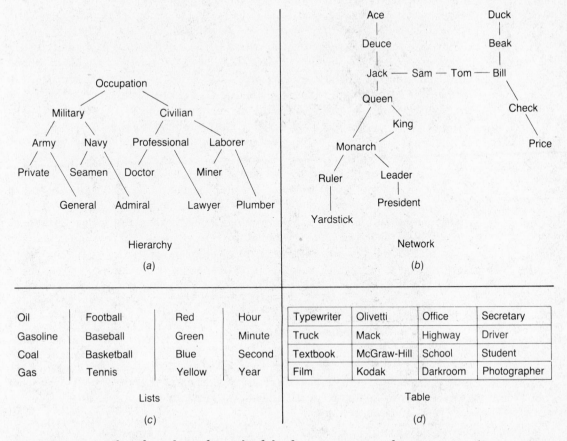

Figure 4.22 Examples of word sets for each of the four major types of organizations. (*Source:* "Data organization" by B. M. Durding, C. A. Becker, and J. D. Gould, 1977, *Human Factors, 19*, p. 4. Copyright 1977 by the Human Factors Society, Inc. Reproduced by permission.)

present complex, multidimensional data in such a way that scientists can physically "explore" and manipulate them. For example, the biochemist may wish to "take a trip" through the network of a complex molecule; the engineer studying fluid dynamics may wish to examine from different perspectives the three-dimensional patterns of motion and density caused by fluid flowing through a certain passage. Or the atmospheric scientist may wish to understand the dynamics of a thunderstorm (Wilhelmson et al., 1989). Such techniques, although computationally expensive, have proven attractive to scientists and are useful in comprehending their data (Wilhelmson et al., 1989, see Figure 4.23). Their expression, however, can often be improved by principles of navigational awareness (Wickens, 1990) and three-dimensional graphic perception (Merwin & Wickens, 1991; Wickens & Todd, 1990), in particular, the principle of visual momentum.

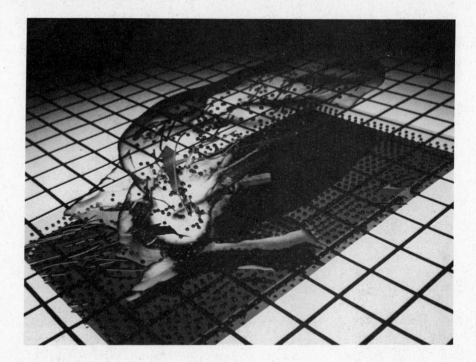

Figure 4.23 Computer-generated visualization of a severe thunderstorm. (*Source:* From Wilhelmson et al, 1989. Courtesy of National Center for Super Computing Applications.)

Visual Momentum

People become confused or disoriented as they examine a sequence of graphs with inconsistent labels, symbols, or orientation. A similar problem is often encountered in multilevel menu systems, multifunction displays, or data-retrieval systems, in which users may get lost in the system and disoriented in their ability to return to a given origin (Billingsley, 1982; Vicente, Hayes, Williges, 1987; Woods, Roth, Stubler, & Mumaw, 1990). Of course, one objective of spatially oriented computer systems is precisely to create a familiar spatial world in which travel can be accomplished without this disorientation. But as we know, it is possible to get lost in spatial environments as well as in symbolic ones.

The concept of *visual momentum* represents an engineering design solution to the problem of becoming cognitively lost as the display user traverses through multiple displays pertaining to different aspects of the same system or data base (Wise & Debons, 1987; Woods, 1984). The concept was originally borrowed from film editors, as a technique to give the viewer an understanding of how successively viewed film cuts relate to one another (Hochberg & Brooks, 1978). When applied to the viewing of successive display frames, either of virtual space (e.g., maps) or of conceptual space (e.g., topologically related components in a process control plant, nodes in a menu or data base, or graphic representations of data), visual momentum involves four basic guidelines.

1. *Use consistent representations.* This guideline of course reiterates a principle that was set forth in the discussion of graphs. Unless there is an explicit rationale for changing some aspects of a display representation, it should not be changed. However, when it is necessary to show new data or a new representation of previously viewed data, display features should show the relationship of the new data to the old. The next three guidelines indicate how this may be accomplished.

2. *Use graceful transitions.* When changes in representation will be made over time, abrupt discontinuities may be disorienting. For example, on an electronic map, the transition from a small-scale, wide-angle map to a large-scale close-up will be cognitively less disorienting if this change is made by a rapid but continuous blowup.

3. *Highlight anchors.* An anchor may be described as a constant invariant feature of the displayed "world," whose identity and location is always prominently highlighted on successive displays. For example, in aircraft attitude displays that might be viewed successively in various orientations, the direction of the vertical (or horizon) should always be prominently highlighted. In map displays, which may be reconfigured from inside-out to outside-in to accommodate different task demands, a salient and consistent color code might highlight both the northerly direction and the heading direction (Andre, Wickens, Moorman, & Boschelli, 1991). As noted earlier, Aretz (1991) successfully used an anchor by portraying the angle subtended by the forward field of view on a top-down north-up map. In displays to examine components of a complex chemical or electrical process, the direction of causal flow (input-output) could be prominently highlighted. In the design of YAH maps (Figure 4.21), a visually prominent landmark in the forward field of view that is highlighted on the map, offers such an anchor (Levine, 1982).

A corollary principle is that when successive display frames are introduced, each new frame should include overlapping areas or features with the previous frame, and these common landmarks should be prominently highlighted (here again, color is an effective highlight).

4. *Display continuous world maps.* Here we refer to a continuously viewable map of the world, always presented from a fixed and compatible perspective. Within this map the current identity of the active display is always highlighted. This is a feature of the topographic maps produced by the U.S. Geological Survey, in which a small map of the state is always viewable in the upper left-hand corner, with the currently displayed quadrant highlighted in black. As discussed earlier in this section, Knepp, Barrett, and Sheridan (1982), developed a display concept to support information search through, and retrieval from, a multidimensional data base: The display featured a small but consistently oriented picture of the full data base in the form of a cube, and the currently examined item was highlighted. A study by Vicente and Williges

(1988) supported the concept of world maps in information retrieval from a hierarchically organized data base in which users had experienced problems of cognitive disorientation. Vicente and Williges found that the presence of a map of file organization, and of a cursor highlighting the momentary position within the file structure, provided significant benefits to user orientation.

In conclusion, visual momentum, like visualization and direct manipulation, although an intuitively appealing concept and supported by initial validation, awaits strong empirical support as a useful engineering concept in system design.

TRANSITION

This chapter has described the perception of space and the relationship between that perception and our understanding of space and spatial relations (spatial cognition). We emphasized the concept of compatibility between the display and the cognitive domain. We will address these topics again when we discuss spatial working memory in Chapter 6, spatial or analog manual control in Chapter 11, and the compatibility between a display and working memory and response in Chapters 6 and 8, respectively. However, as we are well aware, spatial information plays only a partial role in our interactions with other systems, including people. In the next chapter, then, we will discuss the complementary role of verbal and linguistic information in this interaction.

REFERENCES

Alexander, J., & Winarsky, N. (1989, April). Interactive data visualization. *Information Display*, pp. 14–22.

Anderson, J. R. (1979). *Cognitive psychology*. New York: Academic Press.

Andre, A. D., Wickens, C. D., Moorman, L., & Boschelli, M. M. (1991). Display formatting techniques for improving situation awareness in the aircraft cockpit. *International Journal of Aviation Psychology*.

Antsis, S. (1986). Motion perception in the frontal plane. In K. R. Boff, L. Kaufman, & J. P. Thomas (eds.), *Handbook of perception and human performance* (vol. 1). New York: Wiley.

Aretz, A. J. (1991). The design of electronic map displays. *Human Factors, 33*(1), 85–101.

Barfield W. (1987). Computer aided design. In G. Salvendy (ed.), *Handbook of human factors*. New York: Wiley.

Bartram, D. J. (1980). Comprehending spatial information: The relative efficiency of different methods of presenting information about bus routes. *Journal of Applied Psychology, 65*, 103–110.

Baty, D. L., Wempe, T. E., & Huff, E. M. (1974). A study of aircraft map display location and orientation. *IEEE Transactions on Systems, Man & Cybernetics, SMC–4*, 560–568.

Bemis, S. V., Leeds, J. L., & Winer, E. A. (1988). Operator performance as a function of type of display: Conventional versus perspective. *Human Factors*, *30*(2), 163–169.

Billingsley, P. A. (1982). Navigation through hierarchical menu structures: Does it help to have a map? *Proceedings of the 26th Annual Meeting of the Human Factors Society* (pp. 103–107). Santa Monica, CA: Human Factors Society.

Boff, K. R., Kaufman, L., & Thomas, J. P. (eds.). (1986). *Handbook of perception and human performance* (vol. 1). New York: Wiley.

Braunstein, M. L. (1990). Structure from motion. In J. I. Elkind, S. K. Card, J. Hochberg, & B. M. Huey (eds.), *Human performance models for computer-aided engineering* (pp. 89–105). Orlando, FL: Academic Press.

Braunstein, M. L., & Andersen, G. J. (1984). Shape and depth perception from parallel projects of three-dimensional motion. *Journal of Experimental Psychology: Human Perception and Performance*, *10*, 749–760.

Bruno, N., & Cutting, J. E. (1988). Minimodularity and the perception of layout. *Journal of Experimental Psychology: General*, *117*(2), 161–170.

Carroll, J. M., & Olson, J. R. (1987). *Mental models in human-computer interaction*. Washington, DC: National Academy Press.

Carswell, C. M. (1990). Graphical information processing: The effects of proximity compatibility. *Proceedings of the 34th Annual Meeting of the Human Factors Society* (pp. 1494–1498). Santa Monica, CA: Human Factors Society.

Carswell, C. M., & Wickens, C. D. (1987). Information integration and the object display: An interaction of task demands and display superiority. *Ergonomics*, *30*(3), 511–527.

Carswell, C. M., & Wickens, C. D. (1988). *Comparative graphics: History and applications of perceptual integrality theory and the proximity compatibility hypothesis* (University of Illinois Technical Report ARL-88-2/AHEL-88-1/AHEL Technical Memorandum 8-88). Savoy, IL: Aviation Research Laboratory.

Chase, W., & Chi, M. (1979). *Cognitive skill: Implications for spatial skill in large-scale environments* (Technical Report No. 1). Pittsburgh: University of Pittsburgh Learning and Development Center.

Cleveland, W. S., & McGill, R. (1984). Graphical perception: Theory, experimentation, and application to the development of graphic methods. *Journal of the American Statistical Association*, *70*, 531–554.

Cleveland, W. S., & McGill, R. (1985). Graphical perception and graphical methods for analyzing scientific data. *Science*, *229*, 828–833.

Cole, D. I. (1982). Human aspects of office filing: Implications for the electronic office. *Proceedings of the 26th annual meeting of the Human Factors Society* (pp. 59–63). Santa Monica, CA: Human Factors Society.

Coren, S., & Girgus, J. S. (1978). *Seeing is deceiving: The psychology of visual illusions*. Hillsdale, NJ: Lawrence Erlbaum.

Denton, G. G. (1980). The influence of visual pattern on perceived speed. *Perception*, *9*, 393–402.

Dosher, B. A., Sperling, G., & Wurst, S. A. (1986). Tradeoffs between stereopsis and proximity luminance covariance as determinants of perceived 3D structure. *Vision Research*, *26*(6), 973–990.

Durding, B. M., Becker, C. A., & Gould, J. D. (1977). Data organization. *Human Factors, 19*, 1–14.

Eberts, R. E., & MacMillan, A. G. (1985). Misperception of small cars. In R. E. Eberts & C. G. Eberts (eds.), *Trends in ergonomics/human factors II* (pp. 33–39). North Holland, Netherlands: Elsevier Science Publishers B. V.

Eley, M. G. (1988). Determining the shapes of landsurfaces from topographic maps. *Ergonomics, 31*, 355–376.

Ellis, S. R., McGreevy, M. W., & Hitchcock, R. J. (1984). Influence of a perspective cockpit traffic display format on pilot avoidance maneuvers. *Human considerations in high performance aircraft.* Nevilly-sur-Seine, France, Advisory Group for Aerospace Research & Development (AGARD) Proceedings 371.

Ellis, S. R., McGreevy, M. W., & Hitchcock, R. J. (1987). Perspective traffic display format and air pilot traffic avoidance. *Human Factors, 29*, 371–382.

Fogel, L. J. (1959). A new concept: The kinalog display system. *Human Factors, 1*, 30–37.

Franklin, N., & Tversky, B. (1990). Searching imagined environments. *Journal of Experimental Psychology: General, 119*,(1), 63–76.

Gentner, D., & Stevens, A. L. (1983). *Mental models.* Hillsdale, NJ: L. Erlbaum.

Gibson, J. J. (1979). *The ecological approach to visual perception.* Boston: Houghton Mifflin.

Goettl, B., Wickens, C. D., & Kramer, A. (in press). The object display in the perception of graphical data. *Ergonomics.*

Gregory, R. L. (1977). *Eye and brain.* London: Weidenfeld & Nicolson.

Grether, W. F. (1949). Instrument reading I: The design of long-scale indicators for speed and accuracy of quantitative readings. *Journal of Applied Psychology, 33*, 363–372.

Hart, S. (1988). Helicopter human factors. In E. Wiener & D. Nagel (eds.), *Human factors in aviation.* San Diego, CA: Academic Press.

Hart, S. G., & Wempe, T. E. (1979, August). *Cockpit display of traffic information: Airline pilot's opinion about content, symbology, and format* (NASA Technical Memorandum No. 78601). Moffett Field, CA: NASA Ames Research Center.

Harwood, K., & Wickens, C. D. (1991). Frames of reference for helicopter electronic maps: The relevance of spatial cognition and componential analysis. *International Journal of Aviation Psychology, 1*, 5–23.

Hawkins, F. H. (1987). *Human factors in flight.* Brookfield, VT: Gower Technical Press.

Helander, M. G. (1987). Design of visual displays. In G. Salvendy (ed.), *Handbook of human factors* (pp. 507–548). New York: Wiley.

Henderson, D. A., & Card, S. K. (1986). Rooms: The use of multiple virtual workspaces to reduce spatial contention in a window-based graphical use interface. *ACM Transactions on Graphics, 5*, 211–243.

Hintzman, D. L., O'Dell, C. S., & Arndt, D. R. (1981). Orientation in cognitive maps. *Cognitive Psychology, 13*, 149–206.

Hochberg, J., & Brooks, V. (1978). Film cutting and visual momentum. In J. W. Senders, D. F. Fisher, & R. A. Monty (eds.), *Eye movements and the higher psychological functions.* Hillsdale, NJ: Erlbaum.

Holyoak, K. J., & Mah, W. A. (1982). Cognitive reference points in judgments of symbolic magnitude. *Cognitive Psychology, 14,* 328–352.

Howard, J. H., & Kerst, R. C. (1981). Memory and perception of cartographic information for familiar and unfamiliar environments. *Human Factors, 23,* 495–504.

Hutchins, E. L., Hollan, J. D., & Norman, D. A. (1985). Direct manipulation interfaces. *Human-Computer Interaction. 1*(4), 311–338.

Jacob, R. J. K. (1989). Direct manipulation in the intelligent interface. In P. A. Hancock & M. H. Chignell (eds.), *Intelligent Interfaces: Theory, research and design* (pp. 165–212). North Holland, Netherlands: Elsevier Science Publishers B.V.

Johnson, S. L., & Roscoe, S. N. (1972). What moves, the airplane or the world? *Human Factors, 14,* 107–129.

Johnson, W. W., Tsang, P. S., Bennett, C. T., & Phatek, A. V. (1987). The visual control of simulated altitude. In R. Jensen (ed.), *Proceedings of the 4th International Symposium on Aviation Psychology.* Columbus: Ohio State University, Department of Aviation.

Kim, W. S., Ellis, S. R., Tyler, M., Hannaford, B., & Stark, L. (1987). A quantitative evaluation of perspective and stereoscopic displays in three-axis manual tracking tasks. *IEEE Transactions on Systems, Man, and Cybernetics, SMC–17,* 61–71.

Knepp, L., Barrett, D., & Sheridan, T. B. (1982). Searching for an object in four or higher dimensional space. *Proceedings of the 1982 IEEE International Conference on Cybernetics and Society* (pp. 636–640). New York: IEEE.

Kolata, G. (1982). Computer graphics comes to computers. *Science, 217,* 919–920.

Kosslyn, S. M., Ball, T. M., & Reiser, B. J. (1978). Visual images preserve metric spatial information: Evidence from studies of image scanning. *Journal of Experimental Psychology: Human Perception and Performance, 4,* 47–60.

Kraft, C. (1978). A psychophysical approach to air safety. Simulator studies of visual illusions in night approaches. In H. L. Pick, H. W. Leibowitz, J. E. Singer, A. Steinschneider, & H. W. Stevenson (eds.), *Psychology: From research to practice.* New York: Plenum Press.

Kroemer, K. H. E. (1987). Engineering anthropometry. In G. Salvendy (ed.), *Handbook of human factors* (pp. 154–168). New York: Wiley.

Lappin, J. S., Bell, H. H., Harm, O. O., & Kottas, B. (1975). On the relation between time and space in the visual discrimination of velocity. *Journal of Experimental Psychology: Human Perception and Performance, 1,* 383–394.

Larish, J. F., & Flach, J. M. (1990). Sources of optical information useful for perception of speed of rectilinear self-motion. *Journal of Experimental Psychology: Human Perception and Performance, 16*(2), 295–302.

Leibowitz, H. (1988). The human senses in flight. In E. Wiener & D. Nagel (eds.), *Human factors in aviation.* San Diego, CA: Academic Press.

Leibowitz, H., & Post, R. (1982). The two modes of processing concept and some implications. In J. Beck (ed.), *Organization and representation in perception.* Hillsdale, NJ: Erlbaum.

Levine, M. (1982). You-are-here maps: Psychological considerations. *Environment and Behavior, 14,* 221–237.

Loftus, G. R. (1978). Comprehending compass directions. *Memory & Cognition, 6,* 416–422.

McGreevy, M. W., & Ellis, S. R. (1986). The effect of perspective geometry on judged direction in spatial information instruments. *Human Factors*, 28(4), 439–456.

McMillan, G., Beevis, D., Salas, E., Strub, M., Sutton, R., & Van Breda, L. (1989). *Applications of human performance models to system design*. New York: Plenum Press.

Maki, R. H., Maki, W. S., & Marsh, L. G. (1977). Processing locational and orientational information. *Memory & Cognition*, 5, 602–612.

Merwin, D. H., & Wickens, C. D. (1991). Comparison of 2D planar and 3D perspective display formats in multidimensional data visualization. *Proceedings of the International Society for Optical Engineering*. Bellingham, WA: SPIE.

Milgram, S., & Jodelet, D. (1976). Psychological maps of Paris. In H. M. Proshansky, W. H. Itelson, & L. G. Revlin (eds.), *Environmental psychology*. New York: Holt, Rinehart & Winston.

Milroy, R., & Poulton, E. C. (1978). Labeling graphs for increasing reading speed. *Ergonomics*, 21, 55–61.

Morello, S. A., Knox, C. E., & Steinmetz, G. G. (1977, December). *Flight test evaluation of two electronic display formats for approach to landing under instrument conditions* (NASA TP-1085).

Nataupsky, M., & Crittendon, L. (1988). Stereo 3-D and non-stereo presentations of a computer-generated pictorial primary flight display with pathway augmentation. *Proceedings of the AIAA/IEEE 8th Digital Avionics Systems Conference*. San Jose, CA: IEEE.

Owen, D. H., & Warren, R. (1987). Perception and control of self-motion: Implications for visual simulation of vehicular locomotion. In L. S. Mark, J. S. Warm, & R. L. Huston (eds.), *Ergonomics and human factors: Recent research*, 40–70. New York: Springer-Verlag.

Peters, R. D., Yastrop, G. T., & Boehm-Davis, D. A. (1988). Predicting information retrieval performance. *Proceedings of the 32nd Annual Meeting of the Human Factors Society* (pp. 301–305). Santa Monica, CA: Human Factors Society.

Playfair, W. (1786). *Commerical and political atlas*. London: Corry.

Poulton, E. C. (1985). Geometric illusions in reading graphs. *Perception & Psychophysics*, 37, 543–548.

Ramachandran, V. S. (1988). Perceiving shape from shading. *Scientific American*, 259, 76–83.

Robbins, W. C., & Fisher, S. S. (1989). *Three dimensional visualization and display technologies*. Bellingham, WA: SPIE.

Rock, I. (1983). *The logic of perception*. Cambridge, MA: MIT Press.

Roscoe, S. N. (1968). Airborne displays for flight and navigation. *Human Factors*, 10, 321–332.

Roscoe, S. N. (1981). *Aviation psychology*. Iowa City: University of Iowa Press.

Roscoe, S. N., Corl, L., & Jensen, R. S. (1981). Flight display dynamics revisited. *Human Factors*, 23, 341–353.

Roscoe, S. N., & Williges, R. C. (1975). Motion relationships and aircraft attitude guidance displays: A flight experiment. *Human Factors*, 17, 374–387.

Rouse, W. B., & Morris, N. M. (1986). On looking into the black box: Prospects and limits in the search for mental models. *Psychological Bulletin*, 100, 349–363.

Sekular, R. (1974). Spatial vision. *Annual Review of Psychology, 25,* 195–232.

Shneiderman, B. (1987). *Designing the user interface: Strategies for effective human-computer interaction.* Reading, MA: Addison-Wesley.

Sholl, M. J. (1987). Cognitive maps as orienting schemata. *Journal of Experimental Psychology: Learning, Memory and Cognition, 13,* 615–628.

Smith, S. (1981). Exploring compatibility with words and pictures. *Human Factors, 23,* 305–316.

Sollenberger, R. L., & Milgram, P. (1989). Stereoscopic computer graphics for neurosurgery. In G. Salvendy & M. J. Smith (eds.), *Designing and using human-computer interfaces and knowledge based systems* (pp. 294–301). North Holland, Netherlands: Elsevier Science Publishers B.V.

Stevens, A., & Coupe, P. (1978). Distortions in judged spatial relations. *Cognitive Psychology, 10,* 422–437.

Stokes, A. F., Wickens, C. D., & Kite, K. (1990). *Display technology: Human factors concepts.* Warrendale, PA: Society of Automotive Engineers.

Streeter, L. A., Vitello, D., & Wonsiewicz, S. A. (1985). How to tell people where to go: Comparing navigational aids. *International Journal on Man-Machine Studies, 22,* 549–562.

Sweller, O., Chandler, P., Tierney, P., & Cooper, M. (1990). Cognitive load as a factor in the structuring of technical material. *Journal of Experimental Psychology: General, 119,* 176–192.

Thomas, J. C., & Gould, J. D. (1975). A psychological study of query by example. In *Proceedings of the National Computer Conference* (pp. 439–445). Arlington, VA: AFIPS Press.

Thorndyke, P. W. (1980, December). *Performance models for spatial and locational cognition* (Technical Report R–2676-ONR). Washington, DC: Rand.

Thorndyke, P. W., & Hayes-Roth, B. (1978, November). Spatial knowledge acquisition from maps and navigation. Paper presented at the meetings of the Psychonomic Society, San Antonio, TX.

Tole, J. R., Stephens, A. T., Harris, R. L., Ephrath, A. R. (1982). Visual scanning behavior and mental workload in aircraft pilots. *Aviation, Space, and Environmental Medicine, 53(1),* 54–61.

Tolman, E. C. (1948). Cognitive maps in rats and men. *Psychological Review, 55,* 189–208.

Tversky, B., & Schiano, D. (1989). Perceptual and cognitive factors in distortion in memory for graphs and maps. *Journal of Experimental Psychology: General, 118,* 387–398.

Vicente, K. J., Hayes, B. C., & Williges, R. C. (1987). Assaying and isolating individual differences in searching a hierarchical file system. *Human Factors, 29,* 349–359.

Vicente, K. J., & Williges, R. C. (1988). Accommodating individual differences in searching a hierarchical file system. *International Journal of Man-Machine Studies, 29,* 647–668.

Warren, R., & Riccio, G. (1985). *Visual cue dominance hierarchies: Implications for simulator design.* Paper presented at the 1985 SAE Aerospace Technology Conference and Exposition. Long Beach, CA.

Warren, D. H., Rossano, M. J., & Wear, T. D. (1990). Perception of map-environment correspondence: The roles of features and alignment. *Ecological Psychology, 2(2),* 131–150.

Way, T. C. (1988). Stereopsis in cockpit display—A part-task test. *Proceedings of the 32nd Annual Meeting of the Human Factors Society.* Santa Monica, CA: Human Factors Society.

Weinstein, L. F. (1990). The reduction of central-visual overload in the cockpit. *Proceedings of the 12th Symposium on Psychology in the Department of Defense.* Colorado Springs, CO: U.S.A.F. Academy.

Weintraub, D. J., Haines, R. F., & Randle, R. J. (1985). Head-up display (HUD) utility. II. Runway to HUD transition monitoring eye focus and decision times. In *Proceedings of the 29th Annual Meeting of the Human Factors Society* (pp. 615–619). Santa Monica, CA: Human Factors Society.

Wetherell, A. (1979). Short-term memory for verbal and graphic route information. *Proceedings of the 23rd annual meeting on the Human Factors Society.* Santa Monica, CA: Human Factors Society.

Wickens, C. D. (1990). Navigational ergonomics. In E. J. Lovesey (ed.), *Contemporary ergonomics 1990* (pp. 16–29). (Proceedings of the Ergonomics Society's 1990 Annual Conference). London: Taylor and Francis Ltd.

Wickens, C. D., Andre, A. D., & Haskell, I. (1990). Compatibility and consistency in crew station design. In E. J. Lovesey (ed.), *Contemporary ergonomics 1990* (pp. 118–122). (Proceedings of the Annual Ergonomics Society Meeting). London: Taylor & Francis Ltd.

Wickens, C. D., Haskell, I., & Harte, K. (1989). Ergonomic perspective of flight path displays. *IEEE Control Systems Magazine.*

Wickens, C. D., & Todd, S. (1990). Three-dimensional display technology for aerospace and visualization. *Proceedings of the 34th Annual Meeting of the Human Factors Society.* Santa Monica, CA: Human Factors Society.

Wickens, C. D., Todd, S., & Seidler, K. (1989). *Three-dimensional displays: Perception, implementation, and applications* (CSERIAC SOAR–89-01). Wright-Patterson AFB, OH: Armstrong Aerospace Medical Research Laboratory.

Wilhelmson, R. B., Jewett, B., Shaw, C., Wicker, L., Arrott, M., Bushell, C., Bajuk, M., & Yost, J. (1990). A study of the evolution of a numerically modeled severe storm. *International Journal of Super Computer Applications, 4(2),* 22–36.

Wise, J. A., & Debons, A. (1987). Principles of film editing and display system design. *Proceedings of the 31st annual meeting of the Human Factors Society* (pp. 121–124). Santa Monica, CA: Human Factors Society.

Woods, D. D. (1984). Visual momentum: A concept to improve the cognitive coupling of person and computer. *International Journal of Man-Machine Studies, 21,* 229–244.

Woods, D. D., Roth, E. M., Stubler, W. F., & Mumaw, R. J. (1990). Navigating through large display networks in dynamic control applications. *Proceedings of the 34th Annual Meeting of the Human Factors Society* (pp. 396–399). Santa Monica, CA: Human Factors Society.

Zenyuh, J. P., Reising, J. M., Walchli, S., & Biers, D. (1988). A comparison of a stereographic 3-D display versus a 2-D display using an advanced air-to-air format. *Proceedings of the 32nd annual meeting of the Human Factors Society.* Santa Monica, CA: Human Factors Society.

Zorpette, G. (1989). The main event. *IEEE Spectrum, 26(1),* 28.

Chapter
5

Language and Communications

The smooth and efficient operation of human-machine systems very often depends on the efficient processing of written and spoken language, whether in reading instructions, comprehending labels, or exchanging information with a fellow crew member. Not all communication is language-based—important information can be exchanged through gestures and nonverbal means—and not all instructions need to be verbal—symbols and icons are sometimes helpful. But the fundamental tie linking the material in this chapter is the role of language and *symbol* representation. The symbol, whether a letter, word, or icon, stands for something other than itself.

In this chapter, we will first consider the perception of printed language—letters, words, and sentences. We will see how these units are processed both hierarchically and automatically, and we will consider the role of context and redundancy in their perception. After considering applications to print format and code design, we will discuss similar principles in the recognition of pictures and iconic symbols. Next we will address cognitive factors involved in comprehending instructions and procedures and consider guidelines that should be followed. After a discussion of the perception of speech, we will conclude with speech communications in multiperson systems.

THE PERCEPTION OF PRINT

Stages in Word Perception

The perception of printed material is hierarchical in nature. When we read and understand the meaning of a sentence (a categorical response), we must first analyze its words. Each word, in turn, depends on the perception of letters, and each letter is itself a collection of elementary features (lines, angles, and

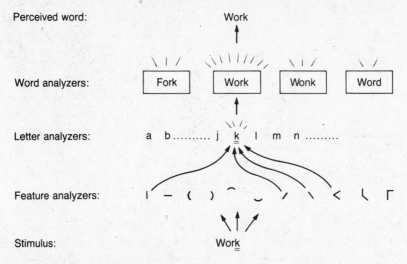

Figure 5.1 Hierarchical process of perception of the visual word *work*.

curves). These hierarchical relations are shown in Figure 5.1, which is based on Neisser (1967). There is good evidence that each level of analysis of the hierarchy (feature, letter, and word) is defined by a set of nodes. The brain manifests a unique, relatively automatic categorical response, characterized by the many-to-one mapping typical of perception. Thus in describing the perception of words we may refer to feature units, letter units, or word units. A given unit at any level will become active if the corresponding stimulus is physically presented and the perceiver has had repeated experience with the stimulus in question.

We will consider, first, the evidence provided for the unit at each level of the hierarchy and the role of learning and experience in integrating higher-level units from experience with the repeated combination of lower-level units. Then we will consider the manner in which our expectancies guide perceptual processing from the "top down." After we describe the theoretical principles of visual pattern recognition, we will then address their practical implications for system design.

The Features as a Unit: Visual Search   The features that make up letters are represented as vertical or diagonal lines, angles, and curved segments of different orientations. Examples of these are shown at the bottom of Figure 5.1. Gibson (1969) has demonstrated that the 26 letters of the alphabet can be economically created by the presence or absence of a limited subset of these features. The importance of features in letter recognition is most clearly demonstrated by the visual search task developed by Neisser, Novick, and Lazer (1964) and discussed in Chapter 3. The researchers demonstrated how the search for a target letter (e.g., *K*) in a list of nontarget "noise" letters was greatly slowed if the latter shared similar features (*N, M, X*), but it was not slowed if the features were distinct (*O, S, U*).

The Letter as a Unit: Automatic Processing   There is strong evidence that a letter is more than simply a bundle of features. That is, the whole letter is greater than the sum of its parts. To illustrate this point, an experiment by LaBerge (1973) revealed that subjects could process letters like *b* or *d* preattentively, or *automatically*, as their attention was directed elsewhere. In contrast, symbols like ↑ or ↓, made up of features that were no more complex but not familiarly associated in past experience, required focal attention in order to be processed. The concept of automaticity — processing that does not require attentional resources — is a key one in understanding human skilled performance. We will encounter it again in our treatment of training in Chapter 6 and again when we discuss attention in Chapter 9. Here we will focus on the development of automaticity in language processing.

What produces the automaticity that we use to process letters and other familiar symbols? Clearly familiarity and extensive perceptual experience is necessary; but research by Schneider and Shiffrin (1977) and Schneider, Dumais, and Shiffrin (1984) suggests that experience is not sufficient. In addition, the symbols must be *consistently* mapped to the same response. Inconsistent responding, when a letter (or other symbol) is sometimes relevant and sometimes not, will be less likely to develop automaticity.

To illustrate this role of consistent mapping in letter processing, Shiffrin, Schneider, Fisk, and their colleagues have done a series of studies in which subjects search an array of letters, in a version of Neisser's visual search task discussed in Chapter 3. Some letters are always consistently responded to and categorized as targets (i.e., by pressing a key if they appear), for sometimes thousands of trials. Other letters are sometimes targets but at other times "distractors" to be ignored. After such training, the consistently mapped target letters reach a special status of automaticity, in which they are perceived without attention and even appear to jump out of the printed page in a way that does not occur with the inconsistent by mapped letters. As we saw in Chapter 2, this automatic processing can produce signals that are much more immune to the vigilance decrement (Schneider & Fisk, 1984).

Subsequent research has expanded the list of categorization processes for which automaticity can be developed through consistency. This research suggests that a higher-level automaticity can be developed that is not bound to repeated exposures to the same physical stimuli. For example, Schneider and Fisk (1984) showed how consistently responding to members of a category (like vehicles) can show the features of automatic processing (fast and preattentive) when each category member is presented, even if that member itself has not been seen frequently. Another example of higher-level automaticity is produced when people consistently respond to *rules*, such as "the higher of two digits is the target." With consistent practice, automatic processing to a given digit will occur if it satisfies the rule, even if that digit has not been seen before in the context of the experiment (Fisk, Ackerman, & Schneider, 1987; Kramer, Strayer, & Buckley, 1990).

The Word as a Unit: Word Shape   Under some circumstances words may be perceived through the analyses of their letters, just as letters were perceived through the analysis of their features. Yet there is also evidence that familiar words can be perceived as units, just as LaBerge's (1973) experiment provided evidence that letters were perceived as units because of the familiarity. Thus the pattern of full-line ascending letters (*h, b*), descenders (*p, g*), and half-line letters (*e, r*) in a familiar word such as *the* forms a global shape that can be recognized and categorized as *the* even if the individual letters are obliterated to such an extent that each is illegible. Broadbent and Broadbent (1977, 1980) propose that the mechanism of spatial frequency analysis, is responsible for this crude analysis of word shape.

The analysis based on word shape is more holistic in nature than the detailed feature analysis described above. The role of word shape, particularly with such frequent words as *and* and *the* for which unitization is likely to have occurred, seems to be revealed in the analysis of proofreading errors (Haber & Schindler, 1981; Healy, 1976). Haber and Schindler had subjects read passages for comprehension and proofreading at the same time. They observed that misspellings of short, function words of higher frequency (*the* and *and*) were difficult to detect. The role of word shape in these shortcomings was suggested because errors in these words were concealed most often if the letter change that created the error was one that substituted a letter of the same class (ascender, descender, or half-line) and thereby preserved the same word shape. An example would be *anb* instead of *and*. If all words were only analyzed letter by letter, these confusions should be as hard to detect in long words as in short ones. As Haber and Schindler observe, they are not. Corcoran and Ween-ing (1967) noted that for words that were longer and less frequent (the two variables are, of course, highly correlated), acoustic or phonetic factors play a more prominent role than visual ones in proofreading errors. In this case, misspellings are concealed if the critical letters are not pronounced in the articulation of the word.

Top-Down Processing: Context and Redundancy

In the system shown in Figure 5.1, "lower-level" units (features and letters) feed into "higher-level" ones (letters and words). Sometimes lower-level units may be bypassed, if higher-level units are unitized. This process then is some-times described as bottom-up or data-driven processing. There is also strong evidence that much of our perception proceeds in a "top-down," context-driven manner (Lindsay & Norman, 1972). More specifically, in the case of reading, hypotheses are formed concerning what a particular word should be, given the context of what has appeared before, and this context enables our perceptual mechanism to guess the nature of a particular letter within that word, even before its bottom-up feature-to-letter analysis may have been com-pleted. Thus the ambiguous word in the sentence "Move the switch to the rxxxx" can be easily and unambiguously perceived, not because of its shape or its features but because the surrounding context limits the alternatives to only a few (e.g., *right* or *left*) and the apparent features of the first letter eliminate all but the first alternative).

In a corresponding fashion, top-down processing can work on letter recognition. Knowledge of surrounding letters may guide the interpretation of ambiguous features, as in the two words "THE CHT." The middle letters of the two words are physically identical. The features are ambiguously presented as parallel vertical lines or as converging lines. Yet the hypotheses generated by the context of the surrounding letters quite naturally force the two stimuli into two different perceptual categories. Top-down processing of this sort, normally of great assistance in reading, can prove to be a source of considerable frustration in proofreading, in which allowing context to fill in the gaps is exactly what is *not* required. All words must be analyzed to their full-letter level to perform the task properly.

The foundations of top-down processing were established in the discussion of redundancy and information in Chapter 2. Top-down processing, in fact, is only possible (or effective) because of the contextual constraints in language that allow certain features, letters, or words to be predicted by surrounding features, letters, words, or sentences. When the redundancy of a language or a code is reduced, the contribution to pattern recognition of top-down, relative to bottom-up, processing is reduced as well. This trade-off of top-down, context-driven processing governed by redundancy against bottom-up, data-driven processing governed by sensory quality is nicely illustrated in an investigation by Tulving, Mandler, and Baumal (1964). They presented subjects with sentences of the form "I'll complete my studies at the ___" and displayed the final word for very brief durations, producing a degraded sensory stimulus. The experimenters could adjust both the duration (and therefore the quality) of the stimulus and the amount of prior word context among eight, four, and zero letters. The results, shown in Figure 5.2, illustrate the almost perfect trade-off in recognition accuracy between stimulus quality and redundancy. As one of these variables increases, the other may be degraded to maintain a constant level of recognition performance.

In addition to redundancy, there is evidence for a second top-down mechanism, in which the letters within a word mutually facilitate one another's analysis, so that a letter that appears in a word can sometimes be processed more rapidly than the letter by itself. This *word superiority effect* (Reicher, 1969) has important implications of models of how people read (Rumelhart & McClelland, 1986), and its implications for engineering psychology are straightforward. The letters in a word are processed faster than a similar number of unrelated letters.

The pattern of analysis of word perception described up to this point may be best summarized by observing that top-down and bottom-up processing are continuously ongoing at all levels in a highly interactive fashion (Navon, 1977; Neisser, 1967; Rumelhart, 1977). Sensory data suggest alternatives, which in turn provide a context that helps interpret more sensory data. This interaction is represented schematically in Figure 5.3. The conventional bottom-up processing sequence of features to letters to words is shown by the upward flowing arrows in the middle of hierarchy. The dashed lines on the left indicate that automatic unitization at the level of the letter and the common word may occur

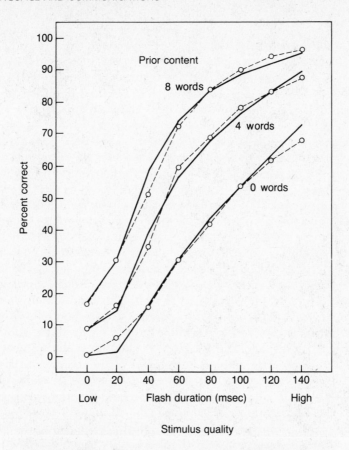

Figure 5.2 Trade-off between bottom-up and top-down processing illustrated by experiment of Tulving, Mandler, and Baumal. (*Source:* E. Tulving, G. Mandler, and R. Baumal, "Interaction of Two Sources of Information on Tachistoscopic Word Recognition." *Canadian Journal of Psychology, 18,* (1964), p. 66. Copyright 1964 by the Canadian Psychological Association. Reproduced by permission.)

as a consequence of the repeated processing of these units. Thus unitization may identify a blurred word by word shape alone, even when features, letters, and context aren't available (Broadbent & Broadbent, 1980). Unitization does not necessarily replace or bypass the sequential bottom-up chain but operates in parallel. Represented on the right of Figure 5.3 are the two forms of top-down processing: those that reduce alternatives through context and redundancy (solid lines) and those that actually facilitate the rate of lower-level analysis (dotted lines).

Although all these factors may be operating simultaneously, two primary dimensions underlie the relative importance of one or the other. The first contrasts sensory quality against context and redundancy, as these trade off in bottom-up versus top-down processing. The second contrasts the relative con-

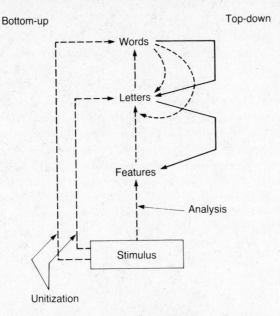

Figure 5.3 Bottom-up processing (analysis and utilization) versus top-down processing.

tribution of higher-level unitization to hierarchical analysis in bottom-up processing. This contribution is determined by the familiarity and consistent mapping of the lower-level units. These two dimensions will be important as a framework for later discussion of the applications of pattern recognition.

Reading: From Words to Sentences

The previous analysis has focused on the recognition of words. Yet more often than not word recognition is reading a string of words in a sentence. We have, of course, implicitly suggested that sentences must be processed to provide the higher-level context that in turn provides top-down processing for word recognition. In normal reading, sentences are processed by visually scanning across the printed page. Scanning occurs by a series of fixations, interspersed by the discrete *saccadic* eye movements discussed in Chapter 3. The fixation has a minimum duration of around 200 msec. During each fixation, as we have seen, there appears to be some degree of parallel processing of the letters within the fixated word. Whereas the meaning of an isolated word can normally be determined during fixations as short as the minimum fixation value of 200 msec, fixations made during continuous reading are sometimes considerably longer (Just & Carpenter, 1980; McConkie, 1983). The extra time is required to integrate the meaning of the word into the ongoing sentence context as well as to process more difficult words. Both the absolute duration of fixations and the frequency of fixations along a line of text vary greatly with the difficulty of the text (McConkie, 1983).

Although a given word is fixated, information understood from the preceding words provides context for top-down processing. A series of investigations conducted by McConkie and his colleagues (see McConkie, 1983) concerned how much information is actually processed from different locations along a line of print during fixations when a given letter is in the center of foveal vision. McConkie's research suggests that different kinds of information may be processed at different regions. As far out as 10 to 14 characters to the right of the fixated letter, very global details pertaining to word boundaries may be perceived for the purpose of directing the saccade to the next fixation. Some processing of word shape may occur somewhat closer to the fixated letter. Individual letters, however, are only processed within a fixation span of roughly 10 letters: 4 to the left and 6 to the right. This span itself is not fixed but varies in width according to the size of the word currently fixated and the position of the fixated letter within the word.

McConkie's research on the use of visual information in normal reading has employed an ingenious technique that we may describe as fixation-guided text display (McConkie & Raynor, 1974). In this technique, subjects read a computer-displayed text while the computer monitors the exact point of fixation of the eye on the printed line. The computer proceeds to "scramble" the text a certain number of characters, N, to the right of fixation. While the eye makes a saccade to the right, the scrambled text is rearranged to its coherent level, and the characters that are now N to the right of the new fixation point are scrambled. McConkie and Raynor found that if N was quite large, subjects were not even aware of the scrambling and they could read at a normal rate. However, when N reached a value of less than 10 (about 2.5 degrees of visual angle to the right, or roughly two words), their reading speed was disrupted. This distance indexes the span of processed peripheral information in normal reading.

In summarizing the results of his research, McConkie (1983) offers a conclusion that is quite consistent with the word superiority effect described previously. It is that the perception of letters within a word are heavily word-driven. When a word lies at the center of fixation, its letters are processed cooperatively and guide the semantic interpretation of the sentence. Letters not in the fixated word may indeed be processed, but the result do not seem to play an active role in comprehension.

APPLICATIONS OF UNITIZATION AND TOP-DOWN PROCESSING

The research on recognition of print is, of course, applicable to system design in contexts in which warning signs are posted or maintenance and instruction manuals are read. These contexts will be discussed later in the chapter. Furthermore, the acquisition of verbal information from computer terminals and video displays is also an area of increasing importance. In designing such displays the goal is to present information in such a manner that it can be read rapidly and accurately. In addition, certain critical items of information (one's own identification code, for example, or critical diagnostic or warning information) should be recognized automatically, with a minimal requirement to invest conscious processing. There are two broad classes of practical implications of the research that generally align themselves with the two dimensions of pattern recognition described: applications that capitalize on unitization and applications that are related to the trade-off between top-down and bottom-up processing.

Unitization

Training and repetition lead to automatic processing. Some of this training is the consequence of a lifetime's experience (e.g., recognition of letters), but as LaBerge (1973) and Schneider and Shiffrin (1977) clearly demonstrated, the special status of automatic processing of critical key targets can also be developed within a relatively short period of practice. These findings suggest that when a task environment is analyzed, it is important to identify critical signals (and these need not necessarily be verbal) that should always receive immediate priority if they are present. Training regimes should then develop their automatic processing. In such training, operators should be presented with a mixture of the critical signals and others and should always make the same consistent responses to the critical signals. A good discussion of applications of automatic processing to training is presented by Schneider (1985).

In this regard, there would appear to be a distinct advantage in calling attention to critical information by developing automatic processing rather than by simply increasing the physical intensity of the stimulus. First, as we will describe in the discussion of alarms in Chapter 12, loud or bright stimuli may be distracting and annoying and may not necessarily ensure a response. Second, physically intense stimuli are intense to all who encounter them. Stimuli that are subjectively intense by virtue of automatic processing may be "personalized," to alert only those who require it.

At any level of perceptual processing it should be apparent that the accuracy and speed of recognition will be greatest if the displayed stimuli are presented in a physical format that is maximally compatible with the visual representation of the unit in memory. For example, the prototypal memory units of letters and digits preserve the angular and curved features as well as the horizontal and vertical ones. As a consequence, "natural" letters that are not distorted into an orthographic grid should be recognized with greater facility than dot matrix letters or letters formed with only horizontal and vertical strokes. These suggestions were confirmed in recognition studies comparing digits constructed in right-angle grids with digits containing angular and curved strokes (e.g., Ellis & Hill, 1978; Plath, 1970). Ellis and Hill, for example, found that five-digit sequences were read more accurately when presented as conventional numerals than in a seven-segment right-angled format. This advantage was enhanced at short exposure durations, as might be typical of time-critical environments.

A similar logic applies to the use of lowercase print in text. Since lowercase letters contain more variety in letter shape, there is more variety in word shape and so a greater opportunity to use this information as a cue for holistic word-shape analysis. Tinker (1955) found that subjects could read text in mixed case better than in all capitals. However, the superiority of lowercase over uppercase letters appears to hold only for printed sentences. For the recognition of isolated words, the stimuli appear to be better processed in capitals than in lowercase (Vartabedian, 1972). These findings would seemingly dictate the use of capital letters in display labeling (Grether & Baker, 1972), where only one or two words are required, but lowercase in longer segments of verbal material.

As a result of unitization, words are both perceived faster and understood better than are abbreviations of acronyms. Therefore, it seems that words should always be used instead of abbreviations, except when space is at an absolute premium. Norman (1981) discusses the difficulties that naive users of computer text-editing systems encounter when confronted with abbreviations like *ln* for *link* or *cat* for *concatenate*. He asks why the full word should not be employed instead. The cost of a few extra letters is surely compensated for by the benefits of better understanding and fewer blunders. Where abbreviations are used, Norman suggests that at a minimum relatively uniform abbreviating principles should be employed (i.e., all abbreviations of common length) and that the abbreviated term should be as logical and meaningful to the user as possible. Moses & Ehrenreich (1981) have summarized an extended evaluation of abbreviation techniques and conclude that the most important principle is to employ *consistent* rules of abbreviation. In particular, they find that truncated abbreviations, in which the first letters of the word are presented, are processed better than contracted abbreviations, in which letters within the word are deleted. For example, *reinforcement* would be better abbreviated by *reinf* than by *rnfnt*. This finding makes sense in terms of our discussion of reading since truncation preserves at least part of any unitized letter sequence. Ehrenreich (1982) concludes that whatever rule is used to generate abbreviations, rule-generated abbreviations are, at least, always superior to subject-generated ones, in which the operator decides the best abbreviations for a given term.

A related recommendation derives from the unitizing influence of gaps between words (Wickelgren, 1979). It appears that this benefit may carry over to the processing of unrelated material such as alphanumeric strings by defining high-order visual "chunks" (see Chapter 6). Klemmer (1969) argues that there is an optimum size of such chunks for encoding unrelated material. In Klemmer's experiment strings of digits were to be entered as rapidly as possible into a keyboard. In this task the most rapid entry was achieved when the chunks between spaces were three or four digits long. Speed declined with either smaller or larger groups. These findings have important implications for deciding on formats for various kinds of displayed material — license plates, identification codes, or data to be entered on a keyboard.

Context-Data Trade-offs

The distinction between bottom-up and top-down processing is important for the design of text displays and code systems. An example of the trade-off of design considerations between bottom-up and top-down processing can be seen when a printed message is to be presented in an aircraft display in which space is clearly at a premium. Given certain conditions of viewing (high stress or vibration), the sensory qualities of the perceived message may be far from optimal. A choice of designs is thereby offered as shown in Figure 5.4: (1) Present large print, thus taking advantage of improving the bottom-up sensory quality but restricting the number of words that can be viewed simultaneously on the screen (and thereby limiting top-down processing). (2) Present more words in smaller print and enhance top-down processing at the expense of bottom-up processing. Naturally the appropriate text size will be determined by an evaluation of the relative contribution of these two factors. For example, if there is more redundancy in the text, smaller text size is indicated. However, if the display contains random strings of digits, there is no opportunity for top-down processing, and larger presentation of fewer digits is advised. If the display or viewing quality is extremely poor, larger size is again suggested. It is essential that the system designer be aware of the factors that influence the trade-off between data-driven and context-driven processing, which determines the optimum point on the trade-off to be selected.

Top-down processing may also be greatly facilitated through the simple technique of restricting a message vocabulary. With fewer possible alternatives

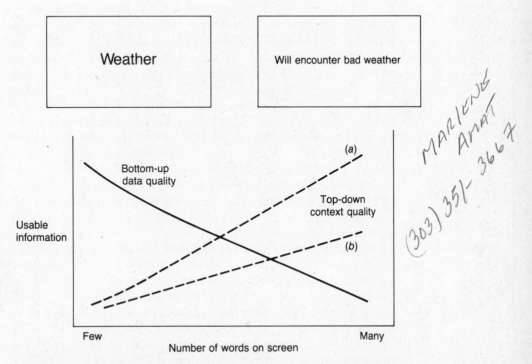

Figure 5.4 An illustration of the trade-off between top-down and bottom-up processing in display of limited size. The two dashed lines represent different amounts of contextual redundancy: (*a*) high context of printed text; (*b*) low context of isolated word strings.

to consider, top-down hypothesis forming becomes far more efficient. This technique is a major factor behind the strict adherence to standard terminology in many message-routing centers or communication systems (Kryter, 1972).

Code Design: Economy versus Security

The trade-off between top-down and bottom-up processing is demonstrated by the fact that messages of greater probability (and therefore less information content formally) may be transmitted with less sensory evidence. We have already encountered one example in the compensatory relation between d' and beta in signal detection theory (Chapter 2). As a signal becomes more frequent, offering less information, and beta is lowered, it can be detected at lower sensitivity (i.e., with less evidence and lower d'). It is fortunate that the trade-off in human performance corresponds quite nicely with a formal specification of the optimum design of codes, referred to as the *Shannon-Fano* principle (Sheridan & Ferrell, 1974). In designing any sort of code or message system in which short strings of alphanumeric or symbolic characters are intended to convey longer ideas, the Shannon-Fano principle dictates that the most efficient, or *economic*, code will be generated when the length of the physical message is proportional to the information content of the message. The principle is violated if all messages are of the same length. Thus high-probability, low-information messages should be short, and low-probability ones should be longer. For example, if the four events that make up a code and their associated probabilities are A (0.4), B (0.4), C (0.1), and D (0.1), a code using binomial symbols that assigns $A = 00$, $B = 01$, $C = 10$, and $D = 11$ violates the Shannon-Fano principle. One that assigns $A = 0$, $B = 1$, $C = 01$, and $D = 10$ does not.

It is interesting to observe that any natural language roughly follows the Shannon-Fano principle. Words that occur frequently (*a*, *of*, or *the*) are short, and ones that occur rarely tend to be longer. This relation is known as Zipf's law (Ellis & Hitchcock, 1986). The relevant finding from the viewpoint of human performance is that adherence to such a code reinforces our natural tendencies to expect frequent signals and therefore requires less sensory evidence for those signals to be recognized. For example, in an efficient code designed to represent engine status, the expected normal operation might be represented by N (one unit), whereas *HOT* (three units) should designate a less-expected, lower-probability overheated condition. Violations of the Shannon-Fano principle are observed in a coding system in which all events, independent of their probability, are specified by a message of the same length. Such a violation was evident in the Navy's system of computerized maintenance records. The operator servicing a malfunctioning component was required to enter a nine-digit malfunction code on a computerized form. This code was of uniform length whether the malfunction was highly probably (overheating engine) or extremely rare.

Numerous other properties of a useful code design have been summarized by Bailey (1989). For example, a code, like an abbreviation or symbol, should be meaningfully related to its referent. Alphabetic codes, because of the greater richness of the alphabet, generally meet this criterion better than *numeric* ones. Also, code strings should be relatively short (fewer than 6 characteristics; The New Jersey driver's license code has 15 digits). It is unwise to make codes a digit or two longer than currently necessary in order to anticipate an expanding population of code vocabulary. For example, a code of the form 00214682, in which the first two 0's anticipate a 100-fold growth in the vocabulary, is not advised. Bailey (1989) points out that the extra processing required of the two place-holding digits will be quite likely to produce a substantial number of copying, memory, and reading errors over the course of the codes' use.

In the context of information theory, there is a second critical factor in addition to efficiency that must be considered when a code or message system is designed. This is security. The security factor illustrates again the trade-offs frequently encountered in human engineering. The Shannon-Fano principle is intended to produce maximum processing efficiency, which is compatible with processing biases. However, it may often be the case that relatively high-frequency (and therefore short) messages of a low information content are in fact very *important*. It is therefore essential that they be perceived with a high degree of security. In these instances the principle of economy should be sacrificed either by enhanced data quality or by including redundancy, as discussed in Chapter 2. Redundancy is accomplished by allowing a number of separate elements of the code to transmit the same information. For voice communications, the use of a communications-code alphabet in which *alpha, bravo,* and *charlie* are substituted for *a, b,* and *c* is a clear example of such redundancy for the sake of security. The second syllable in each utterance conveys information that is highly redundant with the first. Yet this redundancy is advantageous because of the need for absolute security (communication without information loss) in the contexts in which this alphabet is employed. It is possible to look on the trade-off between efficiency and security in code design as an echo of the trade-off discussed in Chapter 2 between maximizing information transmission and minimizing information loss. Certain conditions (orthogonal dimensions and adherence to the Shannon-Fano principle) will be more efficient, and other conditions emphasizing redundancy will be more secure.

RECOGNITION OF OBJECTS

Top-Down and Bottom-Up Processing

The combination of bottom-up and top-down processing involved in word perception characterizes the perception of everyday objects as well. For example, just as letters are perceived, in part, through feature analysis, so Biederman (1987) has proposed that humans recognize objects in terms of combinations of a small number of basic features, which consist of simple geometric

CROSS SECTION

Geon	Edge Straight S Curved C	Symmetry Rot & Ref ++ Ref + Asymm −	Size Constant ++ Expanded − Exp & Cont −−	Axis Straight + Curved −
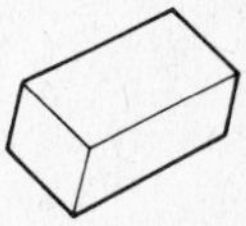	S	++	++	+
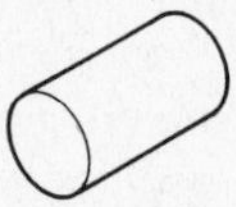	C	++	++	+
	S	+	−	+
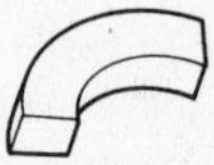	S	++	+	−
	C	++	−	+
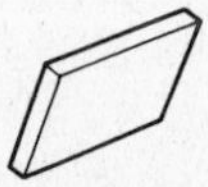	S	+	+	+

Figure 5.5 Proposed set of primitive geometric features, or "geons," used in object recognition. The attributes or dimensions that distinguish each geon from others in the list are shown to the right. (*Source: I. Biederman, "Human Image Understanding." Computer Vision, Graphics and Image Processing, 32* (1985), 29–73. Copyright 1985 by Academic Press. Reprinted by permission.)

solids (e.g., straight and curved cylinders and cones). An example is shown in Figure 5.5. Biederman's theory suggests that the designers of three-dimensional graphics displays might well capitalize on these basic features, by fabricating objects that can be easily recognized without needing to incorporate excessive detail.

The role of top-down processing in object recognition is equally important. In a procedure analogous to that described by Tulving, Mandler, and Baumal (1964), Palmer (1975) presented subjects with a context setting display of a visual scene (e.g., a kitchen). This was followed by tachistoscopic presentation of an object that could be appropriate in the context of that scene (a loaf of bread), appropriate in physical form but out of context (a home mailbox, which looks like a loaf of bread but does not belong in the kitchen), or appropriate in neither form nor context (a drum). Palmer found that the visual recognition threshold was predicted directly by the amount of contextual appropriateness from the loaf of bread (highest appropriateness, lowest threshold) to the mailbox to the drum (lowest appropriateness, highest threshold).

In a related study, Biederman, Mezzanotte, Rabinowitz, Francolin, and Plude (1981) demonstrated top-down processing in rapid photo interpretation of objects. The subjects had to detect objects in a complex visual scene from a rapid 200-msec exposure. The objects were in either appropriate or inappropriate contexts, where appropriateness was defined in terms of several expected properties of the objects (e.g., the object must be supported, and it should be of the expected size given the background). The researchers found that if the object was appropriate, it was detected equally well at visual angles out to 3 degrees of parafoveal vision. If it was not, performance declined rapidly with increased visual angle from fixation.

The role of familiarity in object perception demonstrated by these experiments suggests that familiar objects are coded and represented symbolically (in terms of abstract words and ideas) as well as in analog spatial form. We must assume that presentation of familiar objects rapidly activates both visual and symbolic codes in some cooperative manner (see Chapter 6). Evidence for the speed and efficiency with which the symbolic meaning of objects can be encoded and interpreted is provided by Potter and Faulconer (1975), who found that pictures could be understood at least as rapidly as words.

Pictures and Icons

The fact that pictures can be recognized as rapidly as words leads to the potential application of pictorial symbols or icons to represent familiar concepts. Highway symbols and signs in public buildings are familiar examples of pictures being used to represent or replace words. So increasingly are icons in computer displays (Figure 5.6), where their value over words in allowing rapid processing has been demonstrated (Camacho, Steiner, & Berson, 1990). In spite of the rapid speed of processing pictures that can be shown in ideal circumstances (Potter & Faulconer, 1975) and their status as an international language, caution must be advised against the use of icons as labels for two reasons: legibility and interpretation.

The *legibility* concern relates to the fact that icons and symbols are not always viewed under ideal conditions (neither, of course, are words). Thus it becomes important for symbols to be highly discriminable from one another. The key to discriminability under suboptimal viewing conditions lies not in the fine detailed features but rather in the global shape of the symbol as revealed by spatial frequency analysis typical of that used for word shape (Broadbent & Broadbent, 1980). Thus, if "Previous Page" and "Next Page" in Figure 5.6 were differentiated only by the direction of the arrow, confusion would be

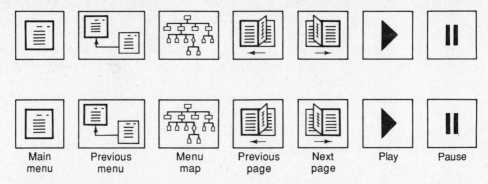

Figure 5.6 Examples of typical icons for a computer display. (*Source:* Brenna and Whitten, 1987.)

likely when viewing under degraded conditions. The difference in the global shape, resulting from the angle of the page, reduces the likelihood of confusion.

The *interpretation* concern relates to the identity of the symbol itself. This concern in turn can be partitioned into two questions: What is depicted on the symbol and what does that depiction mean? A symbol may very clearly depict a recognizable object but remain totally ambiguous concerning the meaning of the object in the context. For example, a clearly defined arrow could be interpreted as pointing to a given region on the display or as commanding an action or movement in a given direction. It is for this reason that Brems and Whitten (1987) have cautioned against the heavy use of icons that are not reinforced by redundant verbal labels. This theme, the redundancy of verbal with pictorial or spatial material, will be repeated later in this chapter as we discuss the presentation of more detailed sequences of information regarding procedures and instructions. Yet redundancy of presentation itself may have a cost. Redundant labels occupy more space and can result in a more cluttered display. Similarly, options to produce better-quality data for superior bottom-up processing may also be costly. These are the sorts of trade-offs that always confront the human factors engineer.

COMPREHENSION

Whether presented by voice or by print, words are normally combined into sentences whose primary function is to convey a message to the receiver. So far our discussion has considered how the meaning of the isolated symbols, words, and word combinations is extracted. In this section we will consider properties of the words themselves, and not just their physical representation, that influence the ease of comprehension (Broadbent, 1977). Instructions and procedures vary dramatically in the ease with which they may be understood. Chapanis (1965), in a delightful article entitled "Words Words Words," provides several classic examples of instructions that are poorly if not incomprehensibly phrased. Some examples are shown in Figure 5.7.

PLEASE

PLEASE
WALK UP ONE FLOOR
WALK DOWN TWO FLOORS
FOR IMPROVED ELEVATOR SERVICE

What it says: (13 words)

IF YOU ARE ONLY GOING
UP ONE FLOOR
OR
DOWN TWO FLOORS
PLEASE WALK
(IF YOU DO THAT WE'LL ALL HAVE BETTER ELEVATOR SERVICE)

What it means: (24 words)

TO GO UP ONE FLOOR
OR
DOWN TWO FLOORS
PLEASE WALK

Would this do? (11 words)

NOTICE
THIS RADIO USES A LONG LIFE PILOT LAMP THAT MAY STAY ON FOR A SHORT TIME IF RADIO IS TURNED OFF BEFORE RADIO WARMS UP AND STARTS TO PLAY

What it said: (29 words)

NOTICE
DON'T WORRY IF THE PILOT LAMP SHOULD STAY ON FOR A LITTLE WHILE AFTER YOU TURN THE RADIO OFF

What is meant: (19 words)

NOTICE
IF YOU TURN THE RADIO ON, AND THEN OFF RIGHT AWAY, THE PILOT LAMP MAY STAY ON FOR A LITTLE WHILE

This is more accurate: (21 words)

NOTICE
THE PILOT LAMP SOMETIMES STAYS ON FOR A LITTLE WHILE AFTER YOU TURN THE RADIO OFF

But would this do? (16 words)

You can Dial LOCAL and TRUNK CALLS

LOCAL CALLS   cost 3d. for 3 mins (cheap rate 6 mins)

for **LONDON** exchanges, dial the first three letters of the exchange name followed by the number you want

LONDON exchanges are shown in the DIALLING CODE BOOKLET and Telephone Directory with the first three letters in heavy capitals

For the following exchanges, dial the code shown followed by the number you want

	Code		Code		Code
Byfleet	BY	Hoddesdon	HO3	St. Albans	LN
Crayford	CY	Hornchurch	HX	Slough	SL
Danford	DA	Ingrebourne	IL	Staines	SW
Erith	ET	Leatherhead	LE7	Uxbridge	UX
Farnborough (Kent)	FN	Northwood	NL	Waltham Cross	WS
Garston	GR7	Orpington	MM	Walton on Thames	WT
Gerrards Cross	GE4	Potters Bar	PR	Watford	WA
Hatfield	HL6	Romford	RO	Weybridge	WR

You can DIAL other LOCAL CALLS shown in the DIALLING CODE BOOKLET (inside the A-D directory)

TRUNK CALLS

For BIRMINGHAM exchanges:	Dial 021	Then the first three letters of the exchange name
EDINBURGH	031	then the number
GLASGOW	041	e.g. for Birmingham Midland 729)
LIVERPOOL	051	dial 021 MID 729)
MANCHESTER	061	

You can DIAL other TRUNK CALLS shown in the DIALLING CODE BOOKLET (inside the A-D directory)

To make a call first check the code (see above)

USE 3d bits, 6d or l/- coins (Not Pennies)

HAVE MONEY READY, but do not try to put it in yet

LIFT RECEIVER, listen for dialling tone and

DIAL — see above — then wait for a tone

Ringing tone (burr-burr) changes, when the number answers, to

Pay tone (rapid pips) — Now PRESS in a coin and speak

(Coins cannot be inserted until first pay tone is heard)

Engaged tone (slow pips) — try again later

N.U. tone (steady note) — check number and redial

INSERT MORE MONEY to prolong the call
at any time during conversation
at once if pay tone returns

Remember—Dial first and when you hear pay tone (rapid pips) press in a coin

For Directory Enquiries dial DIR   For other Enquiries dial INF
For OTHER SERVICES and CALL CHARGES see the DIALLING CODE BOOKLET (inside the A-D directory)

For the Operator – dial 100

Figure 5.7 Three examples of poorly worded instructions. (*Source:* "Words Words Words" by A. Chapanis. Reprinted with permission from *Human Factors* (vol. 7, 1965), pp. 6, 7, 9. Copyright 1965 by The Human Factors Society, Inc. All rights reserved.)

Wordy phrases that are difficult to understand are often encountered in legal documents and instructions. Consider the following set of instructions issued to a deliberating jury:

> You must not consider as evidence any statement of counsel made during the trial; however, if counsel for both parties have stipulated to any fact, or any fact has been admitted by counsel, you will regard that fact as being conclusively proven as to the party or parties making the stipulation or admission. As to any question to which an objection was sustained, you must not speculate as to what the answer might have been or as to the reason for the objection. You must not consider for any purpose any offer of evidence that was rejected or any evidence that was stricken out by the court; such matter is to be treated as through you have never known of it. (Hastie, 1982)

Now consider a modification that attempts to present the same information in a more understandable format.

> As I mentioned earlier, it is your job to decide from the evidence what the facts are. Here are . . . rules that will help you decide what is, and what is not, evidence.
>
> 1. *Lawyer's statement.* Ordinarily, any statement made by the lawyers in this case is not evidence. However, if all the lawyers agree that some particular thing is true, you must accept it as truth.
> 2. *Rejected evidence.* At times during this trial, items or testimony were offered as evidence, but I did not allow them to become evidence. Since they never became evidence, you must not consider them.
> 3. *Stricken evidence.* At times, I ordered some piece of evidence to be stricken, or thrown out. Since that is no longer evidence, you must ignore it, also. (Reed, 1982)

Using this modified text, jurors' comprehension rate was improved by nearly 50 percent.

In writing instructions or procedures that are easy to understand, such as the rewritten set above, it is often sufficient to follow a set of straightforward, common sense principles similar to those outlined by Bailey (1989; see also Dumas & Redish, 1986):

1. State directly what is desired without adding excess words.
2. Use familiar vocabulary.
3. Ensure that all information is explicitly stated, leaving nothing to be inferred.
4. Number and physically separate the different points to be made (or procedural steps to be taken), as has been done here, rather than combining them in a single narrative.
5. Highlight key points or words.

The writing of understandable procedures and instructions may be aided by a number of readability formulas (Bailey, 1989). These formulas take into account such factors as the average word and sentence length to make quantitative assessments of the likelihood that a passage will be correctly understood by a readership with a given educational level. However, useful and necessary as these guidelines may be, they do not consider some other important characteristics of comprehension that are directly related to principles in cognitive psychology and information processing. We will consider four general categories: context, command versus status information, linguistic factors, and the role of pictures.

Context

The important role of context in comprehension is to influence the perceiver to encode the material in the manner that is intended. This is an effect that was considered in two different forms in Chapter 2: the influence of probability on response bias and the influence of context on information. Furthermore, context should provide a framework on which details of the subsequent verbal information may be hung. Bransford and Johnson (1972) have demonstrated the dramatic effect that the context of a descriptive picture or even a thematic title can exert on comprehension. In their experiment, the subjects read a series of sentences that described a particular scene or activity (e.g., the procedures for washing cloths). The subjects were asked to rate the comprehensibility of the sentences and were later asked to recall them. Large improvements in both comprehensibility and recall were found for subjects who had been given a context for understanding the sentences prior to hearing them. This context was in the form of either a picture describing the scene or a simple title of the activity. For those subjects who received no context, there was little means of organizing or storing the material, and performance was poor.

For context to aid in recall or comprehension, however, it should be made available before the presentation of the verbal material (Bower, Clark, Lesgold, & Winzenz, 1969). The important benefits of prior context would account for the results of an investigation by Norcio (1981) of computer program documentation (i.e., commentary on the meaning of the various logical statements). He noted that documentation helped comprehension of the program only if it was given at the beginning and not when it was interspersed throughout. Like a good filing system, context can organize material for comprehension and retrieval if it is set up ahead of time. Even a highly organized filing scheme will be of little assistance if it is made available only after the papers are dumped loosely into a drawer.

Command versus Status

Another issue in the delivery of instructions is related to the distinction between status and command displays. Should a display simply inform the operator of an existing *status*, such as the aircraft attitude display indicators in the previous chapter (Figure 4.8) or verbal instructions ("Your speed is too high"), or should a display *command* an action to be carried out ("Lower your speed")?

Arguments can be made on both sides of the issue, and the data are not altogether consistent. For example, in designing flight path displays to help pilots recover from unusual attitudes, Taylor and Selcon (1990) found that a display that told the pilot what direction to fly in order to recover was more effective than one that showed the aircraft's current status. Barnett (1990) observed no difference in performance on a decision-aiding task between status and command (actually recommended procedure) displays. Finally, a study by Crocoll and Coury (1990) obtained results that favored status displays. When the information was completely reliable in a decision-aiding task, there were no differences in performance; but when information was not totally reliable (as is often the case with automated decision aids, as we will see in Chapter 7 and again in Chapter 12), performance suffered far more when unreliable commands were given than when unreliable status reports were provided.

What conclusions can be drawn from these studies? First, it is probably true that under conditions of high stress and time pressure, a command display is superior to a status display, as the latter will require an extra cognitive step to go from what is to what should be done. Second, Crocoll and Coury's (1990) results suggest that these guidelines might be modified if time pressure is relaxed and/or the source of the status or command information is not fully to be trusted. Finally, as is so often the case in human performance, a strong argument can be made for *redundancy*, presenting both status and command information. This is an approach sometimes advocated in the Threat Collision Avoidance System (TCAS) being introduced in modern aircraft. A command tells the pilot what to do to avoid a collision ("pull up"), while a status display presents the relative location of the threatening traffic (Chappel, 1989). Redundancy of this sort, however, should be introduced only if any possible confusion between what is status and what is command is avoided by making the two sources as different from one another as possible, for example, voice command versus a pictorial status. Otherwise, in the case of such information as directions, a confusion of status ("you are left") with command ("turn left") could lead to disaster. The chance of confusion will be reduced if the verb phrase ("you *are*" versus "turn") is clearly articulated.

Linguistic Factors

Logical Reversals, Negatives, and Falsifications Whenever a reader or listener is required to logically reverse the meaning of a statement to translate from a physical sequence of words to an understanding of what is intended, comprehension is made more difficult. One example is provided by the use of negatives. We comprehend more rapidly that a particular light should be "on" than that it should be "not off." A second example of logical reversals is falsification. It is faster to understand that a proposition should be true than that it should be untrue or false. Experiments by Clark and Chase (1972) and by Just and Carpenter (1971; Carpenter & Just, 1975) suggest that these differences result not simply from the greater number of words or letters that normally occur in reversed statements but from the cognitive difficulties in processing them as well.

The experimental paradigm employed by these investigators was the sentence-picture verification task. It is the experimental analog of an operator who reads a verbal instruction (i.e., "Check to see that valve *X* is closed") to verify it either against the physical state or against his or her own mental representation of the state of the system. In the actual experimental paradigm as implemented by Clark and Chase (1972), the subjects are shown a verbal sentence describing the relationship between two symbols (e.g., "The star is below the plus") along with a picture depicting the two symbols. The symbols are in an orientation that is either true or false relative to the verbal proposition. The subject is to indicate the truth of this statement as rapidly as possible. Sentences vary in their truth value relative to the picture (e.g., whether the correct answer is true or false) and in whether or not they contain a negative ("The star is above the circle" versus "The circle is not above the star"). Four examples of the sentence-picture relationship are shown in Figure 5.8. Beneath each sentence, describing the picture on the left., is the response time needed to verify sentences of that category.

The results of these experiments suggest three important conclusions.

1. Statements that contain negatives always take longer to verify than those that do not, as shown by the greater response latency for the two sentences on the right in Figure 5.8. Therefore, where possible, instructions should contain only positive assertions (i.e., "Check to see that water level is normal") rather than negative ones ("Check to see that water level is not abnormally high"). An added reason to avoid negatives is that the *not* can sometimes be missed or

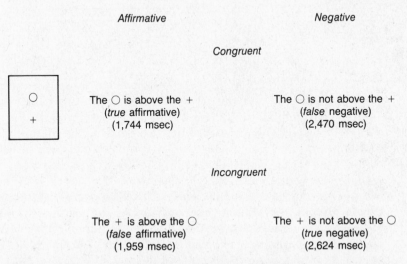

Figure 5.8 The picture-sentence verification task. The four sentences describe the picture on the left. The top two sentences are congruent with the picture relation "circle above plus." The two sentences on the right contain negatives. Response times are shown in parentheses. (*Source:* "On the Process of Comparing Sentences against Pictures" by H. H. Clark and W. G. Chase, 1972, *Cognitive Psychology, 3,* p. 482. Copyright 1972 by Academic Press. Adapted by permission of the authors.)

forgotten if the instructions are read or heard in degraded or hurried circumstances.

2. Whether a statement is verified as true or false influences the verification time in a more complex way. If the statement contains no negatives (is positive), true statements are verified faster than false ones (left side of Figure 5.8). However, if statements contain negatives, false statements are verified more rapidly than true ones (right side). The reason for this reversal relates to the principle of *congruence*, described next.

3. Clark and Chase (1972) and Carpenter and Just (1975) find that there are very predictable differences in latency among the four kinds of picture-sentence relations. In response to this regularity they have modeled the processes involved as a series of very basic constituent comparisons (Carpenter & Just, 1975) of constant duration, each performed in sequence and each taking a constant time. These are performed in series until the final verification is obtained. Comparisons are made concerning the equality, or "congruence," of the propositional form between picture and sentence, disregarding any negatives. For the two sentences at the top, this form is "circle above plus." For the two at the bottom it is "plus above circle." This may be roughly thought of as a comparison of the congruence of the word order on the page with the order of representation in memory of the picture. If there is disagreement in this congruence, extra time is added. Since we normally read from top to bottom, the order of the picture in Figure 5.8, circle–plus, is incongruent for the bottom two sentences. A comparison is also made concerning the existence of negatives (the two right sentences). Negatives also add time. As each of these constituent comparisons is made, units of time are added, and the truth value of the comparison is updated. After all comparisons are made, the final response is given with a latency determined by the number of comparisons. The longest response to the true negative sentence occurs because it alone both is incongruent and contains a negative.

When this model is used to help the designer phrase proper instructions or to predict the time that will be required for operators to respond to instructions, the meaning of true and false must be reconsidered slightly. In Figure 5.8 the relationships are always true or false because the picture never changes. However, in application, a "picture" — the actual state of a system — may take on different values with different probabilities. *True* must thus be defined as the most likely state of a system. Therefore, if a switch is normally in an up position, the instruction should read, "Check to ensure that the switch is up" or "Is the switch up?" Since this position has the greatest frequency, such a statement will normally be verified as a true positive. Furthermore, as long as negatives in wording are avoided, the principle will always hold that affirmations will be processed faster than falsifications.

The results of the basic laboratory work on the superiority of positives over negatives have also been confirmed in applied environments. Newsome and Hochlerin (1989) observed this advantage in computer operating instructions. In highway traffic-regulation signs, experiments have suggested that prohibitive signs, whether verbal ("no left turn") or symbolic, ⊘ are more difficult to comprehend than permissive signs such as "right turn only" (Dewar, 1976; Whitaker & Stacey, 1981). In designing forms to be filled out, such negative phrases as "Do not delay returning this form even if you do not know your insurance number" are harder to comprehend than positive phrases such as "Return this form at once even if you do not know your insurance number (Wright & Barnard, 1975).

Absence of Cues People are generally better at noticing that something unexpected is present than that something expected is missing. The dangers that result when operators must extract information from the absence of cues are somewhat related to the recommendation to avoid negatives in instructions. Fowler (1980) stated this point in his analysis of an airplane crash near the airport at Palm Springs, California. He notes that the *absence* of an R symbol on the pilot's airport chart in the cockpit was the only indication of the critical information that the airport did *not* have radar. Since terminal radar is something pilots come to depend on and the lack of radar is highly significant, Fowler argues that it is far more logical to call attention to the absence of this information by the *presence* of a visible symbol than it is to indicate the presence of this information with a symbol. In general, the presence of a symbol should be associated with information that an operator *needs to know* rather than with certain expected environmental conditions.

Congruence and Order Reversals Many times instructions are intended to convey a sense of ordered events. This order is often in the time domain (procedure *X* is followed by procedure *Y*). When instructions are to convey a sense of order, it is important that the elements in those instructions are *congruent* with the order of events. For example, if subjects are to learn or to verify that the order of elements is $A > B > C$, it is better to say, "*A* is greater than *B*, and *B* is greater than *C*," rather than "*B* is greater than *C*, and *A* is greater than *B*" or "*B* is less than *A*, and *C* is less than *B*" (DeSoto, London, & Handel, 1965). In the first case, the physical ordering of information in the sentence (*A,B,B,C,*) conforms with the intended "true" ordering (*A,B,C*). In the last two cases, it does not (*B,C, A,B* or *B,A,C,B*). Furthermore, in the third case the word *less* is used to verify an ordering that is specified in terms of *greater*. This represents an additional form of cognitive reversal. This finding would dictate that procedural instructions should read, "Do *A*, then do *B*," rather than "Prior to *B*, do *A*," since the former preserves a congruence between the actual sequencing of events and the ordering of statements on the page (Bailey, 1989). For example, a procedural instruction should read, "If the light is on, start the component," rather than, "Start the component if the light is on."

The notion of congruence of ordering appears to be a specific case of the more general finding that people comprehend active sentences more easily than passive ones. Active sentences (e.g., "The malfunction caused the symptoms") preserve an ordering (first the malfunction, then the symptoms) that is more congruent with a mental causal model of the process than a passive sentence is (e.g., "The symptoms were caused by the malfunction"). Research has found that passive sentences both require longer to comprehend and require greater capacity to hold in working memory (Bailey, 1989; Savin & Perchonock, 1965). Many of these principles are relevant to the design of *checklists*, such as those used in aviation. The reader is referred to an interesting human factors evaluation of such checklists by Degani and Wiener (1990).

Multimedia Instructions: Words, Graphics, and Sound

Pictures and Flowcharts versus Words   Earlier in this chapter we discussed the relative merits of pictures versus words for presenting simple concepts, concluding that pictures can sometimes be as effective as words as long as they are easily interpretable. We will now consider the role of pictures and graphics versus words and print in conveying longer series of instructions or procedures, and we find clear advantages for the redundant use of both text and pictures.

In criticizing the use of purely verbal instructions, Kammann (1975) has articulated the so-called two-thirds rule. He summarized a number of investigations of instruction formats and concluded that printed instructions are understood on the average only two-thirds of the time. Kammann then compared the comprehension of a set of dialing instructions for telephone switchboard operators when these were presented entirely in verbal text format and when they were presented in a spatial flowchart format that emphasized the decision-tree aspect of the task. Kammann found that the flowcharts were better understood than the verbal instructions. Furthermore, the superiority in comprehension was maintained in a field trial one to two months after users were initially introduced to the system by the various formats.

The advantage of spatial data displays for comprehension is not, however, universal. In another computer-related study of the relative merits of spatial flowcharts and verbal lists, Wright and Reid (1973) observed that flowcharts were better for immediate comprehension but that information presented in flowchart form showed a greater loss of retention over time. Similar conclusions regarding pictures versus words were reached by Fisk, Scerbo, and Kobylak (1986). A study by Desaulniers, Gillan, and Rudisill (1988) compared flowcharts with text, to help operators diagnose a computer system abnormality. If the number of steps was relatively short, the flowchart was clearly superior in both the speed and the accuracy of diagnosis. But when the procedures were of greater length, thereby preventing all the information from being viewed on the screen at one time, the text provided more accurate performance.

Brooke and Duncan (1980) investigated subjects' abilities to locate faults in a program designed to run a fuel distribution and pricing system. Subjects were given statements about the nature of the error and could then use either a verbal statement listing or a flowchart to assist in their diagnosis. Brooke and Duncan found little difference in performance between the two groups of subjects. A study conducted by Schneiderman, Mayer, McKay, and Heller (1977) also failed to observe any difference between the two alternative display formats in terms of their assistance in computer programming, computer program comprehension, or program debugging.

Ramsey, Atwood, and Van Doren (1978) compared the performance of computer programmers who used flowcharting techniques with those who used a *program design language* (PDL) — a simplified set of verbal statements to help design program formats. Like Schneiderman, Mayer, McKay, and Heller (1977), they noted no difference in comprehension or recall between the two modes but found that use of the PDL generated better-quality programs. Ramsey, Atwood, and Van Doren concluded that the reason for this difference in performance was based on the nature of the information emphasized by the two. Flowcharts emphasize the flow of information, whereas a PDL emphasizes its hierarchical structure. To the extent that efficient programming requires the use of hierarchical relationships in working memory, it would seem that the PDL provides a representation more compatible with this structure. It should be noted, however, that the absence of hierarchical structure is not an inevitable characteristic of all spatial displays. Ramsey, Atwood, and Van Doren did not evaluate a format in which hierarchical structure was displayed spatially, although this alternative is clearly feasible.

Individual Differences   It is apparent from the preceding discussion that the experimental data do not clearly point to one mode as being superior to the other. Some of the ambivalence may result because different modes are preferred by different categories of users. Different users may have different internal models of the system. The advantages of flowcharting found by Kammann (1975), for example, were observed with nonprogrammers, whereas the verbal advantage obtained by Ramsey, Atwood, and Van Doren used experienced computer programmers as subjects.

The question of individual differences in spatial and verbal ability and their interaction with display format was examined in an investigation by Yallow (1980). Subjects with high and low ability in both verbal and spatial aptitude were given material on the topic of economics in a form that emphasized either a verbal or a spatial (i.e., graphic) representation. Not surprisingly, Yallow observed that immediate retention of the material was helped by formats that capitalize on one's strengths (e.g., graphic displays for subjects of high spatial ability). However, she also found that this advantage was markedly short-lived and dissipated when long-term retention was assessed, a finding similar to that of Wright and Reid (1973) with regard to flowcharting.

A second related reason for the lack of pronounced differences between verbal and spatial formats concerns the tremendous sources of variance, often confounding, that enter any experimental comparison of different programming aids (Brooke & Duncan, 1980). Different investigations employ quite different tasks and measures, some using debugging, some programming quality, some programming speed. Performance criteria are often difficult to specify. If problems are formulated in available computer programming languages, subjects must be trained programmers familiar with those languages. These programmers in turn often bring certain biases and differences in familiarity into the laboratory and these biases may confound any comparisons.

Redundancy   A third reason for the ambivalence of results is that different display formats emphasize different properties of the information. Depending on the task, either the spatial, causal flow of information favoring flowcharts or the hierarchical, semantic network relationships emphasized by the verbal statements may be more important (Fitter & Green, 1979). The joint consequences of individual differences and of the different kinds of information emphasized in the two formats suggest a fairly obvious yet important principle: A redundant information format provides the flexibility for different users to capitalize on the format they process best and for the information desired to be represented in whichever mode is most salient.

Three investigations reinforce this conclusion, even as they emphasize the relative strengths of different formats. Booher (1975) evaluated subjects who were mastering a series of procedures required to turn on a piece of equipment. Six different instructional formats were compared: one purely verbal, one purely pictorial, and four combinations of the two codes. Two of these combinations were *redundant:* One code was emphasized and the other provided supplementary cues. Two others were *related:* The nonemphasized mode gave related but not redundant information to the emphasized mode. Booher found the worst performance with the printed instructions and the best with the pictorial emphasis/redundant print format. Although the picture was of primary benefit in this condition, the redundant print clearly provided useful information that was not extracted in the picture-only condition.

Schmidt and Kysor (1987) studied the comprehension of airline passenger safety cards, using samples from 25 of the major air carriers. They found that those cards using mostly words were *least* well understood, those employing mostly diagrams fared better, but the best formats were those in which words were directly integrated with diagrams. The authors describe the value and use of arrows as attention-focusing and attention-directing devices to facilitate this integration.

In the third study, Stone and Gluck (1980) compared subjects' performance in assembling a model using pictorial instruction, text, or a completely redundant presentation of both. Like Booher (1975), Stone and Gluck found the best performance in the redundant condition. The investigators also monitored eye fixation in the redundant condition and found that five times as much time was spent fixating the text as the picture. This finding is consistent with a conclusion drawn by both Booher and by Stone and Gluck: The picture provides an overall context or "frame" within which the words can be used to fill in the details of the procedures or instructions. The importance of context was, of course, emphasized earlier in this chapter.

If pictures and graphics do indeed contribute to the effectiveness of instructions, how realistic should those graphics be? The consensus of research seems to be that more is not better (Spencer, 1988). Simple line drawings appear to do just as well if not better than more elaborate artwork, which captures detail that is not necessary for understanding (Dwyer, 1967). In the study evaluating airline safety codes, Schmidt and Kysor (1987) actually found that photographs, rather than schematic line drawings, led to substantially worse performance. (These findings will have some parallels in our discussion of unnecessary simulator fidelity in Chapter 6.)

Speech Instructions   Another issue regarding redundant presentation concerns the use of speech, a topic that will be discussed in more detail in the following section. It is obvious that speech presentation is not the most effective mode for presenting lengthy instructions, simply because it is transient and cannot be referred to. However, the voice channel can serve as a useful redundant channel, particularly as it will allow information to be conveyed even as the eyes are focused on visual information (see also Chapter 9). A study by Nugent (1987) is illuminating. Investigating various combinations of printed text, graphics, and voice to provide on-line instructions on the use of an oscilloscope, Nugent found that combinations involving both pictures and voice (with or without print) yielded better performance than combinations lacking one or the other of those channels. Similar conclusions were drawn by Baggett (1979) in evaluating the role of film and sound track in conveying instructional information.

In summary, echoing a theme that has recurred in all three of the previous chapters, redundancy of information is of clear benefit to human performance. The extent to which this advantage results because it captures the strengths of different people, because it captures the essence of different kinds of material, because it is less sensitive to fluctuations in attention allocation, or simply because it produces a more firmly anchored and better retrievable knowledge base in long-term memory is not always clear. Probably all four factors work to varying degrees. But the principle stands as one of the more firmly validated in the engineering psychology of instructions.

SPEECH PERCEPTION

In 1977 a tragic event occurred at the Tenerife airport in the Canary Islands: A KLM Royal Dutch Airlines 747 jumbo jet, accelerating for takeoff, crashed into a Pan American 747 taxiing on the same runway. Although poor visibility was partially responsible for the disaster, in which 538 lives were lost, the major responsibility lay with the confusion between the KLM pilot and air traffic control regarding whether clearance had been granted for takeoff. Air traffic control, knowing that the Pan Am plane was still on the runway, was explicit in denying clearance. The KLM pilot misunderstood and, impatient to take off before the deteriorating weather closed the runway, perceived that clearance had been granted. In the terms described earlier, the failure of communications was attributed both to less-than-perfect audio transmission resulting from static and "clipped" messages — poor-quality data or bottom-up processing — and to less-than-adequate message redundancy, so that context and top-down processing could not compensate. The disaster, described in more detail in Hawkins (1987) and fully documented by the Spanish Ministry of Transportation and

Communications (1978), calls attention to the critical role of speech communications in engineering psychology.

In conventional systems, the human operator's auditory channel has been primarily used for transmitting verbal communication from other operators (e.g., messages to the pilot from air traffic control) and for presenting auditory warning signals (tones, horns, buzzers, etc.). Recently, however, rapid advances in microcomputer technology have produced highly efficient speech-synthesis units (Simpson, McCauley, Roland, Ruth, & Williges, 1987). These allow computer-driven displays of auditory verbal messages to be synthesized on-line in a fashion quite analogous to visual information on computer-driven video displays. This capability provides a considerably greater degree of flexibility than that inherent in tape-recorded verbal messages, such as those employed in the cockpit of some aircraft to warn of extreme emergencies. A large and flexible vocabulary of messages can be selected and played instantly without encountering the physical limitations of the tape recorder.

Human perception of speech shares some similarities but also a number of pronounced contrasts with the perception of print, described at the beginning of this chapter. In common with reading, the perception of speech involves both bottom-up hierarchical processing and top-down contextual processing. Corresponding to the reading sequence of features to letters to words, the units of speech go from *phonemes* to *syllables* to *words*. In contrast to reading, on the other hand, the physical units of speech are not so nicely segregated from one another as are the physical units of print. Instead, the physical speech signal, like the cursive line but in contrast to print, is continuous, or analog, in format. The perceptual system must undertake some analog to digital conversion to translate the continuous speech wave form into the discrete units of speech perception. To understand the way in which these units are formed and their relationship to the physical stimulus, it is necessary first to understand the representation of speech. We will consider the difference between the time and frequency representations of continuous analog signals.

Representation of Speech

Physically, the stimulus of speech is a continuous variation or oscillation of the air pressure reaching the eardrum, represented schematically in Figure 5.9a. As with any time-varying signal, the speech stimulus can be analyzed by using the principle of *Fourier analysis* into a series of separate sine wave components of different frequencies and amplitudes. Figure 5.9b is the Fourier-analyzed version of the signal in Figure 5.9a. We may conceptualize the three sinusoidal components in Figure 5.9b as three *features* of the initial stimulus. A more economical portrayal of the stimulus is in the *spectral representation* in Figure 5.9c. Here the frequency value (number of cycles per second, or Hertz) is shown on the abscissa, and the mean amplitude or power (square of amplitude) of oscillation at that particular frequency is on the ordinate. Thus the raw continuous wave form of Figure 5.9a is now represented quite economically by only three points in Figure 5.9c.

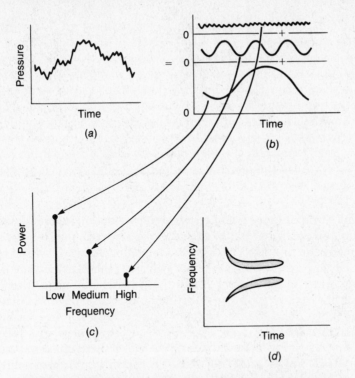

Figure 5.9 Different representations of speech signal: (*a*) time domain; (*b*) frequency components; (*c*) power spectrum; (*d*) speech spectograph.

Unfortunately, the frequency content of articulated speech does not remain constant but changes very rapidly and systematically over time. Therefore, the representation of frequency and amplitude shown in Figure 5.9*c* must also include the third dimension of time. This is done in the *speech spectrograph*, an example of which is shown in Figure 5.9*d*. Here the added dimension of time is now on the abscissa. Frequency, which was originally on the abscissa of the power spectrum in Figure 5.9*c*, is now on the ordinate, and the third dimension, amplitude, is represented by the width of the graph. Thus in the representation of Figure 5.9*d* one tone starts out at a high pitch and low intensity and briefly increases in amplitude while it decreases in pitch, reaching a steady-state level. At the same time a lower-pitched tone increases in both pitch and amplitude to a higher and louder steady level. In fact, this particular stimulus represents the spectrograph that would be produced by the sound *da*. The two separate pitches are called *formants*.

Units of Speech Perception

Phonemes The phoneme, analogous in many respects to the letter unit in reading, represents the basic unit of speech because changing a phoneme in a word will change its meaning (or change it to a nonword). Thus the 38 English phonemes roughly correspond to the letters of the alphabet plus distinctions such as those between long and short vowels and between sounds such as *th* and *sh*. The letters *s* and soft *c* (as in *ceiling*) are mapped into a single phoneme. Although the phoneme in the linguistic analysis of speech is quite analogous to the printed letter, there is a sense in which it is quite different from the letter in its actual perception. The problem is that the physical form of a phoneme is highly dependent on the context in which it appears (the *invariance problem*). The speech spectrograph of the phoneme *k* as in *kid* is quite different from that of *k* as in *lick*. Also, the physical spectrograph of a consonant phoneme differs according to the vowel that follows it.

Syllables Two or more phonemes generally combine to create the *syllable* as the basic unit of speech perception (Massaro, 1975). This definition is in keeping with the notion that although a following vowel (*V*) seems to define the physical form of the preceding consonant (*C*), the syllabic unit (*CV*) is itself relatively invariant in its physical form. The syllable in fact is the smallest unit with such invariance. One line of evidence in support of the syllable unit was provided by Huggins (1964). Huggins' subjects listened to continuous speech that was switched back and forth between the two ears at different rates. Comprehension was somewhat disrupted at all switching rates but was most difficult at a rate of three per second. This is just the rate that would obliterate half of each syllable during the normal rate of speech production (Neisser, 1967). A faster rate obliterating half of each phoneme or a slower one obliterating half of each word was found to be far less disruptive. This finding suggests that people are particularly dependent on the syllable unit in speech perception.

Words Although the word is the smallest cognitive or semantic unit of meaning, like the phoneme it shows a definite lack of correspondence with the physical speech sound. (Actually the morpheme is a slightly smaller cognitive unit than the word, consisting of word stems along with prefixes and suffixes such as *un-* or *-ing*.) This lack of correspondence defines the *segmentation problem* (Neisser, 1967). In a speech spectrograph of continuous speech, there are identifiable breaks or gaps in the continuous record. However, these physical gaps show relatively little correspondence with the subjective pauses at word boundaries that we seem to hear. For example, the spectrograph of the four-word phrase "She uses st∗and∗ard oil" would show the two physical pauses marked, neither one corresponding to the three word-boundary gaps that are heard subjectively. The segmentation issue then highlights another difficulty encountered by automatic speech-recognition systems that function with purely bottom-up processing. If speech is continuous, it is virtually impossible for the recognition system to know the boundaries that separate the words in order to perform the semantic analysis without knowing what the words are already.

Top-Down Processing of Speech

The description presented so far has emphasized the bottom-up analysis of speech. In fact, top-down processing in speech recognition is more essential than it is in reading. The two features that contrast speech perception with reading—the invariance problem and the segmentation problem—make it considerably more difficult to analyze the meaning of a physical unit of speech (bottom-up) without having some prior hypothesis concerning what that unit is likely to be. To make matters more difficult, the serial and transient nature of the auditory message prevents a more detailed and leisurely bottom-up processing of the physical stimulus. This restriction therefore forces a greater reliance on top-down processing.

Demonstrations of top-down or context-dependent processing in speech perception are quite robust. In one experiment, Miller and Isard (1963) compared recognition of degraded word strings between random word lists ("loses poetry spots total wasted"), lists that provided context by virtue only of their syntactic (grammatical) structure but had no semantic content ("sloppy poetry leaves nuclear minutes"), and full semantic and syntactic context ("A witness signed the official document"). The three kinds of lists were presented under varying levels of masking noise. Miller and Isard's data suggested the same trade-off between signal quality and top-down context that was observed by Tulving, Mandler, and Baumal (1964) in the recognition of print. Less context, resulting from the loss of either grammatical or semantic constraints, required greater signal strength to achieve equal performance.

It is apparent that the perception of speech proceeds in a manner similar to the perception of print, through a highly complex, iterative mixture of bottom-up and top-down processing. Neisser (1967) has described this process as one of *analysis by synthesis*. While lower-level analyzers at the acoustic-feature and syllable level progress in a bottom-up fashion, the context provided at the semantic and syntactic levels generates hypotheses and actually synthesizes plausible alternatives concerning what a particular speech sound should be. This synthesis guides and thereby facilitates the bottom-up analysis. The subjective gaps that are heard between word boundaries of continuous speech also give evidence for the dominant role of synthesis. Since such gaps are not present in the physical stimulus, they must result from the top-down processes that decide when each word ends and the next begins.

Applications of Voice Recognition Research

Research and theory of speech perception have contributed to two major categories of applications. First, understanding of how humans perceive speech and employ context-driven top-down processing in recognition has aided efforts to design speech-recognition systems that perform the same task (Lea, 1978). Such systems are becoming increasingly desirable for conveying responses in complex environments, such as high-performance aircraft in which the pilot's hands and arms are continuously occupied.

The second major contribution has been to measure and predict the effects on speech comprehension of various kinds of distortion, which was a source of the Tenerife disaster. Such distortion may be extrinsic to the speech signal—for example, in a noisy environment like an industrial plant. Alternatively, the distortion may be intrinsic to the speech signal when the acoustic wave form is transformed in some fashion, either when synthesized speech is used in computer-generated auditory displays or when a communication channel for human speech is distorted. The following will describe how the disruptive effects of speech distortion are represented and will identify some possible corrective techniques.

As discussed earlier, natural speech is conveyed by the differing amplitudes of the various phonemes distributed across a wide range of frequencies. Thus it is possible to construct a spectrum of the distribution of power at different frequencies generated by "typical" speech. Typical spectra generated by male and female speakers are shown in Figure 5.10. The effects of noise on speech comprehension will clearly depend on the spectrum of the noise involved. A noise that has frequencies identical to the speech spectrum will disrupt understanding more than a noise that has considerably greater power but occupies a narrower frequency range than speech.

Engineers are often interested in predicting the effects of background noise on speech understanding. The *articulation index* (Kryter, 1972) accomplishes this purpose by dividing the speech frequency range into bands and computing the ratio of speech power to noise power within each band. These ratios are then weighted according to the relative contribution of a given frequency band to speech, and the weighted ratios are summed to provide the articulation index (AI). A simplified example of this calculation is shown in Figure 5.11. The spectrum of a relatively low-frequency noise is superimposed on a typical speech spectrum. The speech spectrum has been divided into four bands. The

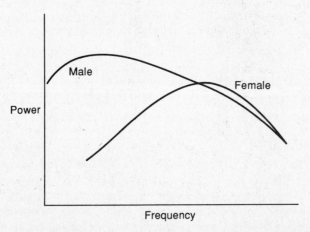

Figure 5.10 Typical power spectra of speech generated by male and female speakers.

MOST IMPORTANT

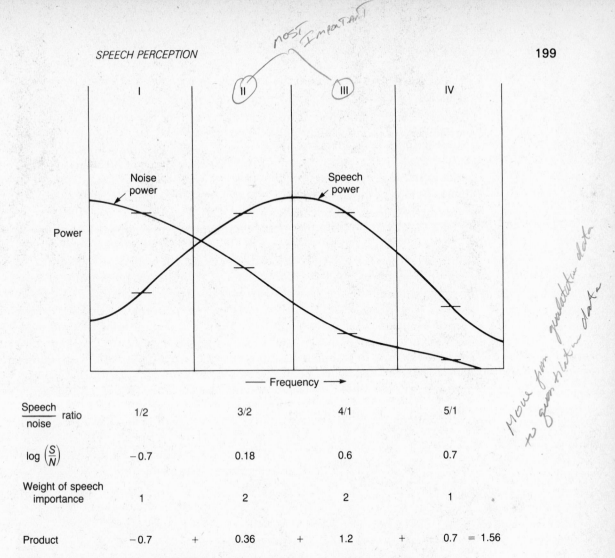

Move from qualitative data to quantitative data

Figure 5.11 Schematic representation of the calculation of an articulation index. The speech spectrum has been divided into four "bands" weighted in importance by the relative power that each contributes to the speech signal.

ratio of speech to noise power and the logarithm of this ratio are shown below each band, and the contributions (weights) are shown below these. To the right is the sum of the weighted products, the AI, reflecting the extent to which the speech signal can be *heard* above the noise.

However, *hearing* is not the same as *comprehension*. From our discussions of bottom-up and top-down processing it is apparent that the AI provides a measure of only bottom-up stimulus quality. A given AI may produce varying levels of comprehension, depending on the information content or redundancy available in the material and the degree of top-down processing used by the listener. To accommodate these factors, measures of *speech intelligibility* are derived by delivering vocal material of a particular level of redundancy over the speech channel in question and computing the percentage of words understood correctly. Naturally, for a given signal-to-noise ratio (defining signal quality and therefore the articulation index) the intelligibility will vary as a function of the redundancy or information content of the stimulus material. A restricted vocabulary produces greater intelligibility than an unrestricted one; words produce greater intelligibility than nonsense syllables; high-frequency words produce greater intelligibility than low-frequency words; and sentence context provides greater intelligibility than no context. Some of these effects on speech understanding are shown in Figure 5.12, which presents data analogous to those in Figure 5.2, concerning print.

The important implications of this discussion are twofold: (1) Either the AI or the speech-intelligibility measures by themselves are inherently ambiguous, unless the redundancy of the transmitted material is carefully specified. (2) To reiterate a point made earlier in this chapter, data-driven, bottom-up processing may trade off with context-driven, top-down processing. For example, as we saw in the discussion of the KLM and Pan American runway disaster, a high level of expectancy for (or motivation to perceive) one meaning of a message, coupled with an ambiguous signal caused by poor-quality data or imprecise wording, can lead to an incorrect interpretation, sometimes with tragic consequences. Several other compelling examples of ambiguity and misinterpretation in aircraft communications are provided by Hawkins (1987), and these are responsible for a large number of pilot errors (Nagel, 1988).

It is important to realize that limitations in signal quality can be compen-

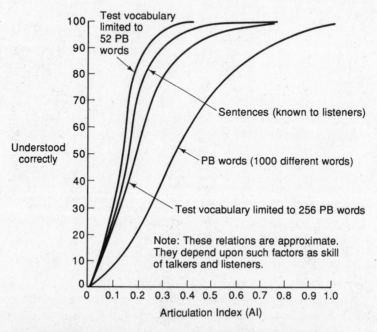

Figure 5.12 The relationship between the articulation index (**AI**) and the intelligibility of various types of speech test materials made up of phonetically balanced (**PI**) words and sentences. (*Source:* Kryter, 1972).

sated for by augmenting top-down processing — creating the ability to "guess" the message without actually (or completely) hearing it. In noisy environments this may be accomplished by restricting the message set size (using standardized vocabulary) or by providing redundant "carrier" sentences to convey a particular message. The latter procedure is analogous to the use of the redundant carrier syllables of the communications-code alphabet (alpha, bravo, charlie, etc.) to convey information concerning a single alphabetic character. A high level of redundancy in the message from air traffic control to the KLM pilot would probably have stopped the premature takeoff and so averted the disaster.

An experiment by Simpson (1976) demonstrated the effect of redundant carrier sentences on comprehension. Pilots listened to synthesized speech warnings presented in background noise. The warnings were either the critical words themselves ("fuel low") or the words embedded in a contextual carrier sentence ("Your fuel is low"). Recognition performance was markedly superior in the latter condition. Simpson also found that the beneficial effects of carrier sentences were greater for one-syllable words than for multisyllabic ones. The greater level of redundancy in the multisyllabic words reduces the need for additional redundancy in the carrier sentence.

Communications

Intuition as well as formal experiments tell us that there is more to communications than simply understanding the words and sentences in speech. The difference between the good and bad lecturer is often the difference between the one who uses gestures, pauses, and voice inflection and the one who reads the lecture in a monotone. A graphic demonstration of the advantages of vision in voice communications is shown in Figure 5.13, showing data in the same

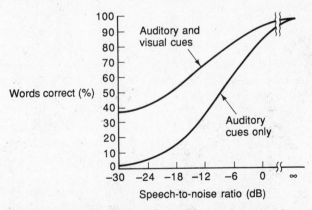

Figure 5.13 Intelligibility of words when perceived with and without visual cues from observing the speaker. (*Source:* W. Sumby and I. Pollack, "Visual Contribution to Speech Intelligibility in Noise." Journal of Acoustical Society of America, 26, 212–215. Reprinted by permission.)

general format as Figure 5.12. As shown in the figure, being able to see the speaker face to face greatly improves communications, particularly when signal quality is low.

Nonverbal Communications   There are four possible causes of differences between the two modes of verbal interaction—face-to-face communications—and voice-only communications, represented in Figure 5.13. All of these causes can influence the efficiency of information exchange.

1. **Visualizing the mouth.** There is little doubt that being able to see a speaker's mouth move and form words is a useful redundant cue—particularly one that can fill in the gaps when voice quality is low. This skill of lipreading is often of critical importance to the hearing impaired. But to understand how important it is to our own speech perception, we need only consider how our understanding of speech in a movie is disrupted when the sound track is poorly dubbed, resulting in desynchronization between sound and mouth movement.

2. **Nonverbal cues.** Being able to see the speaker allows an added range of information conveyance—pointing and gesturing as well as facial cues such as the puzzled look or the nod of acknowledgement that cannot be seen over a conventional auditory channel (e.g., a telephone line). In one study, Chapanis, Ochsman, Parrish, and Weeks (1972) compared problem-solving performance between the two modes and concluded that a face-to-face configuration led to a 14 percent reduction in the time required for a pair of subjects to solve geographical location and equipment assembly problems. The authors also found that this mode actually *increased* the amount of conversation between the members of a team.

3. **Disambiguity.** It is possible, but not clearly validated, that the availability of extra nonverbal cues may resolve ambiguous messages by allowing the speaker to follow up on a puzzled look or other cues suggesting that the listener may have misinterpreted the message. Yet this ability may also be heavily mediated by personality characteristics of the speaker—a point we will address below.

4. **Shared knowledge of action.** In coordinated team performance, such as that typifying the flight crew of an aircraft on a landing approach, a great amount of information is exchanged and shared simply by seeing the actions that a team member has taken (or failed to take), even if this information is totally unrelated to the contents of oral communications (Segal, 1990). For example, the copilot, seeing that the pilot has turned on the autopilot, will be likely to adopt a different mental set as a consequence. The shared knowledge gained by knowing where each member is looking, reaching, and switching potentially contributes a great deal to the smooth functioning of a team (Shaffer, Hendy, & White, 1988).

To the extent that this shared knowledge facilitates communications, changes in the physical configuration of the workspace can affect team performance. For example, cockpit configurations in which the flight engineer looks forward, thereby having activities of pilot and copilot in the direct line of sight, would allow the engineer to share more information than side-looking configurations (Hawkins, 1987). In contrast, the repositioning of flight controls from their position in front of the pilot to the side (the so-called *side-stick controller* used on some modern aircraft) reduces the amount of shared knowledge about control activity between the pilot and copilot since the control activity of one can no longer be easily seen by the other (Beatson, 1989; Segal, 1990). The advances of modern technology, in which spatially distributed dials and keys may be replaced by centralized CRT displays and chording keyboards, may also inhibit the shared knowledge of action by reducing the amount of head and hand movement that can be seen by the coworker (Wiener, 1989).

Resource Management   In spite of the relatively limited amount of research data in the area, the human factors community is becoming very much aware of the importance of effective communications in multioperator environments to support team performance. There is also an increasing awareness that characteristics of the social climate between participants in a dialogue can greatly enhance or degrade that communications pattern and thereby have a strong influence on the efficiency of system performance.

The environment in which this issue has been most directly addressed— because it is potentially relevant to the safety of so many—is in the aircraft flight deck, and here the issue is characterized by the label of cockpit resource management (Foushee & Helmreich, 1988; Office of Technology Assessment, 1988). The concern for social climate and flight deck communications originates from the analysis of several accidents and near accidents in which the copilot or flight engineer, having seen or suspected that the pilot was in error, either failed to call it to the pilot's attention or did so in such a sufficiently ambivalent and deferential manner that the error was not corrected (Foushee, 1984; Hawkins, 1987). Similar incidents are easy to envision in other environments where there is a clear status difference between two operators who must share and exchange information—the surgeon and nurse in the operating room, the pilot and air traffic controller (Wiener, 1977), or the corporate executive and administrative assistant.

The occurrence of such incidents in aviation has led to the establishment of cockpit resource management training programs, which have emphasized the importance of two-way information exchanges to flight deck safety (Office of Technology Assessment, 1988). Although the impact of these programs on flight safety has not yet been clearly documented, there is little doubt that effective cockpit communications has led to less error-prone performance in flight simulator studies. For example, analyzing multicrew simulation studies at NASA Ames Research Center, Foushee and Helmreich (1988) characterized crews that are effective in handling simulated emergencies as sharing more communications, acknowledging communications more frequently, and demonstrating a greater use of commands or assertive statements. However, with the effective crews, these statements are directed both up and down the formal chain of command, from pilot to copilot and from copilot to pilot. It is as if each member is aware of a clearly defined responsibility to communicate and is not hesitant to do so. Foushee and Helmreich also observe better performance in flight crews that have flown together for a longer period of time—as if the communications patterns that have been established for a longer duration have had a chance to develop more effectively (Kanki & Foushee, 1989).

An operational example is provided by the analysis of communications in the cockpit of the severely crippled United Airlines flight 232 before it crashed near Sioux City, Iowa. The study revealed the strong influence of effective communications patterns, which allowed the plane to be guided into a landing with enough control to avoid a complete disaster (Helmreich & Wilhelm, in press). The pilot of the aircraft attributed the success of the landing to the training he received in cockpit resource management (Predmore, 1991).

In summary, the research on communications suggests clearly that the performance of the whole multioperator team is greater than the sum of the parts. This conclusion comes as no surprise to those who have seen a sports team with a collection of superstars fail to meet its expectations because of poor teamwork. The data reemphasize one theme introduced in Chapter 1: The design of effective systems for information display and control with the single operator is a necessary but not sufficient condition for effective human performance.

TRANSITION: PERCEPTION AND MEMORY

Our discussion in the last two chapters has been presented under the categories of spatial and verbal processes in perception. Yet it is quite difficult to divorce these processes from those related to memory. There are four reasons for this close association:

1. Perceptual categorizations, as we saw, were guided by expectancy as manifest in top-down processing. Expectancy was based on both recent experience—the active contents of working memory—and the contents of permanent or long-term memory. Indeed the rules for perceptual categorization themselves are formed only after repeated exposure to a stimulus. These exposures must be remembered to form the categories.
2. In many tasks when perception is not automatic, such as those related to navigation and comprehension, perceptual categorization must operate hand in hand with activities in working memory.
3. The dichotomy that distinguished codes of perceiving into spatial and verbal categories has a direct analog in terms of two codes of working memory.
4. Perception, comprehension, and understanding are necessary precursors for new information to be permanently stored in long-term memory—the issue of learning and training.

REFERENCES

Baggett, P. (1979). Structurally equivalent stories in movie and text and the effect of the medium on recall. *Journal of Verbal Learning and Verbal Behavior, 18,* 333–356.

Bailey, R. W. (1989). *Human performance engineering using human factors/ergonomics to achieve computer system usability* (2nd ed.). Englewood Cliffs, NJ: Prentice Hall.

Barnett, B. J. (1990). Aiding type and format compatibility for decision aid interface design. *Proceedings of the 34th annual meeting of the Human Factors Society* (pp. 1552–1556). Santa Monica, CA: Human Factors Society.

Beatson, S. (1989, April 2). Is America ready to "fly by wire"? *Washington Post,* pp. C3–C4.

Biederman, I. (1987). Recognition-by-components: A theory of human image understanding. *Psychological Review, 94*(2), 115–147.

Biederman, I., Mezzanotte, R. J., Rabinowitz, J. C., Francolin, C. M., & Plude, D. (1981). Detecting the unexpected in photo interpretation. *Human Factors, 23,* 153–163.

Booher, H. R. (1975). Relative comprehensibility of pictorial information and printed words in proceduralized instructions. *Human Factors, 17,* 266–277.

Bower, G. H., Clark, M. C., Lesgold, A. M., & Winzenz, D. (1969). Hierarchical retrieval schemes in the recall of categorical word lists. *Journal of Verbal Learning & Verbal Behavior, 8,* 323–343.

Bransford, J. D., & Johnson, M. K. (1972). Contextual prerequisites for understanding: Some investigations of comprehension and recall. *Journal of Verbal Learning and Verbal Behavior, 11,* 717–726.

Brems, D. J., & Whitten, W. B. (1987). Learning and preference for icon-based interface. *Proceedings of the 31st annual meeting of the Human Factors Society* (pp. 125–129). Santa Monica, CA: Human Factors Society.

Broadbent, D. E. (1977). Language and ergonomics. *Applied Ergonomics, 8,* 15–18.

Broadbent, D., & Broadbent, M. H. (1977). General shape and local detail in word perception. In S. Dornic (ed.), *Attention and performance VI.* Hillsdale, NJ: Erlbaum.

Broadbent, D., & Broadbent, M. H. (1980). Priming and the passive/active model of word recognition. In R. Nickerson (ed.), *Attention and performance VIII.* New York: Academic Press.

Brooke, J. B., & Duncan, K. D. (1980). Flowcharts versus lists as aides in program debugging. *Ergonomics, 23,* 387–399.

Camacho, M. J., Steiner, B. A., & Berson, B. L. (1990). Icons versus alphanumerics in pilot-vehicle interfaces. *Proceedings of the 34th annual meeting of the Human Factors Society* (pp. 11–15). Santa Monica, CA: Human Factors Society.

Carpenter, P. A., & Just, M. A. (1975). Sentence comprehension: A psycholinguistic processing model of verification. *Psychological Review, 82*(1), 45–73.

Chapanis, A. (1965). Words words words. *Human Factors, 7,* 1–17.

Chapanis, A., Ochsman, R. B., Parrish, R. N., & Weeks, G. D. (1972). Studies in interactive communication: I. The effects of four communication modes on the behavior of teams during cooperative problem-solving. *Human Factors, 14*(6), 487–509.

Chappell, S. L. (1989). Avoiding a maneuvering aircraft with TCAS. *Proceedings of the Fifth Symposium on Aviation Psychology.* Columbus, OH: Ohio State University.

Clark, H. H., & Chase, W. G. (1972). On the process of comparing sentences against pictures. *Cognitive Psychology, 3,* 472–517.

Corcoran, D. W., & Weening, D. L. (1967). Acoustic factors in proofreading. *Nature, 214,* 851–852.

Crocoll, W. M., & Coury, B. G. (1990). Status or recommendation: Selecting the type of information for decision aiding. *Proceedings of the 34th annual meeting of the Human Factors Society* (pp. 1524–1528). Santa Monica, CA: Human Factors Society.

Degani, A., & Wiener, E. L. (1990). *Human factors of flight-deck checklists: The normal checklist* (NASA Contractor Report 177549). Moffett Field, CA: NASA Ames Research Center.

Desaulniers, D. R., Gillan, D. J., & Rudisill, M. (1988). The effects of format in computer-based procedure displays. *Proceedings of the 32nd annual meeting of the Human Factors Society* (pp. 291–295). Santa Monica, CA: Human Factors Society.

DeSota, C. B., London, M., & Handel, S. (1965). Social reasoning and spatial paralogic. *Journal of Personal & Social Psychology, 2,* 513–521.

Dewar, R. E. (1976). The slash obscures the symbol on prohibitive traffic signs. *Human Factors, 18,* 253–258.

Dumas, J. S., & Redish, J. (1986). Using plain English in designing the user interface. *Proceedings of the 30th annual meeting of the Human Factors Society* (pp. 1207–1211). Santa Monica, CA: Human Factors Society.

Dwyer, F. M. (1967). Adapting visual illustrations for effective learning. *Harvard Educational Review, 37,* 250–263.

Ehrenreich, S. (1982). The myth about abbreviations. *Proceedings of the 1982 IEEE International Conference on Cybernetics and Society.* New York: Institute of Electrical and Electronic Engineers.

Ellis, N. C., & Hill, S. E. (1978). A comparison of seven-segment numerics. *Human Factors, 20,* 655–660.

Ellis, S. R., & Hitchcock, R. J. (1986). The emergence of Zipf's law: Spontaneous encoding optimization by users of a command language. *IEEE Transactions on Systems, Man, and Cybernetics, SMC–16*(3), 423–427.

Fisk, A. D., Ackerman, P. L., & Schneider, W. (1987). Automatic and controlled processing theory and its applications to human factors problems. In P. A. Hancock (ed.), *Human factors psychology* (pp. 159–197). North Holland, Netherlands: Elsevier Science Publishers B.V.

Fisk, A. D., Scerbo, M. W., & Kobylak, R. F. (1986). Relative value of pictures and text in conveying information: Performance and memory evaluations. *Proceedings of the 30th annual meeting of the Human Factors Society* (pp. 1269–1272). Santa Monica, CA: Human Factors Society.

Fitter, M., & Green, T. R. (1979). When do diagrams make good computer languages? *International Journal of Man-Machine Studies, 11,* 235–261.

Foushee, H. C. (1984). Dyads and triads at 35,000 feet: Factors affecting group process and aircrew performance. *American Psychology, 39,* 885–893.

Foushee, H. C., & Helmreich, R. L. (1988). Group interaction and flightcrew perform-ance. In E. Wiener & D. Nagel (eds.), *Human factors in aviation.* San Diego, CA: Academic Press.

Fowler, F. D. (1980). Air traffic control problems: A pilot's view. *Human Factors, 22,* 645–654.

Gibson, E. J. (1969). *Principles of perceptual learning and development.* Englewood Cliffs, NJ: Prentice Hall.

Grether, W., & Baker, C. A. (1972). Visual presentation of information. In H. P. Van Cott & R. G. Kinkade (eds.), *Human engineering guide to system design.* Washing-ton, DC: U.S. Government Printing Office.

Haber, R. N., & Schindler, R. M. (1981). Error in proofreading: Evidence of syntactic control of letter processing? *Journal of Experimental Psychology: Human Percep-tion & Performance, 7,* 573–579.

Hastie, R. (1982). *An empirical evaluation of five methods of instructing the jury* (Final Report Grant 78-NI-AX-0146). Washington, DC: National Institute of Justice.

Hawkins, F. H. (1987). *Human factors in flight.* Brookfield, VT: Gower Technical Press.

Healy, A. F. (1976). Detection errors on the word "the." *Journal of Experimental Psychology: Human Perception & Performance, 2,* 235–242.

Helmreich, R. L., & Wilhelm, J. A. (in press). Outcomes of crew resource management training. *International Journal of Aviation Psychology.*

Huggins, A. (1964). Distortion of temporal patterns of speech: Interruptions and alter-ations. *Journal of the Acoustical Society of America, 36,* 1055–1065.

Just, M. T., & Carpenter, P. A. (1971). Comprehension of negation with quantification. *Journal of Verbal Learning and Verbal Behavior, 10,* 244–253.

Just, M. T., & Carpenter, P. A. (1980). Cognitive processes in reading: Models based on reader's eye fixation. In C. A. Prefetti & A. M. Lesgold (eds.), *Interactive processes and reading.* Hillsdale, NJ: Erlbaum.

Kammann, R. (1975). The comprehensibility of printed instructions and the flow chart alternative. *Human Factors, 17,* 183–191.

Kanki, B. G., & Foushee, H. C. (1989, May). Communication as group process mediator of aircrew performance. *Aviation, Space, and Environmental Medicine,* pp. 402–410.

Klemmer, E. T. (1969). Grouping of printed digits for manual entry. *Human Factors, 11,* 397–400.

Kramer, A. F., Strayer, D. L., & Buckley, J. (1990). Development and transfer of automatic processing. *Journal of Experimental Psychology: Human Perception & Performance, 16*(3), 505–522.

Kryter, K. D. (1972). Speech communications. In H. P. Van Cott & R. G. Kinkade (eds.), *Human engineering guide to system design.* Washington, DC: U.S. Government Printing Office.

LaBerge, D. (1973). Attention and the measurement of perceptual learning. *Memory & Cognition, 1,* 268–276.

Lea, W. (ed.). (1978). *Trends in speech recognition.* Englewood Cliffs, NJ: Prentice Hall.

Lindsay, P. H., & Norman, D. A. (1972). *Human information processing.* New York: Academic Press.

McConkie, G. W. (1983). Eye movements and perception during reading. In K. Raynor (ed.), *Eye movements in reading.* New York: Academic Press.

McConkie, G. W., & Raynor, K. (1974). *Identifying the span of the effective stimulus in reading* (Final Report OEG 2–71-0537). Washington, DC: U.S. Office of Education.

Massaro, D. W. (1975). *Experimental psychology and information processing.* Chicago: Rand McNally College Publishing.

Miller, G., & Isard, S. (1963). Some perceptual consequences of linguistic rules. *Journal of Verbal Learning and Verbal Behavior, 2,* 217–228.

Moses, F. L., & Ehrenreich, S. L. (1981). Abbreviations for automated systems. In R. Sugarman (ed.), *Proceedings of the 25th annual meeting of the Human Factors Society.* Santa Monica, CA: Human Factors Society.

Nagel, D. (1988). Pilot error. In E. Wiener & D. Nagel (eds.), *Human factors in modern aviation.* Orlando, FL: Academic Press.

Navon, D. (1977). Forest before trees: The presence of global features in visual perception. *Cognitive Psychology, 9,* 353–383.

Neisser, U. (1967). *Cognitive psychology.* Englewood Cliffs, NJ: Prentice Hall.

Neisser, U., Novick, R., & Lazer, R. (1964). Searching for novel targets. *Perceptual and Motor Skills, 19,* 427–432.

Newsome, S. L., & Hocherlin, M. E. (1989). When "not" is not bad: A reevaluation of the use of negatives. *Proceedings of the 33rd annual meeting of the Human Factors Society* (pp. 229–234). Santa Monica, CA: Human Factors Society.

Norcio, A. F. (1981). *Human memory processes for comprehending computer programs* (Technical Report AS–2-81). Annapolis, MD: U.S. Naval Academy, Applied Sciences Department.

Norman, D. A. (1981). The trouble with UNIX. *Datamation, 27*(12), 139–150.

Nugent, W. A. (1987). A comparative assessment of computer-based media for presenting job task instructions. *Proceedings of the 31st annual meeting of the Human Factors Society* (pp. 696–700). Santa Monica, CA: Human Factors Society.

Office of Technology Assessment. (1988). *Safe skies for tomorrow: Aviation safety in a competitive environment* (OTA–SET-381). Washington, DC: U.S. Government Printing Office.

Palmer, S. E. (1975). The effects of contextual scenes on the identification of objects. *Memory & Cognition, 3,* 519–526.

Plath, D. W. (1970). The readability of segmented and conventional numerals. *Human Factors, 12,* 493–497.

Potter, M. C., & Faulconer, B. A. (1975). Time to understand pictures and words. *Nature, 253,* 437–438.

Predmore, S. C. (1991). Micro-coding of cockpit communications in accident analyses: Crew coordination in the United Airlines flight 232 accident. In R. S. Jensen (ed.),

Proceedings of the 6th International Symposium on Aviation Psychology. Columbus, Ohio State University, Department of Aviation.

Ramsey, H. R., Atwood, M. E., & Van Doren, J. R. (1978). Flow charts vs. program design languages. In E. Baise & S. Mitter (eds.), *Proceedings of the 22nd annual meeting of the Human Factors Society.* Santa Monica, CA: Human Factors Society.

Reicher, G. M. (1969). Perceptual recognition as a function of meaningfulness of stimulus material. *Journal of Experimental Psychology, 81,* 275–280.

Rumelhart, D. E., & McClelland, J. L. (1986). *Parallel distributed processing: Explorations in the microstructure of cognition* (Vol. 1). Cambridge, MA: MIT Press.

Rumelhart, D. (1977). *Human information processing.* New York: Wiley.

Savin, H. B., & Perchonock, E. (1965). Grammatical structure and the immediate recall of English sentences. *Journal of Verbal Learning and Verbal Behavior, 4,* 348–353.

Schmidt, J. K., & Kysor, K. P. (1987). Designing airline passenger safety cards. *Proceedings of the 31st annual meeting of the Human Factors Society* (pp. 51–55). Santa Monica, CA: Human Factors Society.

Schneider, W. (1985). Training high-performance skills: Fallacies and guidelines. *Human Factors, 27*(3), 285–300.

Schneider, W., Dumais, S. T., & Shiffrin, R. M. (1984). Automatic and control processing and attention. In R. Parasuraman & D. R. Davies (eds.), *Varieties of attention* (pp. 1–27). Orlando, FL: Academic Press.

Schneider, W., & Fisk, A. D. (1984). Automatic category search and its transfer. *Journal of Experimental Psychology: Learning, Memory, and Cognition, 10,* 1–15.

Schneider, W., & Shiffrin, R. M. (1977). Controlled and automatic human information processing I: Detection, search, and attention. *Psychological Review, 84,* 1–66.

Schneiderman, B. L., Mayer, R., McKay, D., & Heller, P. (1977). Experimental investigations of the utility of detailed flowcharts in programming. *Communications of the Association for Computing Machinery, 20,* 373–381.

Segal, L. D. (1990). Effects of aircraft cockpit design on crew communication. In E. J. Lovesey (ed.), *Contemporary ergonomics 1990* (pp. 247–252). London: Taylor & Francis.

Shaffer, M. T., Hendy, K. C., & White, L. R. (1988). An empirically validated task analysis (EVTA) of low level Army helicopter operations. *Proceedings of the 32nd annual meeting of the Human Factors Society* (pp. 178–183). Santa Monica, CA: Human Factors Society.

Sheridan, T. E., & Ferrell, L. (1974). *Man-machine systems.* Cambridge, MA: MIT Press.

Simpson, C. (1976, May). Effects of linguistic redundancy on pilot's comprehension of synthesized speeds. *Proceedings of the 12th Annual Conference on Manual Control* (NASA TM–X-73, 170). Washington, DC: U.S. Government Printing Office.

Simpson, C. A., McCauley, M. E., Roland, E. F., Ruth, J. C., & Williges, B. H. (1987). In G. Salvendy (ed.), *Handbook of human factors.* New York: Wiley.

Spanish Ministry of Transportation and Communications. (1978). Report of collision between PAA B-747 and KLM B-747 at Tenerife. *Aviation Week & Space Technology, 109* (November 20), 113–121; (November 27), 67–74.

Spencer, K. (1988). *The psychology of educational technology and instructional media.* London: Routledge.

Stone, D. E., & Gluck, M. D. (1980). *How do young adults read directions with and without pictures?* (Technical Report). Ithaca, NY: Cornell University, Department of Education.

Sumby, W., & Pollack, I. (1954). Visual contribution to speech intelligibility in noise. *Journal of Acoustical Society of America, 26,* 212–215.

Taylor, R. M., & Selcon, S. J. (1990). Cognitive quality and situational awareness with advanced aircraft attitude displays. *Proceedings of the 34th annual meeting of the Human Factors Society* (pp. 26–30). Santa Monica, CA: Human Factors Society.

Tinker, M. A. (1955). Prolonged reading tasks in visual research. *Journal of Applied Psychology, 39,* 444–446.

Tulving, E., Mandler, G., & Baumal, R. (1964). Interaction of two sources of information in tachistoscopic word recognition. *Canadian Journal of Psychology, 18,* 62–71.

Vartabedian, A. G. (1972). The effects of letter size, case, and generation method on CRT display search time. *Human Factors, 14,* 511–519.

Whitaker, L. A., & Stacey, S. (1981). Response times to left and right directional signals. *Human Factors, 23,* 447–452.

Wickelgren, W. A. (1979). *Cognitive psychology.* Englewood Cliffs, NJ: Prentice Hall.

Wiener, E. L. (1977). Controlled flight into terrain accidents: System-induced errors. *Human Factors, 19,* 171.

Wiener, E. L. (1989). Reflections on human error: Matters of life and death. *Proceedings of the 33rd annual meeting of the Human Factors Society* (pp. 1–7). Santa Monica, CA: Human Factors Society.

Wright, P., & Barnard, P. (1975). Just fill in this form—A review for designers. *Applied Ergonomics, 6*(4), 213–220.

Wright, P., & Reid, F. (1973). Written information: Some alternatives to prose for expressing the outcomes of complex contingencies. *Journal of Applied Psychology, 57,* 160–166.

Yallow, E. (1980). *Individual differences in learning from verbal and figural materials* (Aptitudes Research Project Technical Report No. 13). Palo Alto, CA: Stanford University, School of Education.

Chapter 6

Memory, Learning, and Training

Failures of memory often plague the human operator. These may be as simple and trivial as forgetting a phone number we have just looked up or as involved as forgetting the procedures to run a word-processing system. Operators may forget to perform a critical item in a checklist (Degani & Wiener, 1990), or an air traffic controller may forget a "temporary" command issued to a pilot (Danaher, 1980). In 1915, a railroad switchman at the Quintinshill Station in Scotland forgot that he had moved a train to an active track, thereby permitting two oncoming trains to use the same track. In the resulting crash over 200 people were killed (Rolt, 1978).

Clearly, then, the success or failure of human memory can have a major impact on the usefulness and safety of a system. As noted in Chapter 1, memory may be thought of as the store of information. In this chapter we will focus on two different storage systems with different durations: working memory and long-term memory. Working memory is the temporary, attention-demanding store that we use to retain new information (like a new phone number) until we use it (dial it). We also use working memory as a kind of a "workbench" of consciousness in which we evaluate, compare, and examine different mental representations. We might use working memory, for example, to carry out mental arithmetic or a mental simulation of what will happen if we schedule jobs in one way instead of in another. Finally, working memory is also used to hold new information until we can give it a more permanent status in memory, that is, until we encode it into long-term memory. Long-term memory thus is our storehouse of facts about the world and about how to do things.

Both of these levels of memory may be thought of in the context of a three-stage representation, shown in Figure 6.1. The first stage, *encoding, learning, or training,* describes the process of putting things into the memory system. The three terms have similar but not overlapping meaning. The latter

211

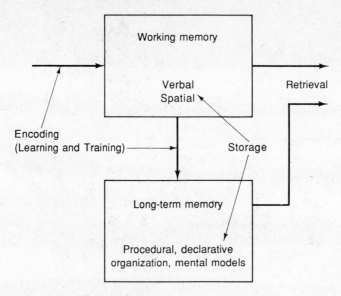

Figure 6.1 A representation of memory functions.

① *encoding*

two typically refer only to the entry into long-term memory. Learning describes the various ways in which this entry can occur, whereas training refers to explicit and intentional techniques used by designers and teachers to maximize the efficiency of learning. Our concern will be primarily with training.

② *Storage*, the second stage, refers to the *way* in which information is held or represented in the two memory systems. The terms that we use to describe it are different for working memory, in which we emphasize spatial versus verbal codes, than it is for long-term memory, in which we emphasize declarative and procedural knowledge and mental models.

③ The third stage, *retrieval*, refers to our ability to get things successfully out of memory. Here we contrast successful retrieval with the various causes of retrieval failure, or *forgetting*. Sometimes material simply cannot be retrieved. At other times it is retrieved incorrectly, as when we mix up the steps in a procedure.

In this chapter, we will first describe the properties of working memory, its spatial and verbal representation and the causes of forgetting. We will then describe long-term memory, focusing heavily on the issue of encoding through a discussion of training. Particular emphasis will be given to the *transfer of training*—how the skills and knowledge acquired in one domain are transferred to another. We will then discuss a number of different ways in which knowledge representation in long-term memory has been described—the storage issue—and finally conclude with a discussion of the causes and consequences of forgetting from long-term memory.

WORKING MEMORY

Codes of Working Memory

By adopting the framework of encoding, storage, and retrieval, it is easy to see that encoding in working memory is not much different from the perceptual processes discussed in the two previous chapters.

As noted earlier, the information retained in working memory can be represented in either of two forms, or *codes. Spatial working memory* represents information in an analog spatial form, often typical of visual images. Baddeley (1985, 1990) defines it as a "visual spatial scratchpad." *Verbal,* or *phonetic, working memory* represents information in linguistic form, typically as words and sounds. It can be rehearsed by articulating those words and sounds, either vocally or subvocally.

Whether represented visually or phonetically, information held in working memory is usually but not always associated with a *semantic* representation (Shulman, 1972). That is, awareness of its meaning is usually activated with its appearance (spatial) or sound (verbal). We say "usually but not always" because it is possible for us to rehearse meaningless sounds (the words of a foreign language we do not know) or to imagine visually meaningless forms.

Although the verbal and spatial categories are of prominent importance, there are actually a number of elaborations that can be and have been made of the working memory system, as described by Baddeley (1990), Posner (1978), and others.

1. As noted in Chapter 1, each sensory system (auditory and visual) has its own *sensory memory,* which retains information very briefly. For visual sensory memory, this duration is so short as to be unimportant in engineering design. The auditory sensory memory, or *echoic memory,* is a longer, more persistent representation, which allows us to "hear" a short utterance even if we were not attending to it as it was spoken. However, despite some distinctions (Wickens, 1984), in most respects the characteristics of echoic memory are similar to those of verbal working memory, and the implications of both for engineering psychology can be treated as one.

2. A model of working memory proposed by Baddeley (1985) actually describes the system as consisting of three subsystems: a master "executive" and two subsidiary "slave systems," a verbal/phonetic rehearsal loop; and a visual, spatial "scratchpad." For most practical applications, however, it appears that the dichotomy of two working memory systems is sufficient, each of which is described below in more detail. Here it can be assumed that both the verbal-phonetic and the visual-spatial system share the conscious, attention-demanding features of Baddeley's central executive.

3. Although sounds are usually associated with verbal memory and visual analog representations with spatial, this coupling is not necessary. Thus, we can represent heard digits, letters, or words by visualizing their appearance, and there is evidence that trying to keep track of the location of sounds around our body space is spatial in its characteristics (Baddeley & Lieberman, 1980).

The practical implications of the distinction drawn between the different working memory codes are provided primarily by three different phenomena: (1) The two kinds of working memory appear to be somewhat independent from one another and therefore are susceptible to interference from different sorts of concurrent activities (Baddeley, Grant, Wight, & Thompson, 1975). (2) The relationship of codes to display modalities has implications for auditory versus visual displays and verbal versus spatial displays. (3) Visual and verbal codes may be differentially effective for different individuals, thereby providing a way to present different operators with materials for which they are best suited.

Code Interference   The verbal-phonetic and visual-spatial codes of working memory appear to function more cooperatively than competitively. Posner (1978), for example, has argued that both may be activated in parallel by certain kinds of material (e.g., pictures of common objects). We also saw examples of the cooperation between codes in our discussion of navigation in Chapter 4. One implication of this cooperation, which will be considered more extensively in Chapter 9, is that the two codes do not entirely compete for the same limited processing resources or attention. That is, if two tasks employ different codes of working memory, they will be time-shared more efficiently than if two tasks share a common code. This is a theme that will be discussed in some detail in Chapter 9. Here we restrict the discussion to time-sharing with working memory.

The research of Baddeley and his colleagues (Baddeley, Grant, Wight, & Thompson, 1975; Baddeley & Hitch, 1974; Baddeley & Lieberman, 1980; Logie, 1989, Logie & Baddeley, 1990) has contributed substantially to the understanding of this dichotomy, in terms of both the kind of material that is manipulated within working memory (spatial-visual or verbal-phonetic) and the separate processing resources used by each. Brooks (1968) performed an important series of experiments that stimulated a good deal of the subsequent research in this area. In one experiment, Brooks required subjects to perform a series of mental operations in spatial working memory. Subjects imagined a capital letter, such as the letter *F*, as depicted in Figure 6.2. They were then asked to "walk" around the perimeter of the letter, indicating in turn whether each corner was or was not in a designated orientation (e.g., facing the lower right). The yes and no answers were indicated either by vocal articulation (verbal response) or by pointing to a column of Y's and N's (spatial response). Brooks found that performance in the verbal condition was reliably better than in the spatial condition, suggesting that the verbal response code in the former used different resources from those underlying the "walking" operations in spatial working memory. The greater resource competition with spatial working memory in the spatial response condition produced the greater degree of interference.

Of course, it is possible that the larger interference with the spatial response mode occurred because this is a more difficult means of responding. However, the results of a second investigation by Brooks (1968) suggest that

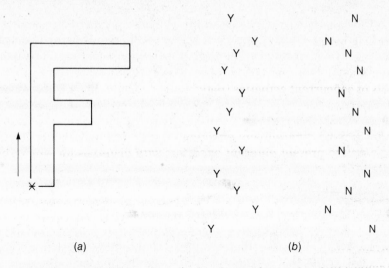

Figure 6.2 (*a*) Example of a stimulus in Brook's study. (*b*) Display used in pointing task. (Source: L. R. Brooks, "Spatial and Verbal Components in the Act of Recall," *Canadian Journal of Psychology*, 22 (1968), pp. 350–351. Copyright 1968 by the Canadian Psychological Association. Reproduced by permission.

this interpretation is unlikely. In this study, the spatial working memory task was replaced with one relying on verbal working memory. Subjects imagined a familiar sentence (e.g., "The quick brown fox jumped over the lazy dog") and then "walked" through the sentence, indicating in turn if each word was or was not a member of a particular grammatical category. Once again, responses were indicated either by speech or by pointing. Using the verbal task, Brooks now found that the reverse pattern of interference was observed. The task was performed better with the spatial-manual response than with the vocal-verbal one. Hence when different codes underlie working memory and response, performance will be improved.

Experiments conducted by Baddeley and his colleagues, Healy (1975) and Klapp and Netick (1988), offer further insight into the distinction between the two memory systems. Klapp and Netick compared two digit recall tasks that used identical input and response systems but varied in their temporal and spatial demands. In the *probe digit* task, the subjects heard a string of eight digits and then were "probed" by one of those digits and asked to recall the digit that had followed it in the list. To respond successfully to the probe, the subjects needed to store the temporal order of the digits. In the missing digit task, the subjects again heard eight digits, but they were now asked to respond with the digit that was missing from the list. A successful response will depend on putting the eight digits in an ordered spatial array as they are heard, and then, at the end of the list, "examining" which "slot" in the imaged array is "empty." Confirming the phonetic and spatial components of these two memory tasks, Klapp and Netick found that the probe digit task was much more disrupted by irrelevant vocalizations (saying, "ba ba ba"), whereas the *missing digit* task was affected by a concurrent task requiring spatial tracking. This contrast is important because the nature of the stimulus and response is identical in both conditions. The only difference between the tasks concerns the attributes of information that need to be retained. Temporal-order information appears to be associated with verbal working memory, spatial-order information with spatial memory (Crowder, 1978).

The relevance of these and related experimental findings may be summarized as follows: We seem to have two forms of working memory. Each is used to process or retain qualitatively different kinds of information (spatial and visual versus temporal, verbal, and phonetic), and each can be disrupted by different concurrent activities. Tasks should be designed, therefore, in such a manner that this disruption does not occur. Tasks that impose high loads on spatial working memory (e.g., air traffic controllers' need to maintain a mental model of the aircraft in their purview) should not be performed concurrently with other tasks that will also use the visual-manual-spatial system. Spatial tasks will be less disrupted by employment of an auditory-phonetic-vocal loop to handle subsidiary information-processing tasks (Wickens & Liu, 1988). Correspondingly, tasks involving heavy demands on verbal working memory, such as editing texts, computing numbers, and using symbolic-based computers, if their memory demands are high, will be more disrupted by concurrent voice input and output than by visual-manual interaction (e.g., control with a mouse).

This guideline does not necessarily suggest that the primary task display should be in a different format from the code of working memory used in the task. In the case of the air traffic controller, this is probably unwise. As we will see, a visual-spatial display will be optimally compatible with the spatial representation in visual working memory. However, there are other circumstances in which it would be best to display information relevant to the primary task by a code different from that used in processing the primary task information. For example, Wetherell (1979) concluded that navigational information was better represented in phonetic route-list format than in map format, in part because the latter interfered more with the spatial characteristics of driving (see Chapter 4).

The implications of code-specific interference have not been extensively followed up with investigations in more applied settings. Hoffman and Mac-Donald (1980) studied how the retention of information on highway traffic signs displayed verbally ("No left turn") or spatially (⊘) was influenced by concurrent verbal or spatial activity during the retention interval. When the subjects' task was to recognize the sign that had been originally presented, their results provide some evidence for code-specific interference. However, when the subjects' task actually involved implementing a response on the basis of the information presented on the sign, the spatial task consistently produced greater interference independent of the code of the original sign. This finding suggests that whether the original sign was verbal or spatial, the implications for action were directly coded in a spatial format, more disrupted by the spatial task. These results suggest in turn that much of driving involves spatial processing (as suggested as well by Wetherell's 1979 results). Since the difference in format had little influence on interference during retention in Hoffman and MacDonald's experiment, it appears that the guidelines for sign formatting can best follow the principle of display compatibility described in Chapter 4; that is, because the implications for action are spatial, the optimum display code will be spatial as well.

Two analogous investigations by Wickens and Weingartner (1985) and Wickens, Sandry, and Vidulich (1983) did find evidence for code-specific interference in working memory. Wickens and Weingartner's subjects engaged in a simulated process-control monitoring task (see Chapter 12) in which several cross-coupled, slowly changing variables were monitored to detect periodic "failures," changes in the dynamic equations governing their relations. Concurrently, subjects performed either a spatial-memory matrix task developed by Brooks (1967), in which they imagined items in different positions of a 4 × 4 matrix, or a verbal task requiring the memory of abstract words. Wickens and Weingartner found that this visual/spatial monitoring task was disrupted more by the spatial than by the verbal matrix task. Since input for both matrix tasks was always auditory, the source of interference was necessarily the code of working memory. Wickens, Sandry, and Vidulich examined how a pilot's task when flying an aircraft simulator was disrupted by either a task requiring memory of verbal navigational information or a task requiring the localization of a target in space. Even when both tasks were presented visually, the localization task, depending on transformations in spatial working memory, disrupted flying more than did the verbal-memory task.

Display Modality and Working Memory Code Figure 6.3 shows four different possible formats of information display defined by the auditory and visual input modalities and the two primary perceptual and memory codes. These can be potentially associated with either of the two different codes of working memory, depending on which is required by the task at hand. Experimental data suggest that the assigment of formats to memory codes should not be arbitrary. Wickens, Sandry, and Vidulich (1983) have described the principle of stimulus/central-processing/response compatibility that prescribes the best association of display formats to codes of working memory. In this principle *S* refers to the four stimulus display formats, *C* to the two possible central-processing codes, and *R* to the two possible response modalties (manual and speech). The following discussion concerns the optimum matching between stimulus and working memory codes (*S-C* compatibility). The compatibility of response modalities will be dealt with in Chapter 8.

In Figure 6.3 the shaded cells indicate the optimum format (combination of mode and code) for each memory task. Thus, although it is possible to employ auditory-spatial displays for tasks that demand spatial working memory, they are usually less effective because the auditory modality is less attuned to processing spatial information than is the visual. For example, Wickens, Vidulich, and Sandry-Garza (1984) used an auditory display of spatial material in which the location and velocity of an aircraft on a two-dimensional grid were

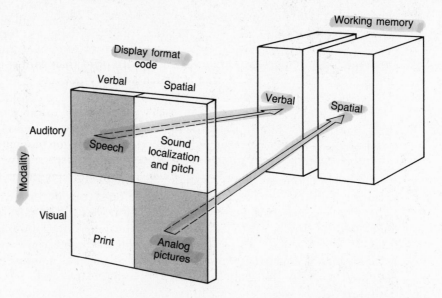

Figure 6.3 Optimum assignment of display format to working memory code.

indicated by tones varying in pitch (corresponding to the vertical dimension) and apparent location (corresponding to the horizontal). Performance with this display was far less proficient than with a visual-spatial display, even in the presence of a concurrent visual task. Further examples will be discussed in the treatment of auditory tracking in Chapter 11.

Taks that demand verbal working memory, in contrast to spatial memory, can be more readily served by either speech or print. However, the fact that echoic memory has a slower decay than iconic memory and that speech displays are more directly compatible with the vocalization used in rehearsal suggests that speech should be employed as a display for verbal tasks, particularly if the verbal material can only be displayed for a short interval (Wickens, Sandry, & Vidulich, 1983). This guideline is supported by a series of laboratory studies that found that verbal material is better retained for short periods when presented by auditory rather than visual means (Murdock, 1968; Nilsson, Ohlsson, & Ronnberg, 1977).

This observation, of course, has considerable practical implications when verbal material is to be presented for temporary storage (e.g., navigational entries presented to the aircraft pilot or the outcome of diagnostic tests presented to the physician or process control monitor by automatic means). Such information will be less susceptible to short-term loss when presented by auditory channels (either spoken or through speech synthesis). Wickens, Sandry, and Vidulich (1983), for example, found that pilots could retain navigational information better from auditory than visual channels, and this advantage was enhanced under conditions of high workload.

An important qualification to the guideline recommending auditory presentation of verbal material occurs when the message is relatively long (i.e., is longer than four to five unrelated words or letters). In this case both auditory and visual channels will be likely to show failures of memory. Neither format will be effective without introducing some means of physically prolonging the message, which of course is more feasible with print than with speech. Hence an optimal format would be one in which auditory delivery is "echoed" by a more permanent printed record. Such systems are being considered for the transmission of critical messages from air traffic control to the pilot, and they will be more feasible by digital rather than radio channel relay (Lee, 1989; Office of Technology Assessment, 1988). The presentation of longer sequences of verbal material in instructions and procedures, as discussed in the previous chapter, is also a circumstance in which the greater length makes a purely auditory presentation unwise. Here the length nearly requires the semipermanent printed record. Yet, as also discussed in the previous chapter, considerable gain can be obtained by the use of redundant auditory and visual display formats in presenting instructions.

WORKING MEMORY RETRIEVAL AND FORGETTING

The rapid rate of decay or loss of availability is one of the greatest limitations of working memory. When verbal information is presented auditorily, the decay may be slightly postponed because of the transient benefits of the echoic code. When information is presented visually, the decay will be more rapid. The consequence of decay before material is used is that retrieval is more difficult and forgetting is more likely. Pilots may forget (or confuse) navigational instructions delivered by air traffic controllers before they are implemented. Computer programmers may lose their place in an interactive data program after having briefly diverted their attention to another task.

As noted, an apparent solution to the problem of such memory failures is to augment the initial transient stimulus (whether visual or auditory) with a more long-lasting visual display — a visual echo of the message a pilot receives from air traffic control, for example (Wiener & Nagel, 1988), or a continuous record of location in a hierarchical computer program. However, such a memory-aiding alternative is not free. Instead, it represents one of the many trade-offs in human engineering design because display augmentation is generally achieved only at the cost of added display clutter. For example, in an air traffic controller's display, computer-generated tags placed next to each aircraft symbol help to off-load working memory by depicting such information as flight number and altitude. But it is likely that much more information presented about each flight might be excessive, so that the added visual clutter would disrupt the controller's overall image of the flight space.

Another example is reflected in deciding whether to equip commercial aircraft cockpits with cockpit displays of traffic information, or CDTI's (Hart & Loomis, 1980; Palmer, Jago, Baty, & O'Connor, 1980). Such a display, while off-loading the pilot's working memory of the surrounding airspace, will also increase the visual work load. Is the increase worth the benefits? Clearly there is no easy answer to this question. The relative benefits of the CDTI will obviously depend on the degree of crowding of the airspace, the pilot's spatial abilities, and a host of other factors. The important point, however, is that simple memory aids that replace memory store with physical store are not an unmixed blessing. At best they will provide a valuable service if they are needed, and they can be perceptually filtered (or physically turned off) by the user if they are not. However, at worst, memory aids may prove unnecessary in most circumstances and either distract or disrupt the perception of more recent events.

Fortunately, other techniques can be used to reduce forgetting. To understand these techniques, we will consider six factors that influence the probability of forgetting in working memory: time; attention; capacity demands; chunking, which serves to improve working memory; interference; and similarity-induced confusion. With knowledge of these variables, the system designer can reduce the effects of those that are harmful and exploit those that are helpful.

Time and Attention Demands

In the late 1950s experiments conducted by Brown (1959) and Peterson and Peterson (1959) used similar techniques to demonstrate two of the most fundamental properties of working memory: its dependence on time and on attention. Using what has subsequently been labeled the Brown-Peterson paradigm, Peterson and Peterson asked subjects to retain a simple sequence of three random letters in memory for short intervals. To prevent subjects from rehearsing the digits, they were asked to count backward aloud by threes from a designated number, presented just after the item to be remembered. On hearing a recall cue, the subject stopped the count and attempted to retrieve the appropriate item. Both Brown and the Petersons found that retention dropped to nearly zero after only 20 seconds when rehearsal was prevented in this manner. This decay function is shown schematically by curve (b) of Figure 6.4.

This highly transient characteristic of short-term memory has been demonstrated repeatedly in numerous variations of the Brown-Peterson paradigm. The various estimates generally suggest that in the absence of attention devoted to continuous rehearsal, little information is retained beyond 10 to 15 seconds. For example, Loftus, Dark, and Williams (1979) obtained decay functions very similar to those in Figure 6.4 for subjects attempting to remember navigational information similar to that which would be delivered to pilots by air traffic controllers. Moray and Richards (Moray, 1986) replicated the same general decay trend of Figure 6.4 for radar controllers attempting to recall displayed information on a radar scope. Their findings suggest that the transience is equally applicable to spatial and to verbal working memory.

Time and Capacity

Working memory is quite limited in its capacity, or the amount of information that it can hold, and this limit directly interacts with time. Curves (a) and (c) of Figure 6.4 represent decay functions in a Brown-Peterson paradigm that would

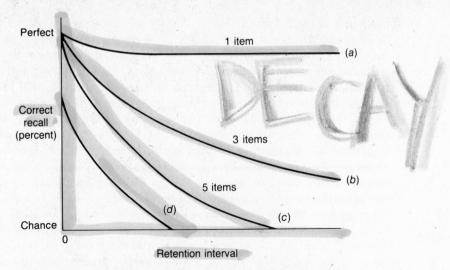

Figure 6.4 Effect of retention interval on recall from working memory with rehearsal prevented. (*Source:* A. W. Melton, "Implications of Short-Term Memory for a General Theory of Memory," *Journal of Verbal Learning and Verbal Behavior*, 2 (1963), p. 9. Copyright 1963 by Academic Press. Reprinted by permission.

be generated by 1- and 5-letter items, respectively (Melton, 1963). Not surprisingly, faster decay is observed when more items are held in working memory, mainly because rehearsal itself (covert speech) is not instantaneous. With more items to be rehearsed in working memory, there will be a longer delay between successive rehearsals of each item. This delay increases the chance that a given item will have decayed below some minimum retrieval threshold before it is next encountered in the rehearsal sequence. In fact, the speed of rehearsal, as dictated either by the length of time it takes to say different items or by differences between people, seems to influence directly the capacity of working memory (Baddeley, 1986, 1990). The faster the speed, the larger the capacity.

The limiting case occurs when a number of items cannot be successfully recalled even immediately after their presentation and with full attention allocated to their rehearsal, as in curve (d) in Figure 6.4. This limiting number is sometimes referred to as the *memory span*. In a classic paper discussed previously in Chapter 2 in the context of absolute judgment, George Miller (1956) identifies the limit of memory span as "the magical number seven plus or minus two" (the title of the paper). Thus somewhere between five and nine items defines the maximum capacity of working memory when full attention is deployed.

To this point, we have spoken somewhat generically of an "item" in working memory, defining it explicitly as a letter in the Brown-Peterson paradigm but not being terribly specific in other cases. Can an "item" (of which 7 ± 2 defines the working memory capacity) also be a number or a word, even if that word consists of 3 letters? Miller (1956) answered this question by proposing the concept of a *chunk*. The capacity of working memory he asserted is 7 ± 2 chunks of information. A chunk can be a letter, a digit, a word, or some other unit. The defining properties of a chunk are a set of adjacent stimulus units that are closely tied together by associations in the subject's long-term memory. Thus seven 3-letter words will define the capacity of working memory, even though this represents 21 letters, because the letter trigrams (*cat, dog,* etc.) are each familiar sequences to the subject—repeatedly experienced together—and so the 3 letters within each are stored together in long-term memory. The 21 letters thereby define seven chunks. Furthermore, if the seven words are combined in a familiar sequence so that the rules that combine the units are also stored in long-term memory ("London is the largest city in England"), the entire string consists only of a single chunk. Thus, the family of decay curves shown in Figure 6.4 would describe equally well a string of 1, 3, 5, or 8 unrelated letters; 1, 3, 5, or 8 unrelated words; or 1, 3, 5, or 8 unrelated but familiar phrases. In each case, the items constituting each chunk within the string are bound together by the "glue" of associations in long-term memory.

It is, of course, important to acknowledge not only the mean (7) but also the variability (± 2) of Miller's limit. For example, some investigators (Conrad, 1957; Jacobs, 1887) have found that unrelated digits have a slightly longer span than unrelated letters, perhaps because we are more familiar with random digit strings or because there are fewer digits and therefore less acoustic confusability between them. Correspondingly, as chunks grow excessively long, the rules for sequencing their elements will be less solid and capacity will be somewhat reduced.

The 7 ± 2 limit is a critically important one in system design. When either presenting auditory or visual information, tasks that encroach on the limits of five to nine chunks should be avoided. In the former category, we might consider the length of strings of navigational information that might be issued to a pilot. For example, the message "Change heading to 179 and speed to 240 knots when you reach flight level 18" approaches or exceeds the limits. As an example of a potential overload of information that must be integrated, consider the number of options to be selected from a menu of a computer on an interactive display terminal. If all alternatives must be compared simultaneously with one another to select the best, the choice will be easier if the number does not exceed working memory limits.

Sometimes the diagnosis of complex electrical system failures imposes an excessive load on working memory. Wohl (1983) offered a model to predict the time needed to troubleshoot system failures as a function of system complexity. In fitting data to his model, he noted that there appears to be a limit on interconnections of about seven or eight components. If this limit is exceeded, producing an overload of working memory, repair time is greatly increased and faults may defy diagnosis.

To provide a quantitative framework for predicting working memory losses, Card, Moran, and Newell (1986) have compiled the data from several experiments to define a "half life" of memory decay. This is the delay after which the probability of correctly recalling the material is estimated to be reduced by one-half, a value of approximately 7 seconds for a memory store of three chunks and 70 seconds for one chunk.

The Exploitation of Chunking

The data just described suggest that one of the best ways to avoid or at least minimize the capacity and decay limitations of working memory is to facilitate chunking of material whenever possible. There is indeed substantial experimental support for this suggestion. A dramatic example is provided by Chase and Ericsson (1981), who trained a subject with a normal memory span to retain and correctly recall strings as long as 82 digits. A second subject was trained to spans of up to 68. Both of these subjects adopted a number of interesting storage and retrieval strategies in the course of the several months of training, and foremost among these was chunking. In particular both subjects, avid runners, learned to encode many of the digit strings as running times for races of a certain distance, and these times became chunks. Thus they might have coded 353431653 as a mile run in 3 minutes, 53.4 seconds, followed by a marathon run in 3 hours, 16 minutes, 53 seconds. Where race times were not appropriate they chunked in terms of familiar dates or ages. Furthermore, to accommodate the longer lists, groups of chunks were combined hierarchically to form "super chunks." Chase and Ericsson argue that subjects do not in fact retain the entire list in working memory. Rather, they use working memory and their chunking strategy to set up a highly efficient storage and retrieval system.

In general, the procedure followed in chunking is either to find or to create a meaningful sequence of stimuli within the total string that because of its meaningfulness has an integral representation already stored in long-term memory. If the physical sequence is close but not identical to the sequence in memory, it may be stored as the memory sequence plus an exception (Chase & Ericsson, 1981). For example, 1676 might be stored as "100 years before Independence Day." As a consequence of this representation, each original string of two to five items that formed the sequence is condensed for storage to a single node. This new set of higher-order nodes is then stored (although these also may be chunked into a second-order node). On recall, only the sequence of nodes must be remembered. When each node is activated in its turn, its contents—the original items—are simply "unpacked" and recalled in their stored order. The issue of how the nodes are recalled in proper sequence will be considered later in this chapter.

Chunking in fact has two correlated benefits: (1) Because it effectively reduces the number of items in working memory, it thereby increases the apparent capacity of this limited store. (2) Because of the reduced load on working memory, material contained therein is more easily rehearsed and is thereby more easily transferred to permanent, long-term memory. This transfer is further facilitated because by definition, the associations between the items that make up a chunk already exist in long-term memory. There appear to be four important characteristics of chunking that may be utilized to advantage in system design. These relate to teaching strategies, familiarity, sequencing data output, and the physical spacing of stimuli.

Teaching Chunking Chunking can be thought of in part as a strategy or mnemonic device that may be taught. The subjects described by Chase and Ericsson (1981) learned to apply grouping principles to facilitate storage. In the same manner, we can learn to search for meaningful subgroups of digits or letters in license plate numbers or other codes that must be learned. In fact, a conclusion from several studies of expert behavior in a variety of disciplines such as computer programming (Norcio, 1981a, 1981b), chess (Chase & Simon, 1973; deGroot, 1965) or decision making (Payne, 1980) is that the expert is able to perceive and store the relevant stimulus material in working memory in terms of its chunks rather than its lowest-level units (Anderson, 1981).

A classic demonstration of this phenomenon was provided by Chase and Simon (1973), comparing recall of a chessboard layout by masters and novices. If the board position was taken from the progression of a logical game, experts recalled far more accurately than did novices. However, if the board position was created by a random placement of pieces, no difference between the two groups was evident. Only in the first case could the expert chunk the various pieces into familiar subpatterns — stored in long-term memory in terms of the evolution of different game strategies. When confronted with the random board that had no correspondence in long-term memory with what a chessboard should look like, the experts' recall regressed to the level of the novice.

Barnett (1989) found similar results when comparing the ability of novice and expert pilots to remember sequences of communications exchanges with air traffic control. When the exchanges flowed in their normal sequence, the experts performed better. When they were randomly scrambled, there was no difference in recall between the groups. We will see in Chapter 7 that expert decision makers seem to encode the relevant pieces of information as a smaller number of chunks if these pieces of information are correlated with one another.

Capitalizing on Familiarity Although chunking is a strategy that is a property of the memorizer, it may also be hindered or helped by properties of the stimulus material that is to be memorized. Obviously, the difference between 21 random letters, which exceed working memory capacity, and the 21 letters of seven 3-letter words, which do not, is that in the second case the sequencing of letters has been constrained to form chunks. Exploiting this difference, system designers should formulate codes in such a way as to facilitate chunking. License plates in many states contain words — 473 HOG — a strategy that takes advantage of this principle. Commercial phone numbers often use familiar alphabetic strings in place of digits ("Dial 263 HELP"). In fact, several years ago the abandonment of the alphabetic prefix (AM 3-7539) for the numeric one (263-7539) in phone numbers seemed to represent a move away from effective chunking. In general, letters induce better chunking than digits because of their greater number and meaningfulness of possible sequential associations.

Parsing Chunking may be facilitated by parsing, or placing physical discontinuities, between subsets that are likely to reflect chunks. Thus the digit sequence 4149283141865 is probably less easily encoded than 4 1492 8 314 1865, which is parsed to emphasize five chunks ("for Columbus ate pie at Appomattox"). For the sufficiently imaginative reader these five chunks in turn may be chunked hierarchically as a single visual image. Loftus, Dark, and Williams (1979) investigated pilots' memory of air traffic control information and observed that four-digit codes were better retained when parsed into two chunks (27 84) than when presented as four digits (2 7 8 4). Bower and Springston (1970) presented sequences of letters that contained familiar acronyms and found that memory was better if pauses separated the acronyms (FBI JFK TV) than if they did not (FB IJF KTV).

When long stimulus sequences are parsed in a way that will facilitate chunking, it is important to consider the optimum size of a chunk for storage. When stimulus-controlled chunks (through parsing) become too long, the likelihood that the units within a chunk will make up a familiar sequence in long-term memory is reduced, to the extent that the parsed unit itself may have to be subdivided by the viewer into two or more chunks. On the other hand, decreasing the size of the parsed unit will, of course, increase the total number of chunks. An experiment by Wickelgren (1964) seems to suggest that the optimum chunk size is three to four. Subjects were shown strings of six to ten digits and asked to rehearse them in groups of one, two, three, four, or five. Recall was best for three-digit rehearsal groups and nearly as good for four-digit groups. This conclusion agrees with the recommendation made by Bailey (1989) concerning the optimum size of grouping for any arbitrary alphanumeric strings used in codes.

The concepts of parsing and chunking have been related to the memory of computer programs by Norcio (1981a and b), who finds that programmers remember sets of lines of codes as chunks. Norcio also suggests that *indentation* of program statements, a form of parsing, can facilitate the chunking process if the indentation is imposed at natural boundaries between chunks (i.e., between logical statements).

In contrast, there is also experimental evidence that in some circumstances indentation may be harmful to comprehension and recall (Schneiderman, 1980). Similar harmful effects are caused by program documentation statements that are inserted within the program (in contrast to the benefits of the initial, context-setting documentation described in Chapter 5). These negative effects may also be attributed to chunking: (1) to the extent that the program aids (indentation or documentation) do not correspond with natural chunks in the programmer's memory or (2) to the extent that either indentation or documentation so expands the physical size of the program that parsimonious groupings can no longer be easily visualized. In the case of indentation, the harm will occur if the shortened lines imposed by indentation require statements to be divided between lines (Schneiderman, 1980). The moral is straightforward. Physical parsing will assist chunking if (1) it conforms to rather than violates subjective chunking boundaries, and (2) it does not greatly increase the physical size of the material.

Sequencing Data Output Memory tasks, such as looking up a phone number and dialing it or hearing instructions and copying them, often involve a sequence of operations in which working memory is first "loaded" and then "emptied" as the information is copied, spoken, or entered onto a keyboard. Normally, we assume that the order of output corresponds to the order of input. But this sequencing does not always reduce the number of chunks that must be stored in working memory as rapidly as possible. Since output sequencing is usually accomplished one *item* at a time, rather than one *chunk* at a time, reducing memory load can be best accomplished by outputting the unchunked items first, followed by the chunked ones.

For example, consider writing down a phone number that you have just heard. First you hear the number 823-7539. Since it is likely that the familiar prefix "823" is a single chunk, your memory load will be unburdened more rapidly if you copy 7539 first, leaving only one chunk remaining (823). An interesting question is why the telephone dialing system itself is not designed to reduce memory load by allowing the four-digit suffix to be entered before the prefix. In fact, this procedure has not been implemented presumably because of the fairly complex changes in switching logic of the switchboard circuiting that would be required. Also, like the redesign of the typewriter keyboard, to be discussed in Chapter 8, any major redesign of a familiar household system such as the telephone would be exceedingly difficult to implement in practice.

Push-button dialing is one technological innovation introduced by the telephone industry that has directly addressed memory load. This simple feature, replacing the old rotary dial, allows the entire sequence of a phone number to be entered with much greater speed, thereby reducing the time during which working memory is loaded.

Output sequencing can be exploited in one other fashion. Whenever any sequence of chunks is encoded, there is a greater likelihood that the first one or two chunks will achieve more permanent storage in long-term memory, and hence will be better recalled at a later time (Chapanis & Moulton, 1990). This is a phenomenon called *primacy* (Adams, 1980). Hence, once a sequence is loaded in working memory, a more efficient strategy is to retrieve those first items last since they, rather than the items in the middle and the end of the sequence, will have the least chance of being lost from working memory (Card, Moran, & Newell, 1986).

Interference

In addition to the forgetting that occurs because of the passage of time, the overload of capacity, and the diversion of attention, material is also lost from working memory through active *interference* from other sources of mental activity. It is important to distinguish the different kinds of interference in terms of the time sequence between their source and the material to be remembered (MTBR).

Figure 6.5 depicts a time sequence during which the operator engages in some activity, is given the MTBR, performs some further activity, and finally retrieves, or "dumps," the MTBR. For example, we might imagine the com-

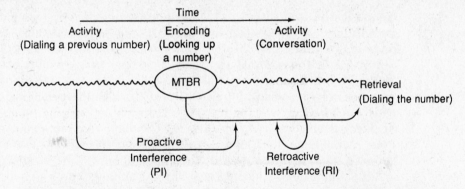

Figure 6.5 Illustrates effects of RI and PI on forgetting of material to be remembered (MTBR). Dialing a previous phone number will produce PI for memory of the next phone number. A conversation after the second number has been looked up will produce RI.

puter user engaged in some program, who then sees a coded error message (the MTBR), must look up the meaning of the error message (in a handbook or through a computer menu), and then finally can retrieve and use the memorized code—to translate its meaning and initiate an action.

As shown in Figure 6.5, *proactive interference* (PI) occurs when activity engaged in prior to encoding the MTBR disrupts its retrieval (Keppel & Underwood, 1962). Its effects can be fairly strong, and when the operator must engage in a whole series of memory tasks with little time between them, its effects are pronounced. Hopkin (1980) suggests that a major potential source of errors for the air traffic controller results when "unwanted baggage" of previously used information disrupts the retention of later material. Using verbal material characteristic of pilots and air traffic controllers, Loftus, Dark, and Williams (1979) found that at least ten seconds' delay was necessary before material remembered in a previous exchange no longer disrupted memory in a subsequent exchange.

Retroactive interference (RI) occurs when material *during* the retention interval disrupts retrieval of the MTBR, for example, forgetting a phone number before it is dialed because someone asked a question during the retention interval. As already stated, increasing similarity (or identity) between *codes* of processing will also increase retroactive interference through resource competition. For example, Brooks's (1968) study showed that working memory for verbal material is more disrupted by concurrent verbal stimulation and activity (listening or talking) than by concurrent spatial activity. The analogous effect is also observed with spatial working memory.

Confusion

Items in working memory are also forgotten because they are confused with other items held at the same time. Intuitively, we can see how this confusion will be most likely to occur if the items are similar to one another. Two examples illustrate this effect. Conrad and Hull (1964) found that working memory capacity for a string of letters is reduced if they are acoustically similar (*E, G, D, T, P,* or *L, X, S, F, N*) than if they are acoustically distinct (*E, U, X, Y, R*). Also, when an air traffic controller must deal with a number of aircraft from one fleet, all having similar identification codes (AI3404, AI3402, AI3401), the interference caused by the similarity of the items makes it difficult for the controller to maintain their separate identity in working memory (Fowler, 1980). Interestingly, although this kind of similarity will disrupt the ordered recall of the items, it may actually enhance their *free recall*—that is, the ability to recall all items independently of their spatial or temporal order (Crowder, 1978). Unfortunately for the air traffic controller, free recall will not be sufficient. The controller must also maintain in working memory the identity of the separate aircraft along some ordered continuum (e.g., projected time of arrival or position in airspace). It is knowing which item goes where that is destroyed by interference from confusion.

Similarity will cause interference at a number of levels. Both acoustic (Conrad & Hull, 1964) and semantic (Shulman, 1972) similarity may cause proactive or retroactive interference, or confusion. In the case of confusion, for example, our memory for a four-digit code will be more disrupted if we must enter it after another four-digit code (4682 1478) rather than after a four-letter code (JBKL 1478) (Bailey, 1989).

The implications of memory interference and confusion for system design are fourfold: When designing coding systems the designer should (1) avoid creating codes with large strings of similar-sounding chunks; (2) use heterogeneous (mixed alphanumeric information), which will reduce similarity-based interference; (3) capitalize on extensive data collected by Chapanis & Moulden (1990) which identify digit strings that are particularly easy to remember (e.g., 333, 888) and are particularly likely to be forgotten or transposed (e.g., 449, 994, 644); and (4) after a task analysis has revealed when information must be stored in working memory, ensure that the intervals before, during, and after storage are free of any unnecessary activity that uses the same code (spatial or verbal) as the stored information.

Running Memory

Our discussion so far has identified the typical estimate of working memory capacity as 7 ± 2 chunks. In many real-world environments, however, this is probably an optimistic figure (Moray, 1980). In the *running memory* task typical of much human-machine interaction, the human's capacity is greatly reduced. A sequence of stimuli is presented to the operator, who neither knows its length nor is expected to retain the entire string. Instead, a different response must be made to each stimulus or series of stimuli at some lag after they occur. For example, a series of aircraft might be "handled" by the air traffic controller, inspected items might be categorized by the quality control inspector, or check readings on instruments of a dynamic system might be stored in order to update an internal model of the system. In this task, if the operator is

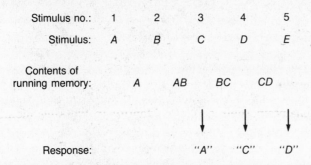

Figure 6.6 Running memory task with $K = 2$.

interrupted and asked to recall the last few items, the memory span will be considerably reduced from the 7 ± 2 value. More difficult still is the situation depicted in Figure 6.6, in which the operator must continuously retrieve information after a lag of a few items. More formally, the operator must hold a certain number, K, of the most recent items in working memory and, as item n is presented, respond with the identity of item $n - K$. In this case, typical of what occurs if an operator falls behind in processing transient stimuli, performance falls off rapidly for values of K greater than 2 (Moray, 1980).

Fortunately, when stimuli are complex and multidimensional in a running memory task, some corrective solutions are available to reduce the effect of the severe limitations. Yntema (1963) has demonstrated some of these solutions in a task analogous to that confronting an air traffic controller. In Yntema's task the subject must keep track of a large number of *objects* (e.g., identified aircraft), each of which varies along a number of *attributes* (e.g., altitude, airspeed, and location), which in turn can take on a number of specific values (e.g., 6000 feet, 400 mph, 60°NW). The value of a particular attribute for one object is periodically updated, and the subject must revise the memory of the current conditions. At random times the subject's memory is probed concerning the nature of one of the objects, and the fidelity of running memory is reflected in the accuracy of report. Yntema investigated different conditions in which the number of objects, the number of attributes per object, and the number of values per attribute were varied. The values were chosen so that each of these manipulations produced formally equivalent changes in the amount of information load.

Three important conclusions were obtained from Yntema's (1963) data. First, subjects performed much better with a few objects that varied on a number of attributes than with many objects that varied on a few attributes, when the total number of variables was the same in both cases. Presumably, the object acts as a sort of integrating "chunk" that readily incorporates its various attributes (Kahneman & Treisman, 1984). This finding suggests that if responsibility in such a monitoring situation is to be divided between operators, it is advisable (from the standpoint of memory limits) to assign each operator to monitor all attributes of a few objects rather than a particular set of attributes for all objects. A similar conclusion was drawn, in an experiment by Harwood et al. (1987), using more representative air-traffic control information.

Second, the number of values per attribute has little influence on the accuracy of memory as long as changes in condition take place at a relatively constant frequency. Thus accuracy will not be greatly sacrificed by increasing the precision with which values are presented and must be retained, despite the fact that more information in the formal sense is required per item. (We should note that Yntema's values were digital readouts and therefore were not susceptible to the absolute judgment limitations described in Chapter 2.)

Third, performance is much better if each attribute has its own unique scale, discriminable from the others. This conclusion is consistent with the confusion caused by item similarity. For example, this guideline would dictate that height and speed must be coded in different scale units—height perhaps in 1,000-foot units and speed in miles per hour. Thus the values 15.2 and 400 would represent an altitude of 15,200 feet and a speed of 400 mph. The discriminability could be enhanced further by presenting the information for each attribute in print of a different size, case, or color (Wickens & Andre, 1990).

In Yntema's (1963) experiment, the information was not continuously available. After an attribute was updated, it was hidden and its value was *required* to be maintained in memory. This state might correspond to the process monitor who is reviewing the status of several plant variables on a selective centralized display. In many real systems, however, information is not so restricted in its availability. For example, the air traffic controller normally has the status of all relevant aircraft continuously visible and so is able to respond on the basis of perceptual rather than memory data. However, even in systems in which there is a continuous display of the updated items, the principles described should still apply. As discussed in Chapter 5, an efficiently updated memory will ease the process of perception through top-down processing and will unburden the operator when perception may be directed away from the display. Furthermore, there is always the possibility of a system failure, in which the perceptual information will no longer exist, not a trivial occurrence in air traffic control. In this case, an accurate working memory becomes *essential* and not just useful.

LEARNING AND LONG-TERM MEMORY

Although limitations of working memory represent a major bottleneck in the operation of many systems, equally important sources of potential failures are the actions people take incorrectly or fail to take because they have forgotten how to do them or simply forgotten to do them at all. For example, the fatal crash of Northwest Airlines flight 255 near Detroit in 1988 was believed to have occurred in part because the pilot forgot to lower the flaps upon takeoff (NTSB, 1988; see also Chapter 10). As more and more powerful computer systems are introduced into the workplace, many hours of wasted time may result because naive users prefer to use simpler but less efficient commands, rather than the more powerful but complex ones they were taught, because the latter have now been forgotten.

The concerns of engineering psychology may be associated with the three stages of memory presented in Figure 6.1: (1) The issue of *training* and how information can be permanently stored in the most efficient manner leads us to the issue of *transfer*: how knowledge learned in one context facilitates the learning of new material. (2) *Storage* or *knowledge representation* leads us to the issues of knowledge organization and mental models. (3) *Retrieval failures, forgetting,* and *retention* leads to what sorts of memory errors people make on retrieval and how forgetting occurs. Although these three categories are treated separately, it is important to realize that they are closely interrelated. For example, the conditions of learning strongly influence both the nature of knowledge representation and the likelihood of forgetting.

Training

Information can be learned through a variety of ways—formal classroom teaching, practice, on-the-job training, focus on principles, theory, and so on. The human factors practitioner who must develop training programs for a new piece of equipment or a job category is concerned with two issues: What method (or device) is most efficient (provides the best learning in the shortest time), leads to the longest retention, and is cheapest? Together these criteria define the issue of *training efficiency*—the greatest level of proficiency per dollar invested.

Naturally any given training technique will not work best for all kinds of learning. Sometimes what we must learn are facts about the domain, which we can easily verbalize or write down. This is called *declarative knowledge* and might describe the safety rules and regulations of a particular company or organization. At other times we must learn how to do something, which is called *procedural knowledge* and is often not easily verbalized (have you ever tried to explain the steps of tying your shoe?). As one might expect, different forms of training are involved for mastery of declarative knowledge (study and rehearsal) versus procedural knowledge (practice and performing). However, a number of other principles, strategies, and considerations may be carefully employed to enhance the learning of either or both kinds of knowledge. We will discuss seven of them and then will consider how prior knowledge influences training: the issue of transfer.

1. *Practice and overlearning.* The expression "practice makes perfect" is one that we are all familiar with, but the issue of *how much* practice is not always obvious. One generalization drawn from the study of skilled performance is that most skills continue to improve after days, months, and even years of practice (Welford, 1968). Such improvement may not be evident in measures of correctness, for with many skills, such as typing or using a piece of equipment, errorless performance can be obtained after a relatively small number of practice trials. However, two other characteristics of performance

continue to develop long after performance errors have been eliminated: The speed of performance will continue to increase at a rate proportional to the logarithm of the number of practice trials (Anderson, 1981), and the attention or resource demand will continue to decline, allowing the skill to be performed in an *automated* fashion (Fisk, Ackerman, & Schneider, 1987; Schneider, 1985). (Overlearning will also decrease the *rate of forgetting* of the skill, as discussed later in this chapter.) These characteristics make it clear that "mastery" training programs, in which training stops after the first or second errorless trial, are short-changing an important part of the automaticity of skill development.

In the previous chapter we discussed briefly the concept of automaticity. It will be recalled that more practice is not sufficient to develop automaticity. For practice to be effective, stimuli or rules must be *consistently* mapped to a response. But when the rules vary from one practice trial to another, as when a computer operator alternates practice between different word-processing programs with different key functions, automaticity will not develop or will do so at a far slower rate (Fisk, Ackerman, & Schneider, 1987).

2. *Elaborative rehearsal.* As noted, rehearsal is an active process, necessary to maintain chunks of information in working memory. In fact, rehearsal appears to be of two types (Craik & Lockhart, 1972). *Rote rehearsal,* involving a pure "recycling" of the phonetic code, is a good way to maintain information in working memory, but it contributes little to the transfer of that material to long-term memory. In contrast, *elaborative rehearsal* involves a greater focus on the semantic and visual code—paying attention to the meaning of the material and, in the process, trying to relate items of that material with one another and with information stored in long-term memory. These operations should be familiar as defining the process of chunking, one of the most crucial skills for creating a permanent, long-term memory representation.

3. *Reducing concurrent task load.* Learning is a task, and it may involve several subcomponents: elaborative rehearsing; perceptual processing, required to notice consistencies in the task environment; processing the feedback about task performance; and formulating strategies. All of these subtasks will be disrupted by competition for resources from concurrent tasks and from one another (see Chapter 9). In short, effective learning will not take place in a high-work-load environment (Nissen & Bullemer, 1987; Schneider, 1985; Schneider & Detweiler, 1988; Sweller, Chandler, Tierney, & Cooper, 1990). Therefore, extraneous task demands, unrelated to the task being learned, should be avoided. Sweller et al. (1990) address those aspects of textbook study of technical material that divert attentional resources from mastery of the underlying concepts, such as separating text from the graphics they describe and imposing difficult problem-solving requirements. Imposing excessively difficult learning demands at the early stages of practice will probably lead to disruption, as the overwhelming challenge of performing a task may divert resources from perceiving and understanding its consistencies (Lintern & Wickens, 1987; Schneider & Detweiler, 1988). This provides a rationale for *adaptive training,* to be discussed below.

4. *Error prevention.* Training that allows errors to be repeated trial after trial will be detrimental, simply because the errors become learned. In addition to creating high workload, very complex learning environments may be so confusing that unwanted errors will occur, leading to further confusion (Carroll & Carrithers, 1984). Error prevention is often accomplished by a technique of *guided training* or *augmented feedback,* which ensures that the learner's performance never strays far from that which is required in the target task. One example of this approach is a "training wheels" program for a word-processing system (Carroll & Carrithers, 1984; Cotrambone & Carroll, 1987). The very brief guided training program was developed in a way that prevents users from making typical mistakes that would affect system performance. In nonguided systems these mistakes might result in wasted time. Instead of allowing the error to affect the system, the training wheels approach offers explicit and immediate feedback about the nature of the error. Compared with a conventional training approach, Cotrambone and Carroll found both faster learning and better transfer to the full word-processing system, concluding that a substantial advantage to training was gained when subjects do not become confused while recovering from errors.

As an example of augmented feedback, consider the difficult task of learning to land an aircraft properly by relying on visual cues from the ground. Feedback might "paint" an ideal flight path through the sky in a visual display; then all the learner must do is track this runway to achieve the proper landing approach (Lintern & Roscoe, 1980). Repeated practice might then "ingrain" the correct sequences of responding necessary to achieve the correct landing. Although such augmented feedback often produces rapid learning of the required skill, it can sometimes lead to fairly *poor* transfer to a more realistic environment (Lintern & Roscoe, 1980; Winstein & Schmidt, 1989). The reason is fairly evident when one considers that the learner may become dependent on the augmented feedback. The learner can perform well by processing the perceptual cues regarding the correct response yet never need to consider the cues that should be learned in the unaided environment when the feedback is withdrawn; nor is the learner forced to make higher-level decisions about the appropriate response. The response to the augmented cues is relatively simple and automatic. When they are withdrawn, the learner may lose a source of information on which he or she is completely dependent, without having acquired the necessary perceptual or decision-making skills to replace it.

Of course a worthwhile goal of the augmented feedback strategy is to prevent errors from occurring, and therefore guard against the possibility that those errors might be repeated and learned. Still, learning may actually be enhanced somewhat if errors are allowed to appear *once,* as long as feedback immediately makes them evident to the learner and they are not repeated (Cotrambone & Carroll, 1987). A unique twist on augmented feedback, which prevents repeated errors but still allows more active perceptual sampling of the environment and choice of responses, is achieved by the strategy of *off-target feedback* (Lintern, Thomley-Yates, Nelson, & Roscoe, 1987). Here supplemental feedback is offered only when performance is outside of some specified criterion. Thus, to stay consistently within the criteria (and thereby perform the skill according to the target level), the learner must be attending to the proper environmental cues and making the appropriate responses. But the feedback, when it appears, ensures that the learner does not err too frequently and understands the appropriate action needed to reduce the error. Studies have revealed the effectiveness of off-target feedback in training aircraft landing skills (Lintern, 1980; Thomley-Yates, Nelson, & Roscoe, 1987).

The two previous guidelines, to reduce workload and to prevent excessive errors during training, are in part the rationale for considering the following two training strategies: adaptive training and part-task training.

5. *Adaptive training.* Adaptive training is one strategy that is available when the target task is a difficult one, so that requirements for its performance may overwhelm the learner. In adaptive training, some component of the task is made simpler to reduce the initial level of difficulty. Then, as training proceeds, this aspect is gradually increased in difficulty until the level of the target task is reached. For example, as we will learn in Chapter 11, the control of dynamic systems with long time lags produced by a higher order is quite difficult. An adaptive training strategy will begin by presenting the learner with a system having no lags (low order), and as proficiency progresses, lags of greater value (higher order) will be introduced (Gopher, Williges, Williges, & Damos, 1975).

Reviews of the adaptive-training literature reveal little if any advantage to these techniques relative to fixed training conditions (Lintern & Gopher, 1978; Lintern & Wickens, 1987). The reason for this ambivalence probably lies in the trade-off of two opposing forces. On the one hand, the simplification imposed by adaptive training probably does release more resources that can be allocated into the learning of task consistencies. On the other hand, the "easy" versions of the task that are presented first may be those that allow (or induce) a response strategy incompatible with the one necessary to perform the task at its final level of difficulty. For example, the human control laws necessary to track a dynamic system without lags are often quite different from those necessary to track lagged systems. Thus, the negative effect that may result from the transfer of incompatible strategies may negate any potential advantage of reduced difficulty. This conclusion suggests that the variable that is adapted must be carefully chosen. One variable that does not distort the response strategy of many tasks is the time stress, or speed with which events occur. Mane, Adams, and Donchin (1989) have successfully demonstrated the value of time pressure as an adaptive variable in training a complex video game. But their experiment also demonstrated the drawbacks of beginning training when the time pressure is too much reduced from the target task level.

6. *Part-task training.* Part-task training describes a strategy in which elements of a complex whole task are learned separately. Wightman and Lintern (1985) have distinguished two different forms of part-task training, shown schematically in Figure 6.7. *Segmentation* defines a training regime in which different *sequential* phases of the skill are practiced extensively by themselves before again being integrated into the whole skill. An example is extensive

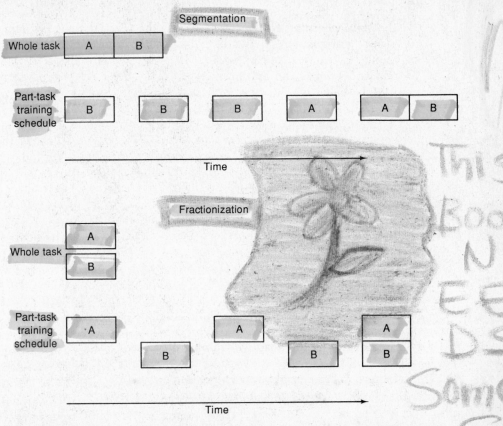

Figure 6.7 Two types of part-task training schedules. (*Source:* D. C. Wightman and G. Lintern, "Part-task Training for Tracking and Manual Control," *Human Factors,* 27(3), 267–283.

practice on a particularly difficult arpeggio of a piano piece. Research indicates that segmentation is a useful strategy, particularly when different segments of the skill vary greatly in their difficulty and therefore in their need for practice. Under these circumstances, it is quite inefficient to spend a great deal of time practicing the easy segments, as would be necessary in a whole-task practice condition (Wightman & Lintern, 1985).

In contrast to segmentation, *fractionization* involves practice on single-task components of two or more tasks (or task components) that must be performed concurrently in the target task. An example is practice on each hand of a piano piece alone or practice on gear shifting without driving. Experimental research indicates that the relative merits of fractionization are much less clear-cut compared to whole-task training because of two competing forces. On the one hand, single-task component training allows full attention to be focused on each component, resulting in more efficient learning of that component, particularly when there are consistencies within the task that must be recognized for automatic processing to develop (Schneider, 1985; Schneider & Detweiler, 1988). This is analogous to the effects of low-work-load training, discussed above. On the other hand, the separation of task components will prevent the development of any *time-sharing skills*, which may be necessary to link and coordinate the two activities (Lintern & Wickens, 1987; Wickens, 1989). Such skills, discussed more in Chapter 9, become particularly crucial if there are dependencies between the different task components—for example, if the events in one component are correlated with those in the other (Naylor & Briggs, 1963). This feature is characteristic of the axes of flight control, in which changes in different instruments depicting vertical and lateral control are correlated. It is also characteristic of working the gear shift and foot pedals in a standard-shift car.

The competition or tension between these two factors helps to explain the ambiguity of experimental results regarding the effectiveness of fractionization (Wightman & Lintern, 1985). However, it also suggests that part-task training may be effective if the whole-task skill is carefully analyzed to identify those components that can be independently broken off (i.e., are not correlated or integrated with the rest of the components) and those that may contain learnable consistencies that can be automated. In one research program, such an approach was taken with a complex video game task known as Space Fortress (Mane, Adams, & Donchin, 1989). Two different task analysis strategies yielded part-task training that was superior to whole-task training (Fabiani et al., 1989; Frederiksen & White, 1989; Mane, Adams, & Donchin, 1989).

One effective form of part-task training, which is also a variant of adaptive training, is *varied priority training*, pioneered by Gopher, Weil, and Siegel (1989). In this technique, training in a multitask skill proceeds by systematically emphasizing one component and deemphasizing the others, even as all are maintained. Hence the integrality of the whole task is not destroyed, and yet full attention does not have to be divided between the various components; rather it can be focused in turn on the consistencies of each component. Training with this strategy produced positive transfer in the Space Fortress game (Fabiani et al., 1989).

7. *Knowledge of results.* Providing the learner with knowledge of results (KR), or feedback about the quality of performance, is useful both for motivation and for correcting and improving performance (Holding, 1987). However, the precise circumstances and factors that optimize KR effectiveness have continued to be the subject of a great deal of inquiry (Winstein & Schmidt, 1989). For example, it is commonly believed that KR should be rich and frequent, but recent research suggests that it can be too detailed. Although some KR is essential for efficient learning, more is not necessarily better.

Two general factors seem to be important in determining the effectiveness of KR: (a) If KR is delayed too long after a practice trial and the interval is filled with other activity, memory for what was done as the skill was executed declines through retroactive interference, and the information regarding what was incorrect about the practiced action becomes less useful for modifying future actions. As we will see in the next chapter, this is an unfortunate characteristic of learning in many decision-making tasks. (b) KR offered while a skill is actually being performed may be processed less well than immediately after the skill's completion, simply because attention must be divided between the two activities (skilled performance and KR processing).

Transfer of Training

One of the most critical factors in skill acquisition is the extent to which learning a new skill, or a skill in a new environment, can capitalize on what has been learned before. This is the issue of *transfer of training* (Holding, 1987; Singley & Andersen, 1989). How well, for example, do lessons learned in a driving simulator transfer to performance on the highway? Or how much does learning one word-processing system help (or even hinder) learning another? Measures of transfer of training are normally used to evaluate the effectiveness of different training strategies, such as some already discussed (Fabiani et al., 1989).

Measuring Transfer   Although there are many ways to measure transfer, the most typical is illustrated in Figure 6.8. The top row represents a *control group*, who learns the target task in its normal setting. This group achieves some satisfactory criterion-level of performance after a certain time — in this example, ten hours. The engineering psychologist then proposes a new training technique, which, it is claimed, will shorten the time needed to learn the target task. A *transfer group* is given some practice with the new training technique and then is transferred to the target task. In the second row, we see that the transfer group trains with the new technique for four hours and then learns the target task faster than the control group, a *savings* of two hours. Hence some information in the training period clearly carried over to the effective performance (or learning) of the target task. Because of this savings in learning time we say that transfer was *positive*. In row 3, we see a second training technique, which apparently had no relevance to the target task (no savings, zero transfer). In row 4, a third training condition was employed, and we see that what was learned before the target task actually *inhibited* learning the target skill. That is, people would have learned the target task faster had they had no prior training. We say here that transfer was *negative*.

Although there are a variety of formulas for expressing transfer, the one that is conventionally used in many settings presents the amount of savings as a percentage of the control group learning time:

$$\% \text{ transfer} = \frac{(\text{control time} - \text{transfer time})}{\text{control time}} \times 100 =$$

$$\frac{\text{savings}}{\text{control time}} \times 100 \tag{6.1}$$

The results of these calculations are shown in Figure 6.8 for the three training conditions.

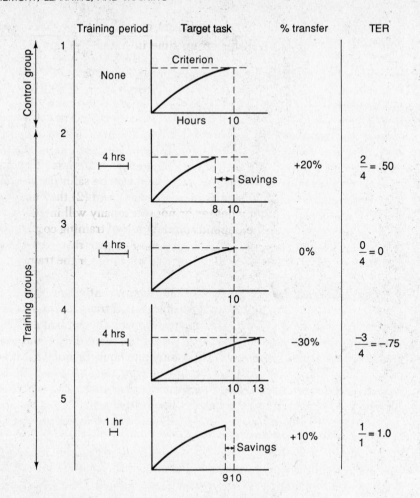

Figure 6.8 The measurement of transfer performance.

Positive transfer is usually a desirable thing, but it is not always clear how much positive transfer is necessary to be effective. For instance, consider the following example, which might have produced the hypothetical data shown in row 2 of Figure 6.8. A driving simulator is developed that produces 20 percent positive transfer to training on the road. That is, learners who use the simulator can reach satisfactory performance on the road in 20 percent fewer road lessons than learners who do all their training on the road. This sounds good, but notice that to get the 20 percent transfer, describing the 2-hour savings, the simulator group had to spend 4 hours in the simulator. Therefore, they actually had to spend 12 total hours of training, compared to the 10 hours spent by the control group. Hence it appears that the simulator, while transferring positively, is less efficient in terms of training time than the actual vehicle.

A measure called the *transfer effectiveness ratio* (TER) expresses this relative efficiency (Povenmire & Roscoe, 1973):

$$\text{TER} = \frac{\text{amount of savings}}{\text{transfer group time in training program}} \qquad (6.2)$$

Examining this formula, we see that if the amount of time spent in the training program (the denominator) is exactly equal to the amount of savings (the numerator), TER = 1. If the total training for the transfer group (training and practice on the target task) is less efficient than for the control group, as is the case with all three groups in Figure 6.8, the TER will be less than 1. If training is more efficient, the TER is greater than 1. A TER that is less than 1 does not mean that the experimental training program is worthless. Two factors may make such programs advantageous: (1) They may be safer (it is clearly safer to train a driver in the simulator than on the road), and (2) they may be cheaper. In fact a major decision of whether or not a company will invest in a particular training program or device depends on the *ratio* of training cost per unit time in the target task environment to that in the training program (e.g., the simulator). This ratio is the *training cost ratio*. In short, the cheaper the training device, the lower can be the TER.

One final characteristic of the TER worth mentioning is the diminishing efficiency of most training devices with increased training time. In the example in row 2 of Figure 6.8, four hours of training were given, and a TER of 0.5 was obtained. But now consider row 5, in which the same device was used for only one hour. Although the savings is now only one hour (half of what it was before) the training time was reduced by 75 percent, and so the TER is 1.0. The difference between rows 2 and 5 expresses a general finding, which is shown in Figure 6.9. Although for very short amounts of training, TERs are typically greater than 1 (Povenmire & Roscoe, 1973), the measure tends to decline as more training is given. The precise point at which TER training should stop and transfer to the target task should begin will depend, in part, on the training cost ratio (TCR). In fact, the amount of training at which TER × TCR = 1 is the point beyond which the training program no longer is cost effective. As noted, however, the training program may still be safety effective for even longer amounts of training.

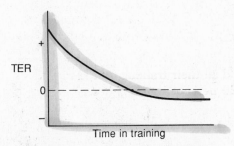

TER

0

Time in training

Figure 6.9 Relationship between time in training and transfer effectiveness ratio, CTER.

What causes transfer to be positive, negative, or zero? To oversimplify the research results somewhat, it is apparent that positive transfer occurs when the training program and the target task are quite similar (in fact, if they are identical, transfer is usually about as positive as it can be, although there are some exceptions). Extreme differences between the training and target task typically produce zero transfer. Learning to type, for example, does not help learning to swim or drive an automobile. Negative transfer occurs from a particularly unique set of circumstances relating to perceptual and response aspects of the task, which we will describe later. We will first consider the goal of the training system design, which is to produce maximum positive transfer by increasing the similarity of the training device to the target task. The issue of *how* similarity should be increased defines training system fidelity. Next we will consider negative transfer between old and new tasks, and finally the role of analogies in transfer.

Training System Fidelity   We stated that maximum positive transfer would generally occur if all elements of a task were identical to the target task. Does this mean that training simulators should resemble the real world as closely as possible? In fact, the answer to this question is no for a number of reasons (Hopkins, 1975; Jones, Hennessy, & Deutsch, 1985; Schneider, 1985). First, highly realistic simulators tend to be very expensive, but their added realism may add little to their TER (Hawkins, 1987). Second, in some cases, high similarity, if it does not achieve complete identity with the target environment, may be detrimental by leading to incompatible response tendencies or strategies. For example, there is little evidence that motion in flight simulators, which cannot approach the actual motion of the aircraft, offers any positive transfer benefits (Hawkins, 1987; Lintern, 1987). Finally, if high realism presents overwhelming complexity, it may so increase workload and divert attention from the critical skill to be learned that effective learning is inhibited.

Instead of total fidelity in training, researchers have emphasized understanding *which* components of training should be made similar to the target task and which may be less important (Holding, 1987; Singley & Andersen, 1989). For example, training simulators for a sequence of procedures may be of very low fidelity to the target task, as long as the sequences of steps are compatible (Hawkins, 1987; Holding, 1987). There is also evidence that a major portion of learning complex skills may be tied to the recognition and use of perceptual consistencies, or *invariants*, in the environment (Lintern, Roscoe, & Sivier, 1990; Schneider, 1985), a concept discussed in Chapter 4. Hence it is important that simulators be designed so that the useful consistencies in the target task are preserved and made visible, and the learner is made aware of them in the training environment (Lintern, Thomley-Yates, Nelson, & Roscoe, 1987). For example, beginning drivers need to recognize the important features of the highway (the heading of the vehicle relative to the vanishing point, more than the momentary deviation) that allow them to steer the vehicle effectively. A training simulator should incorporate this information clearly.

In summary, it appears that some types of departures from full fidelity do not have the detrimental impact on transfer that would be predicted from a straightforward view that maximum similarity produces maximum transfer. Furthermore, some training strategies are based on the assumption that carefully planned departures from similarity can actually enhance transfer if they focus the attention of trainees on critical components of the task.

Negative Transfer The issue of negative transfer is a critical one, as the continued emergence of new technology and different system designs leads operators to switch from one system to another. What causes the skills acquired in one setting to inhibit some aspect of performance in a different one? A history of research in this area (Holding, 1976; Martin, 1965; Osgood, 1949) seems to reveal that the critical conditions for negative transfer are related to the stages of processing. When the two situations have highly similar (or identical) stimulus elements but different response or strategic components, transfer will be negative, particularly if the new and old responses are incompatible with one another (i.e., the new and the old response cannot easily be given at the same time). Stated another way, Singley and Andersen (1989) have argued that a major source of negative transfer is from the positive transfer of *inappropriate* responses. The relationship between the similarity of stimulus and response elements and transfer is shown in Table 6.1.

It is important to realize that many real-world tasks involve the transfer of a large number of different components, and generally more of them produce positive rather than negative transfer. Hence, most transfers from one similar task to another are generally positive. However, the critical design questions may be focused on those aspects of the difference between training and transfer (or between an old and new system) that *do* involve incompatible responses or inappropriate strategies. For example, consider two word-processing systems that present identical screen layouts but require a different set of key presses to accomplish the same editing commands. A high level of skill acquired through extensive training on the first system will show some interference with transfer to the second, even though the overall transfer will be positive. Or consider two control panels that both require a lever movement to accomplish a function. However, in one panel the lever must be pushed up, and in the other it must be pushed down.

Table 6.1 RELATIONSHIP BETWEEN OLD AND NEW TASK

Stimulus Elements	Response Elements	Transfer
Same	Same	++
Same	Difference (incompatible)	−
		−−
Different	Same	+
Different	Different	0

Negative transfer is as much of a concern for an operator who may have to use two systems as it is for a learner who has to transfer from a training device to a target system. In the former case, the designer must be concerned about the kinds of errors that will result from negative transfer, for example, when a company installs a new word-processing system or mandates a new procedure. In the commercial airlines industry, a concern relates to the number of different types of aircraft a pilot may be allowed to fly (transfer between) without undergoing an entirely new training program (Braune, 1989). The lack of standardization in the control arrangement of light aircraft can also lead to serious problems of negative transfer.

There is also a positive side to transfer effects—that differences between two systems are not always harmful for positive transfer. As shown in Table 6.1, two systems may be considerably different in their display characteristics but can involve positive transfer if there is identity in the response elements. For example, there will be high positive transfer between two automobiles with identical control layouts and movements, even if they have very different dashboard displays. Furthermore, if unfortunate circumstances arise in which responses between two systems must be different and incompatible, Table 6.1 suggests that the amount of negative transfer may be reduced by actually *increasing* the display differences. For example, the operator confronting the two control levers with incompatible motion directions will have fewer problems if the appearance of the handles (both their visual and tactile characteristics) are quite distinct.

Transfer of Analogy and Metaphor   The discussion of positive and negative transfer to this point has focused heavily on skill learning. However, the psychologist may also be interested in how best to teach the working of a device or system—the development of a *mental model* (Gentner & Stevens, 1983; Norman, 1988). Some research has found that an effective way of imparting such knowledge is through a metaphor or analogy to another system with which the learner is already familiar (Gentner & Gentner, 1983). Thus, a positive transfer of knowledge is created. For example, one who is familiar with the physical procedures of filing, moving, creating, and discarding in an office can transfer this knowledge effectively to a computer system if it has been designed so that the similarity of such actions is emphasized (Carroll & Olson, 1987). As noted in Chapter 4, the Apple Macintosh computer has capitalized effectively on the office metaphor. Webb and Kramer (1990), successfully taught novice computer users the structure of a data base menu system by analogy to a department store. This analogy provided learning that was superior to strict instruction in procedures. Mayer (1981) found an advantage in teaching computer skills when the training emphasized metaphors, such as storage as a filing cabinet and input as a ticket window.

Analogies, however, may not always be appropriate, and if the learner focuses on characteristics of the analogous system that do not match the system to be learned, problems can arise. For example, the hydrodynamic analogy to electronic power (i.e., water flow to electron flow) is useful to some extent, but it breaks down if learners expect power to keep flowing from a socket if the socket is not in use. Similarly the typewriter analogy to a word processor breaks down when one equates the carriage return to the return key (Carroll & Olson, 1987). Care must be taken to ensure that a system chosen for analogy training matches the target system in its deep structural characteristics and not just its surface features (Gentner, 1983).

Conclusion In summary, the research on training strategies and transfer has produced ambivalent results, many of which suggest that a given training strategy cannot be blindly applied to all tasks. However, a careful analysis of the information-processing requirements of a given task can reveal the conditions under which a given strategy may be more or less effective (Frederiksen & White, 1989; Mane et al., 1989). Concepts of resource competition, response compatibility, automaticity, time-sharing, and processing stages have all been invoked to help explain the success or failure of basic training strategies. In short, a fairly obvious conclusion is reinforced. Learning — the storage of new information in long-term memory — is an information-processing task such as any other discussed in this book. Thus the successful application of principles of training depends in part on an understanding of these information-processing components.

Memory Storage, Organization, and Representation

Once information is encoded into long-term memory through learning and training, its representation in storage can take on a wide variety of forms. As in working memory, the information in long-term memory can have any combination of verbal (sounds), spatial (images), and semantic characteristics. Also, some knowledge is procedural (how to do things), and other knowledge is declarative (knowledge of facts). Another distinction is drawn between general knowledge of things, like word meaning (semantic memory), and memory for specific events (episodic memory). Engineering psychologists have concerned themselves both with the organization of the knowledge in memory and with its representation as a mental model.

Knowledge Organization Researchers have long known that information is not stored in long-term memory as a random collection of facts. Rather, that information has specific structure and organization, defining the ways in which the different factors or items of knowledge are associated with one another. The understanding of knowledge organization has proven useful to engineering psychologists by providing guidelines for designing interfaces for data bases that contain the knowledge to be used. In particular, systems designed to allow the operator to use knowledge from a domain will be well served if their features are congruent with the operator's organization of that knowledge. This principle is the same one discussed in Chapter 4, where we described the advantage of displays that were congruent with operator mental models.

Consider an index or a menu system for retrieval of information from a data base. If the categories and the structures defined by the system do not correspond to the user's mental organization of them, the user's search for a particular item may be time-consuming and frustrating, often needing to follow the linear search procedure outlined in Chapter 3. That is, the user must start at the first item on the list and scan down until the target is reached. This lack of congruence can often result if the system designer organizes knowledge in a way that is quite different from the "typical" user. The psychologist who is writing a book on engineering psychology may index information relevant to display design under the heading "Perception, visual," whereas the engineer who is using the book would want to look under "Visual displays." To support a broad variety of users, indexes should be relatively broad and redundant, with items of information accessible under different categories (Bailey, 1989; Roske-Hofstrand & Paap, 1986). However, with menu systems, in which space may be scarce or time may be critical, this luxury is not always possible. In this case it is particularly important to understand the mental representation of the typical user.

In Chapter 4, we described the study by Durding, Becker, and Gould (1977) in which information about a set of words was best retrieved by users if the words were spatially organized to be compatible with their perceived underlying organization: hierarchical, networked, orthogonal, or listed. A further demonstration of the importance of congruence between menu structure and user knowledge is provided in an experiment by Roske-Hofstrand and Paap (1986), who were interested in determining the proper menu layout for fighter pilots to interact with a computerized data base of aviation systems. Two layouts were compared: one in which the items were organized according to the original designer's principles, and a second in which they were organized according to the pilot's mental representation of the systems. The mental representation described which systems were most closely associated with which other systems in the pilot's long-term memory, and its form was deduced by a technique of *multidimensional scaling,* in which pilots rated the degree of similarity between different terms (Kruskal & Wish, 1978). Roske-Hofstrand and Paap found that the pilot-organized menu structure was easier to use (and gave rise to fewer errors) than the original structure configured by the system designer.

Mental Models   The organization of knowledge about how a system works or operates has been described as a *mental model,* a concept we have discussed briefly at other points in this book (Carroll & Olson, 1987; Gentner & Stevens, 1983; Norman, 1988; Rouse & Morris, 1986; Wilson & Rutherford, 1989). Carroll and Olson define a mental model as follows:

> The user's mental model of a system is here defined as a rich and elaborate structure, reflecting the user's understanding of what the system contains, how it works, and why it works that way. It can be conceived as knowledge about the system sufficient to permit the user to mentally try out actions before choosing one to execute. (p. 12)

Mental models, of course, may be correct or incorrect, and they may be created spontaneously by the user or carefully formed and structured through training, either by explicit use of analogy or direct explanation.

An accurate mental model of a system can be advantageous because it provides the user with knowledge that is useful when other learned procedures fail. Yet mental models may be erroneous, and when they are, they may lead to breakdowns in performance and the commission of errors (Doane, Pellegrino, & Klatzky, 1990). Norman (1988) offers some particularly compelling examples of erroneous mental models of everyday devices. If mental models are sometimes erroneous, it would seem advantageous to create a correct mental model by explicit training that emphasizes the underlying causal structure and principles operating in the system, principles that lurk behind the procedures necessary to operate the system and its visible controls and displays.

Research comparing training through a mental model to that by strict learning procedures has revealed results that are somewhat in favor of the mental model, although not invariably so (Kieras & Bovair, 1984). For example, Halasz and Moran (1983) taught two groups of students how to use a calculator. One was taught procedures and the other given a mental model of the calculator's operations. Although both groups fared equally well in performing standard tasks, those trained with the mental model did better on novel tasks.

An important aid to the formation of a correct mental model of a system is the concept of *visibility*, introduced by Norman (1988). A device is said to have visibility if by looking at it one can immediately tell the state of the device and the alternatives for action. The relation between operator actions and state changes can then be immediately and easily seen (and thereby more easily learned). Switches that occupy different positions when they are activated have visibility; push buttons, such as those found on most digital watches, do not. Multimode systems, when one control can serve more than one function, also tend to obscure visibility. The concept of visibility also refers to the ability of a system to display its intervening variables between an operator's action and the final system response. For example, a thermostatic system that shows the state of the system that is generating or removing heat, as well as the momentary temperature, has visibility. A concern with high levels of automation, to be discussed in Chapter 12, is that they tend to obscure visibility of the system. Norman (1988) offers the example shown in Figure 6.10 as an instance when lack of visibility leads to an incorrect mental model of a refrigerator. The cooling controls in the refrigerator for the freezer and fresh food are shown in (*a*). The two independent arrows suggest the configuration in (*b*) — two independent, noninteracting cooling controls. Yet the correct model is that shown in (*c*), creating an interaction whose effect on the temperature will be difficult to understand when the incorrect, independent mental model is held.

Memory Retrieval and Forgetting

Knowing that a piece of information, a fact, or a skill has been learned and is now stored in long-term memory in no way guarantees that it will be retrieved

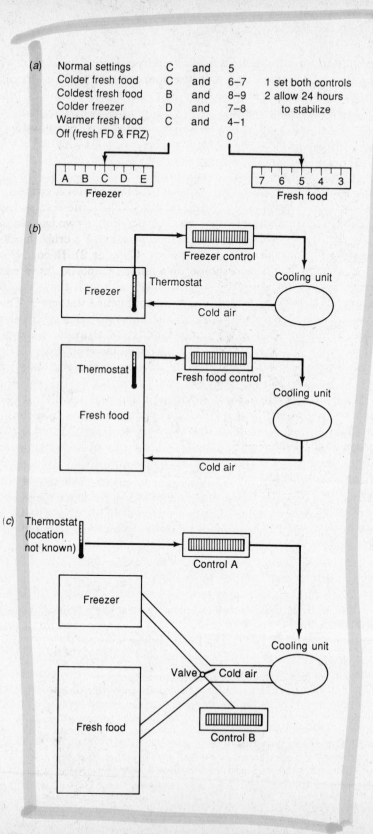

(a)

Normal settings	C	and	5
Colder fresh food	C	and	6–7
Coldest fresh food	B	and	8–9
Colder freezer	D	and	7–8
Warmer fresh food	C	and	4–1
Off (fresh FD & FRZ)			0

1 set both controls
2 allow 24 hours
to stabilize

A B C D E
Freezer

7 6 5 4 3
Fresh food

(b)

Freezer control

Freezer Thermostat

Cooling unit

Cold air

Thermostat

Fresh food control

Fresh food

Cooling unit

Cold air

(c) Thermostat (location not known)

Control A

Freezer

Cooling unit

Valve Cold air

Fresh food

Control B

correctly (or at all) when its use is required. Hence engineering psychologists must be just as concerned with the sources and causes of memory failure as they are with the efficient means of learning. In understanding these causes it is important to distinguish between the two main ways in which memory can be retrieved: recall and recognition. *Recall* describes the situation in which the entire sequence of stored information must be retrieved. For example, the skilled computer user recalls the sequence of steps necessary to load a program or accomplish an operation, or the telephone dialer recalls the phone number. In *recognition*, one must only indicate whether or not a given fact or piece of information is stored in long-term memory. Thus, the lost navigator suddenly recognizes the location because the configuration of two landmarks looks familiar, or the eyewitness recognizes the perpetrator of a crime because the image of the suspect is stored in memory (see Chapter 2). Recognition is typically indicated with a yes-no response or a choice response (as in recognizing the correct answer on a multiple-choice test), although these responses may be augmented by a "strength" estimate or confidence statement. The long history of research on recognition versus recall seems to indicate the relative independence of the two retrieval systems (Mandler, 1980) and to suggest that recognition is a more sensitive measure. That is, even though we may no longer be able to recall things we can often recognize them as familiar once we see or hear them.

What causes the failures of recognition and recall that lead to human performance errors? Research has pointed to the same mechanisms of *retroactive*, *proactive*, and *within-list* interference that disrupt working memory. Also the roll of *similarity* can be identified. As we saw in the discussion of negative transfer, a set of procedures learned for one word-processing system can very easily become confused in memory with a set of procedures for a different system, particularly if many other aspects of the two systems are identical.

Another source of forgetting is the absence of *retrieval cues*, external stimuli, or events that trigger a memory. For example, a piece of equipment

---

Figure 6.10 (at left) (*a*) A refrigerator. Two compartments — fresh food and freezer — and two controls (in the fresh-food unit). The illustration shows the controls and instructions. Your task: Suppose the freezer is too cold, the fresh-food section just right. How would you adjust the controls to make the freezer warmer and keep the fresh food the same? (*Source:* Donald A. Norman, *The Psychology of Everyday Things*. Copyright © 1988 by Basic Books, Inc. Reproduced by permission of Basic Books, a division of HarperCollins Publishers Inc.) (*b* and *c*) Two conceptual models for the refrigerator. The model (*b*) is provided by the system image of the refrigerator as gleaned from the controls and instructions; (*c*) is the correct conceptual model. The problem is that it is impossible to tell in which compartment the thermostat is located and whether the two controls are in the freezer and fresh-food compartment or vice versa.

that has the visual numbers (1), (2), (3) . . . next to the controls to be activated in that sequence when following a procedure is providing retrieval cues. A major source of vulnerability, which we will address again in Chapter 10, is the unprompted checklist when retrieval cues for each action are missing (Degani & Wiener, 1990; Reason, 1990). Finally, the mere *passage of time* causes forgetting. We remember best those things that have happened most recently, an important phenomenon in the discussion in Chapter 7, of information integration in decision making.

We will concern ourselves now with three general applications of research on forgetting: the distinction between recognition and recall in the user-computer interface design, the memory distortions that occur when events are recalled, and the retention of skills.

Recall versus Recognition   Tests of recognition provide either one option or a set of options, and the user picks the appropriate one; recall requires the proper option to be self-generated. The contrast between these two retrieval mechanisms has been considered by the designers of computer interfaces in their choice between *menus* or *keywords* for selectable options. A keyword system requires the operator to recall the correct option in its appropriate context and enter it into a keyboard, whereas the menu system only requires recognition of the correct option from a printed list on the display. To use Norman's (1988) terms, recall requires "knowledge in the head"; recognition places this "knowledge in the world." Not surprisingly, recall failures are often a source of frustration for novice users of keyword systems who would be better served by viewable options on a menu system (Schneiderman, 1987). Yet with the experience and learning of the expert, the appropriate keywords are recalled automatically and are preferred over the more cumbersome and display-consuming menu-based systems. However, since forgetting degrades recall faster than it does recognition, keyword systems are more vulnerable to forgetting when they are not frequently used. The issue of designing systems that will serve both novice and expert users equally well has presented a major challenge to designers. Seemingly an optimum solution is one in which both means are equally available, so that recognition can serve when recall fails.

Event Memory   In Chapter 2, we discussed the application of signal detection theory to recognition memory. Signal was defined as viewing a particular stimulus that had actually been seen before, and *noise* was defined as viewing the stimulus for the first time. We described the sorts of biases that affected the recognition memory of eyewitness testimony. Extending these concepts to broader concerns about the accuracy of episodic memory in real-world environments leads us to consider the many situations in which people are asked to recall events, for example, the witness in a judicial proceeding (Neisser, 1982) or the system operator in a hearing following an accident. An important source of memory failure or bias in both of these cases is the loss of knowledge about the precise *source* of information recalled and the tendency to include erroneously information that was not encoded at the incident. Thus, witnesses are likely to "fill in" details of an event subconsciously to make them plausible with the way the world runs, even though those details might not actually have been observed (Neisser, 1982). This use of one's general knowledge to replace or augment details of a specific event appears to be particularly characteristic of the expert, who will possess a large storage of that knowledge (Arkes & Freedman, 1984). It is as if top-down processing operates on one's memory for reconstructing events.

Of equal concern is the way that events and incidents that have occurred subsequent to an event can be absorbed into one's memory of the event (Loftus, 1979; Loftus & Palmer, 1974; Wells & Loftus, 1984). For example, Brown, Deffenbacher, and Sturgill (1982) had people view staged crimes in poor viewing conditions and then, considerably later, asked them to pick the perpetrator from a lineup. The investigators found that the witnesses were likely to pick an "innocent" face from a lineup if they had seen a picture of that face in the time between the crime and the lineup viewing. What is perhaps disturbing about these shortcomings is not so much that they occur but rather that witnesses tend to be unaware of the fact that they occur. As a result, witnesses' confidence in the accuracy of their memory tends to be high, and this overconfidence is passed on to the jury or inquiry board (Lindsay, Wells, & Rumpel, 1981). We will deal with this form of overconfidence in Chapter 7.

Since human testimony remains a necessary source of information in judicial proceedings or accident investigations, it seems that the appropriate solution is simply for the jury or board of investigation to be made very much aware of (1) the tendency to incorporate subsequent information into the memory of prior information and (2) the general failure of confidence to be calibrated with accuracy (i.e., witnesses who are most confident are not necessarily most accurate).

Skill Retention   In contrast to the recall (or recognition) of specific episodes, operators are more frequently called on to recall how to perform a particular skill. Very often this recall process is accurate and effortless. For example, we do not forget how to drive a car, even though we might not have done so for a few days, weeks, or even years. But sometimes the problem of skill forgetting is a substantial one. The commercial airline industry is sufficiently concerned with pilots forgetting skills that they do not often practice (e.g., recovery from emergencies) that recurrency training is required every six months.

It is critically important for the engineering psychologist to have some way to predict what skills will be forgotten at what rate in order to know how often operators should be required to participate in recurrency training. To develop a predictive model of the decay or forgetting of different skills, Rose (1989) identified four important variables that influenced skill retention.

1. *The retention interval.* The forgetting of an unused skill follows the same general curve illustrated in Figure 6.4, except of course that the rate of forgetting is much, much slower and can be extremely variable, depending on the three following factors.

2. *Degree of overlearning.* Overlearning involves additional trials of prac-

tice after performance is judged to be satisfactory or error free. Practice trials after these criteria have been reached both make the task more automated and reduce the rate of forgetting. This critical factor is sometimes neglected in training programs in which the operator is assumed to be sufficiently practiced once a criterion level of skill is first reached. Of course, in training for skills that are subsequently used on a daily basis (like driving or word processing), such overlearning will occur in the subsequent performance. But because learning skills related to emergency response procedures, for example, will not receive this same level of on-the-job training, their retention will greatly benefit from overlearning. Rose (1989) notes that the effects of overlearning tend to show diminishing benefits with additional overlearning trials. That is, the first few trials of additional practice will give the greatest retention benefits, and each subsequent trial will have a lesser benefit.

3. *Task type.* Rose (1989) identifies two task types that greatly influence the length of skill retention. First, perceptual motor skills, such as driving, flight control, and most sports skills, show very little forgetting over long periods of time. In contrast, procedural skills, which require a sequence of steps, such as how to use a text processor or how to run through a checklist for turning on a piece of equipment, tend to be rapidly forgotten. Second, meaningful material, which the learner has been easily able to relate to old knowledge, tends to be retained longer and better than less meaningful material. This difference attests to the advantages of using analogies in training, previously discussed.

4. *Individual differences.* Faster learners tend to show better retention than slower learners. Rose (1989) suggests that this difference may be related to chunking skills. As we have seen, better chunking will lead to faster acquisition as well as more effective and efficient storage in long-term memory.

TRANSITION

In this chapter we have discussed at length the separate components of verbal and spatial working memory and long-term memory. Each has different properties and different codes of representation, yet all are characterized by stages of encoding, storage, and retrieval. Failures of each of these processes result in forgetting, which is a critical point of breakdown in human-system interaction. Techniques of system and task design and procedures to facilitate memory storage (training) were discussed.

In the next chapter, we discuss decision making, jumping from the memory box in Figure 1.3 into the forward flow of information processing toward response selection. Our treatment of decision making, however, depends closely on our understanding of memory and learning in three respects. First, many decisions place heavy loads on working memory. The costs imposed by these loads often lead to mental shortcuts, or *heuristics*, which produce systematic biases in decision performance. Second, other decisions are more directly tied to long-term memory and experience. We decide to take a particular action because the circumstances match our memory of the circumstances in which we carried out that same decision before, and we remember that its outcome then was successful. Finally, we will learn that the decision-making task has some unique features, which cause learning and expertise in decision making to be somewhat different from that in other skills.

REFERENCES

Adams, J. A. (1980). *Learning and memory: An introduction*. Homewood, IL: The Dorsey Press.

Anderson, J. R. (1981). *Cognitive skills and their acquisition*. Hillsdale, NJ: Erlbaum.

Arkes, H., & Freedman, M. R. (1984). A demonstration of the costs and benefits of expertise in recognition memory. *Memory & Cognition, 12,* 84–89.

Baddeley, A. D. (1986). *Working memory*. Oxford: Clarendon Press.

Baddeley, A. (1990). *Human memory: Theory and practice*. Boston, MA: Allyn and Bacon.

Baddeley, A. D., Grant, S., Wight, E., & Thompson, N. (1975). Imagery and visual working memory. In P. M. Rabbitt & S. Dornic (eds.), *Attention and performance V*. New York: Academic Press.

Baddeley, A. D., & Hitch, G. (1974). Working memory. In G. Bower (ed.), *Recent advances in learning and motivation* (vol. 8). New York: Academic Press.

Baddeley, A. D., & Lieberman, K. (1980). Spatial working memory. In R. S. Nickerson (ed.), *Attention and performance VIII*. Hillsdale, NJ: Erlbaum.

Bailey, R. W. (1989). *Human performance engineering using human factors/ergonomics to achieve computer system usability* (2nd ed.). Englewood Cliffs, NJ: Prentice Hall.

Barnett, B. J. (1989). Information-processing components and knowledge representations: An individual differences approach to modeling pilot judgment. *Proceedings of the 33rd annual meeting of the Human Factors Society* (pp. 878–882). Santa Monica, CA: Human Factors Society.

Bower, G. H., & Springston, F. (1970). Pauses as recoding points in letter series. *Journal of Experimental Psychology, 83,* 421–430.

Braune, R. J. (1989). *The common/same type rating: Human factors and other issues*. Anaheim, CA: SAE.

Brooks, L. R. (1967). The suppression of visualization in reading. *Quarterly Journal of Experimental Psychology, 19,* 289–299.

Brooks, L. R. (1968). Spatial and verbal components in the act of recall. *Canadian Journal of Psychology, 22,* 349–368.

Brown, E., Deffenbacher, K., & Sturgill, W. (1982). Memory for faces and the circumstances of encounter. In U. Neisser (ed.), *Memory observed: Remembering in natural contexts* (pp. 130–138). San Francisco, CA: W. H. Freeman and Company.

Brown, J. (1959). Some tests of the decay theory of immediate memory. *Quarterly Journal of Experimental Psychology, 10,* 12–21.

Card, S., Moran, T., & Newell, A. (1986). The model human processor. In K. Boff, L. Kaufman, & J. Thomas (eds.), *Handbook of perception and human performance* (vol. 2). New York: Wiley.

Carroll, J. M., & Carrithers, C. (1984). Blocking learner error states in a training-wheels system. *Human Factors, 26*(4), 377–389.

Carroll, J. M., & Olson, J. (eds.). (1987). *Mental models in human-computer interaction: Research issues about what the user of software knows.* Washington, DC: National Academy Press.

Chapanis, A., & Moulden, J. V. (1990). Short-term memory for numbers. *Human Factors, 32*(2), 123–137.

Chase, W. G., & Ericsson, A. (1981). Skilled memory. In S. A. Anderson (ed.), *Cognitive skills and their acquisition.* Hillsdale, NJ: Erlbaum.

Chase, W. G., & Simon, H. A. (1973). The mind's eye in chess. In W. G. Chase (ed.), *Visual information processing.* New York: Academic Press.

Conrad, R. (1957). Accuracy of recall using keyset and telephone dial and the effect of a prefix digit. *Journal of Applied Psychology, 42,* 285–288.

Conrad, R., & Hull, A. J. (1964). Information, acoustic confusions, and memory span. *British Journal of Psychology, 55,* 429–432.

Cotrambone, R., & Carroll, J. M. (1987). Learning a word processing system with training wheels and guided exploration. *Proceedings of CHI & GI human factors in computing systems and graphics conference.* New York: ACM, 169–174.

Craik, F. I. M., & Lockhart, R. S. (1972). Levels of processing: A framework for memory research. *Journal of Verbal Learning and Verbal Behavior, 11,* 671–684.

Crowder, R. (1978). Audition and speech coding in short-term memory. In J. Requin (ed.), *Attention and performance VII.* Hillsdale, NJ: Erlbaum.

Danaher, J. W. (1980). Human error in ATC system. *Human Factors, 22,* 535–546.

Degani, A., & Wiener, E. L. (1990). *Human factors of flight-deck checklists: The normal checklist* (NASA Contractor Report 177549). Moffett Field, CA: NASA Ames Research Center.

deGroot, A. D. (1965). *Thought and choice in chess.* The Hague: Mouton.

Doane, S. M., Pellegrino, J. W., & Klatzky, R. L. (1990). Expertise in a computer operating system: Conceptualization and performance. *Human-Computer Interaction, 5,* 267–304.

Durding, B. M., Becker, C. A., & Gould, J. D. (1977). Data organization. *Human Factors, 19,* 1–14.

Fabiani, M., Buckley, J., Gratton, G., Coles, M. G. H., Donchin, E., & Logie, R. (1989). The training of complex task performance. *Acta Psychologica, 71,* 259–299.

Fisk, A. D., Ackerman, P. L., & Schneider, W. (1987). Automatic and controlled processing theory and its applications to human factors problems. In P. A. Hancock (ed.), *Human factors psychology* (pp. 159–197). North-Holland: Elsevier Science Publishers B.V.

Fowler, F. D. (1980). Air traffic control problem: A pilot's view. *Human Factors, 22,* 645–654.

Frederiksen, J. R., & White, B. Y. (1989). An approach to training based upon principled task decomposition. *Acta Psychologica, 71,* 89–146.

Gentner, D. (1983). Structure mapping: A theoretical framework for analogy. *Cognitive Science, 7,* 155–170.

Gentner, D., & Gentner, D. R. (1983). Flowing waters or teeming crowds: Mental models of electricity. In D. Gentner & A. L. Stevens (eds.), *Mental models* (pp. 99–129). Hillsdale, NJ: Erlbaum.

Gentner, D., & Stevens, A. L. (1983). *Mental models.* Hillsdale, NJ: Erlbaum.

Gopher, D., Weil, M., & Siegel, D. (1989). Practice under changing priorities: An approach to the training of complex skills. *Acta Psychologica, 71,* 147–177.

Gopher, D., Williges, B. H., Williges, R. L., & Damos, D. (1975). Varying the type and number of adaptive variables in continuous tracking. *Journal of Motor Behavior, 7(3),* 159–170.

Halasz, F. G., & Moran, T. P. (1983). Mental models and problem solving in using a calculator. In A. Janda (ed.), *Human factors in computing systems: Proceedings of CHI 1983 Conference* (pp. 212–216). New York: Association for Computing Machinery.

Hart, S. G., & Loomis, L. L. (1980). Evaluation of the potential format and content of a cockpit display of traffic information. *Human Factors, 22,* 591–604.

Harwood, K., Wickens, C. D., Kramer, A., Clay, D., & Liu, Y. (1986). Effects of display proximity and memory demands of the understanding of dynamic multidimensional information. *Proceedings of the 30th annual meeting of the Human Factors Society.* Santa Monica, CA: Human Factors Society.

Hawkins, F. H. (1987). *Human factors in flight.* Brookfield, VT: Gower Technical Press.

Healy, A. F. (1975). Temporal-spatial patterns in short-term memory. *Journal of Verbal Learning and Verbal Behavior, 14,* 481–495.

Hoffman, E. R., & MacDonald, W. (1980). Short-term retention of traffic turn restriction signs. *Human Factors, 22,* 241–252.

Holding, D. H. (1976). An approximate transfer surface. *Journal of Motor Behavior, 8(1),* 1–9.

Holding, D. H. (1987). In G. Salvendy (ed.), *Handbook of human factors.* New York: Wiley.

Hopkin, V. S. (1980). The measurement of the air traffic controller. *Human Factors, 22,* 347–360.

Hopkins, C. O. (1975). How much should you pay for that box? *Human Factors, 17(6),* 533–541.

Jacobs, J. (1887). Experiments in prehension. *Mind, 126,* 75–79.

Jones, E. R., Hennessy, R. T., & Deutsch, S. (1985). *Human factors aspects of simulation.* Washington, DC: National Academic Press.

Kahneman, D., & Treisman, A. (1984). Changing views of attention and automaticity. In R. Parasuraman & R. Davies (eds.), *Varieties of attention.* New York: Academic Press.

Keppel, G., & Underwood, B. J. (1962). Proactive inhibition in short-term retention of single items. *Journal of Verbal Learning and Verbal Behavior, 1,* 153–161.

Kieras, D. E., & Bovair, S. (1984). The role of a mental model in learning to operate a device. *Cognitive Science, 8,* 255–273.

Klapp, S. T., & Netick, A. (1988). Multiple resources for processing and storage in short-term working memory. *Human Factors, 30*(5), 617–632.

Kruskal, J. B., & Wish, M. (1978). *Multidimensional scaling.* Sage University Paper Series on Quantitative Applications in the Social Sciences, 07–011. Beverly Hills and London: Sage Publications.

Lee, A. T. (1989). Data link communications in the national airspace system. *Proceedings of the 33rd annual meeting of the Human Factors Society* (pp. 57–60). Santa Monica, CA: Human Factors Society.

Lindsay, R. C., Wells, G. L., & Rumpel, C. M. (1981). Can people detect eyewitness identification accuracy within and across situations? *Journal of Applied Psychology, 67,* 79–89.

Lintern, G. (1980). Transfer of landing skill after training with supplementary visual cues. *Human Factors, 22,* 81–88.

Lintern, G. (1987). Flight simulation motion systems revisited. *Human Factors Society Bulletin, 30*(12), 1–3.

Lintern, G., & Gopher, D. (1978). Adaptive training of perceptual-motor skills: Issues, results, and future directions. *International Journal of Man-Machine Studies, 10,* 521–551.

Lintern, G., & Roscoe, S. N. (1980). Visual cue augmentation in contact flight simulation. In S. N. Roscoe (ed.), *Aviation Psychology.* Ames: Iowa State University Press.

Lintern, G., Roscoe, S. N., & Sivier, J. (1990). Display principles, control dynamics, and environmental factors in pilot performance and transfer of training. *Human Factors, 32,* 299–317.

Lintern, G., Thomley-Yates, K. E., Nelson, B. E., & Roscoe, S. N. (1987). Content, variety, and augmentation of simulated visual scenes for teaching air-to-ground attack. *Human Factors, 29*(1), 45–59.

Lintern, G., & Wickens, C. D. (1987). *Attention theory as a basis for training research* (Technical Report ARL–87-2/NASA-87-3). Savoy: University of Illinois, Institute of Aviation.

Loftus, E. F. (1979). *Eyewitness testimony.* Cambridge, MA: Harvard University Press.

Loftus, E. F., & Palmer, J. C. (1974). Reconstruction of automobile destruction: An example of the interaction between language and memory. *Journal of Verbal Learning and Behavior, 13,* 585–589.

Loftus, G. R., Dark, V. J., & Williams, D. (1979). Short-term memory factors in ground controller/pilot communications. *Human Factors, 21,* 169–181.

Logie, R. H. (1989). Characteristics of visual short-term memory. *European Journal of Cognitive Psychology, 1*(4), 275–284.

Logie, R. H., & Baddeley, A. D. (1990). Imagery and working memory. In P. J. Hampson, D. E. Marks, & J. T. E. Richardson (eds.), *Imagery: Current developments* (pp. 103–128). London: Routledge.

Mandler, G. (1980). Recognizing: The judgment of previous occurrence. *Psychological Review, 87,* 252–271.

Mane, A. M., Adams, J. A., & Donchin, E. (1989). Adaptive and part-whole training in the acquisition of a complex perceptual-motor skill. *Acta Psychologica, 71,* 179–196.

Martin, E. (1965). Transfer of verbal paired associates. *Psychological Review, 72*(5), 327–343.

Mayer, R. E. (1981). The psychology of how novices learn computer programming. *Computer Surveys, 13,* 121–141.

Melton, A. W. (1963). Implications of short-term memory for a general theory of memory. *Journal of Verbal Learning and Verbal Behavior, 2,* 1–21.

Miller, G. A. (1956). The magical number seven plus or minus two: Some limits on our capacity for processing information. *Psychological Review, 63,* 81–97.

Moray, N. (1980, May). *Human information processing and supervisory control* (Technical Report). Cambridge: MIT, Man-Machine System Laboratory.

Moray, N. (1986). Monitoring behavior and supervising control. In K. R. Boff, L. Kaufman, & J. P. Thomas (eds.), *Handbook of perception and human performance.* New York: Wiley.

Murdock, B. B. (1968). Modality effects in short-term memory: Storage or retrieval? *Journal of Experimental Psychology, 77,* 79–86.

National Transportation Safety Board (1988). *Northwest Airlines, Inc. McDonnell Douglas DC-9-82 N312RC, Detroit Metropolitan Wayne County Airport, Romulus, Michigan, August 16, 1987* (Report No. NTSB-AAR-88-05). Washington, DC: Author.

Naylor, J. C., & Briggs, G. E. (1963). Effects of task complexity and task organization on the relative efficiency of part and whole training methods. *Journal of Experimental Psychology, 65,* 217–224.

Neisser, U. (1982). *Memory observed: Remembering in natural contexts.* San Francisco, CA: W. H. Freeman and Company.

Nilsson, L. G., Ohlsson, K., & Ronnberg, J. (1977). Capacity differences in processing and storage of auditory and visual input. In S. Dornick (ed.), *Attention and performance VI.* Hillsdale, NJ: Erlbaum.

Nissen, M. J., & Bullemer, P. (1987). Attentional requirements of learning: Evidence from performance measures. *Cognitive Psychology, 19,* 1–32.

Norcio, A. F. (1981a). *Comprehension aids for computer programs* (Technical Report AS-1-81). Annapolis, MD: U.S. Naval Academy, Applied Sciences Department.

Norcio, A. F. (1981b). *Human memory processes for comprehending computing programs* (Technical Report AS-2-81). Annapolis, MD: U.S. Naval Academy, Applied Sciences Department.

Norman, O. (1988). *The psychology of everyday things.* New York: Harper & Row.

Office of Technology Assessment (1988). *Safe skies for tomorrow: Aviation safety in a competitive environment* (OTA-SET-381). Washington, DC: U.S. Government Printing Office.

Osgood, C. E. (1949). The similarity paradox in human learning: A resolution. *Psychological Review, 56,* 132–143.

Palmer, E., Jago, S., Baty, D., & O'Connor, S. (1980). Perception of horizontal aircraft separation in a cockpit display of traffic information. *Human Factors, 22,* 605–620.

Payne, J. W. (1980). Information processing theory: Some concepts and methods applied to decision research. In T. S. Wallsten (ed.), *Cognitive processes in choice and decision behavior.* Hillsdale, NJ: Erlbaum.

Peterson, L. R., & Peterson, M. J. (1959). Short-term retention of individual verbal items. *Journal of Experimental Psychology, 58*, 193–198.

Posner, M. I. (1978). *Chronometric explorations of the mind.* Hillsdale, NJ: Erlbaum.

Povenmire, H. K., & Roscoe, S. N. (1973). Incremental transfer effectiveness of a ground-based general aviation trainer. *Human Factors, 15*, 534–542.

Reason, J. (1990). *Human error.* Cambridge, England: Cambridge University Press.

Rolt, L. T. C. (1978). *Red for danger.* London: Pan Books.

Rose, A. M. (1989). Acquisition and retention of skills. In G. MacMillan (ed.), *Applications of human performance models to system design.* New York: Plenum Press.

Roske-Hofstrand, R. J., & Paap, K. R. (1986). Cognitive networks as a guide to menu organization: An application in the automated cockpit. *Ergonomics, 29,* 1301–1311.

Rouse, W. B., & Morris, N. M. (1986). On looking into the black box: Prospects and limits in the search for mental models. *Psychological Bulletin, 100*, 349–363.

Schneider, W. (1985). Training high-performance skills: Fallacies and guidelines. *Human Factors, 27*(3), 285–300.

Schneider, W., & Detweiler, M. (1988). The role of practice in dual-task performance: Toward workload modeling in a connectionist/control architecture. *Human Factors, 30*(5), 539–566.

Schneiderman, B. (1987). *Designing the user interface.* Reading, MA: Addison-Wesley.

Schneiderman, B. (1980). *Software psychology.* Cambridge, MA: Winthrop.

Shulman, H. G. (1972). Semantic confusion errors in short-term memory. *Journal of Verbal Learning and Verbal Behavior, 11*, 221–227.

Singley, M. K., & Andersen, J. R. (1989). *The transfer of cognitive skill.* Cambridge, MA: Harvard University Press.

Sweller, O., Chandler, P., Tierney, P., & Cooper, M. (1990). Cognitive load as a factor in the structuring of technical material. *Journal of Experimental Psychology: General, 119*, 176–192.

Webb, J. M., & Kramer, A. F. (1990). Maps or analogies: A comparison of instructional aids for menu navigation. *Human Factors, 32*, 251–266.

Welford, A. T. (1968). *Fundamentals of skill.* London: Methuen.

Wells, G. L., Lindsay, R. C., & Ferguson, T. I. (1979). Accuracy, confidence, and juror perceptions in eyewitness testimony. *Journal of Applied Psychology, 64*, 440–448.

Wells, G. L., & Loftus, E. F. (1984). *Eyewitness testimony: Psychological perspective.* New York: Cambridge University Press.

Wetherell, A. (1979). Short-term memory for verbal and graphic route information. *Proceedings of the 23rd annual meeting of the Human Factors Society.* Santa Monica, CA: Human Factors Society.

Wickelgren, W. A. (1964). Size of rehearsal group in short-term memory. *Journal of Experimental Psychology, 68*, 413–419.

Wickens, C. D. (1984). *Engineering psychology and human performance.* New York: Harper & Row.

Wickens, C. D. (1989). Attention and skilled performance. In D. Holding (ed.), *Human skills* (2nd ed.) (pp. 71–105). New York: Wiley.

Wickens, C. D. & Andre, A. D. (1990). Proximity compatibility and information display: Effects of color, space, and objectness of information integration. *Human Factors, 32*, 61–77.

Wickens, C. D., & Liu, Y. (1988). Codes and modalities in multiple resources: A success and a qualification. *Human Factors, 30*, 599–616.

Wickens, C. D., Sandry, D., & Vidulich, M. (1983). Compatibility and resource competition between modalities of input, central processing, and output: Testing a model of complex task performance. *Human Factors, 25*, 227–248.

Wickens, C. D., Vidulich, M., & Sandry-Garza, D. (1984). Principles of S-C-R compatibility with spatial and verbal tasks: The role of display-control location and voice-interactive display-control interfacing. *Human Factors, 26*, 533–543.

Wickens, C. D., & Weingartner, A. (1985). Process control monitoring: The effects of spatial and verbal ability and current task demand. In R. Eberts & C. Eberts (eds.), *Trends in ergonomics and human factors*. North Holland Pub. Co.

Wiener, E., & Nagel, D. (1988). *Human factors in aviation*. San Diego, CA: Academic Press.

Wightman, D. C., & Lintern, G. (1985). Part-task training for tracking and manual control. *Human Factors, 27*(3), 267–283.

Wilson, J. R., & Rutherford, A. (1989). Mental models: Theory and application in human factors. *Human Factors, 31*(6), 617–634.

Winstein, C. J., & Schmidt, R. A. (1989). Sensorimotor feedback. In D. H. Holding (ed.), *Human skills* (2nd ed.). Chichester, Eng.: Wiley.

Wohl, J. (1983). Cognitive capability versus system complexity in electronic maintenance. *IEEE Transactions on Systems, Man, and Cybernetics, 13*, 624–626.

Yntema, D. (1963). Keeping track of several things at once. *Human Factors, 6*, 7–17.

Chapter 7

Decision Making

OVERVIEW

The previous chapters have described how information is perceived and then sometimes stored in memory. But perceptual processes often trigger the need to decide on a course of action, which may be covert or overt and visible. The general area of decision making considered in this chapter has the following characteristics: (1) The operator must select one of a number of possible choices or courses of action when presented with stimulus information bearing on the choice. (2) The time frame for the choice is relatively long, more than a second (distinguishing the present discussion from that of reaction time in Chapter 8). (3) The probability that the choice of action will be the correct or "best" one, over the long run, is considerably less than 1.0, either because of the probabilistic nature of the stimulus information or because of the operator's own cognitive limitations.

Two particularly tragic examples have recently illustrated the failures and shortcomings of human decision making. In 1985, a decision was made to launch the space shuttle *Challenger* in especially cold weather. This decision was taken in spite of prior knowledge that the cold conditions made the bindings on the o-rings of the booster rocket subject to failure. The subsequent midflight explosion of the *Challenger* resulted in the loss of the entire crew. In 1987 the crew of the U.S.S. *Vincennes*, sailing in potentially hostile waters of the Persian Gulf, erroneously classified a radar signal as belonging to a hostile, attacking fighter aircraft. The aircraft was actually an Iranian civilian airliner, and the decision to launch a defensive missile resulted in the tragic loss of 290 lives (*APA Monitor*, 1988; U.S. Navy, 1988). In hindsight, both of these were unfortunate decisions because their outcome led to tragedy that would have been avoided had another choice of action been made. But whether the decisions were actually "bad" or not, given the information available, is less clear (Klein, 1989a), and to understand the reasons for this uncertainty, we must examine in detail the nature of the decision-making task. This is the objective of this chapter.

Decision making is found in almost any operational environment, including *medical diagnosis:* the physician considers a number of symptoms (stimuli) to diagnose a disease; *fault diagnosis:* the supervisor of a complex dynamic system (e.g., a nuclear plant) processes a number of warning indications to diagnose the nature of the underlying malfunction; *treatment:* the physician (or nuclear operator) weighs the costs and benefits of a number of possible treatments to remedy the diagnosed disease (or malfunction) before deciding on a particular course of action; *weather forecasting:* the forecaster integrates a number of sources of information (time of year, pressure, and temperature) to make a forecast for tomorrow; *pilot judgment:* the aircraft pilot considers a number of factors (visibility, height, fuel consumption, and schedule) to decide whether or not to continue a landing approach in bad weather; *factory production control:* the industrial manager decides which products should be emphasized, given the availability of raw materials and personnel and the demands of the market; *selection:* the admissions committee weighs a number of attributes of an applicant before deciding to admit or reject; *judicial procedures:* the jury (or judge) weighs a number of sources of evidence about the suspect before deciding on a verdict of guilt or innocence or on the length of sentence; and *consumer behavior:* the buyer considers a number of products and compares several attributes before deciding to purchase one. These examples of decision making are diverse, yet they all have certain common elements. More important, all encounter limitations of human information-processing capabilities that prevent decisions from being correct as often as is possible.

In this chapter, a framework is provided for analyzing the prominent features of the decision-making task. Two major classes of limitations in decision making will then be considered: those relating to the estimation of statistics from data and those relating to actual diagnosis decisions and the inference process. The latter involves a lengthy discussion of the manner in which human limitations in memory, attention, and logic affect behavior in a variety of decision-making and troubleshooting situations. A third section considers the role of costs and values in decision making, and a fourth addresses the role of learning and experience. The final section on decision-making aids identifies areas in which efforts have been made to improve decision-making capabilities.

Features of the Decision-Making Task

There are several types of decision-making tasks and a number of different decision phenomena, most of which can be accounted for within the framework of the general decision model shown in Figure 7.1. According to this model, the decision maker first samples a number of cues or information sources, from the environment. These may be perceived all at once or sequentially, but they are often related to the true state of the world (on which the decision should be based) through a shroud of uncertainty. That is, information can be known to be true only with a probability of less than 1.0.

Sampling and integrating this information, the decision maker will usually attempt to formulate a hypothesis about the true state of the world and use it as

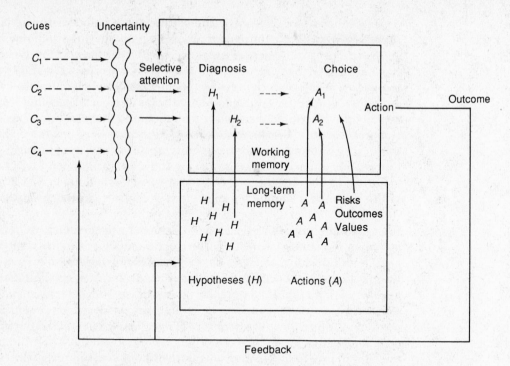

Figure 7.1 Information-processing model of decision making. Uncertain cues about the state of the world are selectively sampled (on the left) to form a hypothesis, which is the basis for a choice of action (on the right). Possible actions, their risks, and the values of their outcomes are also stored in long-term memory. The outcome of the decision may affect both the environment and what is learned (i.e., long-term memory).

a basis for further selections. Sometimes the hypothesis concerns the current state of affairs, as when the troubleshooter hypothesizes why a computer will not operate correctly; this is *diagnosis*. At other times the hypothesis is a *prediction* or *forecast* of what will happen in the future. The hypothesizing process involves an interplay between *long-term memory*, where plausible hypotheses are stored after they have been learned, and *working memory*, where alternative hypotheses are entertained, compared, and evaluated against the information provided by the cues. As we will see, different diagnoses may vary in their reliance on one or the other of the two memory systems.

An initial diagnosis may well be followed by a further search for cues, indicated by the arrow at the top of Figure 7.1, which will either confirm or refute it. Furthermore, even the final diagnosis may not be absolute: It may instead be an expression of degree of belief that one hypothesis rather than another is true (this is the meteorologist's forecast of "70 percent chance of thunderstorms"), but it is usually followed by a *choice of actions*. Given the 70 percent chance of thunderstorms, the pilot may choose to go ahead with the flight plans or cancel. Given the 80 percent chance that a problem lies in the computer hardware and not in the software, the choice is to call the maintenance service rather than to continue debugging. In these and other examples, choice requires the consideration of costs and benefits: What are the costs of canceling the flight? What are the potential consequences of going ahead with the flight if there will be thunderstorms? Hence, choice usually involves the evaluation of *risk*, when there is uncertainty about the state of the world that will affect the consequences of the choice. Much of our information about risk, along with our repertoire of possible actions, is stored in long-term memory. There are situations, however, in which choice is made with relative certainty, as when you choose to purchase one car instead of another. This choice is not based on a diagnosis of present or future states. Rather, it is based on an evaluation of preference, or *values*, for different features or attributes of the cars.

Whether the choice is made with certainty or uncertainty, people typically try to increase gains or reduce losses, and an important factor in predicting and modeling decision making relates to human understanding of gains and losses in an uncertain, probabilistic world. Finally, in any information-processing model, an important element is *feedback*. The outcome of the decision may affect the next round of sampled cues, and knowledge of the outcome, whether successful or not, can influence future decisions through learning or through modification of the current plans.

Decision-making tasks vary greatly in their difficulty, and throughout this chapter we will discuss features that affect that difficulty—both the perceived effort to make the decision and the likelihood that the decision will be correct. One of the most important influences is the number of sources of information that must be considered in making the choice. How many symptoms does the physician consider before making a diagnosis? Or how many features of a college did you consider before your final choice? It turns out that the decision-making problem is greatly eased if there is *correlation* between these cues as they have been experienced in the past. In medical diagnosis, for example, high temperature usually goes hand in hand with a change in skin color. Decision making is also easier if the operator can make the decision on the basis of direct recall of the response from long-term memory, rather than using mental calculations.

Many aspects of decision making are not as accurate as they could be. The limitations of information processing and memory, previously discussed, restrict the accuracy of diagnosis and choice. In addition, limits of attention and cognitive resources lead people to adopt decision-making *heuristics*, or "mental shortcuts," which produce decisions that are often adequate but not usually as precise as they could be (Kahneman, Slovic, & Tversky, 1982; Tversky & Kahneman, 1974). Finally, we will sometimes refer to general *biases* in the decision-making process. These biases are either described as *risky*—leading to a course of action based on insufficient information—or *conservative*—leading to the use of less information or less confidence in a decision than is warranted.

HUMAN LIMITS IN STATISTICAL ESTIMATION

In discussing human limits in decision making, an analogy with the classical procedure of statistical inference may be helpful. This procedure often is viewed as a two-stage process. First, the statistician or psychologist computes some *descriptive statistics* of the data at hand (e.g., mean, proportion, standard deviation). Then these estimated statistics are used to draw some *inference* about the sample under consideration (e.g., whether it is from the same population as another sample or a different population or whether two samples are correlated). When examining human limitations in decision making, two issues that are analogous with these stages must be considered: a person's ability to perceive and store probabilistic data accurately and the ability to draw inferences (and thereby make decisions) on the basis of those data. Peterson and Beach (1967) adopt this framework by considering the human operator as an "intuitive statistician." We will examine data bearing on the ability to estimate four basic descriptive statistics: means, proportions, standard deviations, and parameters of exponential growth. Then we will address human limits in the inferential processes of diagnosis.

Perception of the Mean

When presented with a single number (e.g., 27), human perception proceeds fairly automatically in accordance with principles outlined in Chapter 5. However, when the operator is presented with several numbers or several marks on a measurement scale and asked to estimate their mean value (without resorting to mental arithmetic), very different processes are engaged (Pitz, 1980). Experimental evidence suggests, however, that the estimation of the mean is done reasonably well (Peterson & Beach, 1967; Pitz, 1980; Sniezek, 1980). Although the estimated mean does not necessarily correspond precisely to the true mean of a set of numbers, it does not appear to be systematically biased in one direction or another.

Perception of Proportions

When a quality control monitor is asked to make an intuitive estimate of a defect rate, he or she is estimating the value of a proportion from a sample of data. Estimates of proportions, unlike those of means, tend to show some small but systematic biases. Toward the midrange (e.g., from 0.10 to 0.90), estimates are fairly accurate. However, with more extreme values, subjects seem to "hedge their bets" conservatively away fom extreme values (Sheridan & Ferrell, 1974; Varey, Mellers, & Birnbaum, 1990). In a typical example (Erlick, 1961), subjects saw a rapidly displayed series of two kinds of events (the letters *a* and *c*). Later, they were asked to report the proportion of letters of one kind. On a trial in which only 10 *c* events out of 100 occurred, $P(c) = 0.10$, subjects might report that $P(c) = 0.15$. The same sort of bias against perceiving (or reporting) extremely rare or frequent events was noted in Chapter 2 to be one potential cause of the sluggish beta in signal detection theory, and it seems to influence decision-making tasks such as purchasing insurance or choosing between pairs of gambles having different probabilities of winning.

Why is this conservatism observed? One explanation may be subjects' reluctance to report extreme values, believing that it is safer to err on the side of caution. A second may be that event probability is not just coded in terms of relative event frequency but also weighted by event *salience*. Psychologists investigating the orienting response have long recognized that novel stimuli attract attention and are more salient than those occurring with greater frequency (Sokolov, 1969). Frequent events, on the other hand, induce adaptation, or a failure to attend. The relatively greater salience of the rare event may cause the estimates of its frequency to be biased upward. This finding illustrates an important general point that will recur later. A major source of bias in decision making and judgment results when operators direct attention to the most *perceptually* salient aspects of the environment rather than to those that are most relevant to the task or judgment at hand.

Estimating Variability

Human estimates of variability are used in different contexts. For instance, the task of estimating variability is important when operators must assess the contribution of random noise to a process (e.g., to a meter reading) to determine how large a deviation constitutes a noteworthy signal. In other contexts, the desirability of a product or outcome may be based on its consistency (low variability) on some dimension. Finding that a batch of sheet metal consistently has a precise number of acceptable flaws, for example, may be more desirable than finding that the number of flaws in a batch is highly variable, even though the mean flaw rate in both batches is equivalent. In the former case, a constant corrective action can be taken, whereas in the latter it cannot.

When asked to estimate the variance or standard deviation of a set of numbers, people do not perform as well as when they are estimating means. Given the more complex computation required to compute variance, this difference in accuracy is perhaps not surprising. In fact, two sorts of biases have been reported, one related to the mean and the other to the extremes. Lathrop (1967) noted that the estimation of variability is inversely related to the mean value of the quantities. Of two sets of numbers or analog readings with equal variability, the set with the greater mean will be estimated to have the smaller variability. This finding seems to be a special case of Weber's law in psychophysics, which states that as stimulus magnitude increases, the amount of variability in stimulus magnitude that can just be perceived also increases. Thus, 20-foot variability in the altitude of an airplane flying at 1000 feet may be perceived as small, but the same variability when flying near the ground will be perceived as large. Pitz (1980) observed that estimates of variability are influenced greatly by the members of the data set that are most *salient* to the variability estimation task—namely, the two extreme values. The amount of dispersion of data points between these two is relatively discounted. In fact, for some subjects this dispersion is entirely ignored, and the range of values is the only determination of the estimated variability.

Extrapolating Growth Functions

People are often required to predict or forecast future trends on the basis of a series of past and present data points. The chemical process control monitor must examine the past history of temperature recordings and decide if the process temperature is level or increasing. If it is increasing, is it doing so at a constant rate or is it accelerating? The economic forecaster or business investor must decide if an economic indicator is stable or is going out of control. A point that has and will be discussed elsewhere in this book is that humans do not generally perform well at this prediction task. Waganaar and Sagaria (1975) provide graphic evidence that a systematic *conservative* bias occurs when people are asked to extrapolate the future course of an exponential or accelerating growth function (Figure 7.2). Their future predictions (dashed line) typically underestimate the growth that is predicted by mathematical extrapolation of the observed function into the future (dotted line). Interestingly, Waganaar and Sagaria observed the same magnitude of conservative bias in a group of

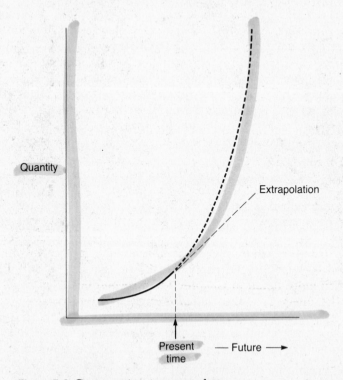

Figure 7.2 Conservatism in extrapolation.

"experts," subjects from the Joint Conservation Committee of the Pennsylvania legislature, who should be better trained in making such predictions than naive subjects. Gottsdanker and Edwards (1957) and Runeson (1975) observed similar conservative trends in extrapolating the future motion of accelerating objects. Subjects tended to describe a constant velocity path that simply extended the velocity at the time of the most recent observation.

Three possible causes may account for this conservative bias in extrapolation: (1) Exponential growth functions are cognitively more complex, requiring more parameters to express verbally, or more analog manipulations to represent spatially, than linear ones. Therefore, the linear representation serves as a simplifying heuristic employed to reduce cognitive effort (Moray, 1980; Rasmussen, 1981; Tversky & Kahneman, 1974). (2) The underestimation may represent a conservative resistance to acknowledging the extreme values that an exponential extrapolation generates. This explanation is similar to one that was postulated to account for conservatism in extreme probability estimations. (3) There may, in fact, be a legitimate rational reason for conservatism in extrapolation. The subject simply goes beyond the data given and changes an estimation task into an inference task. An extrapolation task carries the assumption that the mathematical function generating the first portion of the curve will continue in effect for the latter portion. That is, an extrapolation task assumes a stable environment in which the growth function continues unaltered to infinity. In real-life experiences, however, most processes are not unchanging but have built-in limits or self-correcting feedback loops that will arrest the process as it goes out of control. Countries where populations expand exponentially may adopt family-plannning policies. Temperatures or pressures that increase exponentially will trigger fire extinguishers or safety valves. The expectation that such self-correcting procedures will "linearize" or at least reduce the speed of growth may be such a powerful and automatic bias (and also a legitimate and highly rational one) that it is imposed even in the pure laboratory environment when the subject is told to assume that an unchanging environment does exist (Einhorn & Hogarth, 1981).

Whatever the underlying cause and degree, the essential finding remains that humans do not extrapolate growth functions according to the mathematical laws governing them. If, in an applied setting, it is necessary to extrapolate that function to determine what the future status will be *if self-correcting processes do not occur*, some design innovations must be introduced to counter this human bias. One solution is for computer-generated "best fit" extrapolations of the function to be displayed explicitly. These "predictive" displays will be considered in more detail in Chapter 11. A second possible design change to attenuate the effect of conservatism is suggested by Waganaar and Sagaria (1975). They propose a graphic presentation in which the dependent variable in the extrapolated function is transformed so that the function is displayed linearly rather than as an accelerating curve of the linear function. For example, an exponential population growth of people per square mile will become a decreasing linear function if expressed as square mile per person.

DIAGNOSIS AND HYPOTHESIS FORMATION: INTEGRATING INFORMATION

We will now discuss the human being as an inference maker. As we saw in Figure 7.1, these inferences or hypotheses about the current or future state of the world are based on cues, sampled from the environment. The cues are then interpreted with regard to one or more of the hypothesized diagnostic states until one of three situations is obtained: (1) A state is diagnosed with certainty, (2) a state is diagnosed with uncertainty (as the weather forecaster's "80 percent chance of rain"), or (3) more information is sought. We will first consider the problems that result when cues become available over time, presenting this process in the context of one classical model of optimal decision making—Bayes's theorem. Then we will address other problems relating to integrating information from several *different* sources, whether simultaneous or sequential.

Integrating Information over Time: Conservatism and Anchoring

A medical doctor receives the results of a series of tests over the course of a day, trying to establish whether the patient does or does not have a serious disease. A quality control inspector divides products delivered by an automatic manufacturing device into two groups, defective and normal. On the basis of the frequency of defectives (an estimated proportion), the inspector must decide if the manufacturing process is operating normally (hypothesis H_1) or if it is malfunctioning in some respect (H_2). Both of these scenarios may be represented as those in which an operator samples pieces of information concerning the likelihood that one hypothesis or the other is true. The odds are revised after each sample. Formally, the odds may be represented as the ratio of the two probabilities. That is, the odds that the process is malfunctioning is the ratio of the probability of malfunction to the probability of normal operation, or $P(H_1$ is true$)/P(H_2$ is true$)$. (This formula assumes that the two hypotheses exhaust all possibilities.)

During the 1960s a large quantity of research was performed within this general experimental paradigm (e.g., Edwards, 1962, 1968; Edwards, Lindman, & Phillips, 1965; Edwards, Lindman, & Savage, 1963). The research by Edwards and his colleagues explored this phenomenon within the framework of an optimal model of hypothesis revision known as Bayes's theorem. Briefly, this theorem states that the odds in favor of a given hypothesis after the acquisition of a piece of data should equal the prior odds (i.e., the odds before the datum was collected) multiplied by the *likelihood ratio*. The likelihood ratio is the probability of observing that particular datum if the favored hypothesis is true, divided by the probability of observing the same datum if the other hypothesis is true. This relationship is described formally in Equation 7.1, which expresses the odds in favor of hypothesis 1, given the observation of a datum, D:

$$\frac{P(H_1/D)}{P(H_2/D)} = \frac{P(H_1)}{P(H_2)} \times \frac{P(D/H_1)}{P(D/H_2)}$$ (7.1)

Accordingly, a weather forecaster wishing to predict the odds for rain on a given day should multiply the prior odds for rain (the number of times it has rained on that day of the year divided by the number of times it has not) by the likelihood ratio of a particular piece of meteorological data (the probability of observing that datum if it is going to rain divided by the probability if it is not).

Edwards and his colleagues contrasted human performance when revising odds with the performance of an optimum statistical model following Bayes's theorem. The universal conclusion of their research is that the human operator is generally *conservative*. That is, in revising a hypothesis (or adjusting the odds), the operators do not extract as much information from each diagnostic observation of data as they should do (Edwards, 1968; Edwards, Lindman, & Phillips, 1965).

These findings are described by a more general heuristic, which Tversky and Kahneman (1974) have labeled *anchoring*. The initial piece of evidence provides a cognitive "anchor" for the decision maker's belief in one of several hypotheses. Subsequent sources of evidence are not given the same amount of weight in updating beliefs but are used only to shift the anchor slightly, particularly if those sources provide evidence for the other hypothesis (Einhorn & Hogarth, 1982). One clear implication of the anchoring heuristic is that the strength of belief in one hypothesis over another will be different, and may even reverse, depending on the *order* in which evidence is perceived. Allen (1982) has observed such reversals as weather forecasters study meterological data on the probability of precipitation, and Hogarth and Einhorn (1989) have considered similar reversals as people hear evidence that is either supporting or damaging to a particular hypothesis about an event, for example, jurors hearing different pieces of evidence for the guilt or innocence of a suspect.

A study by Tolcott, Marvin, and Bresoick (1989) of professional Army intelligence analysts clearly demonstrates anchoring. The analysts were given varying pieces of information regarding the intent of an enemy force. After establishing an initial hypotheses, the analysts gave considerably more weight to evidence consistent with that initial hypothesis than to evidence that was contrary. The anchoring heuristic is closely related to the *confirmation bias*, which will be discussed later in this chapter.

Several researchers have developed mathematical models to describe the conservatism and anchoring involved in updating hypotheses on the basis of sequential data (Hogarth & Einhorn, 1989; Lopes, 1982), and others have offered explanations for the phenomenon (Du Charme, 1970; Edwards, Phillips, Hays, & Goodman, 1968; Navon, 1979). Whatever its cause and however much it represents rational or irrational behavior, its existence is generally indisputable. So also is the assertion that because of limits in memory, people encounter a number of problems when aggregating evidence over time. These may be attributed to the tendency to give undue weight to early cues in a sequence (primacy) and the initially formulated hypothesis (anchoring), as well as to the tendency to overweight those cues that have occurred most recently and therefore are fresh in working memory (recency).

In arguing for such innovations as integrated graphics displays for decision support (Bettman, Payne, & Staelin, 1986; Moray, 1981; Woods et al., 1987; MacGregor & Slovic, 1986; see also Chapter 12) or simultaneous displays of unit/price information of a number of comparable products (Russo, 1977), researchers have made a convincing case that where possible, evidence that is available simultaneously should be presented simultaneously and not sequentially (Einhorn & Hogarth, 1981). This format cannot guarantee that simultaneous processing will occur, which of course depends on the limits of attention and the operator's own processing strategies. At least, however, it gives the operator the option of dealing with the information in parallel, if attentional limitations allow, or of alternating between different information sources, if they do not. In this manner, one information source is not given automatic primacy over others.

Integrating Information from Several Sources

Research on decision making and choice suggests that added limitations are imposed when information is derived from different sources, whether presented simultaneously or sequentially. Examples include the physician who incorporates different symptoms in a diagnosis or the technician who must troubleshoot a mechanical component on the basis of several different tests.

When several different information sources, each with less than perfect reliability, are available to support or refute a hypothesis, it can be shown that the likelihood of a correct diagnosis can increase as more cues are considered. In practice, however, as the number of sources grows beyond two, people generally do not use the greater information to make better, more accurate decisions (Allen, 1982; Dawes, 1979; Dawes & Corrigan, 1974; Malhotra, 1982; Schroeder & Benbassat, 1975). Oskamp (1965), for example, observed that when more information was provided to psychiatrists, their confidence in their clinical judgments increased but the accuracy of their judgments did not. Allen (1982) observed the same finding with weather forecasters. The limitations of human attention and working memory seem to be so great that an operator cannot easily integrate simultaneously the diagnostic impact of more than a few sources of information. In fact, Wright (1974) found that under time stress, decision-making performance deteriorated when more rather than less information was provided. Despite these limitations, people have an unfortunate tendency to seek far more information than they can absorb adequately. The admiral or executive, for example, will demand all the facts (Samet, Weltman, & Davis, 1976).

To account for the finding that more information may not improve decision making, we must assume that the human operator employs a selective filtering strategy to process informational cues. When few cues are initially presented, this filtering is unnecessary. When several sources are present, however, the filtering process is required, and it competes for the time (or other resources) available for the integration of information. Thus, more information leads to more time-consuming filtering at the expense of decision quality.

The Salience Bias   Any filter must have a tuning mechanism that determines which information will influence the decision and which will be rejected. Payne (1980) argues that the filter is strongly tuned to the *salience* of the information or cue. Salience itself is a somewhat fuzzy concept that incorporates loud sounds, bright lights, abrupt onsets of intensity or motion, and spatial positions in the front or top of a visual display. Thus, Wallsten and Barton (1982) showed that subjects under time pressure in decision making selectively processed those cues that were presented at the top of an information display. Top locations presumably were more salient to the subjects (as we read from top to bottom), despite the fact that the information presented there was of no greater diagnostic value than that at lower locations.

These findings lead us to expect that in any diagnostic situation, the brightest flashing light or the meter that is largest, is located most centrally, or changes most rapidly will cause the operator to process its diagnostic information content over others. It is important for a system designer to realize, therefore, that the goals of alerting (high salience) are not necessarily compatible with those of diagnosis, in which equal salience of a variety of information sources should be maintained (or salience should be directly related to the value of the information).

In contrast to salience, which may lead to "overprocessing," research also suggests that information that is difficult to interpret or integrate, because it requires arithmetic calculations or contains confusing language, will tend to be ignored, or at least underweighted (Bettman, Johnson, & Payne, 1991; Johnson, Payne, & Bettman, 1988).

Impact of Information Value: The "As If" Heuristic   The effect of salience in producing bias toward some information sources would be of no great harm if all sources were equally informative concerning the truth of one hypothesis or the other. If this were the case, it would not matter which cues were selected and which ignored. However, this equality is not always present. Certain cues may be very diagnostic of a particular hypothesis (e.g., the air pressure 200 miles to the west predicts tomorrow's weather with a high degree of accuracy), whereas other cues (e.g., the temperature 100 miles to the east) convey very little diagnostic information. Experimental results suggest that a salient but uninformative cue often will be weighted heavily relative to an unsalient but informative one, at the cost of ultimate decision-making accuracy. In this light it is important to consider to what extent the operator attends to differences in the informativeness of cues.

Formally, a decision-making cue may be uninformative with regard to a hypothesis for one of two reasons: (1) The particular cue information may be equally likely under each of two hypotheses. In this case we say it is *undiagnostic*. For example, low pressure on a gauge does not discriminate a leaking pipe from a failed pump. (2) The information itself may be *unreliable*. For example, an eyewitness testifying at a trial has a less than perfect probability of reporting a crime accurately because of memory failure (Ellison & Buckhout, 1981; Loftus, 1979, 1987). A remote sensor that provides information concerning weather patterns may occasionally be faulty and give inaccurate readings to the weather forecaster. An auditor may distrust the bookkeeping records of a shady business as an index of the financial health of the business.

The two causes of lack of informativeness—low diagnosticity and low reliability—are logically independent (Schum, 1975). However, both reduce the information that *should* be gained about a hypothesis, given the observed stimulus data. Formal rules may be specified to indicate how much one's belief in a hypothesis should be reduced, given the lack of informativeness (Johnson, Cavanagh, Spooner, & Samet, 1973). Yet in both prediction and diagnostic decision making there is good evidence that people fail to make these optimal conservative adjustments in confidence. When processing several cues that may lack both perfect reliability and diagnosticity, they tend to apply an *"as if"* *heuristic*, treating all evidence "as if" it were equally informative. This heuristic demonstrates a sort of *risky* processing bias in the sense that decision makers extract more implications from the unreliable data than are warranted (Johnson, Cavanagh, Spooner, & Samet, 1973; Schum, 1975).

Numerous examples may be cited in which the "as if" heuristic has been applied. Kahneman & Tversky (1973) have demonstrated that even those well trained in statistical theory do not down-weight unreliable predictions of a criterion variable when making "intuitive" predictions. In Figure 7.3, the

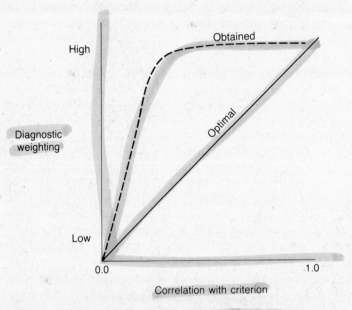

Figure 7.3 Demonstration of the "as if" heuristic. The function shows the relationship of the validity of cues to the optimal and obtained weighting of cues in prediction.

optimal diagnostic weighting of a predictive variable is contrasted with the weights as indicated by subjects' predictive performance. Optimally, the information extracted, or how much weight is given to a cue, should vary as a linear function of the variable's correlation with the criterion. In fact, the weighting varies in more of an "all or none" fashion, as shown in the figure.

The insensitivity to differences in predictive validity or cue reliability should make people ill suited for performing tasks in which prediction is involved. In fact, a large body of evidence (e.g. Dawes & Corrigan, 1974; Dawes, Faust, Meehl, 1989; Kahneman & Tversky, 1973; Kleinmuntz, 1990; Meehl, 1954) does indeed suggest that humans, compared to machines, make relatively poor intuitive or clinical predictors. In these studies, subjects are given information about a number of attributes of a particular case. The attributes vary in their weights, and the subjects are asked to predict some criterion variable for the case at hand (i.e., the likelihood of success in a program or the diagnosis of a patient). Compared with even a crude statistical system that knows only which way a given variable predicts a criterion (e.g., higher tests scores will predict higher criterion scores) and assumes equal weights for all variables, the human predicts relatively poorly. This observation has led Dawes, Faust, and Meehl (1989) to propose that the optimum role of the human in prediction should be to identify relevant predictor variables, determine how they should be measured and coded, and identify the direction of their relationship to the criterion. At this point a computer-based statistical analysis should take over and be given the exclusive power to integrate information and derive the criterion value.

The "as if" heuristic has been amply demonstrated in other decision-making and diagnosis tasks. Kanarick, Huntington, and Peterson (1969) observed that subjects preferred to purchase cheap, unreliable information over more expensive but reliable information when performing a simulated military diagnostic task. This "cheapskate mentality" was exhibited despite the fact that a greater amount of total information per dollar spent could be attained by purchasing the reliable information. Rossi and Madden (1979) found that trained nurses were uninfluenced by the degree of diagnosticity of symptoms in their decision to call a physician. This decision was based only on the total number of symptoms observed.

A particularly dangerous situation occurs when less than perfectly informative information is passed from observer to observer. The lack of perfect reliability or diagnosticity may become lost as the information is transmitted, and what originated with uncertainty might end with conviction. There is some feeling, for example, that in the *Vincennes* incident described at the beginning of this chapter, the uncertain status of the identity of the radar contact may have become lost as the fact of its presence was relayed up the chain of command (APA Monitor, 1988; U.S. Navy, 1988).

Another potential cause of unreliable data results when the sample size of data used to draw an inference is limited. A political poll based on 10 people is a far less reliable indicator of voter preferences than one based on 100. Yet these differences tend to be ignored by subjects when contrasting the evidence for a hypothesis (Fischoff & Bar-Hillel, 1984; Tversky & Kahneman, 1971,

1974). In evaluating two polls, one favoring candidate *A* in 8 out of 10 voters sampled [$P(A) = 0.80$], and the other favoring candidate *B* in 30 out of 50 voters [$P(B) = 0.60$], the "intuitive statistician" might report the net evidence provided by the two polls as favoring *A*, despite the fact that the lower sample size (and therefore lower reliability) of the first poll should in fact lessen its impact. Obviously if the evidence of the two polls were lumped together and then tabulated, candidate *B* would be favored by 32 out of 60 voters. However, when the outcomes are computed separately and reliability must be inferred from sample size information, the impact of differential reliability is diminished. Tversky and Kahneman (1971) observe that even trained scientists who are familiar with principles of statistical inference are not immune to these biases.

Why do subjects demonstrate the "as if" heuristic in prediction and diagnosis? The heuristic seems to be another example of cognitive simplification, in which the operator reduces the load imposed on working memory by treating all data sources as if they were of equal reliability. Thus, a person avoids the differential weighting or multiplication across cue values that would be necessary to implement the most accurate decision. When subjects are asked to estimate differences in reliability of a cue directly, they can clearly do so. However, when this estimate must be used as part of a larger aggregation, the values become distorted. Of course, as we see with other heuristics, the "as if" heuristic usually works adequately. But the occasions when it does lead to biases can be serious.

CHOOSING HYPOTHESES

The process of choosing hypotheses or making forecasts obviously is not independent of the process by which an operator perceives multiple cues of information. In fact, as shown in Figure 7.1, the two operations are closely related, and the limitations of perception affect hypotheses entertainment. We will focus on that portion of the diagnosis problem that is "in the head" and consider how the hypotheses that are entertained influence information-seeking behavior. An analogy can be drawn with top-down processing in perception, described in Chapter 5, in which such expectations guide our search for and interpretation of sensory data.

It is useful to consider the formal representations of diagnosis in more detail, as shown in Figure 7.4. In the example, two symptoms are observed. These might represent the blood pressure and temperature of a patient when the physician is considering two possible hypotheses (disease *A* and disease *B*). To diagnose the most likely disease, the physician must mentally multiply the value, *X*, of each symptom by a *diagnostic weight*, *W*, that represents for a given symptom value the likelihood that disease *A* (or *B*) is present. A high temperature, for example, might be more indicative of *A* than of *B*. Naturally, if the weights for a particular symptom are identical between the diseases, whatever the symptom value may be, its information does not help the physician to make

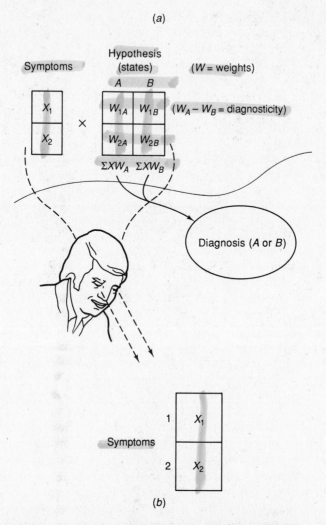

Figure 7.4 Diagnosis: The optimal mode. The observer perceives symptoms and mentally "multiplies" them by diagnostic weights for each of the hypotheses under consideration.

a diagnosis. If the weights differ substantially, they are highly *diagnostic*. That is, the presence of that symptom is quite useful in choosing between the two diseases. After performing a mental multiplication of symptom values across diagnostic weights and aggregating the products across each of the diseases, the physician should diagnose the disease as the one with the highest score. The confidence that the diagnosis is correct should be related to the difference between the two products.

Filtering and Entertaining Hypotheses

Just as limitations of human memory and attention restrict the perceptual cues that are processed, so they may also substantially limit the process of entertaining hypotheses. To begin with, people appear to have limited ability to entertain more than a few (three or four) hypotheses at once, whether these pertain to electronics troubleshooting (Rasmussen, 1981), automotive troubleshooting (Mehle, 1982), medical decision making (Lusted, 1976), or a variety of other decision-making tasks. The limitations of working memory and of dividing attention between hypotheses and cue evaluation appear to be too restrictive to do otherwise. Because of these cognitive limitations, human decision makers seek to avoid, when possible, the optimal but demanding *compensatory* decision-making strategy in which the implications of all symptoms for all hypotheses are considered at once and in which favorable evidence for one hypothesis from one symptom may be canceled or "outvoted" by unfavorable evidence from another (Einhorn & Hogarth, 1981; Payne, 1982).

Therefore, when a large number of alternative hypotheses are initially available and plausible in a diagnosis task or when a large number of objects are presented initially in a choice task, it is possible that relevant hypotheses will not get considered. Faced with an overabundance of alternatives, subjects often employ a simplifying "elimination by aspects" heuristic. This heuristic reduces the demand on cognitive space by reducing the number of hypotheses or options to the restricted number that can be considered more easily with one of the optimal compensatory strategies portrayed in Figure 7.4 (Payne, 1982; Tversky, 1972). In following this heuristic in diagnosis, the operator might focus attention on the symptoms that most discriminate the hypotheses under consideration, and eliminate from consideration any hypothesis that is not consistent with the observed value of that symptom. In diagnosing a process plant failure, for example, the operator might focus on the cue "low pressure at point X" as being most diagnostic and then eliminate from consideration all hypotheses that would not be likely to produce low pressure at X.

This simplifying heuristic may be distorting. On the one hand, cues not processed in the initial reduction of the problem space may point very strongly to a hypothesis that was eliminated from consideration. The weight of evidence of these "secondary symptoms" may even be strong enough to outweigh the evidence of the attended attribute and make one of the eliminated hypotheses the most likely one. However, because the cue is not processed, the likely hypothesis is not entertained. On the other hand, with complex processes, some hypotheses eliminated from consideration may be associated with the value on the relevant attribute (low pressure at X), although with a lower likelihood than the hypotheses that are retained. Yet as we shall see, once discarded from preliminary consideration, further biases in the decision-making process make it unlikely that a hypothesis will be reconsidered.

Once the hypotheses (or objects) under consideration have been restricted to a manageable set, the human decision maker must now choose a hypothesis in light of the symptoms. Following the logic of Bayes's theorem described earlier and presented formally in Equation 7.1, two factors should enter into the choice of a hypothesis: its probability in light of the observed data (data information) and its prior probability, or the absolute frequency with which a given hypothesis appears to be true according to past data. Examples of the latter include the frequency of a certain kind of system failure obtained from system reliability data or the prevalence rate of a disease obtained from medical records. Furthermore, as reflected in Bayes's theorem, these two factors of data information and prior probability should be complementary, so that if the prior probability of a particular hypothesis decreases, more data information will be required for the hypothesis to be chosen with an equal degree of confidence (see Figure 7.5). This complementarity between probability and data evidence in affecting confidence is analogous to the trade-off described in Chapter 2 between signal probability and signal intensity in influencing the likelihood of responding "yes." There we saw that as signals became more probable, the detector should say "yes" on the basis of less signal evidence.

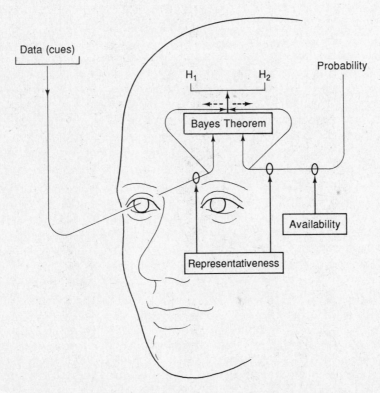

Figure 7.5 Illustration of the trade-off between data and probability in forming beliefs about a hypothesis. The figure illustrates the biases of representativeness (which eliminates the effect of probability) and availability in affecting these beliefs.

There are two important heuristics, however, that often influence a diagnosis in a way that contradicts the optimal balance between prior probability
and data information. The two heuristics, *representativeness* (the degree to
which the data "look like" those of a hypothesis) and *availability* (the ease of
recalling the hypothesis), are simplifying techniques that the decision maker
uses intuitively and automatically to approximate information concerning the
data and prior probabilities, respectively (Figure 7.5) (Kahneman, Slovic, &
Tversky, 1982; Tversky & Kahneman, 1974). Often these heuristics are accurate, but as simplifying shortcuts they are sometimes inaccurate and thereby
lead to some fairly systematic nonoptimal biases.

The Representativeness Heuristic   *Representativeness* refers to the extent to
which a set of observed symptoms is physically similar to or representative of
the symptoms that would be generated if a particular, familiar hypothesis were
true. If this similarity is present, the hypothesis is selected. Thus a physician
who observes a set of temperature and blood pressure readings, blood tests,
x-ray plates, and patient reports that are close to those typical of a given disease
is likely to diagnose that disease. This is an intuitively appealing strategy, and
often there is nothing wrong with it. However, a major danger is that information concerning prior probabilities or "base rate," the other critical element in
hypothesis choice, is ignored, as shown schematically in Figure 7.5.

Consider the example of two diseases with a very similar set of symptoms
but quite different base rates, like cystic fibrosis (disease S: quite rare) and
pneumonia (disease P: relatively frequent). The hypothetical symptoms of
these diseases are shown in Figure 7.6. The physician examines a patient with
the set of symptoms shown in the center. This set is more representative of
disease S, on the left, than of disease P, on the right, matching all but one
symptom of disease S but mismatching two symptoms (2 and 4) of disease P.
The representative heuristic would lead the physician to diagnose disease S.
But is this strategy always correct? Suppose that disease S is extremely rare,
occurring in 1 case out of 100,000. Disease P, which does not match the
symptoms quite as closely (its probability, *given the data*, is less) but which
occurs with 100 times greater frequency in the population, might be far more
likely than S when both prior probabilities and symptomatic information are
combined in an optimal compensatory fashion.

Representativeness reflects another example of the distorting effects of
salience in decision making. Symptoms are salient and visible; probability is
abstract and mental. If probability is to be used, its importance should be
explicitly taught, as discussed later. For example, Christenssen-Szalanski and
Bushyhead (1981) have observed that physicians are insufficiently aware of
disease prevalence rates in making diagnostic decisions. Balla (1980, 1982)
confirmed the limited use of prior probability information by both medical
students and senior physicians in a series of elicited diagnoses of hypothetical
patients. Fischhoff and Bar-Hillel (1984) report that prior probability information effectively enters into subjects' categorization judgments only when the
pattern of symptoms presented by a particular case are not representative of

Symptoms	Disease S (1/100,000)	Patient's observed symptoms	Disease P (1/1,000)
1	+	+	+
2			+
3	+	+	+
4	+	+	
5		+	+
6	+	+	+

Figure 7.6 Illustrates the representatives heuristic. A plus (+) indicates the presence of a symptom, either in the textbook description of the disease (left and right column) or in the patient (middle column). The patient's symptoms are more representative of disease S, despite its greater rarity.

any category. Finally, the sluggish beta adjustment in response to signal probability, described in Chapter 2, in which decision-making criteria are not adjusted sufficiently on the basis of signal frequency information, is another example of this failure to account for base-rate information.

The Availability Heuristic Availability refers to "the ease with which instances or occurrences [of a hypothesis] can be brought to mind" (Tversky & Kahneman, 1974, p. 1127). This heuristic can be employed as a convenient means of approximating prior probability, in that more frequent events or conditions in the world generally are recalled more easily. Therefore, people typically entertain more available hypotheses. Unfortunately, other factors strongly influence the availability of a hypothesis that may be quite unrelated to their absolute frequency or prior probability. As we noted in our discussion of long-term memory, recency is one such factor. An operator trying to diagnose a malfunction may have encountered a possible cause recently in a true situation, in training, or in a description just studied in an operating manual. This recency factor makes the particular hypothesis or cause more available, and thus it may be the first one to be considered. Availability also may be influenced by hypothesis *simplicity*. For example, a hypothesis that is easy to represent in memory (e.g., a single failure rather than a compounded double failure) will be entertained more easily and naturally than one that places greater demands on working memory.

Another factor influencing availability is the elaboration in memory of the past experience of the event. For example, in an experiment simulating the job of an emergency service dispatcher, Fontenelle (1983) found that those emergencies that were described in greater detail to the dispatcher were recalled as having occurred with greater frequency.

It should be reemphasized that the heuristics of availability and representativeness are often effective in the decision-making process. That is, representative hypotheses often are the most likely in terms of the data, and more available hypotheses usually do have higher probabilities. If they did not work well, people probably would not employ them as heuristics. In fact, heuristics may even be considered optimal if the concept of optimality also includes conserving cognitive resources—the optimality that characterizes efficiency. Furthermore, Kleinmutz (1985) has argued that in some situations people may adopt flexible "closed-loop" decision-making strategies, which can minimize the biases caused by representativeness. In these strategies, a diagnosis is made, an action is taken on the basis of that diagnosis, and the outcome is used to modify or correct the diagnosis. We will consider this process of hypothesis testing, which appears to have its own biases, later. For now, because of the important consequences of choosing and acting on an incorrect hypothesis, the concern of engineering psychologists should focus on the situations in which heuristics fail. For example, in the incident at the Three Mile Island nuclear plant, described in Chapter 1, the incorrect initial formulation of a hypothesis was a major cause of the crisis (Rubenstein & Mason, 1979).

Testing Hypotheses: Problems in Logical Inference

Typically the decision process is sequential (Payne, 1980). Once an initial hypothesis is formulated, further evidence is sought to confirm or refute it, as shown in Figure 7.1. Sequential processing is particularly typical of troubleshooting or fault diagnosis, in which it is somewhat unlikely that a particular set of symptoms will instantly trigger a final diagnostic decision (Rasmussen, 1981). Unfortunately, the search process required to gain further information is itself characterized by limitations in cognitive performance that lead to less than optimal decision making.

System Complexity and Working Memory In the previous chapter we discussed at some length the limited capacity of working memory, and we have seen here how it can restrict the number of hypotheses that may be entertained. As we discussed also in Chapter 6, Wohl (1983) has specified working memory limitations as a primary cause of the difficulties troubleshooters have in diagnosing faults in complex systems. His model of fault detection times suggests that when the number of interconnections between components grows beyond the limits of working memory, faults may become exceedingly difficult to detect.

Confirmation Bias The confirmation bias is a manifestation of the anchoring heuristic previously discussed, and it describes an inertia that favors the hypothesis initially formulated. Operators tend to seek (and therefore find) information that confirms the chosen hypothesis and to avoid information or tests whose outcome could disconfirm it (Einhorn & Hogarth, 1978; Mynatt, Doherty, & Tweney, 1977; Schustack & Sternberg, 1981; Wason & Johnson-Laird, 1972). This bias produces a sort of "cognitive tunnel vision" (Sheridan, 1981), in which operators fail to encode or process information that is contradictory to or inconsistent with the initially formulated hypothesis. Such tunneling seems to be enhanced under conditions of high stress and workload (Sheridan, 1981; see also Chapter 10).

The confirmation bias can be described within the framework of Figure 7.7. A simplified set of two hypotheses is considered (although there may be others, as indicated). In an industrial plant these may be a broken pump (H_1) and a clogged relief valve (H_2). A symptom refers to a particular abnormal or unusual observation. A plus sign indicates that the presence of the symptom is evidence for the particular hypothesis; a minus sign means that the symptom is inconsistent with the hypothesis in whose column it appears; a zero indicates that the presence of the symptom is irrelevant to the hypothesis. In this example, the presence of symptom 1 favors H_1 and provides evidence against H_2. Symptoms 2 and 4 provide evidence for H_2 (this contradiction is possible since the symptoms are neither perfectly reliable nor perfectly diagnostic), and symptom 3 favors both hypotheses. That is, it is undiagnostic. Assume that the operator has originally chosen H_1 as the working hypothesis. The confirmation bias will lead the operator to seek (and therefore be likely to find) evidence in favor of H_1 (i.e., S_1 and S_3). The operator would be less likely to attend to S_2 and S_4 since they support the alternative. The confirmation bias clearly leads to a kind of anchoring, as discussed previously, since it describes the reduced impact of contradictory evidence that becomes available after an initial hypothesis is formulated.

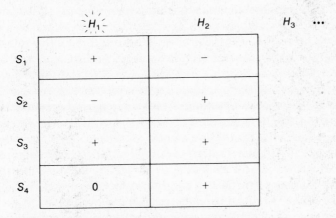

Figure 7.7 The confirmation bias (symptoms 1–4) and the use of negative evidence (symptom 4). H_1 is the hypothesis chosen for consideration. S_1 and S_3 will be attended; S_2, S_4.

The tendency to focus only on the initial hypothesis magnifies the potential danger incurred when biases and heuristics generate an initial hypothesis that is incorrect. It is normally quite easy for the hypothesis tester to obtain information, such as symptom 3 in Figure 7.7, that is consistent with but does not prove a particular formulated hypothesis (i.e., the information also may be consistent with other hypotheses). In theory, it is just as easy to perform critical diagnostic tests that could refute the formulated hypothesis or to seek information contrary to the hypothesis (i.e., symptom 2). Yet this is not typically done.

The investigation into the *Vincennes* incident reveals the confirmation bias at work. Operators of the radar system hypothesized early on that the approaching aircraft was hostile, and they did not interpret the contradictory (and as it turned out, correct) evidence offered by the radar system about the aircraft's neutral status (Tolcott, Marvin, & Bresoick, 1989; U.S. Navy, 1988). The analysis of the Three Mile Island incident also reveals a confirmation bias to accept the erroneous hypothesis of a high-water level in the reactor. Finally, the study of military intelligence analysts by Tolcott, Marvin, and Bresoick (1989), which was described in the context of anchoring, also illustrates the confirmation bias. In one part of their study, analysts were offered their choice of information to seek that would confirm or refute their initial hypothesis about the hostile situation of the enemy. Consistently analysts sought that information that would confirm the hypothesis rather than that which would give them the greatest certainty (i.e., the *most* information, in the formal sense of the word).

Three possible reasons for this failure to seek disconfirmatory evidence may be proposed: (1) People have greater cognitive difficulty dealing with negative information than with positive information (Clark & Chase, 1972; see Chapter 5). (2) To change hypotheses—abandon an old one and reformulate a new one—requires a higher degree of cognitive effort than does the repeated acquisition of information consistent with an old hypothesis (Einhorn & Hogarth, 1981). Given a certain "cost of thinking" (Shugan, 1980) and the tendency of operators, particularly when under stress, to avoid troubleshooting strategies that impose a heavy workload on limited cognitive resources (Rasmussen, 1981), operators tend to retain an old hypothesis rather than go to the trouble of formulating a new one. (3) In some instances it may be possible for operators to influence the outcome of actions taken on the basis of the diagnosis, which will increase their belief that the diagnosis was correct. This is the idea of the "self-fulfilling prophecy" (Einhorn & Hogarth, 1978). It might describe a teacher who, diagnosing a child as "gifted," will provide that child with sufficient extra opportunities and motivation so that high intellectual performance will be almost guaranteed. It might also describe the scientist who, believing a theory to be correct, will now design and carry out experiments that are most likely to produce confirming evidence.

There is some evidence that the confirmation bias is not altogether a bad strategy (Klayman & Ha, 1987). Indeed, maintaining a working hypothesis is valuable because it provides a guide to the search for new information that is more efficient than a random search. The issue of how to force a diagnostician simultaneously to entertain alternative hypotheses and to seek disconfirming evidence—in short, to break through the cognitive tunnel—represents a major challenge to the designer of systems in which troubleshooting will be required.

Investigations by Arkes and Harkness (1980), Bower and Trabasso (1963), and Levine (1966) have studied what information is processed from cues that support an alternate hypothesis (i.e., S_2 and S_4 in support of H_2) while the decision maker is seeking to confirm the chosen hypothesis (H_1). This information is important if the operator ever does overcome the confirmation bias, abandons H_1, and must then choose among the remaining hypotheses. In laboratory studies of concept learning, Levine finds that subjects process far less of this information than is optimal. Bower and Trabasso reach the more pessimistic conclusion that subjects will sometimes fail to process and retain *any* of the information concerning hypotheses that are not entertained. In a more realistic diagnostic setting, Arkes and Harkness also demonstrated the selective biasing of memory induced by the confirmation bias. They presented subjects with several symptoms related to a particular clinical abnormality (experiment 1) or to the state of a hydraulic system (experiment 2). Arkes and Harkness found that if the subject held a hypothesis or made a positive diagnosis, the symptoms they had observed that were consistent with that diagnosis were readily remembered, whereas inconsistent symptoms were more easily forgotten. Furthermore, subjects erroneously reported seeing symptoms that they actually had not seen but that were consistent with the diagnosis.

Absence of Cues   Symptom 4 in Figure 7.6 can also be used to illustrate the second bias often encountered in troubleshooting and hypotheses testing, a bias against the use of the *absence* of information (Wason & Johnson-Laird, 1972). Notice that whether symptom 4 does or does not occur has no bearing on hypothesis 1. But its presence is evidence for hypothesis 2, so that if it is absent, this state provides indirect evidence for H_1, by eliminating one of its competitors. For example, a process operator confronted by a malfunction in a nuclear power system who hypothesizes that the cooling-water level at some point is high could find confirming evidence by observing that a meter is *not* indicating that temperature is excessive. Excessive temperature would be an expected symptom of low cooling-water level. Even though the meter might fail to differentiate normal from high water, it would eliminate one competing hypothesis and thereby reduce uncertainty in informational terms.

In a simulated troubleshooting task, Rouse (1981) found that subjects derived very little diagnostic use from the absence of potentially valuable information that could be used to narrow the competing hypothesis set. Balla (1980) observed that neither medical students nor trained physicians were proficient in using the absence of symptoms to assist in diagnosis. Although Balla found that expertise does not improve the use of negative information, Hunt and Rouse (1981) found that people can be trained to use the absence of cues more efficiently in the simulated troubleshooting task employed by Rouse.

Mental Models and Causal Reasoning   In the real world, troubleshooters rarely embark on a task in the absence of some knowledge of the system, knowledge of how the components are interconnected and their causal links. We may describe this knowledge as a mental model of the system (Gentner & Stevens, 1983), a concept we discussed in Chapter 4 and 6 will explore again in some detail in Chapter 12. Unless the system is extremely simple or the operator is exceptionally well trained, the mental model may not be perfect, and these imperfections can be a further source of problems in troubleshooting, as illustrated in a study by Sanderson and Murtagh (1989). In this study, subjects were taught the nature of the simulated electronic switching system shown in Figure 7.8. After training, the subjects' mental model of the system components and their causal linkages was good but not perfect—a good description of the knowledge possessed by many troubleshooters of complex systems. For example, a subject might fail to realize the existence of a particular link between two components. Following this training period, the subjects were then asked to troubleshoot failures in the system, which were created by "breaking" certain links. Failures that occurred at points in the system where the mental model was incomplete proved to be extremely difficult, if not impossible, to detect.

What are the sources of these incorrect mental models? Sometimes they are simply the result of insufficient training. However, Einhorn and Hogarth (1982) have also described a number of "cues to causality," which people use to infer that a causal relationship exists even when one is absent. For example,

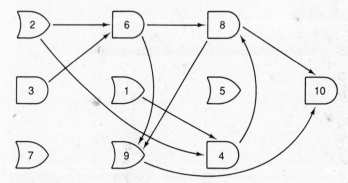

Figure 7.8 Network used in the experiment. The numbers on the components are out of sequence because the components have been moved from a simple left-to-right flow pattern to the present, more complex pattern. Logical components with square left ends are *and* components, and those with curved left ends are *or* components. Arrows, which are not seen by subjects, illustrate how the components are linked by the true wiring of the network. (*Source:* P. M. Sanderson and J. M. Murtagh, "Troubleshooting with an Inaccurate Mental Model," in *Proceedings of the 1989 IEEE International Conference on Systems, Man, and Cybernetics,* 1238–1243. Copyright © 1989 IEEE.)

two components that are displayed close together in space or two events that occur close together in time will be perceived as having a causal link. Einhorn and Hogarth propose that even the simple ordering of two statements or facts on the page leads people to see a causal relationship even when none may exist or when the actual causal relationship may be reversed. Thus, the statement "The water pressure is high and the process is unstable" will be perceived as indicating that high water pressure caused the instability. These inferences may be considered as *causal heuristics* since such factors as closeness in space and time or sequential ordering are appropriate cues to causality much of the time, but not always; and when they are misperceived, an erroneous mental model may be formed.

Reversing Causal Reasoning Another problem in causal and diagnostic reasoning involves the misunderstanding of the asymmetry of probability between cause and effect. Thus, the probability that *A* will cause *B* is generally *not* the same as the probability that *B* was caused by *A*, although people often assume that they are equal. This situation would occur if, for example, *B* could be caused by several possible events and therefore had a higher absolute probability than *A*. In a hypothetical process control example, the relationship might be "If the pump fails, then the water level will be low." Although this causal relationship is unidirectional, people often err by assuming that it is bidirectional (Taplin, 1971; Taplin & Staudenmayer, 1973). That is, they also assume "If *B*, then *A*" ("If the water level is low, then the pump must be broken"). This is not a valid inference, for simple logic informs us that there are numerous other causes for low pressure (e.g., broken pipes or leaking valves). Yet people apparently have a difficult time generating a large set of plausible hypotheses that could have produced the observed symptoms. Mehle (1982), for example, observed this difficulty for subjects trying to troubleshoot problems in automobile engines.

Eddy (1982) has identified similar dangers in bidirectional inference in medical diagnosis when the absolute probabilities of cause-and-effect conditions differ. The probability that someone who has a particular disease will show a certain symptom is not the same as the probability that one who shows the symptom will have the disease. The former may be quite high even as the latter is low. Yet Eddy notes that physicians sometimes erroneously attribute the high probability of the symptom to the disease. As a consequence, physicians may be overconfident in diagnosing the disease in a patient who shows the symptom when such diagnosis is unwarranted, with the cost and potential danger of unnecessary surgery or treatments.

Overconfidence in Diagnosis and Forecasting

When decision makers or troubleshooters complete a diagnosis, one of the most critical pieces of information they should provide is their *confidence* in that diagnosis. This may be an expression of the confidence that a forecast is correct (e.g., a "70 percent chance of rain" is a forecast of rain with 70 percent confidence) or a statement of the probabilities that various alternative hypotheses may be correct (e.g., the physician may state with 60 percent probability that the patient has one particular disease, 30 percent probability of another, and 10 percent probability of a third).

Accurate or *well-calibrated* confidence assessment in diagnosis and forecasting is important for two reasons. First, the level of confidence one has in one's own beliefs can implicitly guide the choice between seeking more evidence (if confidence is low) or choosing an action (if confidence is high). Confidence that is unjustifiably high will contribute to the confirmation bias since such certainty will reduce the desire to look at more evidence, even when that evidence should be sought. Second, expressions of confidence in a diagnosis can influence the choice of actions that the decision maker or someone else may perform on the basis of that diagnosis. For example, if a patient is told with 100 percent certainty that he or she has the fatal disease X, there may be little choice regarding the best treatment, which might itself have risky side effects. However, if the confidence in the diagnosis of X is only 60 percent, the patient may consider the risks associated with the treatment too high and choose a different one.

Later in this chapter, we will discuss how people evaluate risks in choosing actions. Here we consider how accurate or *calibrated* people are in their diagnoses and forecasts. For example, if we evaluated the performance of a weather forecaster, would we find that it actually did rain on 80 percent of the days for which an "80 percent chance of rain" had been forecast? On the basis of such evaluations, a general conclusion is that people tend to be *overconfident*, a bias we may categorize as risky (Kleinmuntz, 1990). This overconfidence has been seen in a number of different kinds of judgments. For example, Mehle (1982) observed that subjects engaged in automotive troubleshooting are unjustly confident that they have entertained all possible diagnostic hypotheses. Fischhoff and MacGregor (1982) performed a forecasting study in which subjects were asked to make predictions about future local and national events and estimate their confidence in the predictions. Later, the confidence of prediction was compared with the frequency with which the predicted events actually did occur. A typical set of results, shown in Figure 7.9, indicates how consistently the estimate of the confidence is greater than the actual chance of being right. Fischhoff and MacGregor cite an impressive number of investigations that demonstrate how pervasive overconfidence is in the behavior of a variety of professional forecasters, such as economic analysts and stockbrokers. There is even some evidence that weather forecasters have a tendency to be overconfident in forecasts of the probability of precipitation (Allen, 1982), although such biases are not large (Murphy, Hsu, Winkler, & Wilks, 1985; Murphy & Winkler, 1984), and weather forecasts remain better calibrated than most because of regular and immediate feedback regarding their accuracy. However, in forecasting weather patterns about which less feedback is received, such overconfidence becomes quite evident (Brown & Murphy, 1987; Lusk, Stewart, Hammond, & Potts, 1990).

Overconfidence appears also in our judgments of our own abilities (Pitz &

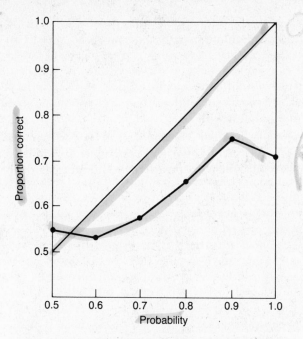

Figure 7.9 Overconfidence in forecasting. The straight line shows an optimal forecast confidence. The actual function plots the forecast probability of occurrence against the proportion of times the event actually did occur. (*Source:* **B. Fischhoff and D. MacGregor, "Subjective Confidence in Forecasts,"** *Journal of Forecasting, 1* (1989), 155–172. Copyright 1989 by John Wiley & Sons, Ltd. Reproduced by permission.)

Sachs, 1984). This bias has been observed in several forms. In Chapter 6 we discussed overconfidence in the accuracy of one's memory in eyewitness testimony. For example, Wells, Lindsay, and Ferguson (1979) simulated a crime and called for eyewitness testimony concerning visual details in front of a mock jury. They found that witnesses were far more confident in the accuracy of their testimony (and therefore their memory of the event) than was warranted by the actual accuracy of their memory. Furthermore, estimates by mock jurors of the witnesses' confidence were totally unrelated to the accuracy of the witnesses' memory.

Fischhoff, Slovic, and Lichtenstein (1977) found that we have the same unwarranted overconfidence in the reliability of our own memory about facts of general knowledge. Subjects reported themselves to be extremely confident in answering general knowledge questions such as the following: "Which is the greater cause of deaths in the United States: abortion, pregnancy and childbirth, or appendicitis?" (The answer is appendicitis.) The disconcerting aspect of their findings is that subjects expressed such overconfidence even on answers that were wrong far more often than would be expected by chance guessing. MacGregor, Fischhoff, and Blackshaw (1987) found that people are overconfident in their ability to retrieve material in data base searches, and Slovic, Fischhoff, and Lichtenstein (1982) reported that people as a whole are overconfident in their abilities to function safely in hazardous environments. For example, people are more confident than they should be that they will not be involved in automobile accidents.

Finally, people's overconfidence in their own judgment leads them to mistrust the outcome of various automated decision aids designed to supplement their judgments (Kleinmuntz, 1990). The overtrust of human, relative to machine, performance has important implications both for decision aiding and for automation, to be discussed in detail in Chapter 12.

In conclusion, people are less accurate than they could be in a variety of diagnosis, forecasting, and troubleshooting tasks as a result of numerous heuristics and biases, which in turn have their origins in properties of attention, memory, and logic. Some have argued that these human limitations call for replacement by computer-based artificial intelligence and expert systems, a topic addressed in Chapter 12. Others argue for the importance of decision aiding, to be described at the end of this chapter. At this point we turn our attention to the processes that take place after a diagnosis is made or a situation is assessed: the decision to carry out an action.

THE CHOICE OF ACTION

Our discussion of diagnosis has focused on understanding the state of the world. As modeled in Figure 7.1 this understanding is usually followed by a *choice* of some action to take, for example, to purchase a particular product, to replace a suspected component, or to abort an intended flight plan. Decision researchers generally make a distinction regarding the consequences of the choice: whether they can be fairly easily anticipated, such as in decisions of what product to buy, or are uncertain such as in decisions about treating a malignancy with surgery, radiation, or chemotherapy. Logically, this distinction is one between decision under *certainty* and decision under *uncertainty*. We will describe the former first, as it provides a framework into which the uncertain element of probability can then be incorporated.

Certain Choice

The choice task is represented in Figure 7.10, which bears some resemblance to the diagnosis task shown in Figure 7.4. In Figure 7.10, we might imagine the consumer observing two objects or products (e.g., two automobiles) differing on two known attributes (e.g., gas mileage and cost). The value, X, of each object on each attribute is known and visible. The operator should then multiply the known value of these attributes by a mental representation of how important they are—their *utilities*, U_A and U_B—before deciding how the two products compare and making a decision. Assume in this example that cost is

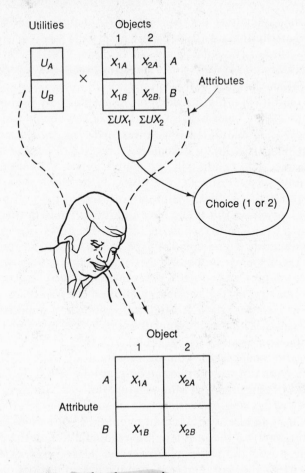

Figure 7.10 The choice task.

twice as important to the consumer as gas mileage. Car 1 has one-third the gas mileage but sells for half the price of car 2. Therefore car 1 would be chosen because of the greater weight placed on the attribute (cost) that favored it. Just as in diagnosis, we again describe this decision-making strategy as *compensatory* because an object that is slightly deficient in one attribute of great utility may be compensated by having higher values on one or more attributes of less utility.

Many of the characteristics of the choice task parallel those of the diagnostic task with multiple cues and multiple hypotheses. In particular, studies of consumer decision making suggest that people perform progressively less well as they are confronted with more objects and more attributes (Malhotra, 1982). Interestingly, however, choice performance is more disrupted by increasing the number of objects than by increasing the number of attributes (Johnson, Bettman, & Payne, 1989), a finding that has a direct parallel to our discussion of working memory in Chapter 6 (Yntema, 1963).

Because of the difficulties encountered with complex choices involving many objects and attributes, people typically first "edit" the decision problem by reducing this number (Bettman, Johnson, & Payne, 1991). Earlier we discussed the *elimination by aspects* (EBA) heuristic, one form of editing in the context of hypothesis entertainment (Tversky, 1972). As applied to choice, the EBA heuristic is employed if the decision maker attends only to the one or few critical attributes that are most heavily weighted in the choice task and eliminates from consideration all but the few objects with the highest values on the critical attributes. For example, in purchasing a car, if price is the most important attribute, the decision maker will eliminate from consideration all but the cheapest automobiles despite the fact that those in the eliminated set may rate quite favorably on attributes that are initially neglected but still may be relatively important. Only after the set has been pared down will other attributes be considered.

Even with reduced problem complexity it is clear that decision makers do not consistently employ the compensatory decision strategy shown in Figure 7.9, in which all attributes of all objects enter into the choice. Instead, a variety of other decision-making rules have been identified (see Bettman, Johnson, & Payne, 1991; Slovic, Lichtenstein & Fischhoff, 1988 for a good review). One example of an alternative rule is *satisficing* (Simon, 1955), in which the decision maker examines each object in turn until encountering one that satisfies some minimum criterion of acceptability on all important attributes. The first such object is selected, even if it may not be the best, as long as it is "good enough."

In their review of consumer decision making, Bettman, Johnson, and Payne (1991) conclude that no single decision rule or model characterizes people's choices; rather the particular model is contingent on a wide number of factors, related to the complexity of the problem, the comparability of the objects, and the format of the information display. From the perspective of engineering psychology, however, the last of these factors has important implications. Choice decisions (e.g., by the consumer) will be made most carefully and accurately if the attribute values (e.g., unit price information or toxicity) are easily interpretable (Johnson, Payne, & Bettman, 1988) and *consistently* expressed in a form that can be compared across objects (Bettman, Payne, & Staelin, 1986; Russo, 1971).

Our discussion of Figure 7.10 has revealed the critical importance of utility, value, or importance in the choice task. It so happens that the concept of utility in decision making is tightly linked with the concepts of uncertainty and probability, so in discussing the role of utility, we will now turn to decision making under uncertainty.

Decision under Uncertainty: Utility and Value

In contrast to consumer choice, in which consequences of purchases are often fairly well known, many other decisions are taken in the face of a good deal of uncertainty, for example, the pilot who chooses to fly into deteriorating weather conditions rather than to turn back (Potter, Rockwell, & McCoy,

1989). Decision theorists define *risk* as taking an action whose outcome is uncertain and whose different possible consequences may have different values or costs to the individual.

To understand the biases that operate on risky decision making, it is important first to consider the way in which costs and benefits should guide decision and choice. In the optimal models of decision making, costs and values are assigned to different potential outcomes. Then the action chosen (which may be based on a diagnostic outcome) should be that which in the long run produces the highest expected gain or the minimum expected loss.

Formally and optimally these relationships may be expressed in the decision matrix shown in Figure 7.11, which is an expansion of the right side of Figure 7.1. Two courses of action (choices and treatments) are specified. Each can be taken in the face of the decision maker's uncertainty about the state of the world. Each state of the world (labeled S) is associated with a subjective probability that it is true (labeled P). For example, after quickly troubleshooting a power turbine that appears to be malfunctioning, the control room operator may believe it quite likely that the turbine has a major malfunction (S_1: $P = 0.70$), possible that the turbine has a trivial malfunction that will not affect its continued running (S_2: $P_2 = 0.20$), and faintly possible that nothing at all is wrong (S_3: $P_3 = 0.10$). Each considered act may have a different outcome (O_{ij}), given that a particular state of the world exists, and these outcomes are shown within each cell. Finally, each outcome is associated with a value or cost, (V_{ij}). For example, consider the act A, in which the operator chooses to keep the turbine running. The outcome of this decision (O_{A1}) could have a large cost

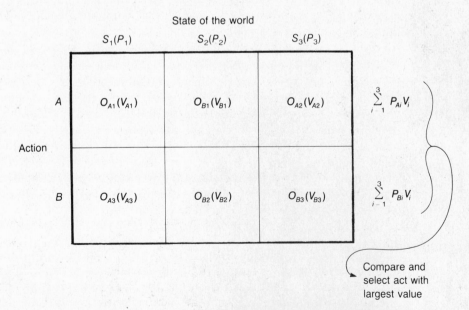

Figure 7.11 Optimal use of expected utility in the selection of action.

$(-V_{A1})$ if, in fact, there is a major malfunction. Within one prescription of optimal decision making, the action that is selected should be that with the highest *expected value*. This quantity is computed by multiplying the value by the probability of each outcome and summing these products across actions. These expected values are shown in the right margin of the figure.

There are, of course, some limitations to this optimal prescription since all costs or values cannot be expressed in monetary units. As discussed in Chapter 2, the cost to the physician of a certain diagnostic treatment that is associated with the possible loss of life obviously cannot be so expressed. This limitation, in fact, is one of the greatest difficulties in applying optimal models of decision making to medical treatment (Lusted, 1976). As a result, decision theorists speak of the subjective *utility* of a given outcome to the decision maker, rather than its objective value (Edwards, 1987). Although the relationship between value and utility is generally monotonic (higher valued outcomes and things have greater utility), the relation is not linear and produces some important biases. For example, according to the prescription, an action that reduces an expected loss by a constant amount should be preferred equally to one that enhances an expected gain by the same amount. Correspondingly, a given change in the expected value should have the same influence on decision behavior, whether this change was produced by a change in probabilities or a change in values. However, these optimal prescriptions do not in fact characterize human decision making.

The distinction between value and utility is important because the relationship between value (expressed in dollars or other currency) and utility (as it influences decision making) is not straightforward. One proposed form is shown in Figure 7.12 (Tversky & Kahneman, 1981), which portrays subjective gains or losses in utility (on the Y axis) as a function of gains or losses in objective value (on the X axis).

The prominent difference in the slope of the positive and negative segments of the function implies that a potential loss of a given amount is perceived as having greater consequences, and therefore exerts a greater influence over decision-making behavior, than does a gain of the same amount. Suppose you are given a choice between refusing or accepting a gamble that offers a 50 percent chance to win or lose $1.00. Most people would typically decline the offer because the potential $1.00 loss is viewed as more negative than the $1.00 gain is viewed as positive. As a result, the expected utility of the gamble (the sum of the probability of outcomes times their utilities) is a loss. Kahneman and Tversky (1984) refer to this characteristic — that losses are preceived as worse than gains are good — as *loss aversion*.

Another characteristic of the function in Figure 7.12 is that both positive and negative limbs are curved toward the horizontal, each showing that equal changes in value produce progressively smaller changes in utility the farther one is from the zero point. This makes intuitive sense. The gain of $10.00 if we have nothing at all is more valued than the gain of the same $10.00 if we already have $100.00. Similarly, we notice the first $10.00 we lose more than an added $10.00 penalty to a loss that is already $100.00.

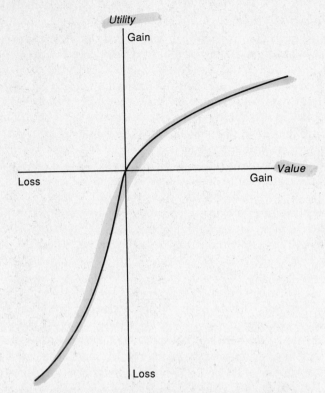

Figure 7.12 The hypothetical relationship between value and utility. (*Source:* A. Tversky and D. Kahneman, "The Framing of Decisions and the Psychology of Choice," *Science, 211* (1981), 453–458. Copyright 1981 by the American Association for the Advancement of Science.)

The utility concept is important in psychology, not only because of its nonlinear relationship with value, but also because it can account for influences on choice and decision that cannot easily be expressed as values. In addition to such areas as pain or happiness, an important concept in many emerging decision models is the utility to a decision maker of expending or conserving *effort*. We noted earlier that heuristics could be viewed as optimal if the reduced effort achieved by their use is taken into account. Similarly, Bettman, Johnson, and Payne (1990), Johnson, Bettman, Payne, and Coupey (1990), and Payne, Bettman, and Johnson (1988) have proposed a model of decision making in which the negative utility of effort is combined with the expected utility of decision accuracy to predict the kind of decision strategy that will be employed to tackle a particular problem. According to their model, if the net gain in expected accuracy is of less utility than the net loss in effort expended on a more complex strategy, when both are expressed as utilities, the more complex strategy will not be chosen.

Risk Seeking and Risk Avoidance: Distortions of Probability

Risky Gains and Losses   In addition to the distorted utility-value function, a second factor, identified by Kahneman and Tversky (1984), also accounts for nonoptimal performance and explains an important difference between risky decisions with positive and with negative outcomes. Recall that a risky decision may be characterized as a choice between a relatively "sure thing" and another option with more uncertain possible outcomes. Kahneman and Tversky observe that when there is a choice between two such actions with *positive* expected outcomes, people tend to avoid the risk. Their behavior is characterized by the phrase "take the money and run."

For example, if given the choice between two alternatives with the same expected values but different possible outcomes—winning $1.00 for sure (no risk) and taking a gamble with a 50/50 chance of winning $2.00 or nothing at all (risky)—subjects typically choose the certain option. However, suppose the word *winning* were replaced by *losing*, so that the choice is between losses. This choice produces a so-called avoidance-avoidance conflict, and subjects would now tend to choose the risky option. They are risk seeking when choosing between losses, and their behavior is characterized by the phrase "throwing good money after bad."

The importance of these differences between perceived losses and gains is that a given change in value (or expected value) may be viewed either as a change in loss or a change in gain, depending on what is considered to be the reference or neutral point. For example, a tax cut may be perceived as a reduction in loss if the neutral point is "paying no taxes" or as a positive gain if the neutral point is "paying last year's taxes" (Tversky & Kahneman, 1981). A meat product may be considered healthy if labeled 75 percent lean or unhealthy if labeled 25 percent fat. As a consequence, different *frames of reference* in a decision problem may produce fairly pronounced changes in decision-making behavior (Carroll, 1980; Tversky & Kahneman, 1981). Puto, Patton, and King (1985) and Schurr (1987) noted that this kind of bias described the behavior of professional buyers, given hypothetical investment decisions, just as aptly as it described the behavior of typical laboratory subjects. McNiel, Pauker, Sox and Tversky (1982) found that it also characterized the choices physicians made between safer and riskier treatments.

The effects of framing in an engineering context can be illustrated by considering a process control operator choosing between two courses of action after diagnosing a potentially damaging failure in a large industrial process: continue to run while further diagnostic tests are performed or shut down the operation immediately. The first action is perceived to lead to a large financial cost (serious damage to the equipment) with some probability less than one (given that the problem is serious and immediate). The second action will produce a cost that is almost certain but of lesser magnitude (lost production time and start-up costs). According to the preceding argument, when the choice is framed in this fashion, as the choice between losses, the operator would tend to select the high-risk alternative (continue to run) over the low-risk alternative (shut down) as long as the expected values of the two actions are seen to be similar. On the other hand, if the operator's perceptions were based on a framework of profits to the company (i.e., gains), the first alternative would be perceived as a probability mix of a full profit if nothing is wrong and a diminished (but still positive) profit if the disastrous event occurs. The second alternative would be perceived as a certain large but not maximum profit. In this case, the choice would more likely be the second, low-risk alternative.

It is apparent that perceived costs and benefits greatly influence most of the decisions that are made. It is important, therefore, for the manager who supervises personnel who make decisions (e.g., process control monitors) to provide them with a relatively uniform set of costs and payoffs concerning the consequences of their various actions. The supervisor then must be aware of what kind of risk-seeking or risk-aversive actions those payoffs will induce.

Probability Perception   Tversky and Kahneman (1981) account for the framing effect by assuming that people's perception of probability (as it influences decision performance) is related to true probability by the function in Figure 7.13 (This function bears a close resemblance to that describing the perception of proportions discussed in this chapter.) The first characteristic that accounts for framing in risk seeking and risk aversion is the fact that for most of its range, the function shows perceived probability as less than actual probability. If the perceived probability that influences one's decision is less than the true probability, then when choosing between two positive outcomes, one risky and one certain, the probability of gain associated with the positive outcome will be underestimated, which will also cause the expected gain of the risky option to be underestimated; therefore the bias will be to choose the sure thing. When choosing between negative outcomes, the probability of the risky negative outcome will also seem less, the expected loss of this option will be underestimated, and it will be chosen over the certain loss.

The second critical part of the function in Figure 7.13 is the way in which the probability of very rare events is *over*estimated, which accounts for two important departures from the typical biases explained by the framing effect. Why do people purchase insurance (choosing a sure loss of money—the cost of the policy—over the risky loss of an accident or disaster, which probably won't happen), and why do people gamble (sacrificing the sure gain of holding on to money for the risky gain of winning)? The answer is that in both cases the risky events are quite rare (the disaster covered by insurance or the winning ticket in the lottery), and hence their probability is overestimated: The image of winning a gamble looms large, as does the possibility of the disaster for which insurance is purchased.

Risky Choices in Everyday Life

The previous discussion has identified some of the biases in the understanding of probabilities, and these biases in turn help us to understand how the decisions we make in everyday life are guided by our conception of risk. But

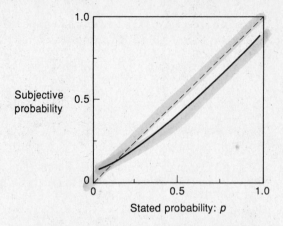

Figure 7.13 A hypothetical weighting function. (*Source:* A. Tversky and D. Kahneman, "The Framing of Decisions and the Psychology of Choice," *Science, 211* (1981), 453–458. Copyright 1981 by the American Association for the Advancement of Science.)

research has identified other factors that influence people's perception of risk, and therefore the choices they make and the policies they support (Slovic et al., 1982; Slovic, 1987). One such factor is another example of the *availability* heuristics discussed earlier in this chapter. People's perception of risk seems to be guided by the availability of examples of the risky event in long-term memory. Why, for example, do people view the riskiness of a nuclear power accident as greater than the risk of death by falling, in spite of the fact that the latter happens much more frequently than the former? Why do people who fear flying have no hesitation in driving an automobile without wearing seatbelts, despite the greater safety record of commercial aviation? Answers to these questions are provided by the research of Slovic, Fischhoff, and Lichtenstein (1981), who assessed people's perception of the risk of various hazards. Their conclusions were, first, that certain risks, like death from cancer, nuclear radiation, or homicide, are considerably overestimated, whereas others, like death from diabetes or accidental falls, are greatly underestimated (Slovic, Fischhoff, & Lichtenstein, 1982). Second, perceived risks are directly correlated with the amount of publicity that the varying hazards receive in the media (Combs & Slovic, 1979). We read a lot about aircraft accidents but relatively little about death by falling. The highly publicized hazards, therefore, become more available to memory.

Personal experience also plays a role in availability. For example, Karnes, Leonard, and Rachwal (1986) found that the estimated risks for all-terrain vehicles decreases for people who have had accident-free experience with the vehicles. Liebowitz (1991) observes that people's tendency to take risks in night driving is in large part the result of a lack of awareness of their own perceptual limitations at night.

Decision theorists have offered various remedies to help people make more rational judgments about risk. For example, Svenson, Fischhoff, and MacGregor (1985) have noted that people are not sensitive to the length of time they are exposed to a given risk, and therefore they underestimate the *cumulative* risks of a repeated activity. They argue that risk figures designed to influence public behavior should explicitly point to cumulative risks, such as the probability of dying from a *lifetime* of not wearing seat belts (which is actually fairly high), rather than the probability of death per trip. Bettman, Payne, and Staelin (1986) have discussed the importance of reducing the cognitive load on consumers in interpreting risk probabilities from product labels. Their perspective relates closely to principles of display formatting discussed in Chapters 3, 4, and 5.

Finally, Keeney (1988) has summarized 12 "facts" about risks that should guide the choices people make in everyday life. Although they will not be given here, one important concept underlying several of them is that decisions to reduce risks, made by individuals, corporations, or governments, usually have risks or costs themselves, which may outweigh the expected gain of the risk reduction. For example, he notes that a decision to buy a heavier, larger car, which will (statistically) reduce the likelihood of fatality in an accident, can lead to faster driving, which will make the probability of an accident more likely. The decision to require very complex safety procedures and all possible hazards to be printed on a piece of equipment might so increase the complexity of instructions that they will be ignored or confused (Bettman, Payne, & Staelin, 1986). The decision to divert large financial resources from a limited budget to reduce certain possible risks (e.g., by imposing stricter emission controls on factories or automobiles) should be taken only after considering how effective those resources could be if applied to reducing risks in other areas.

Keeney's (1988) discussion and others relating to risk perception and public policy (Slovic, Fischhoff, & Lichtenstein, 1984; Sprent, 1988) point out the extent to which moral and ethical issues become involved. These are not always easy to deal with, but a firm understanding of the sources of people's estimates (or misestimates) of probability should be an important component of overall risk analysis.

THE EXPERT DECISION MAKER: LEARNING AND FEEDBACK

Decision Making and Expertise

Like any skill, decision-making performance changes with practice. Of course expertise in decision making does not necessarily guarantee better performance, and this expertise seems to have its costs as well as its benefits. We will focus first on the apparent benefits, which have been identified by examining expertise in such fields as fire fighting (Klein, 1989b), pilot judgment (Wickens et al., 1988, 1991), judicial proceedings (Ebbeson & Konecni, 1980), and medical diagnosis (Lusted, 1976). These benefits may be divided into three categories.

1. *Cue sampling.* Because of the wealth of experience experts have with the cues used in decision making, they will understand the pattern of correlation between cues, thereby reducing the load on perception and working memory through chunking. Thus the expert physician may not need to perceive eight symptoms as eight independent pieces of information; rather, if these symptoms have been correlated in past experience because they define a particular disease, they will be perceived as a single perceptual chunk, or syndrome (Ebbeson & Konecni, 1980; Phelps & Shanteau, 1978). In troubleshooting, an accurate mental model of how a system works, formed through experience, may also guide an efficient search for new information (Sanderson & Murtagh, 1989).

2. *Hypothesis and action generation.* Because of their increased experience, experts have a greater repertoire of possible hypotheses and possible actions stored in long-term memory (see Figure 7.1). Studies of expertise in cognitive skills, such as physics or mathematical problem solving, reveal that hypotheses generation is a major point of contrast with the novice (Chi, Glaser, & Farr, 1988). Coupling this advantage with a greater knowledge of cue correlation, the expert can diagnose by simple pattern matching. A set of correlated symptoms are perceived to match a pattern stored in memory, and the diagnosis is made almost as is the process of perceptual recognition, discussed in Chapter 5. Klein (1989b) refers to this process as *recognition-primed decision making.* Time and effort-consuming cognitive processing are avoided. Similarly the expert may have direct rules that determine that if a certain diagnosis is true, a certain action is most effective, where the action is part of a large store of such actions in long-term memory, learned from past experience. Collectively, these effects will lead the expert to show greater confidence in his or her diagnoses and actions (Wickens et al., 1988).

3. *Risk and probability calibration.* Because of greater experience with a domain, experts should have a better understanding of the actual probabilities of different states and of different outcomes, thereby allowing them to better calibrate their diagnoses and choices to current probabilities and risks. However, as discussed, there is some evidence that experts may be nearly as susceptible to these biases as novices. In this light, we now turn to a discussion of why experience does not always improve decision performance and may, paradoxically, sometimes degrade it.

The Myth of the Expert

Some researchers (e.g., Balla, 1980; Brehmer, 1981; Eddy, 1982; Einhorn & Hogarth, 1978; Slovic, Fischhoff, & Lichtenstein, 1977) assert that even the true expert decision maker, whether medical doctor, investment broker, troubleshooter, or personnel selector, suffers from some fairly severe biases, four of which are discussed below.

Misleading Feedback Since decision making is probabilistic, a correctly made decision (the right cues given appropriate weighting) will often yield an incorrect outcome because of chance factors. (We noted in Chapter 2 that optimal settings of d' and beta will still yield errors.) Correspondingly, one may obtain correct outcomes for the wrong reasons. Thus the "right-wrong" aspect of the feedback may often be misleading, thereby encouraging people to follow incorrect strategies that through chance yielded correct outcomes or to abandon appropriate strategies that yielded incorrect outcomes. Each correct feedback will serve to reinforce the rule that generated the outcome. If this rule is inappropriate but correct by chance, its strength becomes greater and greater, and it is harder to "unlearn" (Einhorn & Hogarth, 1978).

Limited Attention to Delayed Feedback Often the feedback from a decision may be delayed — by a few days perhaps in the case of the physician or by several months or years in the case of an admissions committee (to assess whether an applicant succeeds in a program) or a parole officer (to assess whether the client has successfully adapted to the world outside of prison). This delay may cause many of the factors that went into the decision to be forgotten or distorted by the time the feedback (however limited) becomes available. Furthermore, the decision maker is often quite preoccupied with other matters at the time and thus gives little attention to the feedback. This tendency is exaggerated by a phenomenon that Fischhoff (1977) labels *cognitive conceit.* Through a number of convincing experiments, he demonstrated how much we tend to underestimate the information gained from outcomes and therefore overestimate, retrospectively, the extent of our prior knowledge. If the apparent discrepancy between what we know now (after a decision outcome is observed) and what we thought we knew (retrospectively) before the decision is slight, we see little wrong with our initial decision formulation and thus perceive little need to revise the decision-making process. This process, of course, describes the "20–20 vision of hindsight" (Hawkins & Hastie, 1990).

Fischhoff and MacGregor (1982) identify the lack of attention to feedback as a major cause of overconfidence in forecasting, discussed earlier. As they note, the exception appears to be in weather forecasting, in which forecasters are clearly judged on the long-run accuracy of their predictions. Here, when decision makers are forced to process feedback and thereby calibrate their forecasts, performance appears to be accurate (Murphy & Winkler, 1984).

Selective Perception of Feedback Einhorn and Hogarth (1978) have elegantly demonstrated how the manner of representing the prediction task fosters unwarranted confidence in one's decision-making abilities. This representation in turn leads experts to become progressively poorer, ironically even as their confidence in the correctness of their choice grows. Einhorn and Hogarth consider the situation in which a candidate is to be admitted to or rejected from a program. The basis of selection is the value of a composite predictive variable, aggregated from measures on a number of attributes, as represented in Figure 7.14. If the variable exceeds a criterion, the candidate is selected. Eventually, those candidates selected will be evaluated on the basis of their success in the program (or success of the treatment). Here again the authors assume a dichot-

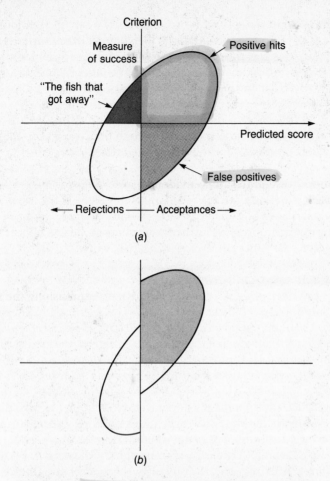

Figure 7.14 Source of unwarranted confidence in prediction. (*Source:* H. J. Einhorn and R. M. Hogarth, "Confidence in Judgment: Persistence of the Illusion of Validity," *Psychological Review*, 85 (1978), p. 397. Copyright © 1978 by the American Psychological Association. Adapted by permission of the authors.)

omy of success and failure. These representations are shown in Figure 7.14a. The core of Einhorn and Hogarth's argument is that a host of factors combine to cause the decision maker to attend to (and therefore overestimate) the *positive hit rate* (applicants who exceeded the selection criterion and succeeded in the program).

First, Einhorn and Hogarth (1978) argue that people tend to encode and thus remember the frequency of positive hits rather than the *probability* of a positive hit (i.e., positive hits divided by the total number of those exceeding the acceptance criterion). This distortion occurs in part because number is a more direct and salient quantity than is probability—an abstraction requiring an extra cognitive step (division). Also the bias against encoding by probability occurs in part because the denominator is dependent on knowing the number of false positives (those who were accepted but failed). As indicated, these entries are data that disconfirm the decision-making strategy or selection rule and therefore, as disconfirming evidence, tend to be ignored.

Perceiving in this fashion the entries in the positive hit cell, the decision maker is spuriously reinforced by the knowledge that "*n* people exceeded the selection criterion based on my decision rule and they succeeded; therefore, my rule must work." In addition to this bias of attention against the false positive numbers (disconfirmatory evidence), the decision maker will rarely possess any data concerning those who were rejected. Even if such data are available, they are unlikely to be in a usable probabilistic form. More likely they will be in the form of a single-case representation, for example, "the fish that got away."

Selective Influence on Outcome Einhorn and Hogarth, (1978) point out that the decision maker often has a vested interest in establishing the success of the rules that were employed. Therefore, treatment or "placebo effects" will increase the likelihood that those selected to a program will be more likely to succeed than those who did not exceed the criterion but might have been admitted to the program for other reasons. This bias is shown by the upward shift in criterion scores to the right of the cutoff in Figure 7.14*b*. Through computer simulation techniques, which assume that the rewarding effects of a positive hit reinforce the existing decision rules, Einhorn and Hogarth show that both increasingly unwarranted confidence and less accurate prediction emerge with increased practice on the prediction task.

In summary, decisions of the expert in a given field may well be superior to those of the novice, as a result of the expert's greater familiarity with the material in the area, greater knowledge of correlations between cues, and better models of causal structure. Yet this knowledge alone is not sufficient to guarantee the accuracy of the decision process. The inherent biases built into the decision-making task will often be sufficient to guarantee that even the confident expert has an inflated estimate of his or her decision-making accuracy.

DECISION-MAKING AIDS

It can be argued that whether decision making by the expert or novice is, in an absolute sense good or bad is somewhat beside the point. What is more important is that careful analysis has identified various influences on the quality of human decision making and has allowed engineering psychologists to identify the circumstances in which decisions are likely to degrade. This knowledge in turn has led psychologists to consider the kinds of decision aids that can support better performance in those circumstances.

Although several techniques have been proposed to assist the decision maker in real-world contexts, many of them have encountered difficulties on their way to implementation (Einhorn & Hogarth, 1981; Pitz & Sachs, 1984; Schrenk, 1969). One problem is that the complexity of many of the aids renders them difficult to use. A second problem is that people's overconfidence in their own decision-making capabilities leave them less than willing to trust the aids, which they view to be less accurate (Kleinmuntz, 1990). A third problem is that it is exceedingly difficult to evaluate the success of decision aids. If a decision is found to be correct with the use of an aid in a real-world context, would the outcome also have been correct if the decision were made without it? In the laboratory such data can easily be brought to bear as decision aids are demonstrated on simulated problems (Samet, Weltman, & Davis, 1976). However, because each real decision problem outside the laboratory is unique, the answer to the question is difficult to obtain. The exact conditions under which the particular decision is made are impossible to re-create once, let alone often enough to obtain a statistically reliable estimate of the probability of a correct (or optimal) decision with and without the aid.

In spite of these difficulties in establishing the usefulness of decision aids, the numerous shortcomings of human decision making implicitly suggest a number of techniques that could assist the decision maker in a wide variety of environments. These aids are divided into two categories: those based on the limits of human memory and attention and those based on training.

Memory and Attention Aids

It is evident that where possible decision-making tasks should be designed in such a manner as to present all diagnostic information concerning all alternatives so that it can be considered simultaneously. Bettman, Payne, and Staelin (1986) and Russo (1977) point to the importance of such display configurations to assist consumers in making product choices. Integrated computer displays in complex process-monitoring stations will be of considerable use (Moray, 1980, 1981; Rasmussen, 1981) and will be considered in more detail in Chapter 12. Naturally some selection may be required to reduce the displayed information to a manageable level. Samet, Weltman, and Davis, (1976) have suggested possible adaptive algorithms. They propose a system in which the computer monitors the operator's choice of information sources and infers the particular *attributes* of those sources that the operator finds most important. For example, does the monitor of a process control plant typically refer to readings of pressure or temperature, and which ones weigh more heavily on diagnostic judgment? After making this inference, the computer selectively filters information sources in order to present primarily those sources that contain the most information on the subjectively important attributes. In this manner the computer reduces some of the problems associated with information overload.

In a similar vein, computer-generated displays may be used to assist the operator in a fault-diagnosis task by considering alternative hypotheses and keeping track of the outcome of sequential tests, so that recency and anchoring on the initial hypotheses do not dominate memory. Rouse (1981) reports that such memory aids substantially improve troubleshooting performance. A computer with sufficient intelligence could provide information relevant to the bearing of test outcomes on alternative hypotheses and could force the operator to entertain alternative hypotheses. This technique could provide an important safeguard against the bias for hypothesis confirmation (Einhorn & Hogarth, 1982; Rasmussen, 1981). Some limited success with this sort of aid has been realized with expert systems in such areas as medical diagnosis when a restricted problem can be very carefully defined (Madni, 1988; Shortcliffe, 1983). These issues will be discussed further in Chapter 12.

Other aids emphasize the computer's assistance in integrating diagnostic information. As described earlier, Dawes (1979; Dawes, Faust, & Meehl, 1989) argues convincingly that computer assistance can be of great use in integrating predictive information to provide statistical prediction. Edwards (1962; Edwards, Phillips, Hays, & Goodman, 1968) proposes a system of cooperation between humans and machines in sequential data-processing tasks. To overcome the inherent human conservatism relative to the "Bayesian optimal," he proposes that the person act as the likelihood-ratio estimator and the machine act as the Bayesian-odds aggregator. Kleinmuntz (1990) discusses other techniques for best combining human and automated statistical components in decision making.

One of the most important and prominent decision-aiding techniques in this area is the application of multiattribute utility theory (Edwards, 1987; Edwards & Newman, 1982; Slovic, Fischhoff, & Lichtenstein, 1977). This technique assists people in carrying out the choice process, outlined in Figure 7.10, by a structural procedure in which objects and attributes are identified, utilities of different attributes are specified, and the rating of each alternative on each attribute is obtained. In this manner it is possible to derive a choice that will most closely satisfy the decision maker's value structure, even if the total information load is high. The technique has been successfully employed in a number of diverse circumstances—for example, assisting land developers and environmentalists in reaching a compromise on coastal development policy in California (Gardiner & Edwards, 1975) and choosing the location of the Mexico City airport (Kenney, 1973).

Training Aids

Training aids can be offered to the decision maker in three forms. One method is simply to make decision makers aware of the nature of limitations and biases of which they may be totally unconscious. For example, it would seem that training operators to consider alternative hypotheses and to be aware of their resistance to doing it under stress might reduce the likelihood of cognitive tunnel vision.

A number of efforts have obtained some success in "debiasing" decision makers from particular forms of overconfidence (Fischhoff, 1982; Murphy & Winkler, 1984). Koriat, Lichtenstein, and Fischhoff (1980) found that forcing forecasters to entertain reasons why their forecasts might not be correct reduced their biases toward overconfidence in the accuracy of the forecast. As noted, Hunt and Rouse (1981) succeeded in training operators to extract diagnostic information from the absence of cues. Lopes (1982) achieved some success at training subjects away from nonoptimal anchoring biases when processing multiple information sources. She called subjects' attention to their tendency to anchor on initial stimuli that may not be informative and had them anchor instead on the most informative sources. When this was done, the biases were reduced. The study of intelligence analysts by Tolcott, Marvin, and Bresoick (1989) included an experimental treatment designed to reduce the influence of anchoring and the confirmation bias by administering a brief training program regarding their effects. This training was partially successful.

A second training aid is to provide more comprehensive and immediate feedback in predictive and diagnostic tasks, so that operators are forced to attend to the degree of success or failure of their rules. We noted that the feedback given to weather forecasters is successful in reducing the tendency for overconfidence in forecasting (Murphy & Winkler, 1984). Jenkins and Ward (1965) demonstrated that providing decision makers simultaneously with data in all four cells of the contingency table in Figure 7.14 instead of simply the hit probability, improves their appreciation of predictive relations. Where selection tasks or diagnostic treatments are prescribed, box scores should be maintained to integrate data in as many cells of the matrix as possible (Einhorn & Hogarth, 1978; Goldberg, 1968). Tversky and Kahneman (1974) suggested that decision makers should be taught to encode events in terms of probability rather than frequency since probabilities intrinsically account for events that did not occur (negative evidence) as well as those that did.

A third way to improve training is to capitalize on the natural efforts of people to seek causal relations between variables and their superiority in integrating cues when correlations between variables are known beforehand (Einhorn & Hogarth, 1981; Medin, Alton, Edelson, & Freko, 1982; Phelps & Shanteau, 1978). To capitalize in this manner, every effort should be made to identify "syndromes" and to emphasize the correlational structure existing in the cues that represent a certain hypothesis. Where possible it seems advantageous to explain these correlations in terms of a causal structure (if one exists) or a process model that can economically represent the nature of the intercorrelations (Moray, 1981). We will cite some examples of the success of this instructional procedure in Chapter 12.

TRANSITION

We will revisit some of the issues of decision and diagnosis in our discussion of process control in Chapter 12. However it is appropriate to end this chapter on a note of optimism. Although human decision making is often far from that which could be achieved by the intelligent computer with all facts at hand, in most (but not all) situations humans far outperform decisions that would be made by the flip of a coin or a roll of the dice. Performance probably lies somewhere midway between chance and optimal, shifting further away from the latter as complexity and stress increase.

But as we have seen, when decisions become more predictable, with certain rather than uncertain outcomes, and when the decision maker becomes very familiar with the structure and correlation of cues in the environment, human decision making may become both rapid and accurate, following the principles of perception and pattern recognition discussed in Chapter 5. In these circumstances the concern of the engineering psychologist is less for the cognitive limitations of integrating information and computing in working memory and more for the response limitations that constrain the speed with which the decision is translated to action. This, then, is the focus of Chapter 8.

REFERENCES

Allen, G. (1982). Probability judgment in weather forecasting. *Ninth Conference in Weather Forecasting and Analysis*. Boston: American Meterological Society.

APA Monitor. (1988). *Vincennes*: Findings could have helped avert tragedy: Scientists tell Hill panel.

Arkes, H., & Harkness, R. R. (1980). The effect of making a diagnosis on subsequent recognition of symptoms. *Journal of Experimental Psychology: Human Learning and Memory, 6,* 568–575.

Balla, J. (1980). Logical thinking and the diagnostic process. *Methodology and Information in Medicine, 19,* 88–92.

Balla, J. (1982). The use of critical cues and prior probability in concept identification. *Methodology and Information in Medicine, 21,* 9–14.

Bettman, J. R., Payne, J. W., & Staelin, R. (1986). Cognitive considerations in designing effective labels for presenting risk information. *Journal of Marketing and Public Policy, 5,* 1–28.

Bettman, J. R., Johnson, E. J., & Payne, J. (1990). A componential analysis of cognitive effort and choice. *Organizational Behavior and Human Performance, 45,* 111–139.

Bettman, J. R., Johnson, E. J., & Payne, J. W. (1991). Consumer decision making. In T. S. Robertson & H. S. Kassarjin (eds.), *Handbook of consumer behavior* (pp. 50–84). New York: Prentice Hall.

Bower, G., & Trabasso, T. (1963). Reversals prior to a solution in concept identification. *Journal of Experimental Psychology, 66,* 409–418.

Brehmer, B. (1981). Models of diagnostic judgment. In J. Rasmussen & W. Rouse (eds.), *Human detection and diagnosis of system failures.* New York: Plenum Press.

Brown, B. G., & Murphy, A. H. (1987). Quantification of uncertainty in fire-weather forecasts: Some results of operational and experimental forecasting programs. *Weather and Forecasting, 2*(3), 190–205.

Carroll, J. S. (1980). Analyzing behavior: The magician's audience. In T. S. Wallsten (ed.), *Cognitive processes in choice and decision making.* Hillsdale, NJ: Erlbaum.

Chi, M. T. H., Glaser, R., & Farr, M. J. (eds.). (1988). *The nature of expertise.* Hillsdale, NJ: Erlbaum.

Christenssen-Szalanski, J. J., & Bushyhead, J. B. (1981). Physicians' use of probabilistic information in a real clinical setting. *Journal of Experimental Psychology: Human Perception and Performance, 7,* 928–936.

Clark, H. H., & Chase, W. G. (1972). On the process of comparing sentences against pictures. *Cognitive Psychology, 3,* 472–517.

Combs, B., & Slovic, P. (1979). Newspaper coverage of causes of death. *Journalism Quarterly, 56*(4), 837–843; 849.

Dawes, R. M. (1979). The robust beauty of improper linear models in decision making. *American Psychologist, 34,* 571–582.

Dawes, R. M., & Corrigan, B. (1974). Linear models in decision making. *Psychological Bulletin, 81,* 95–106.

Dawes, R. M., Faust, D., & Meehl, P. E. (1989). Clinical versus statistical judgment. *Science, 243,* 1668–1673.

Du Charme, W. (1970). Response bias explanation of conservative human inference. *Journal of Experimental Psychology, 85,* 66–74.

Ebbeson, E. D., & Konecni, V. (1980). On external validity in decision-making research. In T. Wallsten (ed.), *Cognitive processes in choice and decision making.* Hillsdale, NJ: Erlbaum.

Eddy, D. M. (1982). Probabilistic reasoning in clinical medicine: Problems and opportunities. In D. Kahneman, P. Slovic, & A. Tversky (eds.), *Judgment under uncertainty: Heuristics and biases.* New York: Cambridge University Press.

Edwards, W. (1962). Dynamic decision theory and probabilistic information processing. *Human Factors, 4,* 59–73.

Edwards, W. (1968). Conservatism in human information processing. In B. Kleinmuntz (ed.), *Formal representation of human judgment* (pp. 17–52). New York: Wiley.

Edwards, W. (1987). Decision making. In G. Salvendy (ed.), *Handbook of human factors* (pp. 1061–1104). New York: Wiley.

Edwards, W., Lindman, H., & Phillips, L. D. (1965). Emerging technologies for making decisions. In T. M. Newcomb (ed.), *New directions in psychology II.* New York: Holt, Rinehart & Winston.

Edwards, W., Lindman, H., & Savage, L. J. (1963). Bayesian statistical inference for psychological research. *Psychological Review, 70,* 193–242.

Edwards, W., & Newman, J. R. (1982). *Multiattribute evaluation.* Beverly Hills, CA: Sage.

Edwards, W., Phillips, L. D., Hays, W. L., & Goodman, B. C. (1968). Probabilistic information processing systems: Design and evaluation. *IEEE Transactions on Systems, Science, and Cybernetics, SSC–4,* 248–265.

Einhorn, H. J., & Hogarth, R. M. (1978). Confidence in judgment: Persistence of the illusion of validity. *Psychological Review, 85,* 395–416.

Einhorn, H. J., & Hogarth, R. M. (1981). Behavioral decision theory. *Annual Review of Psychology, 32,* 53–88.

Einhorn, H. J., & Hogarth, R. M. (1982). *Theory of diagnostic interference I: Imagination and the psychophysics of evidence* (Technical Report no. 2). Chicago: University of Chicago, School of Business.

Einhorn, H. J., & Hogarth, R. M. (1983, January). *Diagnostic inference and causal judgment: A decision making frame* (Research Report). Chicago: University of Chicago, Center for Decision Research.

Ellison, K. W., & Buckhout, R. (1981). *Psychology & criminal justice.* New York: Harper & Row.

Erlick, D. E. (1961). Judgments of the relative frequency of a sequential series of two events. *Journal of Experimental Psychology, 62*, 105–112.

Fischhoff, B. (1977). Perceived informativeness of facts. *Journal of Experimental Psychology: Human Perception and Performance, 3*, 349–358.

Fischhoff, B. (1982). Debiasing. In D. Kahneman, P. Slovic, & A. Tversky (eds.), *Judgment under uncertainty: Heuristics and biases* (pp. 422–444). New York: Cambridge University Press.

Fischhoff, B., & MacGregor, D. (1982). Subjective confidence in forecasts. *Journal of Forecasting, 1*, 155–172.

Fischhoff, B., Slovic, P., & Lichtenstein, S. (1977). Knowing with certainty: The appropriateness of extreme confidence. *Journal of Experimental Psychology: Human Perception and Performance, 3*, 552–564.

Fischhoff, B., & Bar-Hillel, M. (1984). Diagnosticity and the base-rate effect. *Memory and Cognition, 12*(4), 402–410.

Fischhoff, B., & MacGregor, D. (1982). Subjective confidence in forecasts. *Journal of Forecasting, 1*, 155–172.

Fischhoff, B., Slovic, P., & Lichtenstein, S. (1977). Knowing with certainty: The appropriateness of extreme confidence. *Journal of Experimental Psychology: Human Perception and Performance, 3*, 552–564.

Fontenelle, G. A. (1983). *The effect of task characteristics on the availability heuristic for judgments of uncertainty* (Report No. 83–1). Office of Naval Research, Rice University.

Gardiner, P. D., & Edwards, W. (1975). Public values: Multiattribute ability measurement for social decision making. In M. F. Kaplan & B. Schwartz (eds.), *Human judgment and decision processes.* New York: Academic Press.

Gentner, D., & Stevens, A. L. (1983). *Mental models.* Hillsdale, NJ: Erlbaum.

Goldberg, L. R. (1968). Simple models or simple processes? *American Psychologist, 23*, 483–496.

Gottsdanker, R. M., & Edwards, R. V. (1957). The prediction of collision. *American Journal of Psychology, 70*, 110–113.

Hogarth, R., & Einhorn, H. (1989). *Order effects in belief updating: The belief adjustment model* (Research Report no. 155). Chicago: University of Chicago Center for Decision.

Hunt, R., & Rouse, W. (1981). Problem-solving skills of maintenance trainees in diagnosing faults in simulated power plants. *Human Factors, 23*, 317–328.

Jenkins, H. M., & Ward, W. C. (1965). Judgment of contingency between responses and outcomes. *Psychological Monographs: General and Applied, 79* (whole no. 594).

Johnson, E. M., Cavanagh, R. C., Spooner, R. L., & Samet, M. G. (1973). Utilization of reliability measurements in Bayesian inference: Models and human performance. *IEEE Transactions on Reliability, 22,* 176–183.

Johnson, E. J., Payne, J. W., & Bettman, J. R. (1988). Information displays and preference reversals. *Organizational Behavior and Human Decision Processes, 42,* 1–21.

Kahneman, D., Slovic, P., & Tversky, A. (eds.). (1982). *Judgment under uncertainty: Heuristics and biases.* New York: Cambridge University Press.

Kahneman, D., & Tversky, A. (1984). Choices, values, and frames. *American Psychologist, 39,* 341–350.

Kahneman, D., & Tversky, A. (1973). On the psychology of prediction. *Psychological Review, 80,* 251–273.

Kanarick, A. F., Huntington, A., & Peterson, R. C. (1969). Multisource information acquisition with optimal stopping. *Human Factors, 11,* 379–386.

Karnes, E. W., Leonard, S. D., & Rachwal, G. (1986). Effects of benign experiences on the perception of risk. *Proceedings of the 30th annual meeting of the Human Factors Society* (pp. 121–125). Santa Monica, CA: Human Factors Society.

Keeney, R. L. (1988). Facts to guide thinking about life threatening risks. *Proceedings 1988 IEEE Conference on Systems, Man, and Cybernetics.* Beijing, China: Pergamon-CNPIEC.

Kenney, R. L. (1973). A decision analysis with multiple objectives: The Mexico City airport. *Bell Telephone Economic Management Science, 4,* 101–117.

Klayman, J., & Ha, Y. W. (1987). Confirmation, disconfirmation, and information in hypothesis testing. *Journal of Experimental Psychology: Human Learning and Memory,* 211–228.

Klein, G. A. (1989a). Do decision biases explain too much? *Human Factors Society Bulletin, 32,* 1–3.

Klein, G. A. (1989b). Recognition-primed decisions. In W. Rouse (ed.), *Advances in man-machine systems research* (Vol. 5, pp. 47–92). Greenwich, CT: JAI Press.

Kleinmuntz, B. (1990). Why we still use our heads instead of formulas: Toward an integrative approach. *Psychological Bulletin, 107*(3), 296–310.

Kleinmutz, D. (1985). Cognitive heuristics and feedback in a dynamic decision environment. *Management Science, 31,* 680–702.

Koriat, A., Lichtenstein, S., & Fischhoff, B. (1980). Reasons for confidence. *Journal of Experimental Psychology: Human Learning and Memory, 6,* 107–118.

Lathrop, R. G. (1967). Perceived variability. *Journal of Experimental Psychology, 23,* 498–502.

Levine, M. (1966). Hypothesis behavior by humans during discrimination learning. *Journal of Experimental Psychology, 71,* 331–338.

Leibowitz, H. W. (1991). Perceptually induced misperception of risk: A common factor in transportation accidents. In L. Lipsitt and L. Mituick (eds.), *Behavior and risk taking: Causes and consequences.* Norwood, NJ: Ablex.

Loftus, E. F. (1979). *Eyewitness testimony.* Cambridge, MA: Harvard University Press.

Loftus, E. F. (1987). Psychology and the law. In F. Farley and C. Null (eds.), *Using psychological science: Making the public case.* Washington, DC: Federation of Behavioral, Psychological, and Cognitive Science.

Lopes, L. L. (1982, October). *Procedural debiasing* (Technical Report WHIPP 15). Madison: Wisconsin Human Information Processing Program.

Lusk, C. M., Stewart, T. R., Hammond, K. R., & Potts, R. (1990). Judgment and decision making in dynamic tasks: The case of forecasting the microburst. *Weather and Forecasting.*

Lusted, L. B. (1976). Clinical decision making. In D. Dombal & J. Grevy (eds.), *Decision making and medical care.* Amsterdam: North Holland.

MacGregor, D., Fischhoff, B., & Blackshaw, L. (1987). Search success and expectations with a computer interface. *Information Processing and Management, 23,* 419–432.

MacGregor, D., & Slovic, P. (1986). Graphic representation of judgmental information. *Human-Computer Interaction, 2,* 179–200.

McNeil, B. J., Pauker, S. G., Sox, H. C., Jr., & Tversky, A. (1982). On the elicitation of preferences for alternative therapies. *New England Journal of Medicine, 306,* 1259–1262.

Madni, A. M. (1988). The role of human factors in expert systems design and acceptance. *Human Factors, 30,* 395–414.

Malhotra, N. K. (1982). Information load and consumer decision making. *Journal of Consumer Research, 8,* 419–430.

Medin, D. L., Alton, M. W., Edelson, S. M., & Freko, D. (1982). Correlated symptoms and simulated medical classification. *Journal of Experimental Psychology: Learning, Memory, and Cognition, 8,* 37–50.

Meehl, P. C. (1954). *Clinical versus statistical prediction.* Minneapolis: University of Minnesota Press.

Mehle, T. (1982). Hypothesis generation in an automobile malfunction inference task. *Acta Psychologica, 52,* 87–116.

Moray, N. (1980). *Information processing and supervisory control* (MIT Man Machine Systems Laboratory Report). Cambridge, MA: MIT Press.

Moray, N. (1981). The role of attention in the detection of errors and the diagnosis of errors in man-machine systems. In J. Rasmussen & W. Rouse (eds.), *Human detection and diagnosis of system failures.* New York: Plenum Press.

Murphy, A. H., Hsu, W. R., Winkler, R. L., & Wilks, D. S. (1985). The use of probabilities in subjective quantitative precipitation forecasts: Some experimental results. *Monthly Weather Review, 113*(12), 2075–2089.

Murphy, A. H., & Winkler, R. L., (1984). Probability of precipitation forecasts. *Journal of the Association Study of Perception, 79,* 391–400.

Mynatt, C. R., Doherty, M. E., & Tweney, R. D. (1977). Confirmation bias in a simulated research environment: An experimental study of scientific inference. *Quarterly Journal of Experimental Psychology, 29,* 85–95.

Navon, D. (1979). The importance of being conservative. *British Journal of Mathematical and Statistical Psychology, 31,* 33–48.

Oskamp, S. (1965). Overconfidence in case-study judgments. *Journal of Consulting Psychology, 29,* 261–265.

Payne, J. W. (1980). Information processing theory: Some concepts and methods applied to decision research. In T. S. Wallsten (ed.), *Cognitive processes in choice and decision behavior.* Hillsdale, NJ: Erlbaum.

Payne, J. W. (1982). *Contingent decision behavior: A review and discussion of issues* (Technical Report 82–1). Durham, NC: Duke University School of Business Administration.

Payne, J. W., Bettman, J. R., & Johnson, E. J. (1988). Adaptive strategy selection in decision making. *Journal of Experimental Psychology: Learning, Memory, and Cognition, 14,* 534–552.

Peterson, C. R., & Beach, L. R. (1967). Man as an intuitive statistician. *Psychological Bulletin, 68,* 29–46.

Phelps, R. H., & Shanteau, J. (1978). Livestock judges: How much information can an expert use? *Organizational Behavior in Human Performance, 21,* 209–219.

Pitz, G. F. (1980). The very guide of life: The use of probabilistic information for making decisions. In T. S. Wallsten (ed.), *Cognitive processes in choice and decision behavior.* Hillsdale, NJ: Erlbaum.

Pitz, G. F., & Sachs, N. J. (1984). Judgment and decision: Theory and application. *Annual Review of Psychology, 35,* 139–163.

Potter, S. S., Rockwell, T. H., & McCoy, C. E. (1989). General aviation pilot error in computer simulated adverse weather scenarios. In R. S. Jensen (ed.), *Proceedings of the Fifth International Symposium on Aviation Psychology* (pp. 570–575). Columbus: Ohio State University, Department of Aviation.

Puto, C. P., Patton, W. E., III, & King, R. H. (1985). Risk handling stratetgies in industrial vendor selection decisions. *Journal of Marketing, 49,* 89–98.

Rasmussen, J. (1981). Models of mental strategies in process control. In J. Rasmussen & W. Rouse (eds.), *Human detection and diagnosis of system failures.* New York: Plenum Press.

Rossi, A. L., & Madden, J. M. (1979). Clinical judgment of nurses. *Bulletin of the Psychonomic Society, 14,* 281–284.

Rouse, W. B. (1981). Experimental studies and mathematical models of human problem solving performance in fault diagnosis tasks. In J. Rasmussen & W. Rouse (eds.), *Human detection and diagnosis of system failures.* New York: Plenum Press.

Rubenstein, T., & Mason, A. F. (1979, November). The accident that shouldn't have happened: An analysis of Three Mile Island. *IEEE Spectrum,* pp. 33–57.

Runeson, S. (1975). Visual prediction of collisions with natural and nonnatural motion functions. *Perception & Psychophysics, 18,* 261–266.

Russo, J. E. (1977). The value of unit price information. *Journal of Marketing Research, 14,* 193–201.

Samet, M. G., Weltman, G., & Davis, K. B. (1976, December). *Application of adaptive models to information selection in C3 systems* (Technical Report PTR–1033-76-12). Woodland Hills, CA: Perceptronics.

Sanderson, P. M., & Murtagh, J. M. (1989). Predicting fault diagnosis performance: Why are some bugs hard to find? *IEEE Transactions on Systems, Man, and Cybernetics.*

Schrenk, L. P. (1969). Aiding the decision maker—A decision process model. *IEEE Transactions on Man-Machine Systems, MMS–10,* 204–218.

Schroeder, R. G., & Benbassat, D. (1975). An experimental evaluation of the relationship of uncertainty to information used by decision makers. *Decision Sciences, 6,* 556–567.

Schum, D. (1975). The weighing of testimony of judicial proceedings from sources having reduced credibility. *Human Factors, 17*, 172–203.

Schurr, P. H. (1987). Effects of gain and loss decision frames on risky purchase negotiations. *Journal of Applied Psychology, 72*(3), 351–358.

Schustack, M. W., & Sternberg, R. J. (1981). Evaluation of evidence in causal inference. *Journal of Experimental Psychology: General, 110*, 101–120.

Sheridan, T. (1981). Understanding human error and aiding human diagnostic behavior in nuclear power plants. In J. Rasmussen & W. Rouse (eds.), *Human detection and diagnosis of system failures.* New York: Plenum Press.

Sheridan, T. B., & Ferrell, L. (1974). *Man-machine systems.* Cambridge, MA: MIT Press.

Shortliffe, E. H. (1983). Medical consultation systems. In M. E. Sime and M. J. Coombs (eds.), *Designing for human-computer communications* (pp. 209–238). New York: Academic Press.

Shugan, S. M. (1980). The cost of thinking. *Journal of Consumer Research, 7*, 99–111.

Simon, H. A. (1955). A behavioral model of rational choice. *Quarterly Journal of Economics, 69*, 99–118.

Slovic, P. (1987). Facts vs. fears: Understanding perceived risk. In F. Farley & C. H. Null (eds.), *Using psychological science: Making the public case* (pp. 57–68). Washington, DC: The Federation of Behavioral, Psychological and Cognitive Sciences.

Slovic, P., Fischhoff, B., & Lichtenstein, S. (1977). Behavioral decision theory. *Annual Review of Psychology, 28*, 1–39.

Slovic, P., Fischhoff, B., & Lichtenstein, S. (1982). Facts versus fears: Understanding perceived risk. In D. Kahneman, P. Slovic, & A. Tversky (eds.), *Judgment under uncertainty: Heuristics and biases* (pp. 463–489). Cambridge, England: Cambridge University Press.

Slovic, P., Fischhoff, B., & Lichtenstein, S. (1981). Perceived risk: Psychological factors and social implications. In F. Warner & D. H. Slater (eds.), *The assessment and perception of risk* (pp. 17–34). London: Royal Society.

Slovic, P., Fischhoff, B., & Lichtenstein, S. (1984). Behavioral decision theory perspectives on risk and safety. *Acta Psychologica, 56*, 183–203.

Slovic, P., Lichtenstein, S., & Fischhoff, B. (1988). Decision making. In R. C. Atkinson, R. J. Herrnstein, G. Lindzey, & R. D. Luce (eds.), *Steven's handbook of experimental psychology* (2nd ed.). New York: Wiley.

Slovic, P., Fischhoff, B., & Lichtenstein, S. (1977). Behavioral decision theory. *Annual Review of Psychology, 28*, 1–39.

Slovic, P., Fischhoff, B., & Lichtenstein, S. (1981). Perceived risk: Psychological factors and social implications. In F. Warner & D. H. Slater (eds.), *The assessment and perception of risk* (pp. 17–34). London: Royal Society.

Slovic, P., Fischhoff, B., & Lichtenstein, S. (1984). Behavioral decision theory perspectives on risk and safety. *Acta Psychologica, 56*, 183–203.

Sniezek, J. A. (1980). Judgments of probabilistic events: Remembering the past and predicting the future. *Journal of Experimental Psychology: Human Perception & Performance, 6,* 695–706.

Sokolov, E. N. (1969). The modeling properties of the nervous system. In I. Maltzman & K. Cole (eds.), *Handbook of contemporary Soviet psychology.* New York: Basic Books.

Sprent, P. (1988). *Taking risks: The science of uncertainty.* England: Penguin.

Taplin, J. E. (1971). Reasoning with conditional sentences. *Journal of Verbal Learning and Verbal Behavior, 10,* 218–225.

Taplin, J. E., & Staudenmayer, H. (1973). Interpretation of abstract conditional sentences in deductive reasoning. *Journal of Verbal Learning and Verbal Behavior, 12,* 530–542.

Tolcott, M. A., Marvin, F. F., & Bresoick, T. A. (1989). *The confirmation bias in military situation assessment.* Reston, VA: Decision Science Consortium.

Tversky, A. (1972). Elimination by aspects: A theory of choice. *Psychological Review, 79,* 281–299.

Tversky, A., & Kahneman, D. (1971). The law of small numbers. *Psychological Bulletin, 76,* 105–110.

Tversky, A., & Kahneman, D. (1974). Judgment under uncertainty: Heuristics and biases. *Science, 185,* 1124–1131.

Tversky, A., & Kahneman, D. (1981). The framing of decisions and the psychology of choice. *Science, 211,* 453–458.

U.S. Navy. (1988). *Investigation report: Formal investigation into the circumstances surrounding the downing of Iran air flight 655 on 3 July 1988.* Washington, DC: Department of Defense Investigation Report.

Varey, C. A., Mellers, B. A., & Birnbaum, M. H. (1990). Judgments of proportions. *Journal of Experimental Psychology: Human Perception and Performance, 16*(3), 613–625.

Waganaar, W. A., & Sagaria, S. D. (1975). Misperception of exponential growth. *Perception & Psychophysics, 18,* 416–422.

Wallsten, T. S., & Barton, C. (1982). Processing probabilistic multidimensional information for decisions. *Journal of Experimental Psychology: Learning, Memory and Cognition 8,* 361–384.

Wason, P. C., & Johnson-Laird, P. N. (1972). *Psychology of reasoning: Structure and content.* London: Batsford.

Wells, G. L., Lindsay, R. C., & Ferguson, T. I. (1979). Accuracy, confidence, and juror perceptions in eyewitness testimony. *Journal of Applied Psychology, 64,* 440–448.

Wickens, C. D., Barnett, B., Stokes, A., Davis, T., Jr., & Hyman, F. (1989). Expertise, stress, and pilot judgment. *Proceedings of the NATO/AGARD Conference (AGARD-CP-458) on Human Behaviour in High Stress Situations in Aerospace Operations* (pp. 10 = 1–10 = 8). Loughton, Essex, England: Specialised Printing Services.

Wickens, C. D., Stokes, A., Barnett, B., & Hyman, F. (1991, in press). The effects of stress on pilot judgment in a MIDIS simulator. In O. Svenson & J. Maule (eds.), *Time pressure and stress in human judgment and decision making.* Cambridge, England: Cambridge University Press.

REFERENCES 311

Wohl, J. (1983). Cognitive capability versus system complexity in electronic maintenance. *IEEE Transaction on Systems, Man, and Cybernetics, 13,* 624–626.

Woods, D. D., O'Brien, J. F., & Hanes, L. F. (1987). Human factors challenges in process control: The case of nuclear power plants. In G. Salvendy (ed.), *Handbook of human factors* (pp. 1724–1770). New York: Wiley.

Wright, P. (1974). The harassed decision maker: Time pressures, distractions, and the use of evidence. *Journal of Applied Psychology, 59,* 555–561.

Yntema, D. Keeping track of several things at once. *Human Factors 6,* 7–17.

Chapter
8

Selection of Action

OVERVIEW

In most systems the human operator must translate the information that is perceived about the environment into an action. Sometimes the action is an immediate response to a perceived stimulus: We slam on the brake when a car unexpectedly pulls into the intersection ahead; the robot operator must press a button rapidly when the robot arm moves beyond safety limits (Helender, Karwan, & Etherton, 1987); or the monitor in an intensive care ward must make an immediate corrective action to the patient's cardiac arrest. At other times, the action is based more on a thorough, time-consuming evaluation of the current state of the world, integrating information from a large number of sources over a longer period of time. Many examples of this latter process were considered in Chapter 7, including the medical diagnosis and selection of treatment by a physician.

The two types of selection of action represent end points on a continuum related to the degree of *automaticity* with which the action is chosen. This automaticity in turn is determined by the amount of practice that operators have had in applying the rules of action selection. Rasmussen (1980, 1986) has distinguished three distinct categories on this continuum: skill-based, rule-based, and knowledge-based behavior. At the most automated level, *skill-based* behavior assigns stimuli to responses in a rapid automatic mode with a minimum investment of resources (see Chapter 9). Applying the brake on a car in response to the appearance of a red light is skill-based behavior. The stimuli for skill-based behavior need not necessarily be simple. For example, certain combinations of correlated features — a syndrome — may be rapidly classified into a category that triggers an automatic action. This is an example of skill-based action with high stimulus complexity. However, when such complexity exists, the rapid action will occur only after the operator has received extensive training and experience. The skilled physician, for example, may immediately detect the pattern of symptoms indicating a certain disease and identify the appropriate treatment at once. The medical school student or intern with far less medical training will evaluate the same symptoms in a much more time-consuming fashion to reach the same conclusion.

The level typified by the medical student illustrates *rule-based* behavior. Here an action is selected by bringing into working memory a hierarchy of rules: "If *X* occurs, then do *Y*." After mentally scanning these rules the decision maker will initiate the appropriate action. The situation may be familiar, but the processing is considerably less automatic and timely. The final category of action, *knowledge-based* behavior, is invoked when entirely new problems are encountered. Neither rules nor automatic mappings exist, and more general knowledge concerning the behavior of the system, the characteristics of the environment, and the goals to be obtained must be integrated to formulate a novel plan of action. This level is often typical of the kinds of decisions discussed in Chapter 7.

In decision making and diagnosis, more characteristic of rule- and knowledge-based behavior, accuracy is the most important measure of performance. This chapter will focus more on the *speed* of selection of skill-based actions. Although accuracy is still important, it is usually quite high in these tasks, and so response time is usually considered to be the critical measure of the performance quality of a person (or a system).

In the laboratory, the selection of skill-based actions is understood through the study of *reaction time*, or RT. What are the factors that determine the speed with which an operator can perceive a stimulus and translate that perception into a well-learned action? The trade-off between speed and errors will be considered in this chapter, and the nature of errors will be considered in Chapter 10.

Many different variables influence reaction time both inside and outside of the laboratory. One of the most important is the degree of uncertainty about what stimulus will occur and therefore the degree of choice in the action to make. For the sprinter at the starting line of a race, there is no uncertainty about the stimulus—the sound of the starting gun—nor is there a choice of what response to make: to get off the blocks as fast as possible. On the other hand, for the driver of an automobile, wary of potential obstacles in the road, there is both stimulus uncertainty and response choice. An obstacle could be encountered on the left, requiring a swerve to the right; on the right, requiring a swerve to the left; or perhaps at dead center, requiring that the brakes be applied. The situation of the sprinter illustrates the *simple reaction time* task, the vehicle driver the task of *choice reaction time*.

Examples of simple reaction time rarely occur outside of the laboratory—the sprinter's start is a rare exception. Yet the simple RT task is nevertheless important for the following reason: All of the variables that influence reaction time can be dichotomized into those that depend in some way on the choice of a response and those that do not, that is, those that influence only choice RT and those that affect all reaction times. When the simple RT task is examined in the laboratory, it is possible to study the second class of variables more precisely because the measurement of response speed cannot be contaminated by factors related to the degree of choice. We will see that the influences of the latter are considerable. Hence in the following treatment we will consider the variables that influence both choice and simple RT before discussing those variables unique to the choice task.

After both sets of variables are discussed, as well as a stage model of speeded information processing, we will consider what happens when several reaction times are strung together in a series—the *serial reaction time* task and its manifestations beyond the laboratory.

VARIABI   REACTION TIME

                                     estigated by providing the subject
                                     us occurs. The subject may or may
                                     stimulus. Four major variables—
                                     al uncertainty, and expectancy—
                                     aradigm.

Stimulus

                                     le RT to auditory stimuli is about
                                     roughly 130 msec and 170 msec,
                                     65). This difference has been at-
                                     sory processing between the two
                                     ms that the difference is instead
                                     control for intensity differences
                                     y controlled for these differences,

he found that auditory and visual stimuli of equal subjective intensity produced equal simple reaction times, as long as the visual stimulus was intense enough to be processed in foveal vision.

Stimulus Intensity

Simple RT decreases with increases in intensity of the stimulus to an asymptotic value, following a function as shown in Figure 8.1. Simple RT reflects the latency of a *decision* process that something has happened (Fitts & Posner,

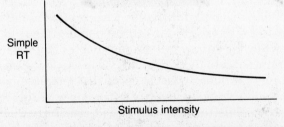

Figure 8.1 Relationship between stimulus intensity and simple reaction time.

1967; Teichner & Krebs, 1972). This decision is based on the aggregation over time of sensory evidence in the sensory channel until a criterion is exceeded.

In this sense, the simple RT is conceived as a two-stage process, as in signal detection theory. Aggregation of stimulus evidence may be fast or slow, depending on the intensity of the stimulus, and the criterion can be lowered or raised, depending on the "set" of the subject. In the example of the sprinter, a lowered criterion might well induce a false start if a random noise in the environment exceeded the criterion. After one false start, the runner will raise the criterion and be slower to start on the second gun in order to guard against the possibility of being disqualified. This model then attributes the only source of uncertainty in simple RT to be *temporal*.

Temporal Uncertainty

The degree of predictability of when the stimulus will occur is called temporal uncertainty. This factor can be manipulated by varying the *warning interval* (WI) occurring between a *warning signal* and the *imperative stimulus* to which the subject must respond. In the case of the sprinter, two warning signals are provided: "Take your mark" and "Set." The gunshot then represents the imperative stimulus. If the warning interval (WI) is short and remains constant over a block of trials, the imperative stimulus is highly predictable in time and the RT will be short. In fact, if the WI is always constant at around 0.5 seconds, the subject can shorten simple RT to nearly 0 seconds by synchronizing the response with the predictable imperative stimulus. On the other hand, if the warning intervals are long or variable, RT will be long (Klemmer, 1957). An experiment of Warrick, Kibler, Topmiller, and Bates (1964) investigated variable warning intervals as long as two and a half days! The subjects were secretaries engaged in routine typing. Occasionally they had to respond with a key press when a red light on the typewriter was illuminated. Even with this extreme degree of variability, simple RT was prolonged only to around 700 msec.

Temporal uncertainty thus results from increases in the variability and the length of the WI. When the variability of the WI is increased, this uncertainty is in the environment. When the mean length of the WI is greater, the uncertainty is localized in the subjects' internal timing mechanism since the variability of their estimates of time intervals increases linearly with the mean duration of those intervals (Fitts & Posner, 1967).

Although warning intervals should not be too long, neither should they be so short that there is not enough time for preparation. This characteristic is illustrated in a real-world analogy: the duration of the yellow light on a traffic signal, the time that a driver has to prepare to make a decision of whether or not to stop when the red signal occurs. In a study of traffic behavior at a number of intersections, Van Der Horst (1988) concluded that the existing warning interval (yellow light duration) was too short to allow adequate preparation. When the duration was lengthened by one second at two selected intersections, over a period of one year the frequency of red-light violations was reduced by half, with the obvious implications for traffic safety. At the same time, Van Der

Horst warns against excessively long warning intervals because of the temporal uncertainty it presents. This uncertainty, he notes, is a contributing cause to the many warning signal violations at drawbridges, where a 30-second warning signal precedes the lowering of the gate.

Expectancy

Whereas basic research reveals that the RT increases as the *average* WI of a block of trials becomes longer (the effect of temporal uncertainty), an opposite effect is observed when RTs for long and short WIs are examined *within* a block having varied WIs. In an illustration of this effect, Drazin (1961) examined individual RTs within a block of variable WIs and found that responses that follow long WIs tend to be faster than those following short WIs. This short-ened reaction time with longer warning intervals is explained by the concept of *expectancy*. The longer the time since the warning signal has passed, the more the subject *expects* the imperative signal to occur. As a consequence, the response criterion is progressively lowered with the passing of time and pro-duces fast RTs when the imperative stimulus finally does appear after a long wait. In terms of the sprinting example, if the delay of the starter's gun is variable from one sprint to the next, those sprinters who must wait an excep-tionally long time between "Set" and the gun will produce fast RTs and fast times. On the other hand, if the gun is delayed too long, there will be an increased number of false starts.

The role of expectancy and warning intervals in reaction time is critical in many real-world situations. Helander, Karwan, and Etherton (1987), for exam-ple, model the RT to respond to an unexpected (and potentially dangerous) move of a robot arm, in terms of an expectancy-driven criterion. Danaher (1980) discusses an incident in which an aircraft gaining altitude in low visibil-ity was flying on a collision course with a second aircraft flying level and directly above. When the upper plane suddenly appeared in view of the lower one, a rapid maneuver (RT response) was required for evasion. This incident is relevant to the present discussion because a *warning signal* was provided by air traffic control just before the visual sighting. It was this warning signal that may well have lowered the pilots' response criterion sufficiently to initiate the timely response. (The question of why the warning was not given sooner to prevent the close pass will be discussed in Chapter 12.)

Finally, in his study of traffic behavior, Van Der Horst (1988) compared constant timed lights to lights with vehicle-controlled timing. The latter lights tend to remain green when an approaching driver is sensed, and hence they maintain a more continuous flow of traffic. However, they also increase the oncoming driver's *expectancy* that the light will remain green. Consistent with the predictions of the underlying expectancy principle, Van Der Horst found that such lights increase by a full second the time at which the driver will stop when a yellow light appears at any point prior to the intersection. That is, lower expectancy of yellow seems to add a full second to the stop-response RT.

VARIABLES INFLUENCING CHOICE REACTION TIME

When actions are chosen in the face of environmental uncertainty, a host of additional variables related to the choice process itself influences the speed of action. In the terms described in Chapter 2, the operator is *transmitting information* from stimulus to response. This characteristic has led several investigators to use information theory to describe the effects of many of the variables on choice reaction time.

The Information Theory Model: The Hick-Hyman Law

It is intuitive that the more complex decisions or choices require longer to initiate. A straightforward example is the difference between simple RT and choice RT, in which there is uncertainty about which stimulus will occur and therefore about which action to take. More than a century ago, Donders (1869, trans. 1969) demonstrated that choice RT was longer than simple RT. The actual function that related the amount of uncertainty or degree of choice to RT was first presented by Merkel (1885). He found that RT was a negatively accelerating function of the number of stimulus-response alternatives. Each added alternative increases RT, but by a smaller amount than the previous alternative.

The theoretical importance of this function remained relatively dormant until the early 1950s, when in parallel developments Hick (1952) and Hyman (1953) applied information theory to quantify the uncertainty of stimulus events. Recall from Chapter 2 that three variables influence the information conveyed by a stimulus: the number of possible stimuli, the probability of a stimulus, and its context or sequential constraints. These variables were also found by Hick and Hyman to affect RT in a predictable manner. First, both investigators found that choice RT increased linearly with stimulus information—$\log_2 N$, where N is the number of alternatives—in the manner shown in Figure 8.2*a*. Reaction time increases by a constant amount each time N is doubled or, alternatively, each time the information in the stimulus is increased by one bit. When a linear equation is fitted to the data in Figure 8.2*a*, RT can be expressed by the equation $RT = a + bH_s$, a relation often referred to as the Hick-Hyman law. The constant b reflects the slope of the function—the amount of added processing time that results from each added bit of stimulus information to be processed. The constant a describes the sum of those processing latencies that are unrelated to the reduction of uncertainty. These would include, for example, the time taken to encode the stimulus and to execute the response. The issue of whether stimulus-response uncertainty affects perception, response selection, or both will be considered later in the chapter.

If the Hick-Hyman law is valid in a general sense, a function similar to that in Figure 8.2 should be obtained when information is manipulated by various means, as described in Chapter 2. Both Hick (1952) and Hyman (1953) manipulated the number of stimulus-response alternatives, N. Thus the points representing 1, 2, and 3 bits of information on the abscissa of Figure 8.2*a* could be

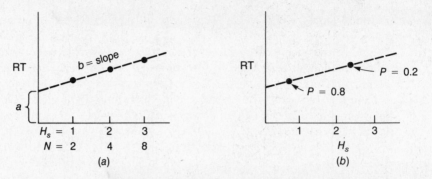

Figure 8.2 The Hick-Hyman law of choice reaction time: $RT = a + bH_s = a + bH_t$. (*a*) RT as a function of the number of alternatives. (*b*) RT for two alternatives of different probabilities.

replaced by the values $\log_2 2$, $\log_2 4$, and $\log_2 8$, respectively. Hyman further demonstrated that the function was still linear when the average information transmitted by stimuli during a block of trials was manipulated by varying the probability of stimuli and their sequential expectancy. If probability is varied, when N alternatives are equally likely, as described in Chapter 2, information is maximum (i.e., four alternatives yield two bits). When the probabilities are imbalanced, the average information is reduced. Hyman observed that the mean RT for a block of trials is shortened by this reduction of information in such a way that the new, faster data point still lies along the linear function of the Hick-Hyman law. When sequential constraints were imposed on the series of stimuli, thus also increasing redundancy and reducing the information conveyed, Hyman found a similar reduction in RT so that the faster point still lies along the function in Figure 8.2*a*.

Hyman's data measured the average reaction time for trial blocks. Fitts, Peterson, and Wolpe (1963) examined the RT to rare and frequent stimuli *within* a block of trials. They found that frequent stimuli, highly expected and conveying little information, produced faster RTs than rare, surprising, high-information stimuli. Adding further generality to the Hick-Hyman Law, Fitts, Peterson, and Wolpe demonstrated that individual stimuli of low (or high) probability, and therefore high (or low) information content respectively, produced RTs that also fall along the linear function describing the average RTs. Figure 8.2*b* presents these results for two stimuli randomly occurring in a series, one with a probability of 0.2 (high information) and the other with a probability of 0.8 (low information). Note that the response time to the low-probability, unexpected stimulus is slower, but it is slowed just enough to fall along the function at the point predicted by its higher information value, $\log_2 (0.2) = 2.2$ bits.

The Speed-Accuracy Trade-off

In reaction time tasks, and in speeded performance in general, people often make errors. Furthermore, they tend to make more errors as they try to respond more rapidly. This reciprocity between latency and errors is referred to as the *speed-accuracy trade-off*. According to the analysis of information transmission in Chapter 2, errors of response will reduce the information transmitted (H_t). We described the concept of a *bandwidth* as H_t/RT expressed as bits/seconds. If a person actually has a constant bandwidth for transmitting information, shifting the speed-accuracy trade-off from accurate to fast performance by changing the instructional emphasis should decrease H_t by the same proportion that RT is decreased, keeping the bandwidth constant.

It turns out, however, that the constant bandwidth model of human performance is not quite accurate. Rather, there is one level of the speed-accuracy trade-off that produces optimal performance in terms of processing bandwidth. For example, Howell and Kreidler (1963, 1964) compared performance on both easy and complex choice RT tasks as the set for speed versus accuracy was varied by different instructions. Different subjects were told to be fast, to be accurate, or to be fast *and* accurate, and finally they were given instructions that explicitly induced them to maximize the information transmission rate in bits per second. With both simple and complex tasks, Howell and Kreidler found that instructions changed latency and error rate in the expected directions, with the speed instructions having the largest effect on both variables. However, performance efficiency was not constant across the different sets for either task. When the choice task was easy, maximum bits per second was obtained by subjects instructed to maximize this quantity. When the task was complex, the highest level of performance efficiency (bits per second) was obtained with the speed set instructions.

Other investigations by Fitts (1966) and Rabbitt (1981, 1989), using reaction time, and by Seibel (1972), employing typing, also conclude that performance efficiency reaches a maximum value at some intermediate level of speed-accuracy set. These investigators conclude furthermore that subjects left to their own devices will seek out and select the level of set that achieves the maximum performance efficiency. This searching and maximizing behavior may be viewed as another example of the optimality of human performance.

The Speed-Accuracy Operating Characteristic Reaction time and error rate represent two dimensions of the efficiency of processing information. These dimensions are analogous in some respects to the dimensions of hit and false-alarm rate in signal detection (Chapter 2). Furthermore, just as operators can adjust their response criterion in signal detection, they can also adjust their set for speed versus accuracy to various levels defining "optimal" performance under different occasions, as the preceding experiments demonstrated. The *speed-accuracy operating characteristic*, or SAOC, is a function that represents RT performance in a manner analogous to the receiver operating characteristic (ROC) representation of detection performance.

Conventionally, the SAOC may be shown in one of two forms. In Figure 8.3, the RT is plotted on the abscissa and some measure of accuracy (the inverse of error rate) on the ordinate (Pachella, 1974). The four different points in the figure represent mean accuracy and latency data collected on four different blocks of trials when the speed-accuracy set is shifted. From the

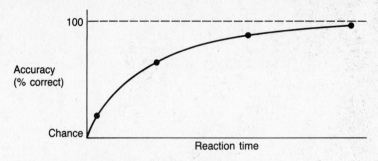

Figure 8.3 The speed-accuracy trade-off.

figure, it is easy to see why information transmission is optimal at intermediate speed-accuracy sets. When too much speed stress is given, accuracy will be at chance, and no information will be transmitted. When too much accuracy stress is given, performance will be greatly prolonged with little gain in accuracy. This characteristic has an important practical implication concerning the kind of accuracy instructions that should be given to operators in speeded tasks such as typing or keypunching. Performance efficiency will be greatest at intermediate levels of speed-accuracy set. It is reasonable to tolerate a small percentage of errors in order to obtain efficient performance, and it is probably not reasonable to demand zero defects, or perfect performance. We can see why this is so by examining the speed-accuracy trade-off plotted in Figure 8.3. Forcing the subject to commit no errors will induce impossibly long RTs.

An important warning to experimenters emphasized by Pachella (1974) and Wickelgren (1977) is also implied by the form of Figure 8.3. If experimenters instruct their subjects to make no errors, they are forcing them to operate at a region along the SAOC in which very small changes in accuracy generate very large differences in latency since the slope of the right-hand portion of Figure 8.3 is almost flat at that level. Hence, reaction time will be highly variable, and the reliable assessment of its true value will be a difficult undertaking.

Pew (1969) has shown that when accuracy is expressed in terms of the measure log $[P(\text{correct})/P(\text{errors})]$, the SAOC is typically *linear*. This relationship, shown in Figure 8.4, indicates that a constant increase in time buys the operator a constant increase in the logarithm odds of being correct. In terms of the SAOC space, one should think of movement from "southeast" to "northwest" as changing performance quality, efficiency, or bandwidth (bits per second). This relationship may be captured by the mnemonic "Northwest is best; southeast is least." Movement *along* an SAOC, on the other hand, represents different cognitive sets for speed versus accuracy. Two such SAOCs are shown in Figure 8.4. An excellent discussion of the speed-accuracy trade-off in reaction time may be found in Pachella (1974) and Wickelgren (1977).

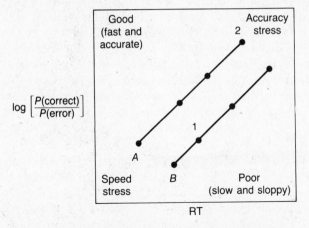

Figure 8.4 The speed-accuracy operating characteristic (SAOC). (*Source:* R. W. Pew, "The Speed-Accuracy Operating Characteristic," *Acta Psychological, 30* (1969), 18. Copyright 1969 by the North Holland Publishing Company. Reproduced by permission.)

From an engineering psychology perspective, one important aspect of the speed-accuracy trade-off is its usefulness in deciding what is best. Suppose, for example, that lines *A* and *B* in Figure 8.4 described the performance of operators on two data entry devices. From the graph, there is no doubt that *A* supports better performance than *B*. But suppose the evaluation had only compared one level on the SAOC of each device and produced the data of point 1 (for system *B*) and point 2 (for system *A*). If the evaluator examined only response time (or data-entry speed), he or she would conclude that *B* is the better device because it has shorter RT. Even if the evaluator looked at both speed and accuracy, any conclusion about which is the superior device would be difficult because there is no way of knowing how much of a trade-off there is between speed and accuracy, unless the trade-offs are actually manipulated. If SAOCs are not actually created, it is critical to keep the error rate or the latency of the two systems at equivalent levels to one another and to the real-world conditions in which the systems (and their operators) are expected to operate.

System designers should also be aware that certain design features seem automatically to shift performance *along* the SAOC. For example, auditory rather than visual presentation often leads to more rapid but more error-prone processing, a fact that in part leads aircraft designers to use auditory displays only for the most critical alerts, where speed of response is vital. Presenting more information, of greater precision, on a visual display will often lead to more accurate performance (as that information is used by the operator) but at a greater cost of time. Using SAOC analysis, Strayer, Wickens, and Braune (1987) showed that older adults were less efficient in responding than younger ones, but they also operated at a more conservative, accuracy-stress portion of the SAOC. As we will discuss in more detail in Chapter 10, the stress induced by emergency conditions sometimes leads to a speed-accuracy trade-off such that operators are disposed to take rapid but not always clearly conceived actions. It is for this reason that regulations in some nuclear power industries require controllers to stop and take no action at all for a specified time following a fault, thereby encouraging an accuracy set on the speed-accuracy trade-off.

The Speed-Accuracy Micro-trade-off The general picture presented suggests that conditions or sets in which speed is emphasized tend to produce more errors. A different way of looking at the speed-accuracy relationship is to compare the accuracy of fast and slow responses *within* a block of trials, using the same system (or experimental condition). (Alternatively one can compare the reaction time of correct and error responses.) This comparison describes the speed-accuracy *micro-trade-off*. Although the details of the micro-trade-off and its relationship to the models of human information processing are beyond the scope of this book (see Coles, 1988; Gratton et al., 1988; Pachella, 1974; Rabbitt, 1989; Wickens, 1984), the most important general point that can be made is that whether errors are faster or slower than correct responses seems to depend on the particular nature of the RT task, which in turn determines what varies from trial to trial, thereby causing some trials to be correct and others to be in error.

Thinking of the RT task as a decision task, with a criterion for how much perceptual evidence is required to respond, it is easy to see that variability in the setting of the response criterion will cause the micro-trade-off to have the same form as the macro-trade-off: When the criterion is conservative, full information will be processed, taking a longer time, and accuracy will be high. When the criterion is short, a response will be initiated rapidly, on the basis of little evidence, and errors will be likely to occur. In the extreme, the subject may emit "fast guesses," in which a random response is initiated as soon as the stimulus is detected (Gratton et al., 1988; Pachella, 1974). This positive micro-trade-off between reaction time and accuracy seems to be characteristic of most speeded tasks when RTs are generally short and stimulus quality is good.

In contrast, Wickens (1984) concludes that when stimulus evidence is relatively poor (as in many signal detection tasks) or processing is long and imposes working memory load (as in many decision tasks), the opposite form of the micro-trade-off is more likely to be observed. Fast responses are no longer more error-prone and may even be more likely to be correct. When there is generally poor signal quality, the responses on some trials will be longer because the variable quality of the stimulus drops, which also makes an error more likely. When decision tasks impose memory load, anything that delays processing imposes a greater (longer) memory load, which yields poorer decision quality. Hence the "inverted," or negative, form of the speed-accuracy trade-off is observed.

DEPARTURES FROM INFORMATION THEORY

The Hick-Hyman law has proven in general to be quite successful in accounting for the changes in RT with informational variables. The linear relationship between reaction time and information, in fact, led a number of investigators to conclude that people have a relatively constant bandwidth of information processing, provided by the inverse slope of the Hick-Hyman law function. Yet it soon became evident that information theory was not entirely adequate to describe RT data. We have noted already that bandwidth is not constant across wide ranges of speed and accuracy set. More crucially, five additional variables will be discussed that influence reaction time but are not easily quantified by information theory. These relate to stimulus discriminability, the repetition effect, response factors, practice, and compatibility. The existence of these variables does not invalidate the Hick-Hyman law but merely restricts somewhat its generality and requires some degree of caution when it is applied.

Stimulus Discriminability

Reaction time is lengthened as a set of stimuli are made less discriminable from one another (Vickers, 1970). This "noninformation" factor has some important implications. Tversky (1977) has argued that we judge the similarity or difference between two stimuli on the basis of the ratio of shared features to total features within a stimulus, and not simply on the basis of the absolute number of shared (or different) features. Thus, the numbers 4 and 7 are quite distinct, but the numbers 721834 and 721837 are quite similar, although in each case only one digit differentiates the pair. Discriminability difficulties in RT, like confusions in memory (see Chapter 6), can be reduced by deleting shared and redundant features where possible. In the context of nuclear power plant design, Kirkpatrick and Mallory (1981) have emphasized the importance of avoiding confusability between display items by minimizing the feature similarity between separate labels.

The Repetition Effect

A number of investigators have noted that in a random stimulus series, the repetition of a stimulus-response (S-R) pair yields a faster reaction time to the second stimulus than does an alternation. For example, if the stimuli were designated A and B, the response to A following A will be faster than to A following B (e.g., Hyman, 1953; Kirby, 1976). The advantage of repetitions over alternations, referred to as the *repetition effect*, appears to be enhanced by increasing N (the number of S-R alternatives), by decreasing S-R compatibility (see below), and by shortening the interval between stimuli and responses (Kornblum, 1973). Research by Bertelson (1965) and others (see Kornblum for a summary) suggests that the response to repeated stimuli is speeded both by the repetition of the stimulus and by the repetition of the response.

There are two important circumstances in which the repetition effect is *not* observed. (1) As summarized by Kornblum (1973), the repetition effect declines with long intervals between stimuli and may sometimes be replaced by an alternation effect (faster RTs to a stimulus change). In this case, it appears that the "gambler's fallacy" takes over. Subjects do not expect a continued run of stimuli of the same sort, just as gamblers believe that they are "due for a win" after a string of losses. (2) In some tasks, such as typing, rapid repetition of the same digit or even digits on the same hand will be slower than alternations (Sternberg, Kroll, & Wright, 1978). The nature of this reversal of the repetition effect in fast-responding (transcription) tasks will be discussed more fully at the end of this chapter.

Response Factors

Two characteristics of the response appear to influence reaction time. (1) Reaction time is lengthened as the *discriminability* between the responses is decreased. Thus, for example, Shulman and McConkie (1973) found that two choice RTs executed by the fingers on opposite hands were faster than those executed by two fingers on the same hand, the latter pair being less discriminable from one another. Similarly, distinct shape and feel of a pair of controls will reduce the likelihood of their being confused. (2) Reaction time is lengthened by the *complexity* of the response. For example, Klapp and Irwin (1976) showed that the time to initiate a vocal or manual response is directly related to the duration of the response. Sternberg, Kroll, and Wright (1978) found that it takes progressively longer to *initiate* the response of typing a string of characters as the number of characters in the string is increased.

Practice

Consistent results suggest that practice, a noninformational variable, decreases the slope of the Hick-Hyman law function relating RT to information. In fact, compatibility (to be discussed below) and practice appear to trade off reciprocally in their effect on this slope. This trade-off is nicely illustrated by comparing three studies. Leonard (1959) found that no practice was needed to obtain a flat slope with the highly compatible mapping of finger presses to tactile stimulation. Davis, Moray, and Treisman (1961) required a few hundred trials to obtain a flat slope with the slightly lower compatibility task of naming a heard word. Finally, Mowbray and Rhoades (1959) examined a mapping of slightly lower (but still high) compatibility. The subject depressed keys adjacent to lights. For this unusually stoic subject, 42,000 trials were required to produce a flat slope.

S-R Compatibility

In June 1989, the pilots of a commercial aircraft flying over the United Kingdom detected a burning engine but mistakenly shut down the good engine instead. When their remaining engine (the burning one) eventually lost power, the plane crashed, with a large loss of life. Why? Analysis suggests that a violation of *stimulus-response compatibility* in the display control relation may have been a contributing factor (*Flight International*, 1990). We have already encountered the concept of compatibility in earlier chapters—in Chapter 3, the compatibility of *proximity* between display elements and information processing; in Chapter 4, compatibility between a display and the static or dynamic properties of the operator's mental model of the displayed elements. We have also discussed in Chapter 4 the direct manipulation interface, whose advantages, in part, relate to the compatibility between actions and display changes. Now we will discuss compatibility between a display location or movement and the location or movement of the associated operator response. We devote a fair amount of space to this topic because of its historic prominence in engineering psychology research and because of its tremendous importance in system design.

As suggested, S-R compatibility has both static elements (where response devices should be located to control their respective displays) and dynamic elements (how response devices should move to control their displays). We refer to these as *locational* and *movement* compatibility, respectively. Much of compatibility describes spatially oriented actions (e.g., switches arranged in space and moving in space), but it can also characterize other mappings between displays and responses. More compatible mappings require fewer mental transformations from display to response. We will also examine compatibility in terms of *modalities* of control and display. What is common about all of these different types of S-R compatibility, however, is the importance of *mapping*. There is no single best display configuration or control configuration. Rather, each display configuration will be compatible only when it is mapped to certain control configurations.

Location Compatibility The foundations of location compatibility are provided in part by the human being's intrinsic tendency to move or orient toward the source of stimulation (Simon, 1969). Given the predominance of this effect, it is not surprising that compatible relations are those in which controls are located next to the relevant displays, a characteristic that defines the *colocation principle*. Touch-screen CRT displays are an example of designs that maximize S-R compatibility through colocation (but see Chapter 11 for some limitations of this concept). However, many systems in the real world often fail to adhere to this principle, for example, the location of stove burner controls (Chapanis & Lindenbaum, 1959; Shinar & Acton, 1978). Controls colocated beside their respective burners (Figure 8.5a) are compatible and will of course eliminate the possible confusions caused by arrays shown in Figure 8.5(b and c), which are more typical.

Unfortunately the principle of colocation is not always possible to achieve. Operators of some systems may need to remain seated, with controls at their fingertips that activate a more distant array of displays. In combat aircraft, the high gravitational forces encountered in some maneuvers may make it impossi-

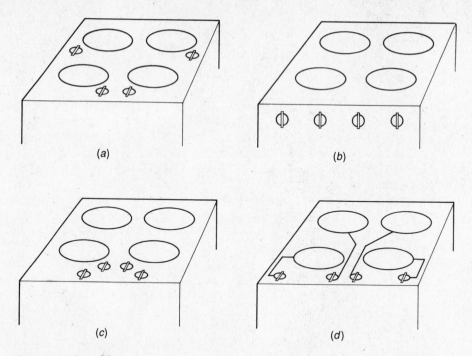

Figure 8.5 Possible arrangements of stove burner controls. (*a*) Adheres to coloca-tion principle. (*d*) Solves the compatibility problem by the visual linkages.

ble to move the hands far to reach colocated controls. Even the colocation of Figure 8.5*a* may require the chef to reach across an active (hot) burner to adjust a control. Where colocation cannot be obtained, two important compati-bility principles are congruence and rules.

The general principle of *congruence* is based on the idea that the spatial array of controls should be congruent with the spatial array of displays. This principle was illustrated in a study by Fitts and Seeger (1953), who evaluated RT performance when each of the three patterns of light stimuli on the left in Figure 8.6 was assigned to one of the three response mappings (moving a lever) indicated across the top. In each case an eight-choice RT was imposed. In stimulus array S_a, any one of the eight lights could illuminate (and for R_a the eight lever positions could be occupied). In S_b, the same eight angular positions could be defined by the four single lights and the four combinations of adjacent lights. In R_b, the eight shaded lever positions could be occupied. In S_c, the eight stimuli were defined by the four single lights and four pairwise combinations of one light from each panel. In R_c, each or both levers could be moved to either side. Fitts and Seeger found that the best performance for each stimulus array was obtained from the spatially congruent response array: S_a to R_a, S_b to R_b, and S_c to R_c. This advantage is indicated by both faster responses and greater accuracy.

A stove-top array such as that shown in Figure 8.5*c* would also achieve this congruence. Notice in *b* and *d* that there is no possible congruent mapping of

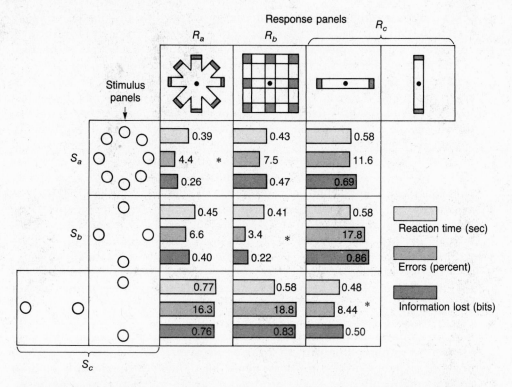

Figure 8.6 Each of the three stimulus panels on the left was assigned to one of the three response panels across the top. The natural compatibility assignments are seen down the negative diagonal and indicated by an asterisk (°). (*Source:* P. M. Fitts and C. M. Seeger, "S-R Compatibility: Spatial Characteristics of Stimulus and Response Codes" *Journal of Experimental Psychology, 46* (1953), 203.)

the linear array of controls to the square array of burners (displays). The only way to bypass this lack of compatibility is through the drawn links in Figure 8.5d (Osborn & Ellingstad, 1987).

Congruence is often defined in terms of an ordered array (e.g., left-right or top-down). A study by Hartzell et al. (1982) revealed a marked departure of left-right congruence in the design of the military helicopter, which is configured in such a way that altitude information is presented to the right of the instrument display and controlled with the left hand, whereas airspeed information, presented to the left, is controlled with the right hand. In a simulation experiment, this configuration produced considerably worse performance in adjusting the two variables than a more congruent design in which the two display elements are reversed. In the Boeing 737 crash, a violation of location compatibility resulted because the relevant indicator of malfunction of the burning engine, which was the left engine, was located on the right side of the cockpit midline.

Why are incongruent systems difficult to map? In a theoretical analysis of S-R compatibility effects, Kornblum, Hasbroucq, and Osman (1990) argue that if the response dimension can be physically mapped to any dimension along which the stimuli are ordered (e.g., both are linear arrays), the onset of a stimulus in an array automatically activates a tendency to respond at the associated location. If this is not the correct location, a time-consuming process is required to suppress this tendency and activate the rule for the correct response mapping instead. This discussion brings us to the second feature of location compatibility—the importance of *rules* when congruence is not obtained. Simple rules should be available to map the set of stimuli to the set of responses (Kornblum, Hasbroucq, & Osman, 1990). This feature is illustrated in a study by Fitts and Deininger (1954), who compared three mappings between a linear array of displays and a linear array of controls. One mapping was congruent; the second was reversed, so that the leftmost display was associated with the rightmost control and so forth; the third mapping involved a random assignment of controls to displays. Not altogether surprising, but important, performance was best in the first array but considerably better in the reversed than in the random array. In the reversed array, a single rule can provide the mapping. A study by Haskell, Wickens, and Sarno (1990) revealed that the number of rules necessary to specify a mapping between linear arrays of four displays and four controls was a strong predictor of RT.

There are times, however, when even congruence is difficult to achieve. Consider a linear array of switches that must be positioned along an armrest to control (or respond to) a vertical array of displays. Since a congruent, vertical array of switches on the armrest would be difficult to implement (and an anthropometrically poor design), the axis of switch orientation must be incongruent with the display axis. However, there are rules to guide the designer. These rules describe a mapping of ordered quantities from least to most in space, which specifies that increases move from left to right, aft to forward, clockwise (for a circular array), and (to a lesser extent) from bottom to top. Hence, a far right control should be mapped to a top display when a left-right array is mapped to a vertically oriented display (Weeks & Proctor, 1990). It is unfortunate, however, that the top-down ordering is not strong. On the one hand, high values are compatible with top locations (as noted in Chapter 4; see also the typical calculator keyboard). On the other hand, the order of counting (1, 2, 3 . . .), following the order of reading in English, is from top to bottom (see the push-button telephone). These conflicting stereotypes suggest that vertical display (or control) arrays that are not congruent with control (display) arrays should only be used with caution. However, an important design solution that can resolve any potential mapping ambiguity is to put a slight cant, or angling, of one array in a direction that is congruent with the other, as shown in Figure 8.7. If this cant is as great as 45°, then reaction time can be as fast as if the axes are parallel (Andre, Haskell, & Wickens, 1991).

Finally, there is some evidence that, in cases where the hand must reach to activate the appropriate response, the most compatible mappings for incongruent (right-angle) arrays may be different from those cases where fingers are dedicated to (and resting on) the appropriate key. In the latter case, there is evidence that different rules govern the two different hands. For each hand,

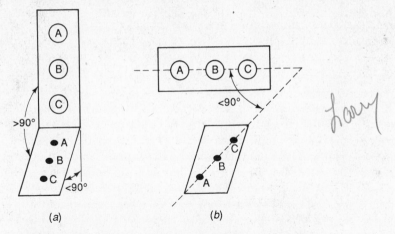

Figure 8.7 Solutions of location compatibility problems by using cant. (*a*) The control panel slopes downward slightly (an angle greater than 90 degrees), so that control A is clearly above B, and B is above C, just as they are in the display array. (*b*) The controls are slightly angled from left to right across the panel, creating a left-right ordering that is congruent with the display array.

the compatible mapping to a left-right stimulus array is the finger assignment that would hold if the hands were held in the palms down, fingers forward orientation—that is, thumb to the right stimulus for the left hand, and to the left stimulus for the right hand. But this same finger assignment would continue to be most compatible if the hands were rotated inward or vertically (Andre, Haskell, & Wickens, 1991; Ehrenstein, Schroeder-Heister, & Heister, 1989), even though this would now produce an opposite spatial mapping for the two hands.

Movement Compatibility When an operator moves a position switch, rotary, or sliding control, it often changes the state of a displayed variable. *Movement compatibility* defines the set of expectancies that an operator has about how the display will respond to the control activity. Technically we may define this as C-R-S compatibility because a *cognitive* intention to change a variable leads to a *response* on the control, which in turn produces a *stimulus* change in the display. When C-R-S compatibility is violated, an operator may move the control, then perceive the system responding in what he or she thinks is the opposite direction, triggering a further unnecessary and possibly disastrous control action.

The congruence principle of location compatibility also applies to the compatibility of movement. Thus linear controls should move in an axis and direction parallel to display movement (and as discussed in Chapter 4, this movement should also be parallel to the operator's mental model of the displayed variable). Dial displays are more compatible or congruent with rotary controls and linear displays with sliding controls. And when congruence must be violated, a common mapping of increase (up, right, forward, and clockwise), should be used to identify how the display responds to the control. Thus in Figure 8.8a, the operator would expect a clockwise rotation to move the linear pointer upward over the fixed scale.

Movement compatibility, however, is also governed by a principle of *movement proximity*, which is not related to congruence. This principle, also known as the Warrick principle (Warrick, 1947), asserts that the closest part of the moving element of a control should move in the same direction as the closest part of the moving element of a display, as if the operator has a mental model of a mechanical linkage between the two. To illustrate, consider Figure 8.8b, in which the clockwise-to-move-upward dial has been placed to the left of the linear display. Here the clockwise stereotype conflicts with the movement proximity stereotype. In these circumstances, Loveless (1963) concludes that the movement proximity stereotype is dominant. That is, counterclockwise rotation would be expected to increase. However, as shown in Figure 8.8c, a simple relocation of the rotary control to the right of the linear scale is the sort of engineering psychology solution that is so gratifying because it captures both principles at once.

A study by Hoffman (1990) has examined how these (and other) principles of compatibility interact in different display-control layouts. He finds that the overall compatibility of a given layout can be modeled as the sum of the forces exerted by the different components of compatability as they either work together (Figure 8.8c) or in opposition (Figure 8.8b).

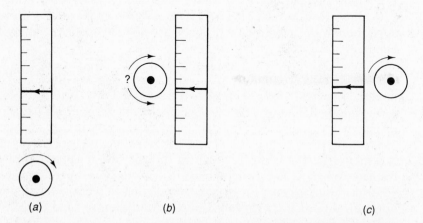

(a)   (b)   (c)

Figure 8.8 Three control display layout configurations illustrating movement compatibility principles. The arrow indicates the expected direction to *increase* the indicator. (b) The display is ambiguous because the clockwise-to-increase, and proximity of movement principles are in opposition.

Compatibility Ambiguities Applying compatibility principles to control-display movement relations is not always straightforward because the operator's *mental model* of what a control is doing may sometimes make the relation ambiguous, as seen in three examples. First, Hoffman (1990) found that the movement proximity principle was far less pronounced for students of psychology than students of mechanical engineering, suggesting that only the latter had the strong mental model of the mechanical linkage. Second Figure 8.9*a* is a control design for the vertical speed of an aircraft. Immediately, the incompatibility appears evident: Move the control up to set the speed downward? However, if the operator (in this case the pilot) conceives of the control motion not as a linear slide but as a rotating wheel whose front surface is exposed, and the pilot has a mental model that this rotation is directly coupled to the rotation of the aircraft around the axis of its wings (upward rotation for increased vertical velocity), the relation now appears to be quite compatible (Figure 8.9*b*).

A third compatibility ambiguity relates to the distinction between status and command displays. In Chapter 5, we saw that a status display presents the current status of a variable (e.g., "Your power setting is too high"). A command display tells you what to do to correct the situation ("Lower your power setting"). From the strict viewpoint of S-R compatibility, it is apparent that a command display satisfies the principle that movement of the display is com-

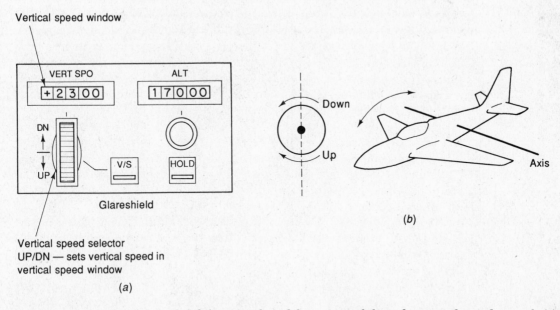

Figure 8.9 Example of the role of the mental model in compatibility of a vertical speed control. (*a*) Vertical speed selector control, suggesting an apparent incompatibility. (*Source:* Courtesy of Boeing Commercial Airplane Co., Seattle.) (*b*) Mental model of the aircraft, rotating around the axis parallel to its wings. Here the rotary control is quite compatible with the desired change of state.

patible with the required action. But this mapping bypasses the display of a representation of the state for which the command is required; and as we discuss in Chapter 12, it is not always good design to tell the operator what should be done without displaying the reasons why. A status display achieves this requirement; but now, if the display is S-C compatible, it requires a movement that is S-R *in*compatible (e.g., the upward display movement "too high" triggers the downward control movement "reduce"). In most circumstances, it would appear that the need for status information would override a command configuration that is more S-R compatible, and indeed when operators have a good mental model of the system under control, they perform as well if not better with status than with command displays (Andre, Wickens, & Goldwasser 1990). As noted in Chapter 5, however, high levels of time pressure may favor command displays, and discriminable presentation of both status and command information may be the most effective format.

Transformations and Population Stereotypes Not all compatibility relationships are spatially defined. Any S-R mapping that requires some *transformation*, even if it is not spatial, will be reduced in its compatibility. Hence a mapping between stimulus and response digits of $1-1$, $2-2$, and $3-3$ is more compatible than $1-2$, $2-3$, and $3-4$, which imposes the transformation "add one." Similarly, the relationship between stimulus digits and response letters $(1-A, 2-B, 3-C,$ etc.) is less compatible than digits–digits or letters–letters mappings. Also any S-R mapping that is many-to-one will be less compatible than a one-to-one mapping (Norman, 1988; Posner, 1964). Consider, for example, the added cognitive difficulty of dialing alphabetic phone numbers, like $437-HELP$, resulting from the $3-1$ mapping of letters (stimuli) to keys (responses). Ironically, in Chapter 6 we identified this form of phone number as better from the standpoint of memory load. As we continuously see, human engineering is always encountering such trade-offs.

Population stereotypes define mappings that are more directly related to *experience*. For example, consider the relationship between the desired lighting of a room and the movement of a light switch. In North America, the compatible relation is to flip the switch up to turn the light on. In Europe, the compatible relation is the opposite (up is off). This difference is clearly unrelated to any difference in the psychological hardware between Americans and Europeans but rather is a function of experience. S. Smith (1981) has evaluated population stereotypes in a number of verbal-pictorial relations. For example, he asks whether the "inside lane" of a four-lane highway refers to the center-most lane on each side or to the driving lane. Smith finds that the population is equally divided on this categorization. Any mapping that bases order on reading patterns (left-right and top-bottom) will also be stereotypic, and thereby not applicable, say, to Hebrew or Oriental readers. Finally, as noted in Chapter 3, color coding is strongly governed by population stereotypes: Red for danger, stop, and so on.

Modality S-R Compatibility Stimulus-response compatibility appears to be defined by stimulus and response modality as well as by spatial correspondence. Brainard, Irby, Fitts, and Alluisi (1962) found that if a stimulus was a light, choice RT was faster for a pointing (manual) than a voice response, but if the stimulus was an auditorily presented digit, RT was more rapid with a vocal naming response than with a manual pointing one. In a thorough review of the factors influencing choice RT, Teichner and Krebs (1974) summarized a number of studies and concluded that the four S-R combinations defined by visual and auditory input and manual and vocal response produced reaction times in the following order: A voice response to a light is slowest, a key-press response to a digit is of intermediate latency, and a manual key-press response to a light and naming of a digit are fastest.

Greenwald (1970) discusses the related concept of *ideomotor compatibility*, which will occur if a stimulus matches the sensory feedback produced by the response. Under these conditions, RT will be fast and relatively automatic (Greenwald, 1979). Thus Greenwald observes fast RTs in the ideomotor compatible conditions when a written response is given to a seen letter and when a spoken response is given to a heard letter. Incompatible mappings in which a written response is made to a heard letter and a spoken response to a seen letter are slower. Ideomotor compatible mappings not only are fast but also appear to be influenced neither by the information content of the RT task (*N*) nor by dual-task loading (Greenwald & Shulman, 1973).

Wickens, Sandry, and Vidulich (1983) and Wickens, Vidulich, and Sandry-Garza (1984) proposed that these modality-based S-R compatibility relations may partially depend on the central processing code (verbal-spatial) used in the task. In both the laboratory environment and in an aircraft simulator, they demonstrated that tasks that use verbal working memory are served best by auditory inputs and vocal outputs, whereas spatial tasks are better served by visual inputs and manual outputs. In the aircraft simulation, Wickens, Sandry, and Vidulich found that these compatibility effects were enhanced when the flight task became more difficult.

As noted in Chapter 5, these guidelines would hold only when the material is short. Furthermore, for the voice control, they would hold only when the vocal response does not disrupt rehearsal of the retained information (Wickens & Liu, 1988). The particular advantages of voice control in multitask environments such as the aircraft cockpit (Henderson, 1989) or the computer design station (Martin, 1989) will be further discussed in the next chapter and again in Chapter 11.

Consistency and Training Compatibility is normally considered to be an asset in system design. However, to reiterate a point made in Chapter 4, the designer should always be wary of any possible violation of *consistency* across a set of control-display relations that may result from trying to optimize the compatibility of each. For example, Duncan (1984) found that subjects actually had a more difficult time responding to two RT tasks if one was compatibly mapped and one incompatible than responding when both were incompatible. In other words, the consistency of having identical (but incompatible) mappings in both tasks outweighed the advantages of compatibility in one. Correspondingly, a designer who needs to add another function to a system that already contains a lot of control-display mappings should be wary of whether the compatible addition proposed (e.g., status display) is in disharmony with the existing set (e.g., several command displays) (Andre, Wickens, & Goldwasser, 1990).

We have seen how training and experience form the basis for population stereotypes. Training can also be used to formulate or enhance correct mental models, as in Figure 8.9. It is also evident that training will improve performance on both compatible and incompatible mappings. In fact the rate of improvement is actually faster with the incompatible mappings because they have more room to improve (Fitts & Seeger, 1963). However, the data are also clear that extensive training of an incompatible mapping will never fully catch up to a compatible one. And when the operator is placed under stress, performance with the incompatible mapping will regress further than with the compatible one (Fuchs, 1962; Loveless, 1962). Hence the system user should be wary of a designer who excuses an incompatible design with the argument that the problem can be "trained away."

Is Information Theory Still Viable?

Five important factors have been identified that influence RT and yet cannot be accounted for in informational terms. Teichner and Krebs (1974) assert that two of these variables, practice and S-R compatibility, along with the informational variable, N, are the most potent influences on choice RT. In spite of this fact, the information metric still appears to be an important and useful concept in the RT paradigm. For example, investigators generally find that within a constant level of any practice, repetition, discriminability, and compatibility variable, the linearity of the Hick-Hyman law still seems to hold. Correspondingly, the effects of these five variables seem generally to be exerted on the slope of the constant function determined by RT and H_t (Fitts & Posner, 1967; Teichner & Krebs, 1974).

It is reasonable to assume that high-compatibility mappings and/or extensive practice merely allows processing to bypass the time-consuming response-selection stages (Fitts & Posner, 1967), so that the stimulus automatically activates its associated response. Furthermore, when S-R compatibility is low, a stimulus will automatically activate an incompatible response. Then a time-consuming process of suppressing the competing response tendency is required. As described in Chapter 3, reaction time is delayed.

It seems that as long as the literal assumption is abandoned that the reciprocal of the slope of the Hick-Hyman law function reflects the bandwidth of human performance (this leads to the indefensible position that the human has infinite bandwidth when the slope is zero), the information concept still has a lot to offer. Information theory will clearly not account for all effects on choice reaction time, but there is no reason to expect that it should. It does do a good job of describing some.

STAGES IN REACTION TIME

A number of variables have been discussed that influence or prolong reaction time. The model of information processing, presented in Chapter 1, also assumes that total reaction time equals the sum of the duration of a number of component processing stages (e.g., perceptual encoding, and response selection). In a series of experiments and theoretical papers, psychologists have attempted to establish where these variables have their effects and, in fact, if processing really does proceed by discrete stages or sequential mental operations. Pachella (1974) has contrasted two approaches used to justify the existence of processing stages and to examine the influences of stage duration on total RT latency. These are the *subtractive method* and the *additive factors* technique. A third approach based on the *event-related brain potential* will also be discussed.

The Subtractive Method

In the subtractive method an experimental manipulation is used to delete a mental operation entirely from the RT task. The decrease in RT that results is then assumed to reflect the time required to perform the absent operation. More than a century ago, Donders (1869, trans. 1969) first used the subtractive method in RT to provide evidence for a response selection stage. He compared reaction time in which several stimuli were assigned to several responses (conventional choice RT) with reaction time in which there were several possible stimuli but only one demanded a response. This is known as the disjunctive or "go-no-go" RT. The difference between these two conditions was assumed by Donders to be the time taken to select a response since response selection is not necessary when only one response is required. By similar logic, when disjunctive RT was compared with simple RT, Donders assumed that the difference reflected the latency of a stimulus discrimination stage since only in disjunctive RT is it necessary to discriminate the stimuli. In simple RT, detection, not recognition, is sufficient to initiate the response.

Additive Factors Technique

The subtractive method is intuitive, and its application leads to a number of plausible findings, although some of its assumptions have been criticized (Pachella, 1974). Fortunately, confirming evidence for the existence and identity of processing stages has been provided by the *additive factors* method (Sternberg, 1969, 1975). Although all of the specific assumptions underlying the applications will not be described here, the intent of additive factors is to define the existence and distinctiveness of different stages by manipulating variables that are known to lengthen reaction time.

In a typical application, two variables are manipulated independently. It is assumed that if the two influence a common stage of processing, their effects on reaction time will interact; that is, the extent to which one variable lengthens RT will be *greater* at the more difficult level of the other. This situation, shown in Figure 8.10*b*, is similar to one in real life, when troubles from one source disrupt our ability to deal with troubles from another source, particularly if both sources affect the same general aspect of our life (i.e., both relate to trouble with classwork). If, however, the two manipulated variables influence different stages of processing, their effects will be *additive*, as shown in Figure 8.10*c*. That is, the influence of one variable will not be affected by the level of the other.

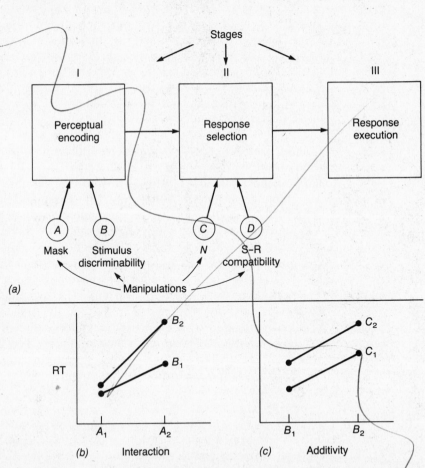

Figure 8.10 (*a*) Hypothetical relationship between orthogonal manipulation of four variables interpreted in the framework of additive factors. (*b*) Variables *A* and *B* interact, effecting a common stage, as do *C* and *D*. (*c*) Variables *B* and *C* are additive, effecting different stages.

Experimental Techniques A specific example using the additive factors method is represented in Figure 8.10. The experimenter wishes to make inferences about what manipulations influence what stages of processing, as shown in Figure 8.10a. The experimenter first identifies four experimental variables to manipulate, each identified by a letter and each run at an easy and difficult level. In our examples, these variables are the absence or presence of a mask over the stimulus (A_1 and A_2, respectively), which degrades stimulus quality; the discriminability of the stimuli in a set (B_1: high; B_2: low); the number of S-R pairs $N(C_1$: small; C_2: large); and stimulus-response compatibility (D_1: high; D_2: low). (Notice that the subscript 1 always designates the easy level of the variable.) The experimenter then measures reaction time as each pair of variables is manipulated orthogonally. For example, RT is measured when stimuli of high and low discriminability are presented with and without a mask. The four RT measures are shown in Figure 8.10b, indicating an interaction. The effect of the mask (A_1 versus A_2) is more pronounced when discriminability is low (B_2) than when it is high (B_1). Hence the experimenter concludes that discriminability and stimulus quality must influence the same perceptual processing stage, as shown in Figure 8.10a.

Next, the experimenter manipulates discriminability and set size together by presenting stimuli of high and low discriminability in, for example, a two- and four-choice RT task. Here the results are shown in Figure 8.10c. The two variables are additive. The set-size effect is uninfluenced by the ease of discriminating the stimuli. The experimenter concludes that the two variables influence different processing stages. Since, therefore, N cannot affect perceptual encoding, it is likely instead to influence response selection. This finding would be confirmed if the experimenter manipulated N and S-R compatibility together. These variables would be found to interact like those in Figure 8.9b, although this graph is not shown in the figure.

As experimenters have performed a large number of these orthogonal manipulations of RT difficulty, the additive factors data have provided a fairly consistent picture of processing stages. Certain pairs of variables consistently interact, and others are consistently additive. The experimenters use a certain amount of intuition to infer the stage affected by a cluster of interacting variables. For example, it is clear intuitively that S-R compatibility must influence response selection and that stimulus quality must influence perceptual processing. These anchors help interpret the locus of effect of other factors.

Figure 8.11 presents the pattern of additivity and interactions that have been collectively aggregated from a number of RT investigations. Four processing stages are portrayed across the bottom. Experimental manipulations are circled, and additive or interactive relations obtained between these manipulations are indicated by narrow or thick lines, respectively. Each line is coded by the investigation in which the manipulation was performed, and the codes are identified by the list below the figure. The dashed arrows point to the stages inferred to be affected by the manipulation in question.

Generally, the relationships that are shown in Figure 8.11 are consistent. Clusters of variables, such as S-R compatibility, N, and the repetition effect, are consistently found to interact with one another by a number of investigators and to be additive with variables that logically should affect other stages.

In addition to the pattern of interactions and additivity, the picture of results shown in Figure 8.11 allows two general conclusions to be drawn. First, the seat of the "action" of most of the variables seems to be in response selection, emphasizing a point that will be addressed again later in this chapter: The response selection process is a major bottleneck in speeded information processing. Second, stimulus probability appears to affect two stages: Improbable stimuli require longer to be recognized, and their associated responses take longer to be selected.

Applications of Additive Factors Methodology   Although the additive factors methodology is an important theoretical method for investigating information processing, it has also served as a useful applied tool for establishing how the speed of information processing is influenced by different environmental and organismic factors, such as aging (Strayer, Wickens, & Braune, 1987); poisoning (Smith & Langolf, 1981); and mental work load, as discussed in Chapter 9 (Crosby & Parkinson, 1979; Wickens, Hyman, Dellinger, Taylor, & Meador, 1986).

When the technique is applied for these purposes, latency in the RT task is measured as the demands of different processing stages are manipulated, in both the low and high level of the factor (e.g., aging) to be investigated. This factor is then treated just like any other manipulation. If it interacts with a variable that influences a known stage, the environmental variable is assumed to affect the stage in question. For example, Smith and Langolf (1981) used a Sternberg memory search task (an RT task described in Chapter 9) to measure information processing of subjects who had been exposed to different amounts of mercury in industrial environments. They found an interaction between memory load and the amount of mercury poisoning in the bloodstream. Therefore, the behavioral effects of such toxins were localized at the stage of short-term memory retrieval.

Problems with Additive Factors   The additive factors technique has had a remarkable history of success in accounting for RT data, and it has been employed in a number of applied contexts. One primary criticism, however, is directed at the assumption that stages proceed strictly in series—like a factory assembly line—and that the effects that either slow down or speed up an earlier stage have no effect on the speed of processing at a later stage. By now there is convincing evidence that information processing does not strictly proceed in a serial fashion (Coles, Gratton, & Donchin, 1988; McClelland, 1979). For example, the process of preparation based on expectancy may overlap different stages in time, so that an increase in preparation that is made for a particular, frequent response (hence, reducing response selection time) can proceed while perceptual recognition is still taking place (Coles, Gratton, & Donchin, 1988). This overlap in time can occasionally do strange things to the RT relationship shown in Figure 8.10, such as producing an *underadditive*

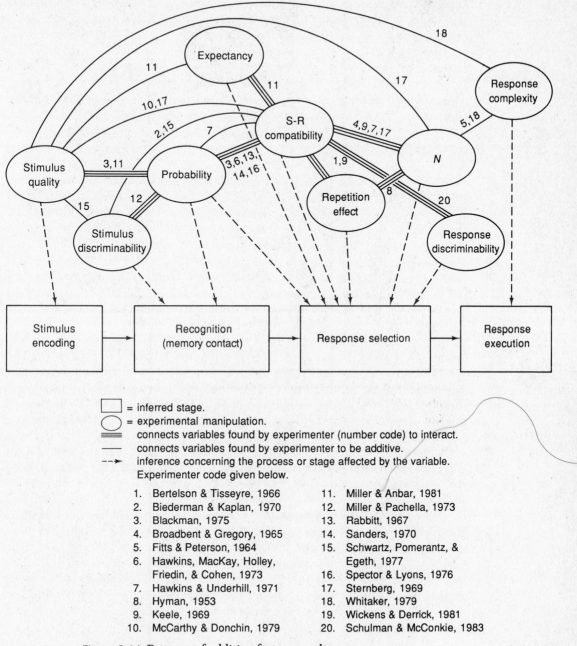

Figure 8.11 Patterns of additive factors results.

= inferred stage.

= experimental manipulation.

connects variables found by experimenter (number code) to interact.

connects variables found by experimenter to be additive.

inference concerning the process or stage affected by the variable. Experimenter code given below.

1. Bertelson & Tisseyre, 1966
2. Biederman & Kaplan, 1970
3. Blackman, 1975
4. Broadbent & Gregory, 1965
5. Fitts & Peterson, 1964
6. Hawkins, MacKay, Holley, Friedin, & Cohen, 1973
7. Hawkins & Underhill, 1971
8. Hyman, 1953
9. Keele, 1969
10. McCarthy & Donchin, 1979

11. Miller & Anbar, 1981
12. Miller & Pachella, 1973
13. Rabbitt, 1967
14. Sanders, 1970
15. Schwartz, Pomerantz, & Egeth, 1977
16. Spector & Lyons, 1976
17. Sternberg, 1969
18. Whitaker, 1979
19. Wickens & Derrick, 1981
20. Schulman & McConkie, 1983

relationship, where the delay caused by increasing the difficulty at one stage of processing is actually smaller at the more difficult level of the other stage (Schwartz, Pomerantz, & Egeth, 1977).

The Event-Related Brain Potential as an Index of Mental Chronometry

Despite some shortcomings, Sternberg's additive-factors approach provides a reasonably good approximation of the processing mechanisms involved in reaction time. In fact, the inferences concerning the influence of manipulations on stages that are made from additive and interaction data are still sound in many instances even when processes do overlap in time (McClelland, 1979). However, no matter what theory is adopted, all RT investigations must make inferences about the processes between stimulus and response by looking at the final product of the response and not examining directly those intervening processes. To augment this mental chronometry, the evoked brain potential has been used to provide a direct estimate of the timing of processes up to the intermediate stage of stimulus categorization (Coles, 1988; Donchin, 1981; Gratton, Coles, Sirevaag, Eriksen, & Donchin, 1988; McCarthy & Donchin, 1979). The evoked potential is a series of voltage oscillations or components that are recorded from the surface of the scalp to indicate the brain's electrical response to discrete environmental events.

An investigation by McCarthy and Donchin (1979) suggests that the late positive, or P300, component of the evoked potential seems to covary with the duration of perceptual processing but not with the duration of response selection. Their subjects performed the RT task in which the displayed words *left* and *right* were responded to by pressing left and right keys. The difficulty of stimulus encoding was varied by increasing noise on the display. Response selection difficulty was manipulated by changing S-R compatibility. As shown in Figure 8.12, when RT was examined, an additive effect of both variables was observed, as would be expected by the data shown in Figure 8.11. The latency of the P300 component elicited by the stimuli was also delayed by the display mask but was *unaffected* by S-R compatibility. This finding suggested that compatibility influenced a stage *after* perceptual categorization. An interesting observation from their data was that the delay produced by the display noise was greater for RT than for P300. This finding also provides support for a model of information processing, which assumes that the effects of degrading one stage (perception) can carry through to prolong processing at a subsequent stage (response selection). Longer processing at encoding will delay P300 somewhat. The final manual response will be delayed even more because it includes slower response selection as well as slower encoding.

Conclusion: The Value of Stages

Collectively, the data from the subtractive method and additive factors and from the event-related potential are quite consistent with the model of information processing described in Chapter 1. However, these data also suggest

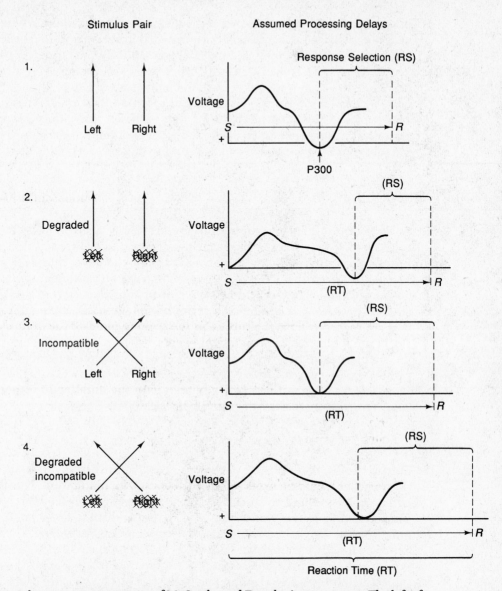

Figure 8.12 Schematic representation of McCarthy and Donchin's experiment. The left of each row shows the stimulus pair presented in a condition, along with the mapping to the response (arrows). To the right is the time line of the evoked potential and response for a particular experimental condition. The evoked potential is conventionally plotted with positive voltages at the bottom. Changes in encoding delay are reflected in P300 latency and reaction time (compare row 1 to 2 and 3 to 4). Changes in response selection delay are reflected in the difference between RT and P300 latency (compare conditions 1 to 3 and 2 to 4). The effects of masking and S-R compatibility in this experiment show an additive relationship.

that the separation of processing stages should not be taken too literally. There undoubtedly is some overlap in time between processing in successive stages, just as the brain in general is capable of a good deal of parallel processing (see Chapters 3 and 9). However, as with other models and conceptions discussed in this book, the stage concept is a useful one that is consistent with dichotomies made elsewhere between sensitivity and response bias in detection, between early and late selection in attention, and between early and late processing resources in time-sharing (see Chapter 9). The integrating value of the stage concept more than compensates for any limitations in its complete accuracy.

SERIAL RESPONSES

So far we have discussed primarily the selection of a single discrete action in the RT task. Many tasks in the real world, however, call for not just one but a series of repetitive actions. Typing and assembly-line work are two examples. The factors that influence single reaction time are just as important in influencing the speed of repetitive performance. However, the fact that several stimuli must be processed in sequence brings into play a set of additional influences that relate to the timing and pacing of sequential stimuli and responses.

In the discussion of serial or repeated responses, we will focus initially on the simplest case: only two stimuli presented in rapid succession. This is the paradigm of the *psychological refractory period*. Next we will examine response times to several stimuli in rapid succession, the serial RT task. This discussion will lead us to an analysis of transcription skills, such as typing.

The Psychological Refractory Period

The psychological refractory period, or PRP (Telford, 1931), describes a situation in which two RT tasks are presented close together in time. The separation in time between the two stimuli is called the *interstimulus interval*, or ISI. The general finding is that the response to the second stimulus is delayed by the processing of the first when the ISI is short. Suppose, for example, a subject is to press a key (R_1) as soon as a tone (S_1) is heard and speak (R_2) as soon as a light (S_2) is seen. If the light is presented a fifth of a second or so after the tone, the subject will be slowed in responding to the light (RT_2) because of the response to the tone. However, reaction time to the tone (RT_1) will be unaffected. The PRP delay in RT_2 is typically measured with respect to a single-task control condition, in which S_2 is responded to without any requirement to respond to S_1.

The most plausible account of the PRP is a model that proposes the human being to be a *single-channel processor* of information. The single-channel theory of the PRP was originally proposed by Craik (1947) and has subsequently been expressed and elaborated on by Bertelson (1966), Davis (1965), Pashler (1989), and Welford (1967). It is compatible with Broadbent's (1958) conception of attention as an information-processing bottleneck that can only process one stimulus or piece of information at a time (see Chapter 3). In explaining the PRP effect, single-channel theory assumes that the processing of S_1 temporarily "captures" the single-channel bottleneck of the decision-making/response-selection stage. Thus until R_1 has been released (the single channel has finished processing S_1), the processor cannot begin to deal with S_2. The stimulus S_2 must therefore wait at the "gates" of this single-channel bottleneck until they open. This waiting time is what prolongs RT_2. The sooner S_2 arrives, the longer it must wait. According to this view, anything that prolongs the processing of S_1 will increase the PRP delay of RT_2. Reynold's (1966), for example, found that the PRP delay in RT_2 was lengthened if the RT task of RT_1 involved a choice rather than a simple response.

This bottleneck in the sequence of information-processing activities does not appear to be located at the peripheral sensory end of the processing sequence (such as blinders over the eyes that are not removed until R_1 has occurred). If this were the case, no processing of S_2 whatsoever could begin until RT_1 is complete. However, as described in Chapter 5, much of perception is relatively automatic. Therefore the basic *perceptual* analysis of S_2 can proceed even as the processor is fully occupied with selecting the response to S_1 (Karlin & Kestinbaum, 1968; Keele, 1973; Pashler, 1989). Only after its perceptual processing is completed does S_2 have to wait for the bottleneck to dispense with R_1. These relations are shown in Figure 8.13.

Imagine as an analogy a kindergarten teacher (who is the bottleneck) who must get two children (S_1 and S_2) ready for recess. Both are able to put on their coats by themselves (this is the automatic "early" processing that does not require the teacher — the single channel — to function), but both need the

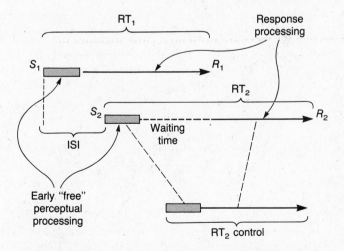

Figure 8.13 Single-channel theory explanation of the psychological refractory period. The figure shows the delay (waiting time) imposed on RT_2 by the presence of RT_1.

teacher to button the coats (select and execute a response). Therefore, how long $child_2$ will have to wait is a joint function of how soon he or she arrives after $child_1$ (the ISI) and how long it takes the teacher to button $child_1$. But this waiting time will not include the time it takes $child_2$ to put on the coat since this can be done in parallel with the buttoning of $child_1$. Only once $child_2$'s coat is on, must the child wait. Therefore, the total time required for $child_2$ to get the coat on and buttoned (analogous to RT_2) is equal to the time it normally takes to put the coat on and have the teacher button it, plus the waiting time. The latter may be predicted by the ISI and by the buttoning time of $child_1$.

Returning to the PRP paradigm, we see that the *delay* in RT_2, beyond its single-task baseline, will increase linearly (on a one-to-one basis) with a decrease in ISI and an increase in the complexity of response selection of RT_1 since both increase the waiting time. This relationship is shown in Figure 8.14. Assuming that the single-channel bottleneck is perfect (i.e., postperceptual processing of S_2 will not start at all until R_1 is released), the relationship between ISI and RT_2 will look like that shown in Figure 8.14. When ISI is long (much greater than RT_1), RT_2 is not delayed at all. When ISI is shortened to about the length of RT_1, some temporal overlap will occur and RT_2 will be prolonged because of a waiting period. This waiting time will then increase linearly as ISI is shortened further.

The relationship between ISI and RT_2 as shown in Figure 8.14 describes rather successfully a large amount of the PRP data (Bertelson, 1966; Kantowitz, 1974; Pashler, 1989; M. Smith, 1967). There are, however two important qualifications to the general single-channel model as it has been presented so far.

(1) When the ISI is sufficiently short (less than about 100 msec), a qualitatively different processing sequence occurs; both responses are emitted *together* (grouping) and *both* are delayed (Kantowitz, 1974). It is as if the two stimuli are occurring so close together in time that S_2 gets

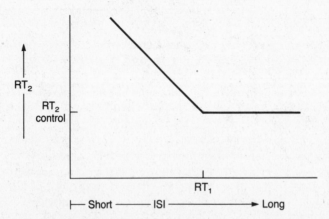

Figure 8.14 Relationship between ISI and RT_2 predicted by single-channel theory.

through the gate while it is still accepting S_1 (Kantowitz, 1974; Welford, 1952).

(2.) Sometimes RT_2 suffers a PRP delay even when the ISI is greater than RT_1. That is, S_2 is presented after R_1 has been completed. This delay occurs when the subject is monitoring the feedback from the response of RT_1 as it is executed (Welford, 1967).

In the real world, operators are more likely to encounter a *series* of stimuli that must be rapidly processed than a simple pair. In the laboratory the former situation is realized in the serial RT paradigm. Here a series of RT trials occurs sufficiently close to one another in time that each RT is affected by the processing of the previous stimulus in the manner described by the single-channel theory. A large number of factors influence performance in this paradigm, typical of tasks ranging from quality control inspection to typewriting to assembly-line manufacturing. Many of these variables were considered earlier in this chapter. For example, S-R compatibility, stimulus discriminability, and practice influence serial RT just as they do single-trial-choice RT. However, some of these variables interact in important ways with the variables that describe the sequential timing of the successive stimuli. Other variables describe that timing itself. Five of these factors will be considered: decision complexity, number of information sources, pacing, response, and preview factors.

Decision Complexity

The Decision Complexity Advantage   Earlier we described how the linear relationship between choice reaction time and the amount of information transmitted—the Hick-Hyman law—was seen to reflect a capacity limit of the human operator. The slope of this function expressed as seconds per bit could be inverted and expressed as bits per second. Early interpretations of the Hick-Hyman law assumed that the latter figure provided an estimate of the bandwidth of the human processing system. As decisions become more complex, decision rate slows proportionately.

If the human being really did have a constant fixed bandwidth for processing information, in terms of bits per second, this limit should be the same, whether we make a small number of high-bit decisions per unit time or a large number of small-bit decisions. For example, if one 6-bit decision/sec was our maximum performance, we should also be able to make two 3-bit decisions/sec, three 2-bit decisions/sec, or six 1-bit decisions/sec.

In fact, this trade-off does not appear to hold. The most restricting limit in human performance appears to relate to the absolute number of decisions that can be made per second rather than the number of bits that can be processed per second. People are better able to process information delivered in the format of one 6-bit decision per second than in the format of six 1-bit decisions per second (Alluisi, Muller, & Fitts, 1957; Broadbent, 1971). Thus the frequency of decisions and their complexity do *not* trade off reciprocally. The advantage of a few complex decisions over several simple ones may be defined as a *decision complexity advantage*. This finding suggests that there is some fundamental limit to the central-processing or decision-making rate, independent of decision complexity, that constrains the speed of other stages of processing. This limit appears to be about 2.5 decisions/sec for decisions of the simplest possible kind (Debecker & Desmedt, 1970). Such a limit might well explain why our motor output often outruns our decision-making competence. The *uhs* or *uhms* that we sometimes interject into rapid speech are examples of how our motor system fills in the noninformative responses while the decision system is slowed by its limits in selecting the appropriate response (Welford, 1976).

The most general implication of the decision complexity advantage is that greater gains in information transmission may be achieved by calling for a few complex decisions than by calling for many simple decisions. Several investigators suggest that this is a reasonable guideline. For example, Deininger, Billington, and Riesz (1966) evaluated push-button dialing. A sequence of 5, 6, 8, or 11 letters to be dialed was drawn from a vocabulary of 22, 13, 7, and 4 alternatives, respectively (22.5 bits each). The total dialing time was lowest with the shortest number of units (5 letters), each delivering the greatest information content per letter. Thus, again, frequency of decisions and complexity of decisions did not trade off reciprocally.

A general guideline emerging in computer menu design is that people work better with broad-shallow menus — each choice is among a fairly large number of alternatives, but there are only a few layers — than with narrow-deep menus — choices are simple, but several choices must be made to get to the bottom of the menu (Miller, 1981; Schneiderman, 1987).

A property of working memory that was discussed in Chapter 6 is seemingly related to the decision complexity advantage. This is Yntema's (1963) finding that memory capacity is expanded when subjects must retain a few items with several attributes (i.e., items of greater complexity), as opposed to several items with a few attributes.

Implications for Keyboard Design   The decision complexity advantage has implications for any data-entry task, such as typewriting. For example, Seibel (1972) concluded that making text more redundant (less information per key stroke) will increase somewhat the rate at which key responses can be made (decisions per second) but will decrease the overall information transmission rate (bits per second). It follows from these data that processing efficiency could be increased by allowing each key press to convey more information than the 1.5 bits provided on the average by each letter (see Chapter 2). One possibility is to allow separate keys to indicate certain words or common sequences such as *and*, *ing*, or *th*. This "rapid type" technique has indeed proven to be more efficient than conventional typing, given that the operator receives a minimal level of training (Seibel, 1963). However, if there are too many of these high-information units, the keyboard itself will become overly large, like the keyboard of a Chinese character typewriter. In this case efficiency may decrease because the sheer size of the keyboard will increase the time it takes to locate keys and to move the fingers from one key to another (see Chapter 11).

One obvious solution to this motor limitation is to allow chording, in which simultaneous rather than sequential key presses are required. This approach would increase the number of possible strokes without imposing a proportional increase in the number of keys. Thus, with only a five-finger keyboard, it is possible to produce $2^5 - 1$, or 31, possible chords without requiring any finger movement to different keys. With ten fingers resting on ten keys the possibilities are $2^{10} - 1$, or 1023.

A number of studies indeed suggest that the greater information available per key stroke in chording provides a more efficient means of transmitting information. For example, Seibel (1963) found that increasing the number of possible chords beyond five had little effect on processing speed (chords per minute) but increased overall information transmission. Another study by Seibel (1964) examined skilled court typists using a chording "court writer," or stenotype. Although these operators responded at a third of the rate of skilled typists in the number of keys per unit time, they also succeeded in transmitting twice as much information (bits per second), again consistent with the decision complexity advantage. Lockhead and Klemmer (1959) observed successful performance with a chord typewriter, and Conrad and Longman (1965) found that chording was better than conventional typing as a means of sorting letters in automatic mail-sorting consoles. A ten-key chording board was selected by the British post office for foreign mail sorting, having demonstrated faster performance than a sequential keying device (Barton, 1986).

Besides capitalizing on the decision complexity advantage, chording keyboards are also useful because they can be easily operated while vision is fixated elsewhere. A major problem with chording keyboards, however, is that the sometimes arbitrary finger assignments take a long time to learn (Richardson et al., 1987). One solution is to capitalize on visual imagery, assigning t' chording fingers in a way that "looks" like the image of the letters. S' chording keyboard was designed by Sidorsky (1974), following the sc' Figure 8.15. Using three fingers, the operator presses twice for e "painting" it from the top row to the bottom. In the figure, the that are not pressed. Once the operator remembers the particu' shapes of the letters, little learning is required, and Sidors' jects were able to type from 60 to 110 percent as fast w' with the conventional keyboard. See also Gopher and only one hand is required, the chording keyboard can mouse, controlled by the other hand.

Load Stress

Conrad (1951) used the term *load stress* to refer t channels, as defined in Chapter 3) along which sti from our discussion of divided attention in Ch decrement will probably occur if the operator is r more than one channel at a time. If increasing the

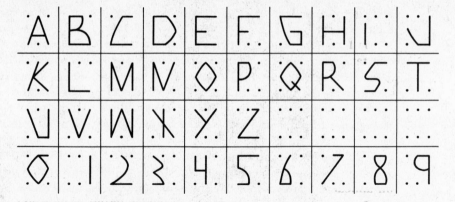

Figure 8.15 The letter-shape keyboard devised by Sidorsky uses visual imagery to specify the form of the key press for an alphanumeric character. There are three keys, and one to three of them must be pressed twice. The small dots indicate the keys that are not pressed. The top row of each letter represents the first key press; the bottom row represents the second. The keys that are successively pressed have a movement pattern that approximates the visual pattern of the letter. (*Source:* C. Sidorsky, *Alpha-dot: A New Approach to Direct Computer Entry of Battlefield Data* (Arlington, VA: U.S. Army Research Institute for the Behavioral and Social Sciences, 1974), Figure 1.)

a corresponding increase in the number of signals to be processed per unit time, the decrement is not surprising. An investigation by Goldstein and Dorfman (1978) determined that there is also a decrease in performance resulting from increasing load stress when the total rate of information delivery across all channels remains constant. That is, although an operator can successfully monitor or respond to one channel that provides X stimuli/min, that operator may not be able to monitor N channels each with X/N stimuli/min.

In their experiment Goldstein and Dorfman (1978) required subjects to monitor one, two, or three meters that had to be responded to with a key press if the indicators entered a danger zone. These events could vary in their frequency. The subjects' response reversed the indicated movement, but only ʳer some inertial delay. This inertia required subjects to anticipate and re- an indicator before it entered the danger zone. Goldstein and Dorfman at the effect of *load* was prepotent over the effect of *speed* in degrading nce. That is, when the number of events/unit time was constant, nce was better on one source than several. More dramatically, sub- better job of processing 72 events/min on one meter than only 24 distributed across three meters, even though the latter condition a mere 8 events/meter.

ʳctor defines the circumstances under which the operator pro- e stimulus to the next. Pacing can be characterized in terms of s, one dichotomous and one continuous, both of which generally describe the degree of time constraints placed on the human operator. The dichotomous dimension contrasts a *force-paced* with a *self-paced* stimulus rate. In the force-paced schedule each stimulus follows the preceding stimulus at a constant interval. The critical parameter is the interstimulus interval, or ISI. The frequency of stimulus presentation in the force-paced schedule is thus independent of the operator's responses. Work on an assembly line in which the items continuously move past the operator on a conveyor belt is force-paced. The speed of the belt defines the ISI. This schedule is typical of semiautomated letter-sorting consoles in the post office. In the self-paced schedule each stimulus follows the previous *response* by a constant interval known as the *response-stimulus interval*, or RSI. In this case, the frequency with which stimuli appear depends on the latency of the operator's response. This schedule characterizes manual letter sorting, in which letters are tossed into the appropriate bins.

The continuous dimension in pacing defines the value of the timing parameters. Either self-paced or force-paced schedules may be perceived as leisurely if the RSI or ISI, respectively, is long, so that a long time passes between a response and a subsequent stimulus. However, if the RSI is reduced to near zero or the ISI to a value near the average reaction time, the speed stress can be quite intense indeed. In fact, the self-paced schedule with a zero RSI can seem just as forced as the force-paced schedule if the operator is responding rapidly. No matter how short the response, the subsequent stimulus will always be waiting.

The differences between these two schedules and the ease with which one ˜ the other may be implemented in such systems as automatic postal sorting, a[sse]mbly-line work, or industrial inspection have led investigators repeatedly verse which is better. Welford (1968, 1976) argues on intuitive grounds in found t[h]e self-paced schedule because as long as there is any variability in the perform[n]se (because of fluctuations in decision complexity, stimulus quality, performa[nce] efficiency), the self-paced schedule will allow long responses at one jects did a[c]ompensated for by shorter responses at another. Welford argues events/min[]e-paced schedule allows no such flexibility. Either the ISI must be generated [th]an the longest expected RT to avoid temporal overlap, in which [numb]er of stimuli processed will be fewer than in the self-paced rate, [may] be set shorter (e.g., at the mean response rate of the self-paced [] this case a long RT to one stimulus will cause processing to [] with the subsequent stimulus. A PRP effect will then result. []d not be damaging if the subsequent stimulus was permanently [] quired a short latency response. However, if the stimulus was []ppeared before the process of S_1 was dispensed with, an error []he processing of S_2 was also lengthy, the PRP effect would [] and so on.

Pacing

The pacing fa[ctor]
ceeds from on[]
two dimensions[]

Unfortunately, the empirical data appear to be so ambiguous that firm conclusions are difficult to draw concerning which schedule is better (e.g., Drury & Coury, 1981; Knight & Salvendy, 1981; Waganaar & Stakenberg, 1975; Wickens, 1984). One reason is that differences between schedules depend so much on the timing parameters, the RSI and ISI, that are chosen. If these are short, then with long-duration tasks, subjects in a force-paced schedule will be unable to rest, unless these too are built into the system. With long RSIs, however, rest will occur often. It seems safe to assert that as the variability of processing latency increases, for whatever reasons, the relative merits of self-paced as opposed to force-paced schedules will improve. The exact level of variability at which a superiority of self- over force-paced schedules will be observed cannot be stated with confidence.

Response Factors

Response Complexity   As shown in Figure 8.11, more complex responses require longer to initiate. In the serial RT task, one important consequence of increased response complexity is the requirement for more monitoring of the response. As noted in the discussion of the psychological refractory period, monitoring the execution of and feedback from a response will sometimes delay the start of processing a subsequent stimulus (Welford, 1976).

Response Feedback   The feedback from a response can have two effects on performance, depending on the sensory modality in which it is received. Consider first the case in which the feedback is an *intrinsic* part of the response — the perceived sound of one's voice or the visualization of one's moving hand. These are linkages described as ideomotor compatible earlier in the chapter (Grunwald, 1970). Delays, distortions, or elimination of the intrinsic feedback can produce substantial deficits in performance (K. V. Smith, 1962). For example, consider the difficulty one has in speaking in a controlled voice when listening to loud music over headphones so that one's voice cannot be perceived, or the difficulty in controlling one's own hand movements if they are perceived through a TV image with a delay (see Chapter 11).

Less serious are distortions of secondary feedback, such as the click of a depressed key or the appearance of a visual letter on a screen after the keystroke. Delays or degradation of this feedback can be harmful (Miller, 1968). However as expertise on the skill develops, and the operator becomes less reliant on the feedback to ensure that the right response has been executed, such feedback can be ignored; hence the harmful effects of its delays (or elimination) are themselves reduced (Long, 1976).

Response Repetition   Earlier in this chapter, we saw that a response that repeated itself was more rapid than if it followed a different response (Kornblum, 1973). However, there is a trend in many serial response skills such as typewriting for the opposite effect to occur, in which a response is slowed by its repetition.

The differing effects of response repetition between the single-trial RT paradigm (in which repetition is good) and the typing task described here (in which it is harmful) is worthy of note. In the single-trial RT the repetition effect is considered to be a kind of shortcut, which eliminates the repeated engagement of the response selection stage. If a stimulus repeats, the same response is activated as before, and time is saved. Reaction times in this case are relatively long (around 200–300 msec), however, compared to the interresponse times in typing, which may be less than half that amount. This produces a speed of five to eight responses/sec. In light of Debecker and Desmedt's (1970) data, which showed a limit to serial response speed of two to three decisions/sec, the high speeds in typing indicate that the separate responses may be selected without engaging a higher-level decision process. As a consequence there is no longer any benefit to repetitions by shortcutting this process. In fact, with the shorter interresponse times, processing instead begins to impose on the refractoriness of the motor system when repeated commands are issued to the same muscles (Fitts & Posner, 1967). In this case, it becomes advantageous to employ separate muscle groups for successive responses — hence, the advantage for alternations.

The slowing of repeated responses is greatest when a single finger is repeated, particularly when the repeated finger must strike two different keys because movement time is now required. However, slowing is also evident when two different digits on the same hand are repeated (Rummelhart & Norman, 1982). For example, Sternberg, Kroll, and Wright (1978) observed that words that needed key strokes of alternate hands were typed faster than those in which all letters were typed with one hand. One characteristic of the keyboard layout on the conventional Sholes, or QWERTY, keyboard is that keys are placed so that common sequences of letters will be struck with keys that are far apart and therefore, on the average, likely to be struck with different hands.

Unfortunately, the QWERTY design fails to include some other characteristics that would lead to more rapid performance. For example, more letters are typed on the row above than on the home keys, thus adding extra movement. Also the amount of effort between the two hands is not balanced. Dvorak (1943) designed a keyboard that reflects these two considerations and that attempts as well to maximize between-hand alternations. There is some consensus that with proper training, the Dvorak keyboard could lead to improved typing performance. However, the Dvorak board will probably never prove to be practical because the 5 to 10 percent estimated improvement (Alden, Daniels, & Kanarick, 1972; Norman & Fisher, 1982) is small relative to the familiarity that users have with the conventional keyboard and to the great inertia against change.

Preview and Transcription

We have noted that the limits of serial RT performance are around 2½ decisions per second. Yet skilled typists can execute key strokes at a rate of more than 15 per second for short bursts (Rummelhart & Norman, 1982). The major difference here is in the way in which typing and, more generally, the class of *transcription tasks* (e.g., typing, reading aloud, and musical performance) are structured to allow the operator to make use of *preview*, *lag*, and *parallel processing.* These are characteristics that allow more than one stimulus to be displayed at a time (preview is available) and therefore allow the operator to lag the response behind perception. Thus at any time the response executed is not necessarily relevant to the stimulus that was most recently encoded but is more likely to be related to a stimulus encoded earlier in the sequence. Therefore perception and response are occurring in parallel. Whether one speaks of this as *preview* (seeing into the future) or *lag* (responding behind the present) obviously depends on the somewhat arbitrary frame of reference one chooses to define the present.

Preview and lag are both possible in either self-paced or force-paced tasks. Thus, in typing (a self-paced activity), the typist typically encodes letters (as judged from visual fixation) approximately one second before they are entered into the keyboard. Similarly, in reading aloud, the voice will lag well behind the eye fixations. Oral translation is a force-paced task because the auditory speech flows at a rate that is not determined by the translator. There is a lag of a few words between when a word is heard and when it is spoken. In general, this lag in transcription is beneficial to performance. Yet the physical constraints of the task determine the extent to which a lag is possible. In a self-paced task a lag can be created only if a preview of two or more stimuli is provided (e.g., two or more stimuli are displayed simultaneously). This typifies the cases of typing or reading aloud, in which preview is essentially unlimited and is subject only to the constraints of visual fixation (see Chapter 3). In a force-paced task, in which a stimulus does not wait for a response to occur—for example, taking oral dictation or translating spoken languages—the operator need only build up a slight lag or queue before responding.

When operators use preview and lag, they must maintain a running "buffer" memory of encoded stimuli that have not yet been executed as responses. It is therefore interesting to consider why the lag is beneficial in transcription, in light of the fact that task-induced lags between input and response were shown to be harmful in the running memory tasks discussed in Chapter 6. A major difference between the two cases concerns the size of the lag involved. The lag typically observed in transcription is short—around one second—relative to the decay of working memory; so the contents of memory are readily available at the time that output is called for. Resource-demanding rehearsal processes need not be invoked. The problems of running memory are typically observed when the delay is greater.

Since the costs encountered in running memory tasks are not present in the lags of transcription, it is possible for the transcriber to realize the two important benefits of these lags: allowance (1) for variability and (2) for chunking (Shaffer, 1973; Shaffer & Hardwick, 1970).

1. *Allowance for variability.* In a nonlagged system, if an input is encountered that is particularly difficult to encode, this delay will be shown at the response as well. Correspondingly, a particularly difficult response will slow down processing of the subsequent input in a PRP-like fashion. However, if there is a lag between input and output, accounted for by a buffer of three or four items, a steady stream of output at a constant rate can proceed (for at least a short while) even if input is temporarily slowed. The buffer is just "emptied"

at a constant rate. A prolonged input will, of course, temporarily reduce the size of the buffer until it may be "refilled" with a more rapidly encoded input. If easily encoded stimuli do not appear, the operator may eventually be forced to adopt a less efficient nonlagged mode of processing, which is what occurs when the text on a page being typed becomes degraded.

  2. *Allowance for chunking.* There is good evidence in typing that inputs are encoded in chunks, so that the letters within each chunk are processed more or less in parallel (McClelland & Rummelhart, 1986; Reicher, 1969). The output, however, must be serial (assuming that a "rapid type" is not employed). The creation of a lag, therefore, allows the steady flow of serial output to proceed even as the buffer is suddenly increased in the number of required output units (key presses stored) as a result of the parallel perception of a chunk. Furthermore, factors affecting the speed of individual responses (i.e., the reach time for letters on a keyboard) are totally unrelated to encoding difficulty. A buffer will prevent slow responses from disrupting subsequent encoding. Figure 8.16 provides a schematic example in which a relatively constant output stream (response per unit time) is maintained in spite of variations in encoding speed

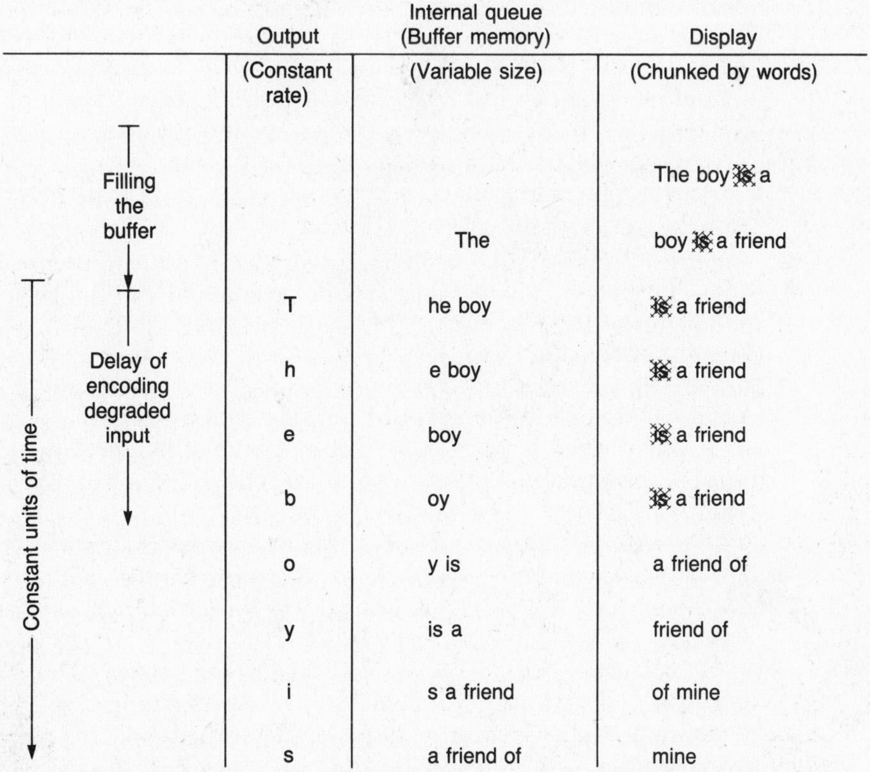

	Output	Internal queue (Buffer memory)	Display
	(Constant rate)	(Variable size)	(Chunked by words)
Filling the buffer			The boy 💥 a
		The	boy 💥 a friend
	T	he boy	💥 a friend
Delay of encoding degraded input	h	e boy	💥 a friend
	e	boy	💥 a friend
	b	oy	💥 a friend
	o	y is	a friend of
	y	is a	friend of
	i	s a friend	of mine
	s	a friend of	mine

(left vertical axis label: Constant units of time)

Figure 8.16 Schematic representation of transcription, showing variable input rate (caused by chunking and stimulus quality) providing constant output rate. Note the long delay when encoding the degraded word *is*.

and buffer contents caused by variations in input chunks and input quality. These variations will cause the size of the buffer to vary yet allow an even, "rhythmic" flow of responses.

One remarkable characteristic of all of these activities—encoding, buffer storage, and response—is that they appear to proceed more or less in parallel, with little mutual interference, and are even time-shared with a fourth mental activity, the monitoring of errors in response (Rabbitt, 1981). This success stands in general contrast to the failures that are often observed in multiple-task performance, a topic that will be addressed in the next chapter. It suggests that parallel processing of mental operations within a task is easier than parallel processing between tasks.

Use of Preview   The availability of preview in transcription tasks does not, of course, mean that preview will necessarily be used. The unskilled typist will still type one letter at a time and will go no faster if preview is available than it if it is not. The skilled typist depends heavily on preview for efficient transcription. Investigations by Hershon and Hillix (1965), Shaffer (1973), and Shaffer and Hardwick (1970) suggest that preview helps performance. These data make clear that there are two benefits of preview: making available more advance information and giving the operator an opportunity to perceive chunks (see Chapter 6). Hershon and Hillix had subjects type a written message, of which various numbers of letters could be displayed in advance. They increased the number of preview letters from one to two, then to three, then to six, and finally to an unlimited number. Greater preview provided some benefits to typing random-letter strings but much greater benefits to typing random words. The greater benefit in the latter condition, of course, resulted because of the chunkability of words.

Shaffer (1973) conducted a systematic investigation of preview effects using one highly skilled typist. Shaffer examined the differences in interresponse times (IRT) as the subject typed varying kinds of text with different amounts of preview. The slowest typing was obtained while typing random letters with no preview, a condition equivalent to a self-paced serial RT task with negligible response-stimulus interval. Here the IRT was 500 msec. Progressively shorter IRTs (faster typing) were obtained while typing random letters with unlimited preview (IRT = 200 msec) and typing random words with preview (IRT = 100 msec). In a fourth condition, words in coherent text were typed with unlimited preview. The results of this condition were interesting because no further gain in typing speed was observed over the random-word condition. This finding suggests that the benefits of preview are *not* related to the semantic level of processing but rather to the fact that preview allows the letters within chunk-sized units (i.e., words) to be processed in parallel (see Chapter 5).

Further support for the conclusion that word chunks and not semantic content is the critical factor in preview is provided by another of Shaffer's findings. In typing either coherent prose or random words, eight letters of preview are sufficient to accomplish all necessary gains in performance. Eight letters would be sufficient to encompass the great majority of words but generally not enough to extract coherent semantic meaning from word strings. The absence of heavy semantic involvement in transcription would thereby explain how skilled typists may be able to carry on a conversation or perform other verbal activity while typing (Shaffer, 1975; see also Chapter 9).

A second conclusion of Shaffer's (1975) research is that the benefits of chunking are primarily perceptual and may be seen in storage but not in response. Groups of letters are perceived as a unit, perhaps are stored in the buffer memory as a unit, but are rarely output to the keyboard as an integrated motor program; that is, they are not a highly overlearned response pattern, such as one's signature, that is executed as an open-loop motor chunk or motor program (Keele, 1973; Summers, 1981; see Chapter 11). Shaffer's arguments are based primarily on an analysis of the interresponse times between letters within a word. If these are, in fact, parts of a motor chunk or motor program, the sum of the IRTs within a chunked word should be less than a similar sum of mean IRTs derived from all words. For example, since the IRT averaged across typing words with preview is 100 msec, to be typed as a chunk the total IRT in the word *cat* should be *less* than 200 msec (i.e., less than two 100-msec IRTs). Shaffer observes that this motor-packaging effect holds true for only a few extremely common sequences (*and*, *the*, and *-ing*) but not for the general class of common words. This conclusion is supported by investigations of typing by Gentner (1982) and Sternberg, Kroll, & Wright (1978). Gentner, for example, performed a detailed analysis of finger movement in typing and concluded that they were determined entirely by constraints of the hands and keyboard interacting with the letter sequence. Higher-level word units played no role in response timing.

In summary, people are remarkably efficient at performing transcription tasks, overcoming many of the limits of reaction time. Only relatively small margins for improvement are probably feasible in the redesign of the typewriter (Norman & Fisher, 1982). Furthermore, it is likely that future developments in automatic speech recognition devices may eventually make a person's manual function in typing obsolete. The read or created message may simply be spoken to the computer and a printed text prepared from the automatically recognized speech. The major benefit of research on typing would not then appear to be on issues of keyboard redesign. Instead, its value seems to be in terms of how models of optimal typing performance, such as those of Card, Moran, and Newell (1983; see Chapter 1); Gentner (1982); Rummelhart and Norman (1982); and Shaffer and Hardwick (1970), can be applied to design interfaces for other transcription and data-entry tasks such as those involved in more complex human-computer interactions.

TRANSITION

With the discussion of discrete actions in this chapter, we have now completed our presentation of the different stages and units of human performance in Chapters 2 through 8. We have also seen how stages of perception, memory, and action can be assembled by the performer to create complex decision and

RT tasks, and how these tasks in turn can be analyzed by the psychologist into their stagelike components. In our discussion of transcription skills, we saw evidence for parallel processing of different mental processes within a single task. Much of our performance in daily life, however, does not involve performance of a single task at a time but rather the time-sharing of multiple activities. As you read these words, you may well be listening to music or tapping your fingers. The focus of the next chapter will be on this issue of dual-task performance, or divided attention, in a much broader context than that addressed in Chapter 3, when we considered attention as a perceptual phenomenon.

We also discussed the trade-off between the speed and the accuracy of performance, but we did not examine in detail the nature of the errors produced when performance is rapid. The discussion of these and other sorts of errors, the models for predicting their occurrence, and techniques for preventing their damaging effects will form a major component of Chapter 10, in which our focus is on the relationship between stress and human error.

REFERENCES

Alden, D. G., Daniels, R. W., & Kanarick, A. F. (1972). Keyboard design and operations: A review of the major issues. *Human Factors, 14,* 275–293.

Alluisi, E., Muller, P. I., & Fitts, P. M. (1957). An information analysis of verbal and motor response in a force-paced serial task. *Journal of Experimental Psychology, 53,* 153–158.

Andre, A. D., Wickens, C. D., & Goldwasser, J. B. (1990). *Compatibility and consistency in display-control systems: Implications for decision aid design.* University of Illinois Institute of Aviation Technical Report (ARL-90-13/NASA-A³I-90-2). Savoy, IL: Aviation Research Laboratory.

Andre, A. D., Haskell, I., & Wickens, C. D. (1991). S-R compatibility effects with orthogonal stimulus and response dimensions. *Proceedings of the 35th annual meeting of the Human Factors Society.* Santa Monica, CA: Human Factors Society.

Barton, P. H. (1986). The development of a new keyboard for outward sorting foreign mail. *IMechE,* 57–63.

Bertelson, P. (1965). Serial choice reaction-time as a function of response versus signal-and-response repetition. *Nature, 206,* 217–218.

Bertelson, P. (1966). Central intermittency twenty years later. *Quarterly Journal of Experimental Psychology, 18,* 153–163.

Bertelson, P., & Tisseyre, F. (1966). Choice reaction time as a function of stimulus versus response relative frequency of occurrence. *Nature, 212,* 1069–1070.

Biederman, I., & Kaplan, R. (1970). Stimulus discriminability and S-R compatibility: Evidence for independent effects in choice reaction time. *Journal of Experimental Psychology, 86,* 434–439.

Blackman, A. (1975). Test of the additive-factor method of choice reaction time analysis. *Perceptual Motor Skills, 41,* 607–613.

Brainard, R. W., Irby, T. S., Fitts, P. M., & Alluisi, E. (1962). Some variable influencing the rate of gain of information. *Journal of Experimental Psychology, 63,* 105–110.

Broadbent, D. E. (1971). *Decision and stress.* London: Academic Press.

Broadbent, D. E., & Gregory, M. (1965). On the interaction of S-R compatibility with other variables affecting reaction time. *British Journal of Psychology, 56,* 61–67.

Card, S., Moran, T. P., & Newell, A. (1983). *The psychology of human-computer interactions.* Hillsdale, NJ: Erlbaum.

Chapanis, A., & Lindenbaum, L. E. (1959). A reaction time study of four control-display linkages. *Human Factors, 1,* 1–14.

Coles, M. G. H. (1988). Modern mind-brain reading: Psychophysiology, physiology, and cognition. *Psychophysiology, 26*(3), 251–269.

Coles, M. G. H., Gratton, G., & Donchin, E. (1988). Detecting early communication: Using measures of movement-related potentials to illuminate human information processing. *Biological Psychology, 26,* 69–89.

Conrad, R. (1951). Speed and load stress in sensori-motor skill. *British Journal of Industrial Medicine, 8,* 1–7.

Conrad, R., & Longman, D. S. A. (1965). Standard typewriter vs. chord keyboard: An experimental comparision. *Ergonomics, 8,* 77–88.

Craik, K. W. J. (1947). Theory of the human operator in control systems I: The operator as an engineering system. *British Journal of Psychology, 38,* 56–61.

Crosby, J. V., & Parkinson, S. R. (1979). A dual task investigation of pilot's skill level. *Ergonomics, 22,* 1301–1313.

Danaher, J. W. (1980). Human error in air traffic control systems operations. *Human Factors, 22,* 535–545.

Davis, R. (1965). Expectancy and intermittency. *Quarterly Journal of Experimental Psychology, 17,* 75–78.

Davis, R., Moray, N., & Treisman, A. (1961). Imitation responses and the rate of gain of information. *Quarterly Journal of Experimental Psychology, 13,* 78–89.

Debecker, J., & Desmedt, R. (1970). Maximum capacity for sequential one-bit auditory decisions. *Journal of Experimental Psychology, 83,* 366–373.

Deininger, R. L., Billington, M. J., & Riesz, R. R. (1966). The display mode and the combination of sequence length and alphabet size as factors of speed and accuracy. *IEEE Transactions on Human Factors in Electronics, 7,* 110–115.

Donchin, E. (1981). Surprise! . . . Surprise? *Psychophysiology, 18,* 493–513.

Donders, F. C. (1869, trans. 1969). On the speed of mental processes (trans. W. G. Koster). *Acta Psychologica, 30,* 412–431.

Drazin, D. (1961). Effects of fore-period, fore-period variability and probability of stimulus occurrence on simple reaction time. *Journal of Experimental Psychology, 62,* 43–50.

Drury, C., & Coury, B. G. (1981). Stress, pacing, and inspection. In G. Salvendy & M. J. Smith (eds.), *Machine pacing and operational stress.* London: Taylor & Francis.

Duncan, J. (1979). Divided attention: The whole is more than the sum of the parts. *Journal of Experimental Psychology: Human Perception and Performance, 5,* 216–228.

Dvorak, A. (1943). There is a better typewriter keyboard. *National Business Education Quarterly, 12,* 51–58.

Ehrenstein, W. H., Schroeder-Heister, P., & Heister, G. (1989). Spatial S-R compatibility with orthogonal stimulus-response relationship. *Perception & Psychophysics, 45,* 215–220.

Fitts, P. M. (1966). Cognitive aspects of information processing III: Set for speed versus accuracy. *Journal of Experimental Psychology, 71,* 849–857.

Fitts, P. M., & Deininger, R. L. (1954). S-R compatibility: Correspondence among paired elements within stimulus and response codes. *Journal of Experimental Psychology, 48,* 483–492.

Fitts, P. M., & Peterson, J. R. (1964). Information capacity of discrete motor responses. *Journal of Experimental Psychology, 67,* 103–112.

Fitts, P. M., Peterson, J. R., & Wolpe, G. (1963). Cognitive aspects of information processing II: Adjustments to stimulus redundancy. *Journal of Experimental Psychology, 65,* 423–432.

Fitts, P. M., & Posner, M. A. (1967). *Human performance.* Pacific Palisades, CA: Brooks Cole.

Fitts, P. M., & Seeger, C. M. (1953). S-R compatibility: Spatial characteristics of stimulus and response codes. *Journal of Experimental Psychology, 46,* 199–210.

Flight International. (1990, October 31). Lessons to be learned, pp. 24–26.

Fuchs, (1962). The progression regression hypothesis in perceptual-motor skill learning. *Journal of Experimental Psychology, 63,* 177–192.

Gentner, C. R. (1982). Evidence against a central control model of timing in typing. *Journal of Experimental Psychology: Human Perception and Performance, 9,* 793–810.

Goldstein, I. L., & Dorfman, P. W. (1978). Speed stress and load stress as determinants of performance in a time-sharing task. *Human Factors, 20,* 603–610.

Gopher, D., & Raij, D. (1988). Typing with a two hand chord keyboard—will the QWERTY become obsolete? *IEEE Transactions in System, Man, and Cybernetics, 18,* 601–609.

Gratton, G., Coles, M. G. H., Sirevaag, E., Eriksen, C. W., & Donchin, E. (1988). Pre- and post-stimulus activation of response channels: A psychophysiological analysis. *Journal of Experimental Psychology: Human Perception and Performance, 14,* 331–344.

Greenwald, A. (1970). A double stimulation test of ideomotor theory with implications for selective attention. *Journal of Experimental Psychology, 84,* 392–398.

Greenwald, A. G., (1979). Time-sharing, ideomotor compatibility and automaticity. In C. Bensel (ed.), *Proceedings of the 23rd annual meeting of the Human Factors Society.* Santa Monica, CA: Human Factors Society.

Greenwald, H., & Shulman, H. (1973). On doing two things at once: Eliminating the psychological refractory period affect. *Journal of Experimental Psychology, 101,* 70–76.

Hartzell, E. J., Dunbar, S., Beveridge, R., & Cortilla, R. (1982). Helicopter pilot response latency as a function of the spatial arrangement of instruments and controls. *Proceedings of the 28th Annual Conference on Manual Control.* Dayton, OH: Wright Patterson AFB.

Haskell, I., Wickens, C. D. & Sarno, K. (1990). Quantifying stimulus-response compatibility for the Army/NASA A[3] I display layout analysis tool. *Proceedings of the 5th Mid-Central Human Factors/Ergonomics Conference.* Dayton, OH.

Hawkins, H., MacKay, S., Holley, S., Friedin, B., & Cohen, S. (1973). Locus of the relative frequency effect in choice reaction time. *Journal of Experimental Psychology, 101,* 90–99.

Hawkins, H., & Underhill, L. (1971). S-R compatibility and the relative frequency effect in choice reaction time. *Journal of Experimental Psychology*, 91, 280–286.

Helander, M. G., Karwan, M. H., & Etherton, J. (1987). A model of human reaction time to dangerous robot arm movements. *Proceedings of the 31st annual meeting of the Human Factors Society* (pp. 191–195). Santa Monica, CA: Human Factors Society.

Henderson, B. W. (1989, May 22). Army pursues voice-controlled avionics to improve helicopter pilot performance. *Aviation Week & Space Technology*, p. 43.

Hershon, R. L., & Hillix, W. A. (1965). Data processing in typing:.Typing rate as a function of kind of material and amount exposed. *Human Factors*, 7, 483–492.

Hick, W. E. (1952). On the rate of gain of information. *Quarterly Journal of Experimental Psychology*, 4, 11–26.

Hoffmann, E. R. (1990). Strength of component principles determining direction-of-turn stereotypes for horizontally moving displays. *Proceedings of the 34th annual meeting of the Human Factors Society* (pp. 457–461). Santa Monica, CA: Human Factors Society.

Howell, W. C., & Kreidler, D. L. (1963). Information processing under contradictory instructional sets. *Journal of Experimental Psychology*, 65, 39–46.

Howell, W. C., & Kreidler, D. L. (1964). Instructional sets and subjective criterion levels in a complex information processing task.' *Journal of Experimental Psychology*, 68, 612–614.

Hyman, R. (1953). Stimulus information as a determinant of reaction time. *Journal of Experimental Psychology*, 45, 423–432.

Kantowitz, B. H. (1974). Double stimulation. In B. H. Kantowitz (ed.), *Human information processing*. Hillsdale, NJ: Erlbaum.

Karlin, L., & Kestinbaum, R. (1968). Effects of number of alternatives on the psychological refractory period. *Quarterly Journal of Experimental Psychology*, 20, 160–178.

Keele, S. W. (1969). Repetition effect: A memory dependent process. *Journal of Experimental Psychology*, 80, 243–248.

Keele, S. W. (1973). *Attention and human performance*. Pacific Palisades, CA: Goodyear.

Kirby, P. H. (1976). Sequential affects in two choice reaction time: Automatic facilitation or subjective expectation. *Journal of Experimental Psychology: Human Perception and Performance*, 2, 567–577.

Kirkpatrick, M., & Mallory, K. (1981). Substitution error potential in nuclear power plant control rooms. In R. Sugarman (ed.), *Proceedings of the 25th annual meeting of the Human Factors Society*. Santa Monica, CA: Human Factors Society.

Klapp, S. T., & Irwin, C. I. (1976). Relation between programming time and duration of response being programmed. *Journal of Experimental Psychology: Human Perception and Performance*, 2, 591–598.

Klemmer, E. T. (1957). Simple reaction time as a function of time uncertainty. *Journal of Experimental Psychology*, 54, 195–200.

Knight, J., & Salvendy, G. (1981). Effects of task stringency of external pacing on mental load and work performance. *Ergonomics*, 24, 757–764.

Kohlberg, D. L. (1971). Simple reaction time as a function of stimulus intensity in decibels of light and sound. *Journal of Experimental Psychology*, 88, 251–257.

Kornblum, S. (1973). Sequential effects in choice reaction time: A tutorial review. In S. Kornblum (ed.), *Attention and performance IV*. New York: Academic Press.

Kornblum, S., Hasbroucq, T., & Osman, A. (1990). Dimensional overlap: Cognitive basis for stimulus-response compatibility—A model and taxonomy. *Psychological Review, 97*, 253–270.

Leonard, J. A. (1959). Tactile choice reactions I. *Quarterly Journal of Experimental Psychology, 11*, 76–83.

Lockhead, G. R., & Klemmer, E. T. (1959, November). *An evaluation of an 8-key wordwriting typewriter* (IBM Research Report RC–180). Yorktown Heights, NY: IBM Research Center.

Long, J. (1976). Effects of delayed irregular feedback on unskilled and skilled keying performance. *Ergonomics, 19*, 183–202.

Loveless, N. E. (1963). Direction of motion stereotypes: A review. *Ergonomics, 5*, 357–383.

McCarthy, G., & Donchin, E. (1979). Event-related potentials: Manifestation of cognitive activity. In F. Hoffmeister & C. Muller (eds.), *Bayer Symposium VIII: Brain function in old age*. New York: Springer.

McClelland, J. L. (1979). On the time-relations of mental processes: An examination of processes in cascade. *Psychological Review, 86*, 287–330.

Martin, G. (1989). The utility of speech input in user-computer interfaces. *International Journal of Man-Machine System Study, 18*, 355–376.

Merkel, J. (1885). Die zeitlichen Verhaltnisse der Willensthatigkeit. *Philosophische Studien, 2*, 73–127.

Miller, R. B. (1968). Response time in non-computer conversational transactions. In *Proceedings of 1968 Fall Joint Computer Conference*. Arlington, VA: AFIPS Press.

Miller, D. P. (1981). The depth/breadth trade-off in hierarchical computer menus. In R. Sugarman (ed.), *Proceedings, 25th Annual Meeting of the Human Factors Society*. Santa Monica, CA: Human Factors.

Miller, J., & Anbar, R. (1981). Expectancy and frequency effects on perceptual and motor systems in choice reaction time. *Memory & Cognition, 9*, 631–641.

Miller, J., & Pachella, R. (1973). On the locus of the stimulus probability effect. *Journal of Experimental Psychology, 101*, 501–506.

Mowbray, G. H., & Rhoades, M. V. (1959). On the reduction of choice reaction time with practice. *Quarterly Journal of Experimental Psychology, 11*, 16–23.

Norman, D. (1988). *The psychology of every day things*. New York: Harper & Row.

Norman, D. A., & Fisher, D. (1982). Why alphabetic keyboards are not easy to use: Keyboard layout doesn't matter much. *Human Factors, 24*, 509–520.

Osborne, D. W., & Ellingstad, V. S. (1987). Using sensor lines to show control-display linkages on a four burner stove. *Proceedings of the 31st annual meeting of the Human Factors Society* (pp. 581–584). Santa Monica, CA: Human Factors Society.

Pachella, R. (1974). The use of reaction time measures in information processing research. In B. H. Kantowitz (ed.), *Human information processing*. Hillsdale, NJ: Erlbaum.

Pashler, H. (1989). Dissociations and contingencies between speed and accuracy: Evidence for a two-component theory of divided attention in simple tasks. *Cognitive Psychology, 21*, 469–514.

Pew, R. W. (1969). The speed-accuracy operating characteristic. *Acta Psychologica, 30,* 16–26.

Posner, M. I. (1964). Information reduction in the analysis of sequential tasks. *Psychological Review, 71,* 491–504.

Rabbitt, P. M. A. (1967). Signal discriminability, S-R compatibility and choice reaction time. *Psychonomics Science, 7,* 419–420.

Rabbitt, P. M. A. (1989). Sequential reactions. In D. H. Holding (ed.), *Human skills* (2nd ed.). New York: Wiley.

Rasmussen, J. (1980). The human as a system's component. In H. T. Smith & T. R. Green (eds.), *Human interaction with computers.* London: Academic Press.

Rasmussen, J. (1986). *Information processing and human-machine interaction: An approach to cognitive engineering.* New York: North Holland.

Reicher, G. M. (1969). Perceptual recognition as a function of meaningfulness of stimulus material. *Journal of Experimental Psychology, 81,* 275–280.

Reynolds, D. (1966). Time and event uncertainty in unisensory reaction time. *Journal of Experimental Psychology, 71,* 286–293.

Richardson, R. M. M., Telson, R. U., Koch, C. G., & Chrysler, S. T. (1987). Evaluation of conventional, serial, and chord keyboard options for mail encoding. *Proceedings of the 31st annual meeting of the Human Factors Society* (pp. 911–915). Santa Monica, CA: Human Factors Society.

Rumelhart, D. E., McClelland, J. L., & The PDP Research Group (1986). Parallel distributed processing, Vol. I: Foundations. Cambridge, MA: MIT Press.

Rumelhart, D., & Norman, D. (1982). Simulating a skilled typist: A study of skilled cognitive-motor performance. *Cognitive Science, 6,* 1–36.

Sanders, A. F. (1970). Some variables affecting the relation between relative stimulus frequency and choice reaction time. In A. F. Sanders (ed.), *Attention and performance III.* Amsterdam: North Holland.

Schwartz, S. P., Pomerantz, S. R., & Egeth, H. E. (1977). State and process limitations in information processing. *Journal of Experimental Psychology: Human Perception and Performance, 3,* 402–422.

Seibel, R. (1963). Discrimination reaction time for a 1,023-alternative task. *Journal of Experimental Psychology, 66,* 215–226.

Seibel, R. (1964). Data entry through chord, parallel entry devices. *Human Factors, 6,* 189–192.

Seibel, R. (1972). Data entry devices and procedures. In R. G. Kinkade & H. S. Van Cott (eds.), *Human engineering guide to equipment design.* Washington, DC: U.S. Government Printing Office.

Shaffer, L. H. (1973). Latency mechanisms in transcription. In S. Kornblum (ed.), *Attention and performance IV.* New York: Academic Press.

Shaffer, L. H. (1975). Multiple attention in continuous verbal tasks. In S. Dornic (ed.), *Attention and performance V.* New York: Academic Press.

Shaffer, L. H., & Hardiwck, J. (1970). The basis of transcription skill. *Journal of Experimental Psychology, 84,* 424–440.

Shinar, D., & Acton, M. B. (1978). Control-display relationships on the four burner range: Population stereotypes versus standards. *Human Factors, 20,* 13–17.

Shneiderman, B. (1987). *Designing the user interface: Strategies for effective human-computer interaction.* Reading, MA: Addison-Wesley.

Shulman, H. G., & McConkie, A. (1973). S-R compatibility, response discriminability and response codes in choice reaction time. *Journal of Experimental Psychology, 98,* 375–378.

Sidorsky, R. C. (1974, January). *Alpha-dot: A new approach to direct computer entry of battlefield data* (Technical Paper 249). Arlington, VA: U.S. Army Research Institute for the Behavioral and Social Sciences.

Simon, J. R. (1969). Reaction toward the source of stimulus. *Journal of Experimental Psychology, 81,* 174–176.

Smith, K. U. (1962). *Delayed sensory feedback and balance.* Philadelphia: Saunders.

Smith, M. (1967). Theories of the psychological refractory period. *Psychological Bulletin, 19,* 352–359.

Smith, P., & Langolf, G. D. (1981). The use of Sternberg's memory-scanning paradigm in assessing effects of chemical exposure. *Human Factors, 23,* 701–708.

Smith, S. (1981). Exploring compatibility with words and pictures. *Human Factors, 23,* 305–316.

Spector, A., & Lyons, R. (1976). The locus of the stimulus probability effect in choice RT. *Bulletin of the Psychonomics Society, 7,* 519–521.

Sternberg, S. (1969). The discovery of processing stages: Extension of Donders' method. *Acta Psychologica, 30,* 276–315.

Sternberg, S. (1975). Memory scanning: New findings and current controversies. *Quarterly Journal of Experimental Psychology, 27,* 1–32.

Sternberg, S., Kroll, R. L., & Wright, C. E. (1978). Experiments on temporal aspects of keyboard entry. In J. P. Duncanson (ed.), *Getting it together: Research and application in human factors.* Santa Monica, CA: Human Factors Society.

Strayer, D. L., Wickens, C. D., & Braune, R. (1987). Adult age differences in the speed and capacity of information processing. II. An electrophysiological approach. *Psychology and Aging, 2,* 99–110.

Summers, J. J. (1981). Motor programs. In D. H. Holding (ed.), *Human skills.* New York: Wiley.

Teichner, W., & Krebs, M. (1972). The laws of simple visual reaction time. *Psychological Review, 79,* 344–358.

Teichner, W., & Krebs, M. (1974). Laws of visual choice reaction time. *Psychological Review, 81,* 75–98.

Telford, C. W. (1931). Refractory phase of voluntary and associate response. *Journal of Experimental Psychology, 14,* 1–35.

Tversky, A. (1977). Features of similarity. *Psychological Review, 84,* 327–352.

Van Der Horst, R. (1988). Driver decision making at traffic signals. In *Traffic accident analysis and roadway visibility* (pp. 93–97). Washington, DC: National Research Council.

Vickers, D. (1970). Evidence for an accumulator model of psychophysical discrimination. *Ergonomics, 13,* 37–58.

Waganaar, W. A., & Stakenberg, H. (1975). Paced and self-paced continuous reaction time. *Quarterly Journal of Experimental Psychology, 27,* 559–563.

Warrick, M. J. (1947). *Direction of movement in the use of control knobs to position visual indicators* (USAF AMC Report no. 694–4C). Wright AFB: U.S. Airforce.

Warrick, M. S., Kibler, A., Topmiller, D. H., & Bates, C. (1964). Response time to unexpected stimuli. *American Psychologist, 19*, 528.

Weeks, D. J., & Proctor, R. W. (1990). Salient features coding in the translation between orthogonal stimulus and response dimensions. *Journal of Experimental Psychology: General, 119*, 355–366.

Welford, A. T. (1952). The psychological refractory period and the timing of high speed performance. *British Journal of Psychology, 43*, 2–19.

Welford, A. T. (1967). Single channel operation in the brain. *Acta Psychologica, 27*, 5–21.

Welford, A. T. (1968). *Fundamentals of skill*. London: Methuen.

Welford, A. T. (1976). *Skilled performance: Perceptual and motor skills*. Glenview, IL: Scott, Foresman.

Whitaker, L. A. (1979). Dual-task interference as a function of cognitive processing load. *Acta Psychologica, 43*, 71–84.

Wickelgren, W. (1977). Speed accuracy tradeoff and information processing dynamics. *Acta Psychologica, 41*, 67–85.

Wickens, C. D. (1984). Processing resources in attention. In R. Parasuraman & R. Davies (eds.), *Varieties of attention* (pp. 63–101). New York: Academic Press.

Wickens, C. D., & Derrick, W. (1981, March). *The processing demands of higher order manual control: Application of additive factors methodology* (Technical Report EPL–80-1/ONR-80-1). Champaign: University of Illinois, Engineering Psychology Research Laboratory.

Wickens, C. D., Hyman, F., Dellinger, J., Taylor, H., & Meador, M. (1986). The Sternberg memory search task as an index of pilot workload. *Ergonomics, 29*, 1371–1383.

Wickens, C. D., & Liu, Y. (1988). Codes and modalities in multiple resources: A success and a qualification. *Human Factors, 30*, 599–616.

Wickens, C. D., Sandry, D., & Vidulich, M. (1983). Compatibility and resource competition between modalities of input, central processing, and output: Testing a model of complex task performance. *Human Factors, 25*, 227–248.

Wickens, C. D., Vidulich, M., & Sandry-Garza, D. (1984). Principles of S-C-R compatibility with spatial and verbal tasks: The role of display-control location and voice-interactive display-control interfacing. *Human Factors, 26*, 533–543.

Woodworth, R. S., & Schlossberg, H. (1965). *Experimental psychology*. New York: Holt, Rinehart & Winston.

Yntema, D. (1963). Keeping track of several things at once. *Human Factors, 6*, 7–17.

Chapter
9

Attention, Time-Sharing, and Workload

OVERVIEW

As described in Chapter 3, as the searchlight illuminates the visual world, so attention guides our perception of the environment. The concept of attention, however, is relevant to a much broader range of human performance, for example, dividing attention between tasks that may not be perceptual. In a recent decision, a train engineer could not recite poetry to his passengers because of concern that diversion of attention to poetry recitation (response selection and execution) would disrupt the primarily perceptual/monitoring aspects of train driving. Our discussions of working memory revealed how vulnerable this system was to the competing demands of other activities, like driving, speech production, or encoding. To account for the role of attention in time-sharing between tasks, in this chapter we adopt a different metaphor, that of the resource (Normal & Bobrow, 1975), which has great functional utility in predicting workload and task interference.

Whereas the searchlight metaphor emphasizes the unity of attention, the *resource metaphor* emphasizes its divisibility. When performing any task, different mental operations must be carried out (responding, rehearsing, perceiving, etc.), and performance of each requires some degree of the operator's limited processing resources. Since these resources are limited, the resource metaphor readily accounts for our failures of time-sharing. Two activities will demand more resources than a single activity, and so there will be a greater deficiency between supply and demand. The resource metaphor also proposes that some operations may require resources that are different from others (just

364

as some furnaces utilize oil and others natural gas or coal). As a consequence, there is less competition between these processes for their enabling resources, and time-sharing between them may be more successful. An example of separate resources in Chapter 6 showed the distinction between verbal and spatial working memory (Baddeley & Hitch, 1974). Different resources underlie the two memory codes, and so there is relatively effective time-sharing between two tasks using the two types of memory. This chapter will focus on resources as the most important but not the only mechanism that determines how effectively two or more tasks can be time-shared.

How do we time-share when we must perform two or more activities in a short period of time? There are probably several mechanisms that determine our successes and failures at this endeavor. First, it is clear that *scheduling* and efficient *switching* between activities are important ingredients to success. If given ten minutes to perform two five-minute tasks, the operator will achieve success if full use is made of the available time (efficient scheduling) and no time is wasted in switching from one activity to the next. However, if given only seven minutes to complete the two five-minute tasks, the person may be forced to engage in *concurrent* processing. If this is the case, three further factors will influence the effectiveness of multiple-task performance: confusion of task elements, cooperation between task processes, and competition for task resources.

Confusion results when elements for one task become confused with the processing of another task because of their similarity. Thus, if we are trying to listen to two speakers at once, their voices are of similar quality, and they are discussing a similar topic, it is quite likely that we will mistakenly link the words of one to the message of the other. As noted in our discussion of auditory attention in Chapter 3, these confusions will diminish if the physical and semantic characteristics of two messages are more different.

Sometimes the high similarity of processing routines can result in *cooperation* or even an integration of the two task elements into one. Thus, for example, when the pianist or drummer must implement two separate response streams, it is easier if they share a common rhythm (Klapp, 1978).

Switching, confusion, and cooperation are all important components of multiple task performance. Indeed we have discussed the role of optimal task switching and scheduling in Chapter 3 and will address it again later in this chapter. However, a critical element of concurrent task time-sharing is related to task *difficulty*. We can time-share driving and conversation on an open freeway, but on a crowded freeway our conversation will deteriorate because driving in heavy traffic is more difficult. The resource metaphor of attention emphasizes the quantitative, intensive aspects of attention demands. The demand for resources is determined by the difficulty of the task confronting the operator as well as by its priority. We first discuss a resource model of time-sharing that does not deal with the structural aspects of the tasks (e.g., whether the voice or hands are used to respond) and then a multiple-resource model that does account for the structural characteristics of both the task and of the human operator (Wickens, 1984, 1989, 1991).

MECHANISMS OF TIME-SHARING

Emphasis on the quantitative properties of attention owes much to an important paper by Moray (1967), who proposed that attention was like the limited
processing capacity of a general-purpose computer. This capacity could be
allocated in graded amounts to various activities depending on their difficulty
or demand for that capacity. The capacity concept emphasizes both the flexible
and the sharable nature of attention or processing resources. Tasks demand
more of these hypothetical resources (attention or mental effort) as they become more difficult or their desired level of performance increases. With fewer
resources available for other tasks, performance will deteriorate.

The concept of attention as a flexible, sharable, processing resource of
limited availability was extended by Kahneman (1973), Navon and Gopher
(1979), and Norman and Bobrow (1975).

Single-Resource Theory

Kahneman (1973) states some of the predictions of the resource or capacity
concept of human attention. In the early chapters of his comprehensive book,
he proposes that there is a single undifferentiated pool of such resources,
available to all tasks and mental activities, as shown in Figure 9.1a. (This is a
position that he qualifies later in the book.) As task demands increase either by
making a given task more difficult or by imposing additional tasks, physiological
arousal mechanisms produce an increase in the supply of resources. However,
this increase is insufficient to compensate entirely for the increased resource
demands (as shown by the supply-demand curve in Figure 9.1b). Thus performance falls off as the supply-demand shortfall increases, and physiological manifestations of increased arousal such as heart rate and pupil diameter are evident
as indexes of resources mobilization. These physiological indicators of resource
demand, or *mental workload*, will be discussed in more detail later in the
chapter.

The Performance-Resource Function   Norman and Bobrow (1975) introduced the important concept of the *performance-resource function*. If two tasks
do in fact interfere with one another (are performed less well) because they are
sharing resources to which each previously had exclusive access, there must be
some underlying function that relates the quality of performance to the quantity of resources invested in a task. This hypothetical function is the performance-resource function, or PRF, an example of which is shown in Figure 9.2.
Single-task performance occurs when all resources are invested in the task
(point *A*) and is the best that can be obtained. Diverting a large amount of
resources away from the task to a concurrent one will depress performance
accordingly, as indicated by point *B*. As more resources are then reinvested
back into the task, performance will improve up to the point *C* at which no
further change in performance is possible. To the right of point *C*, the task is
said to be *data-limited* (limited by the quality of data, not by the resources

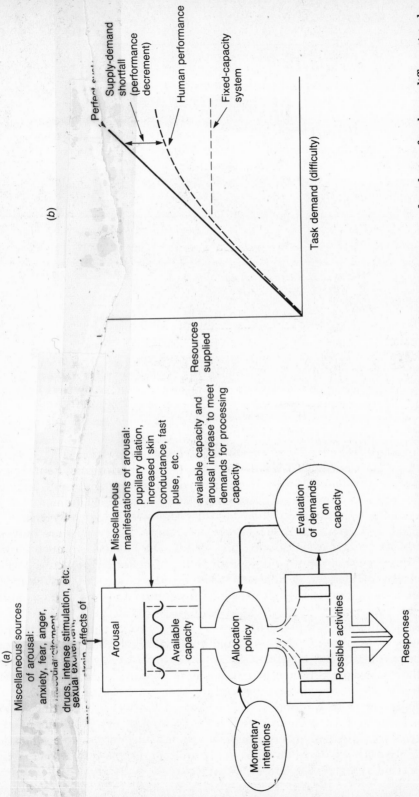

Figure 9.1 (a) Relationship among capacity demands, arousal, and performance. This conception is a foundation for the undifferentiated capacity view of resources. (b) Hypothetical relationship between resources demanded with increasing task difficulty and resources supplied. (*Source:* Daniel Kahneman, *Attention and Effort* (Englewood Cliffs, NJ: Prentice-Hall, 1973), pp. 10, 15. Copyright 1973 by Prentice-Hall. Reprinted by permission.)

367

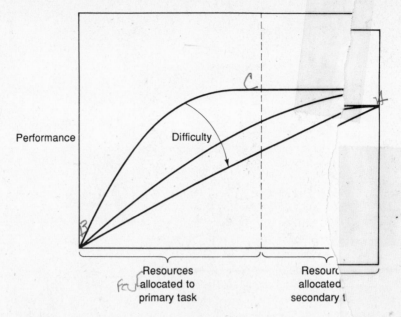

Figure 9.2 A hypothetical performance-resource function. Poin
and C are described in the text. (Source: D. Norman and D. B, *B*,
"On Data-Limited and Source-Limited Processing," *Cognitive P.* ow,
ogy, 7 (1975), p. 49. Copyright 1975 by the American Psychol *ol-*
Association. Adapted by permission of the authors.   cal

invested). Data limits may occur at any level of performance
two-digit number is data-limited since perfect performance Remembering a
with few resources and further effort will lead to no improv an be obtained
    But data limits also occur in vigilance with low-intens ent.
understanding a conversation in a language with which you signals or in
familiar. In both cases, no matter how hard you try (how re only faintly
invest), perfect performance is impossible, and beyond a partich effort you
formance gains will not be realized with more effort. When pelar level, per-
change with added or depleted resources, the task is said to formance does
ited, the region to the left of point C in Figure 9.2.   e resource-lim-

Time-Sharing and the PRF   If all resources are presumed t
same reservoir (i.e., they are undifferentiated), the amount ome from the
between two tasks is determined by the form of the two PRFs. interference
the PRF for the bottom task (*B*) is plotted backward so that a giv Figure 9.3*a*
through both functions will represent a single policy of alloca vertical slice
resources to task *A* and (100 − *X*) percent to task *B*. Maxim g *X* percent
performance on each task is given by 100 percent resource single-task
resources are divided between tasks, performance must drop off llocation. If
unless both are data-limited as in Figure 9.3*b*. In this case the a one or both
percent resources to task *A* and 60 percent to task *B* will location of 40
                                                             vide perfect

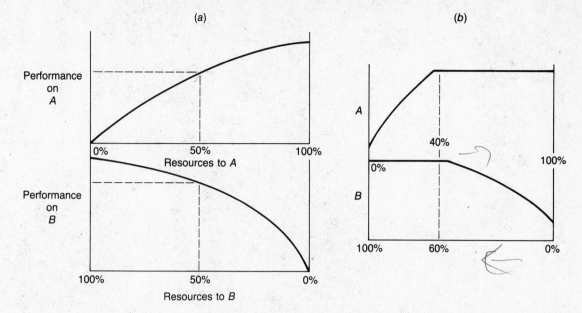

Figure 9.3 Performance-resource functions of two time-shared tasks. (*a*) Two resource-limited tasks with 50/50 allocation policy. (*b*) Two data-limited tasks with 40/60 allocation policy, demonstrating perfect time-sharing.

time-sharing. That is, both tasks performed concurrently can be done as well as either task performed alone.

Allocation of Resources: The Performance Operating Characteristic In the discussion of Figure 9.3 we have assumed that subjects can *allocate* their resources flexibly between tasks in any proportion desired. That is, they need not adopt a 50/50 split but can choose any other proportional allocation. This ability has been well documented in a number of investigations (e.g., Gopher, Brickner, & Navon, 1982; Gopher & Navon, 1980; Schneider & Fish, 1982; Sperling & Dosher, 1986; Wickens, Sandry, & Vidulich, 1983). In fact, when two tasks are time-shared and the subject is asked to adopt different strategies of allocation on successive trials, it is possible to cross-plot performance of the two tasks on a single graph. This curve is called a *performance operating characteristic, or POC* (Norman & Bobrow, 1975; Wickens & Yeh, 1985). A POC constructed from the two PRFs of Figure 9.3*a* is shown in Figure 9.4*a*. The POC has proven to be a useful way of summarizing a number of characteristics of two time-shared tasks. Hence it is important to describe certain important landmarks, or characteristics, of the POC.

1. *Single-task performance* is shown by points on the two axes (*A* and *B*) that have a hypothetical intersection in the POC space at *P*. This point represents perfect time-sharing. As shown in Figure 9.4, the single-task points may

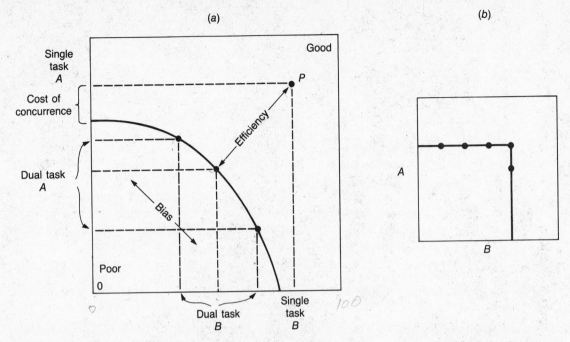

Figure 9.4 The performance operating characteristic (POC). (*a*) Two resource-limited tasks shown in Figure 9.3*a* with three allocation policies. (*b*) Two data-limited tasks shown in Figure 9.3*b* with five allocation policies.

not be continuous with the extension of the POC to the single-task axes. If, as in Figure 9.4, these single-task points are higher (better performance), there is, in the words of Navon and Gopher (1979), a "cost of concurrence." The act of time-sharing itself pulls resources away from both tasks above and beyond the resources that each task demands by itself. Thus, time-sharing, even with no resources allocated to the other task, produces worse performance than the single-task condition. This cost of concurrence will be imposed, for example, if the two tasks are displayed at different locations in the visual field. Thus both cannot be fixated simultaneously. The cost of concurrence may also reflect the resource demands of an executive time-sharing mechanism that is responsible for coordinating responses, sampling display locations, and deciding how to allocate resources. Such a mechanism will be called into play only in time-sharing conditions (Allport, 1980; Hunt & Lansman, 1981; McLeod, 1977).

2. The *time-sharing efficiency* of the two tasks is indicated by the average distance of the curve from the origin (*O*); obviously the farther from the origin, the more nearly dual-task performance is close to the single-task performance point *P*. This is efficient time-sharing.

3. The *linearity*, or smoothness, of the function indicates the extent of shared or exchangeable resources between the tasks (Sperling & Dosher, 1986). A curve such as shown in Figure 9.4*a* is smooth, and so indicates that a

given number of hypothetical units of resources removed from task A (thereby decreasing its performance) can be transferred to and efficiently utilized to improve the performance of task B. A discontinuous or "boxlike" POC (Figure 9.4b) suggests that resources are not as interchangeable. When resources are withdrawn from one task, they cannot be used to improve performance of the other one. This situation may occur when there are data limits in the tasks. The POC shown in Figure 9.4b would result from the shift in resource allocation across the PRFs of the two data-limited tasks shown in Figure 9.3b. A second cause of the boxlike POC is the possibility that different resources may be used in the two tasks. This possibility will be considered in some detail later when we discuss multiple resources.

4. Since the POC is actually a series of points, each one collected in a different time-sharing trial, the *allocation bias* of a given condition is indicated by the closeness of a given point on the POC to one axis over the other. A point on the positive diagonal indicates an equal allocation of resources between tasks.°

Automaticity and Difficulty   The effects of both practice and task difficulty can be easily represented by the resource metaphor. Figure 9.5 shows the PRF represented by two tasks, A and B. Task B demands fewer resources to reach

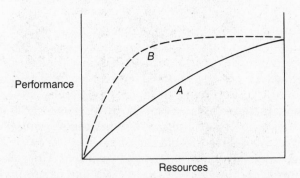

Figure 9.5 The performance resource function and practice. Task B: practiced or easy; task A: unpracticed or difficult.

°When the two tasks in a POC are the same, with the same performance measure, there is no problem with the assumption that equal allocation of resources lies on the positive diagonal. However, when the tasks are different, with different dependent variables (e.g., tracking and reaction time), a question arises concerning how to scale the two axes into common units so that spatial relations along the two axes may be meaningfully compared (Kantowitz & Weldon, 1985). One possible solution is to convert both to common dimensionless units such as standardized scores (Wickens, Mountford, & Schreiner, 1981; Wickens & Yeh, 1985).

performance levels that are equivalent to A. Task B also contains a greater data-limited region. Task B then differs from A by being of lesser difficulty or having received more practice (being more automatic). Note that task B may not necessarily be performed better than task A if full resources are invested into A but will simply be performed at that level with more spare capacity. Hence task B will be less disrupted by diverting resources from its performance than will task A. We saw in Chapters 3 and 5 that the best way to produce this automaticity is through repeated practice on tasks with consistently mapped characteristics.

The implication of the difference between PRFs such as those shown in Figure 9.5 is that the extent of differences between two tasks or two situations may not be appreciated by examining the performance on each task alone. Only when the primary task (i.e., the task of interest) is time-shared with a concurrent task will the differences be realized. When the primary task is emphasized in a dual-task situation, the concurrent task is called a *secondary task.* Thus, for example, Bahrick, Noble, and Fitts (1954) employed the secondary task to index differences in learning of a perceptual-motor task that were not revealed by primary-task performance. Dornic (1980) compared comprehension of first and second languages by bilingual speakers and observed differences in the secondary tasks and none in the primary. The use of secondary tasks as one means to assess mental work load will be considered in greater detail later in this chapter.

Performance Strategies and the PRF   The concept of the PRF—the relationship between effort and performance—has been encountered before in the discussion of decision-making heuristics in Chapter 7. Heuristics are seen as mental shortcuts that can provide reasonably good performance without the investment of too much effort. For example, in choosing between several options, a heuristiclike elimination by aspects might show a PRF such as that shown by the solid line in Figure 9.6. An optimal *compensatory* strategy, in which all attributes of all options are considered, might show a PRF such as that shown by the dashed line. Which strategy will be chosen? The answer is given by considering the role of *utility* in relation to effort and performance. Assuming that there is a positive utility to good performance, but also a positive utility to conserving effort (prolonged investment of high effort is fatiguing), we may think of the PRF space as having a "good" region (low effort, high performance at the upper left) and a "bad" region (high effort, low performance in the lower right). Navon and Gopher (1979) speak of lines of constant utility radiating from the origin of the PRF (these are the light lines shown in the figure). Accordingly, people will choose to operate at points farther to the upper left in the PRF space. As we can clearly see in Figure 9.6, the heuristic PRF has a region farther to the upper left than does the optimal strategy PRF—which can explain why people will choose to use heuristics rather than optimal strategies.

Some models of decision making have directly incorporated concepts of effort into their computations (Bettman, Johnson, & Payne, 1990; Payne, Bett-

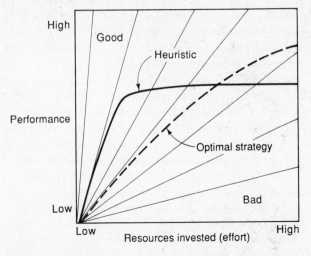

Figure 9.6 Relationship between heuristic and optimal strategies for performing the same task, as represented in the PRF space.

man & Johnson, 1988). But the utility of the trade-off between performance and effort is also relevant to other domains as well. For example, Soede (1980) and Shingledecker (1989) discuss the role of effort in the use of prosthetic devices for the handicapped. Very effective devices (yielding high performance) may not be chosen because of their high demand for effort. The trade-off is equally relevant for understanding why certain powerful but complex features in computing systems are not used. Indeed, the rationale behind many of the direct manipulation techniques discussed in Chapter 4 was not that they offered greater potential performance levels but rather that their performance could be obtained with reduced effort. The issue of how to measure the effort required for task performance—the measurement of mental workload—will be discussed later in this chapter.

Limitation of the Single Resource Theory   A major limitation with single-resource theory is that it cannot account for several aspects of the data from dual-task interference studies (Wickens, 1980, 1984). As noted, the concept of a resource translates closely to that of difficulty. Tasks of greater difficulty performed at the same level of performance demand more resources. Yet examples abound in which interference between tasks is predicted not by their difficulty but by their structure (e.g., the stages, codes, and modalities of processing required). For example, Wickens (1976) found that performance on a manual tracking task (a task similar to flying an aircraft, discussed in detail in Chapter 11) was more disrupted by a concurrent task requiring a pure response (maintaining constant pressure on a stick) than by auditory signal detection, even though the latter was judged by subjects to be the more difficult task and therefore presumably demanded more resources. We may describe this example as a case of *difficulty-structure uncoupling*.

There are other examples in which increases in the difficulty of one task, which should presumably consume more resources (as allocation is held constant), fail to degrade the performance of a second task. Such *difficulty insensitivity* is seen in a study by Wickens, Sandry, and Vidulich (1983), who asked subjects to time-share a tracking task with a verbal RT task. In one condition, the RT stimuli were presented auditorily and yes-no responses were given by speech. In spite of the effort demands imposed by both tasks, an increase in tracking difficulty produced no increase in mutual interference. However, this difficulty insensitivity disappeared, and the difficulty-performance trade-off predicted by resource theory reappeared, when visual input and manual responses were employed with the RT task. This difference will have important implications for multiple-resources theory, as described below.

Another class of results not readily explained by an undifferentiated capacity viewpoint is that in which two tasks that are both clearly attention demanding can be *time-shared perfectly*. This success is possible because the two tasks have separate structures. Allport, Antonis, and Reynolds (1972), for example, showed that skilled pianists could sight-read music and engage in verbal shadowing with no disruption of either task by the other. Shaffer (1975) reports an experiment in which a skilled typist can transcribe a written message and engage in verbal shadowing, again with perfect time-sharing efficiency.

An explanation for some of these phenomena is that the characteristic that allows for perfect time-sharing and difficulty insensitivity is simply the shape of the performance-resource functions. As shown in Figure 9.3*b*, PRFs with large data limits for one or both tasks should allow either of these phenomena to occur. Indeed, it is true that much of the variation in the efficiency with which any two tasks may be time-shared is due to the automation of one or both of the component tasks (Hunt & Lansman, 1981; Schneider & Fish, 1982; Schneider & Shiffrin, 1977). Yet this argument cannot explain the differences in the amount of interference or the degree of difficulty-performance trade-off when the structure or processing mechanism of one of the paired tasks is altered. These changes are referred to as *structural alteration effects* (Wickens, 1980) and were encountered in Chapter 6 in the discussion of the changes in interference within the same working memory task when voice rather than manual responses were used (Brooks, 1968). In the study described above, Wickens, Sandry, and Vidulich (1983) observed a reduction in interference and in the difficulty-performance trade-off between tracking and reaction time when the input of the latter task was changed from visual to auditory and when the response was changed from manual to vocal. Many other examples of structural alteration effects are identified by Wickens (1980). Collectively, the phenomena of difficulty-structure uncoupling, difficulty insensitivity, perfect time-sharing, and structural alteration effects are consistent with a multiple-resource conception of human attention.

Multiple-Resource Theory

The multiple-resource view argues that instead of one single supply of undifferentiated resources, people have several different capacities with resource properties. Tasks will interfere more and difficulty-performance trade-offs will be more likely to occur, if more resources are shared. This position has been proposed by Allport (1980), Kantowitz and Knight (1976), Kinsbourne and Hicks (1978), and McLeod (1977) but has received the most explicit theoretical development within the framework of the POC by Navon and Gopher (1979). Wickens (1980, 1984), drawing on the results from a large number of dual-task studies, has argued that, at one level, resources may be defined by three relatively simple dichotomous dimensions: two stage-defined resources (early versus late processes), two modality-defined resources (auditory versus visual encoding), and two resources defined by processing codes (spatial versus verbal).

Figure 9.7 presents the dimensional representation of multiple resources. To the extent that any two tasks demand separate rather than common resources on any of the three dimensions, three phenomena will occur: (1) Time-sharing will be more efficient. (2) Changes in the difficulty of one task will be less likely to influence performance of the other; that is, difficulty insensitivity will be observed. (3) The POC constructed between the tasks will

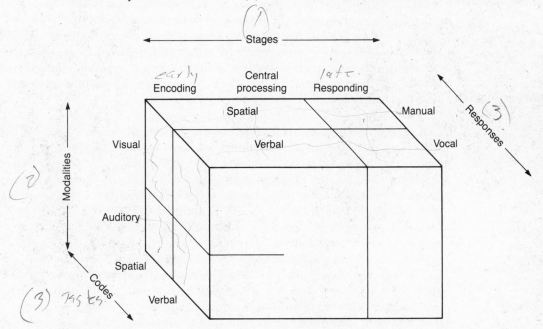

Figure 9.7 The proposed structure of processing resources. From "Processing Resources in Attention" by C. D. Wickens, 1984, in R. Parasuraman and R. Davies (eds.), *Varieties of Attention*, New York: Academic Press. Reproduced by permission.

be more of the boxlike form of Figure 9.4*b* because resources withdrawn from one task cannot be used to advantage by the other since it depends on different resources (Gopher & Navon, 1980; Wickens, 1984, 1991). The nature of the three dimensions that constitute the multiple-resource model will now be considered in more detail.

Stages The resources used for perceptual and central-processing activities appear to be the same, and they are functionally separate from those underlying the selection and execution of responses. Evidence for this dichotomy is provided when the difficulty of responding in a task is varied and this manipulation does not affect performance of a concurrent task whose demands are more perceptual in nature. In a series of experiments by Isreal, Chesney, Wickens, and Donchin (1980) and Isreal, Wickens, Chesney, and Donchin (1980), the amplitude of an evoked brain potential (see Chapter 8) elicited by a series of counted tones is assumed to reflect the investment of perceptual and central-processing resources since the evoked potential can be elicited without requiring any overt responses. The experiments found that the evoked potential is not sensitive to response-related manipulations of tracking difficulty but is influenced by manipulations of display load. Shallice, McLeod, and Lewis (1985) have examined dual-task performance on a series of tasks involving speech recognition (perception) and production (response) and have concluded that the resources underlying these two processes are separate.

As an operational example of separate stage-defined resources, we would predict that the added requirement for an air traffic controller to acknowledge vocally or manually each change in aircraft state (a response demand) would not disrupt his or her ability to maintain an accurate mental model of the airspace (a perceptual-cognitive demand).

Perceptual Modalities It is apparent that we can sometimes divide attention between the eye and ear better than between two auditory channels or two visual channels. That is, cross-modal time-sharing is better than intramodal. Wickens, Sandry, and Vidulich (1983) found advantages to cross-modal over intramodal displays in both a laboratory tracking experiment and a fairly complex flight simulation. Parkes and Coleman (1990) found that discrete route guidance was better presented auditorily than visually while subjects were concurrently driving a simulated vehicle. Wickens (1980) reviews several other studies that report similar advantages.

However, the relative advantage of cross-modal (auditory-visual or **AV**) over intramodal (**VV** and **AA**) time-sharing may not really be the result of separate perceptual resources but rather the result of peripheral factors that place the two intramodal conditions at a disadvantage. Thus, two competing visual channels, if they are far enough apart, will require visual scanning between them—an added cost, discussed in Chapter 3. If too close together, they may impose confusion and masking, just as two auditory messages may mask one another. The degree to which peripheral rather than central factors are responsible for the examples of better cross-modal time-sharing remains uncertain, and when visual scanning is carefully controlled, cross-modal displays do not always produce better time-sharing (Wickens & Liu, 1988). However, in most applied settings, visual scanning is enough of a factor that dual-task interference can be reduced by off-loading some information channels from the visual to the auditory modality. And simultaneous auditory messages are difficult enough to process that an advantage can usually be gained by displaying one of them visually (Rollins & Hendricks, 1980).

Processing Codes The role that the spatial and verbal codes play in defining separate processing resources has been discussed in some detail in previous chapters: spatial and verbal perceptual processes in Chapters 4 and 5, the degree of interference within and between spatial and verbal working memory in Chapter 6, and the compatibility of these two codes with information display in Chapter 8. The data indicate that spatial and verbal processes, or *codes*, whether functioning in perception, working memory, or response, depend on separate resources (Polson & Friedman, 1988). An experiment by Kinsbourne and Hicks (1978) shows that performing a tracking task with the right hand disrupts a verbal task more than performing it with the left hand, and it suggests that the code-related resource dimension may be defined, in part, by the two cerebral hemispheres. Resources underlying spatial processing and left-hand control reside predominantly in the right hemisphere. Resources underlying verbal processing, speech responses, and right-hand control reside more in the left.

The separation of spatial and verbal resources seemingly accounts for the high degree of efficiency with which manual and vocal outputs can be time-shared, assuming that manual responses are usually spatial in nature and vocal ones are verbal. In this regard investigations by Martin (1989); McLeod (1977); Tsang and Wickens (1988); Vidulich (1988); Wickens (1980); Wickens and Liu (1988); and Wickens, Sandry, and Vidulich (1983) have shown that tracking and a discrete verbal task are time-shared more efficiently when the latter employs vocal as opposed to manual response mechanisms. Discrete manual responses using the nontracking hand appear to interrupt the continuous flow of the tracking response, whereas discrete vocal responses leave this flow untouched.

Finally, consider the near perfect time-sharing efficiency with which shadowing and visual-manual transcription tasks (typing and sight-reading) can be carried out, as demonstrated by Shaffer (1975) and by Allport, Antonis, and Reynolds (1972), respectively. This success is clearly due in large part to the separation of codes and modalities of processing of the two tasks. If the auditory shadowing and piano sight-reading task pair investigated by Allport, Antonis, and Reynolds are examined within the framework of Figure 9.7, we see that auditory shadowing is clearly auditory, verbal, and vocal. Piano sight-reading is visual and manual. If the further assumption is made that music involves right-hemispheric processing (Nebes, 1977), the two tasks may be considered to require predominately separate resources.

An important practical implication of processing codes is the ability to predict when it might or might not be advantageous to employ voice versus manual control. As noted by Brooks (1968) and confirmed in a more applied context by Wickens and Liu (1988), manual control may disrupt performance in a task environment imposing heavy demands on spatial working memory (e.g., driving), whereas voice control may disrupt performance of tasks with heavy verbal demands. Thus, for example, the model predicts the potential dangers of manual dialing of cellular phones, given the visual, spatial, and manual demands of vehicle driving, and it suggests the considerable benefits to be gained from voice dialing. This being said, it is important to recall that verbal tasks may be most *compatibly* responded to with voice control and spatial tasks with manual control, as discussed in Chapter 8. Occasionally, then, the choice between control types may be dictated by a consideration of trade-offs between compatibility and resource competition, addressed explicitly by Wickens et al. (1984; Wickens, Sandry, & Vidulich, 1983, Wickens, Vidulich & Sandry-Garza, 1984). Research generally indicates, however, that resource conflict, rather than compatibility, is the more dominant of the two forces in dual-task situations (Goettl & Wickens, 1989).

Task Similarity and Time-Sharing

The three dimensions of the multiple-resource model are not intended to account for all structural influences on dual-task performance and time-sharing efficiency. There are other entities in the human processing system that have some resourcelike processes (better time-sharing between than within). Indeed, multiple-resource theory has sometimes been criticized for the fuzziness of defining what a resource is (Hirst & Kalmer, 1987). As we will see, many of these other similarities in processing can be used to predict task interference differences in system design (see Wickens, 1989, 1991, for further discussion). However, stages, modalities, and codes do represent three major dichotomies that can account for a reasonably large portion of these influences and can be readily used in predicting task interference. Furthermore, the loadings on these three dimensions can be fairly easily established a priori by a system designer. Thus visual, auditory, print, and speech input channels are easy to define, as are voice versus manual controls. Task analysis can usually reveal the extent of linguistic or phonetic requirements in working memory (verbal code) and the extent to which demands are perceptual, cognitive, or response related. In this sense, the model is an effort to gain usability and parsimony by sacrificing some degree of precision (see Chapter 1).

It is important, however, to consider other mechanisms that may influence time-sharing efficiency. As discussed in more detail in Chapter 3, efficient scheduling and switching are important components when two tasks are performed more or less in sequence. Another important component is related to similarity. In fact, in a global sense, the effects of similarity on dual-task performance seem to act as a two-edged sword, which can sometimes enhance the efficiency of time-sharing and sometimes degrade it.

Cooperation The enhancement of time-sharing by increasing similarity results from circumstances in which a common mental set, processing routine, or timing mechanism can be activated in the service of two tasks that are performed concurrently. Thus, there is some evidence that the performance of two tracking tasks is better if the dynamics on both axes are the same than if they are different, even if like dynamics are produced by combining two more difficult tasks (Chernikoff, Duey, & Taylor, 1960) (see Chapter 11). Even when the performance of two identical but difficult tasks is not actually better than the performance of a difficult-easy pair, performance of the difficult pair is less degraded than would be predicted by a pure resource model (Braune & Wickens, 1986; Fracker & Wickens, 1989). That is, there is an advantage for the identity of two difficult dynamics, which compensates for the cost of their difficulty.

A similar phenomenon, discussed also in Chapter 8, has been observed in the domain of choice reaction time by Duncan (1979). He observed better time-sharing performance between two incompatible RT tasks than between a compatible and an incompatible one, in spite of the fact that the average difficulty of the incompatible pair was greater. Here again, the common rules of mapping helped performance. Finally, a series of investigations have pointed to the superior time-sharing performance of two rhythmic activities when the rhythms are the same rather than different (Klapp, 1979; Klapp, 1981; Peters, 1981).

These three examples illustrate that similarity in information-processing routines leads to cooperation and facilitation of task performance, whereas differences lead to interference, confusion, and conflict, an issue we will address in more detail shortly. Other aspects of identity and cooperation are also reflected in a kind of resonance or compatibility between similarity at one stage of processing and similarity at another. This resonance is described by the proximity compatibility principle (Barnett & Wickens, 1988), discussed in the context of object displays in Chapter 3 and graphs in Chapter 4. Thus, for example, Fracker and Wickens (1989) and Chernikoff and Lemay (1963) found that two tracking tasks benefited from an integrated display if they shared similar dynamics, but not if the dynamics were different. Fracker and Wickens also found that two tracking tasks using separate dynamics benefited from having separate control sticks, an issue that will be discussed further in Chapter 11.

Confusion We have discussed ways in which increasing similarity of processing *routines* can bring about improved dual-task performance. A contradictory trend, in which the increasing similarity of processing *material* may reduce rather than increase time-sharing efficiency, is a result of *confusion*. For example, Hirst and Kalmar (1987) found that time-sharing between a spelling and mental arithmetic task is easier than time-sharing between two spelling or two mental arithmetic tasks. Hirst (1986) showed how distinctive acoustic features of two dichotic messages can improve the operator's ability to deal with each separately. Carswell (1990) found increased dual-task interference when the stimuli for the two tasks are part of the same perceptual object rather than of two different objects. Many of these confusion effects may be closely related to interference effects in memory, discussed in Chapter 6. Indeed, Venturino (1991) has shown similar effects when tasks are performed successively, so that the memory trace of one interferes with the processing of the other.

Although these findings are similar in one sense to the concepts underlying multiple-resources theory (greater similarity producing greater interference), it is probably not appropriate to label these elements as resources in the same sense as stages, codes, and modalities since such items as a spelling routine or distinctive acoustic features hardly share the physiologically based energetics characteristic of the dimensions of the multiple-resources model (Wickens, 1986, 1991). Instead, it appears that interference of this sort is more likely based on confusion or a mechanism that Navon (1984; Navon & Miller, 1987) has labeled *outcome conflict.* Responses (or processes) relevant for one task are activated by stimuli for a different task, producing confusion or cross talk between the two. The most notorious example of this phenomenon is in the Stroop task, discussed in Chapter 3, in which the semantic characteristics of a color word (*white* or *blue*) interfere with the subjects' ability to report the color of ink in which the word is printed. The necessary condition for confusion and cross talk to occur is similarity. In the Stroop task, there is similarity both in the common location of the two stimulus properties and in their common reference to color. Thus Stroop interference may be lessened either by reducing the physical similarity (increased distance) between the two attributes or by increasing the "semantic distance" between the color and semantic properties of the word (Klein, 1964). Thus in manipulating physical distance, Stroop interference is reduced when the color word is placed next to a color patch rather than printed in the color ink (Kahneman & Chajczyk, 1983). In semantic distance, color-related words like *sky* or *grass* produce some but reduced Stroop interference, whereas color-neutral words like *will* or *five* produce very little interference at all.

However, the Stroop task is a focused, not a divided-attention, task. Are confusion and cross talk also mechanisms that cause dual-task interference? Experiments by Fracker and Wickens (1989) and by Navon and Miller (1987) suggest that they have at least some role. Navon and Miller studied subjects' abilities to categorize simultaneously words in two visually displayed word sequences. They found that items on one sequence that were similar to those on the other slowed down the response to the other, as if producing an outcome conflict. Fracker and Wickens asked subjects to perform two tracking tasks at the same time. By looking at the time-series analysis between input and output signals of the two tasks, they could measure the degree of cross talk, in which the error of one task was compensated for by an unwanted control response of the hand controlling the other task. Generally this cross talk was small, but when the displays and controls of the two axes were made more similar by integrating them, the cross talk increased. However, this increase was not the source of increased tracking error, nor was there an increase in cross talk when the resource demands of one of the tasks was increased, producing greater interference.

In summary, although confusion due to similarity certainly contributes to task interference in some circumstances, it is not always present nor always an important source of task interference. Its greatest impact probably occurs when an operator must deal with two verbal tasks, requiring concurrently working memory for one and active processing (comprehension, rehearsal, or speech) for the other. In these cases, as discussed in Chapter 6, similarity-based confusions in working memory may play an important role.

A Note on Terminology

In concluding our discussion of the mechanisms of time-sharing, it is worthwhile taking time to define clearly and distinguish some key terms used here, in Chapter 3, and elsewhere to refer to attentional phenomena. Unfortunately, the definitions will not be approved by all readers because of the diversity of meanings in the field. However, it is hoped that the following may provide some standardization.

A *channel* has two characteristics: It defines the flow of information through some or all of the stages of processing, and it is characterized by some distinct perceptual property as its source. That perceptual property (like a location in space or a particular color) is not itself a part of the information conveyed along the channel (Garner, 1974) but rather serves as its identifier, or tag, in a particular time-sharing context (e.g., the right ear channel, the top-center display channel, or the red color channel). For an automobile driver, one channel is the flow of information regarding the car's lateral position on the highway. Two additional channels may be a news broadcast on the radio and conversation with a passenger. A fourth channel may be an analog speedometer on a head-up display.

Information may be processed by the operator along two (or more) channels in parallel or serially. The driver will process conversation and road position in parallel, but the driver must process in series the flow of vocal information from the radio and the passenger. However, parallel (concurrent) processing along two channels does not necessarily imply that the parallel processing is perfect (i.e., that the quality of transformation on one channel is unaffected by the presence of processing on the other). In particular, confusion or cross talk may occur between channels processed in parallel. Such confusion often causes outcome conflict. In driving, confusion might result if a sudden change in position of the HUD speedometer was processed as a lateral deviation of the car on the road or if words on the radio were mistakenly attributed to the conversation of the passenger.

The term *capacity* simply refers to a maximum or upper limit of processing capability. It has been (and will continue to be) used broadly to describe such items as the channel capacity of absolute judgment, the capacity of working memory, or the bandwidth capacity to transmit information along a channel in bits per unit of time.

Capacity should not be confused with *resources,* which represent the mental effort supplied to improve the processing efficiency along a channel or the performance of any other mental operation (e.g., rehearsal or generation of mental images). Resources are characterized by two general properties: Their deployment is under voluntary control and they are scarce, so allocation of resources to one process to increase its efficiency has a cost elsewhere, in either long-term fatigue or reduced performance on a concurrent operation. Thus, the heavy resource demands of driving in poor visibility leave fewer resources available to concentrate on the passenger's voice and will also produce a buildup of fatigue. There is more than one kind of resource, so scarcity will be more obvious between two tasks sharing a common resource—such as driving and mentally rotating a map (both spatial) or steering while tuning a radio (both response)—than between two tasks utilizing separate resources—such as steering while listening to the conversation on the radio.

Separate resources are characteristically defined by distinct anatomical structures or regions in the brain with their own activation mechanisms (Friedman & Polson, 1981; Gopher & Sanders, 1985) in addition to time-sharing criteria. Other aspects of similarity between tasks may also affect their time-sharing efficiency, such as the similarity of the properties that define their channels or the overlap in the vocabulary or semantic content of their information. For example, the extent to which we wish to label French and English processors or mental arithmetic and spelling as "separate resources," rather than "other aspects of similarity," is not clear-cut, but we argue here that this labeling should be based jointly on how consistently the distinction meets the performance criteria of separate resources and how readily different anatomical structures can be associated with the pair of entities in question.

Finally, *similarity* is clearly a multidimensional concept, with *identity* (between two tasks or channels) at one end and many dimensions of difference (and numerous levels along each dimension) varying away from identity. Greater similarity along any and all dimensions will be more likely to increase confusion and cross talk, the consequence being greater interference. For example, the radio and passenger voices will more likely be confused if both are male speakers, using the same language, with messages emanating from the same location in the vehicle than if either or all of these dimensions are different. Changes in speed displayed on a HUD will more likely be confused with changes in lateral position if the former are presented in analog (rather than digital) format and if they are displayed head up, superimposed on the roadway, rather than head down.

These definitions and distinctions will undoubtedly leave some readers uncomfortable because of their fuzziness and lack of precision. The alternative is to strive for precise exclusionary definitions; but the human processing system is sufficiently complex, and behavioral phenomena sufficiently varied, that greater crispness and precision of definitions will only lead to more qualifications.

INDIVIDUAL DIFFERENCES AND LEARNING IN TIME-SHARING

When observing the expert perform a complex task, whether juggling, doing secretarial work, flying an aircraft, or engaging in industrial inspection and assembly, the novice is often overwhelmed at the ease with which the expert can effectively time-share a number of separate activities. Indeed, across people and across age groups there is a considerable diversity in the efficiency with which tasks may be time-shared. Some of these differences may result from differences in practice and others from differences in a basic time-sharing ability.

Attention as a Skill

Continued practice in a dual-task environment will soon lead to improved dual-task performance. Naturally some component of this gain could result simply from an improvement in the single-task component skills. As the two skills demand fewer resources and become more data-limited through practice, their combined resource demand will be diminished, and their dual-task efficiency will improve accordingly (Schneider, 1985; Schneider & Detweiler, 1988). It should also be recalled that the change in single-task performance we call automaticity, discussed in Chapters 5 and 6, need not entail an increased level of single-task performance but only an increase in the data-limited region of a PRF that asymptotes at the same level (see Figure 9.5).

This form of improvement does not actually result from acquisition of an attentional or time-sharing "skill" but from a reduced resource demand of single-task performance. It may be acquired, in short, from extensive single-task practice. Such a mechanism explains the automatic processing of familiar perceptual stimuli such as letters (LaBerge, 1973), consistently assigned targets (Schneider & Fisk, 1982; Schneider & Shiffrin, 1977), and repeated sequences of stimuli (Bahrick, Noble, & Fitts, 1954; Bahrick & Shelley, 1958), as well as the automatic performance of habitual motor acts such as signing one's own name. When describing the response end of processing, this automaticity of performance is referred as the *motor program* (Keele, 1973; Summers, 1989), discussed more in Chapter 11.

Distinct from automaticity is the true skill in time-sharing that results explicitly and exclusively from multiple-task practice. To show that improvements in time sharing efficiency are due to the development of a time-sharing skill and not simply to increased automaticity of the component task, Damos and Wickens (1980) suggest that any of three procedures should be employed: (1) Show that time-sharing performance for a given task pair develops more rapidly with dual- than with single-task performance. (2) Show that a time-sharing "skill" developed with one task combination transfers to a qualitatively different time-sharing task combination. (3) Demonstrate, through a microscopic analysis of the timing of responses in dual-task performance, that changes in strategy develop that directly reflect differences in the manner in which the two tasks are interwoven and coordinated. As Moray and Fitter (1973) argue, such changes would indicate that time-sharing, switching, or sampling behavior is learned, similar to the optimal sampling prescriptions discussed in Chapter 3.

Damos and Wickens (1980) conducted an investigation of time-sharing skill development that emphasized these three procedures. They asked subjects to time-share two speeded tasks: a digit running memory task (see Chapter 6) and a digit categorization task (digit pairs were judged on their similarity of value and physical size). Performance on both tasks was assessed when they were performed together and on periodic single-task trials, spaced throughout roughly 25 trials of dual-task training. When dual-task training was terminated, the subjects then transferred to a dual-axis tracking task, and the same training procedure that was employed on the two discrete tasks was repeated.

Three kinds of analyses suggested that a unique time-sharing skill was developing: (1) The degree to which time-sharing on the dual-axis tracking task benefited from the prior exposure to the dual-task digit task was evaluated in a transfer of training design by comparing the performance of these subjects with that of a control group. The control group had corresponding single-task practice on the two discrete tasks but had never performed them together. The transfer was positive, suggesting that the earlier dual-task training of the transfer group did, in fact, develop a generalizable dual-task skill. (2) Microscopic analysis of dual-task performance of both the discrete task pair and the tracking pair revealed that practiced subjects engaged in more parallel processing of the stimuli. (3) Individual differences between subjects revealed some who alternated rapidly between the discrete tasks and others who processed both in parallel. The latter group consistently performed better.

The learning of perceptual sampling strategies related to time-sharing was discussed in Chapter 3 in regard to the experiments performed by Senders (1964) and Sheridan (1972). The notion that strategies in allocating and switching attention contribute to improved time-sharing performance has received further support from three more recent experiments. Gopher and Brickner (1980) observed that subjects who were trained in a time-sharing regime that successively emphasized different resource allocation policies became more efficient time-sharers in general than did a group trained only with equal priorities. The former group was also better able to adjust performance in response to changes in dual-task difficulty. As noted in Chapter 6, Gopher, Weil, and Siegel (1989) found that training that emphasizes attention control leads to better transfer to a complex multitask video game, and recent evidence suggests that similar training benefits later performance in military aircraft (Gopher, 1991).

Schneider and Fisk (1982) found that subjects could time-share an automatic and a resource-demanding letter-detection task with perfect efficiency if they received training to allocate their attention *away from* the automatic task. In the absence of this training, subjects allocated resources in a nonoptimal fashion by providing more resources to the automatic task than it needed, at the expense of the resource-limited task. In the example in Figure 9.3*b*, this situation was as if subjects initially allocated 60 percent of the resources to *A* and 40 percent to *B* and needed to be trained to adopt the more optimal 40/60 split. This paradigm is interesting in that it shows the contributions of both single-task automaticity (for the automatic detection task) *and* dual-task resource-allocation training to overall time-sharing efficiency.

It appears safe to conclude, therefore, that the very efficient time-sharing performance of the expert results not only from the more automated performance of component tasks but also from a true skill in time-sharing: knowing when to sample what from the display, when to make which response, and how to integrate better the flow of information in the two tasks. To what extent the time-sharing skill acquired in one environment is generalizable to others is not well established. Damos and Wickens (1980), as noted, did find some transfer. The amount, however, was not large relative to the amount of skill learning demonstrated by both groups on the new task. It seems, then, that most time-sharing skills that are learned are probably fairly specific to a given task combination and are not of the generic kind.

Attention as an Ability

Differences between individuals in time-sharing efficiency may to a large extent be related either to differences in automaticity of single-task skills or to the practice-related acquisition of time-sharing skills. Thus the observation made by Damos (1978) that flight instructors have greater reserve capacity than novices, as measured by the performance of a secondary task, was undoubtedly related to the greater degree of automaticity of the flight task for the instructors. However, Hunt and Lansman (1981), using an information-theory approach, provided evidence that some portion of the individual differences in dual-task performance is related directly to the amount of resources available and not to the level of skill on the component tasks. A similar conclusion was offered by Fogarty and Stankov (1982) on the basis of factor analysis of single- and dual-task scores. Gopher (1982); Gopher and Kahneman (1971); and Kahneman, Ben-Ishai, and Lotan (1973) report that measures of the flexibility of attention switching in a dichotic listening task predict both the success in flight training and the frequency of accidents encountered by bus drivers. Keele, Neill, and DeLemos (1977) report tentative evidence for a fairly general skill of attention switching that correlates across a number of different paradigms.

Given that there are individual differences in time-sharing skill, to what extent are they general — applicable across a wide variety of different dual-task combinations? Phrased in other terms, if there is such a "general" time-sharing ability, would it suggest that individuals who perform well on one dual-task combination will be more likely to perform better on a second combination involving entirely different component tasks?

The statistical and analytical problems involved in answering this question are numerous (Ackerman & Wickens, 1982). However, except for the attention-switching ability described by Keele, Neill, and DeLemos (1977) and more tentative data offered by Fogarty and Stankov (1982), there is not much evidence at this point for a general time-sharing ability. Wickens, Mountford, and Schreiner (1981), for example, examined differences in time-sharing ability of 40 subjects performing four different tasks in nine different pairwise combinations. Although there were substantial individual differences in the efficiency of time-sharing a given task pair, they did not correlate highly across the different task combinations. These findings, along with similar results obtained by Braune and Wickens (1986), Brookings (1987), Jennings and Chiles (1977), and Sverko (1977), suggest that what accounts for differences between individuals in time-sharing ability is related either to differences in automaticity of the component tasks or to differences in the ability to time-share a specific task pair. For example, Damos, Smist, and Bittner (1983) have identified a fairly stable difference in the ability with which people can process information in parallel between two discrete tasks. Some can easily process stimuli and responses of the two at the same time (but with some slowing), whereas others must deal with epochs of one task followed by epochs of the other. Once again, however, this dichotomy is probably somewhat task specific. It is not likely that it also accounts for differences in the ability of people to perform dual-axis tracking or to the ability to read one message and shadow another.

PRACTICAL IMPLICATIONS

The practical implications of research and theory on attention and time-sharing are as numerous as the cases in which a human operator is called on to perform two activities concurrently, and his or her limitations in doing so represent a bottleneck in performance. These instances include the pilot of the high-performance aircraft who may have a variety of component tasks simultaneously imposed; the process control or nuclear power plant monitor who is trying to diagnose a fault and is simultaneously deciding, remembering, and scanning to acquire new information; the musical performer who is attending to notes, rhythm, accompanist, and the quality of his or her own performance; the learner of any skill who must concurrently perceive different stimuli associated with a task, make responses, and process feedback; or the vehicle driver who must drive safely while operating a radio or navigational device.

As with any domain of human performance, there are three broad categories of applications of attention theory: to system and task design, to operator training, and to operator selection. The applications to the last two areas will be described only briefly. When operators must be trained for time-sharing in complex environments, such as the aircraft cockpit, equal attention must be given to the development of automaticity and to the training of time-sharing skills, issues that are discussed in considerable detail by Gopher (1991), Schneider (1985), and Schneider and Detweiler (1988). If operators are to be *selected* for task environments in which they will be required to engage in time-sharing activity, at issue is the extent to which tests can be derived that will predict success in time-sharing. As noted, whether such tests of a general sort can be found (e.g., attention switching) remains unresolved, and at present we must assume that the best tests will be those that impose demands similar to those imposed in the target environment.

The third applications category, which will concern us most, is that of predicting multiple-task performance imposed by different task environments or system design features. This issue is also closely tied to the issues of predicting and measuring the mental workload imposed by these design features.

Predicting Multiple-Task Performance

Using the structural and demand characteristics of models like the multiple-resources model, system designers may be concerned with either relative or absolute predictions of task interference (Hart & Wickens, 1990). *Relative predictions* allow the designer to know ahead of time which of two or more configurations will provide better multiple-task performance (this is closely related to the issue of which configuration will provide lower work load, a distinction to be made in the following section). If the computer user wishes to move a cursor while reading a screen, will the two activities interfere least with keyboard, voice, or mouse control (Martin, 1989)? Should automated navigational commands be presented to the automobile driver auditorily or visually (Parkes & Coleman, 1990)?

To these sorts of questions, multiple-resource theory can provide some answers (Wickens, Larish, & Contorer, 1989). For example, the adoption of voice recognition and synthesis technology (a talking, listening computer) may not provide many advantages—and may even be worse than a visual/manual interface if the information exchange is to be carried out in an environment in which the operator must rehearse other verbal material. As we have seen before, voice technology will offer its greatest benefits for reducing task interference if the concurrent task demands are heavily spatial, as in computer-based design (Martin, 1989).

The role of confusion in dual-task performance is also relevant to performance prediction. Where concurrent activity may be required, designers should not impose two different tasks using similar material. For example, the entering or transcription of digital data will be disrupted if others in the surrounding environment are currently speaking digits, but less so if others are speaking normal conversations.

An interesting application here concerns the presence of music as background or entertainment for operators engaged in various tasks. Multiple-resource theory would predict relative independence between the perception of music (more associated with spatial/analog processing) and the involvement with tasks that are manual and/or verbal. This prediction seems to be consistent with the results of a study by Tayyari and Smith (1987), who found no interference between "light orchestral" music listening (at up to 85 decibels) and visual-manual data entry (and in fact a slight improvement in the speed of data entry). Similarly, Martin, Wogalter, and Forlano (1988) found no interference between instrumental music and reading comprehension, but did find that comprehension suffered when lyrics were added, thus imposing a dual load on the verbal code.

Absolute predictions of task interference and task performance are somewhat more complex. This goal calls to mind the kind of question asked by the Federal Aviation Administration before certifying new aircraft: Are the demands imposed on the pilot excessive? If *excessive* is to be defined relative to some absolute standard, such as "80 percent of maximal capacity," an absolute question is being asked. Similarly, to resolve the debate over whether the Amtrak engineer should or should not be allowed to recite poetry, one must implicitly make an absolute workload assumption—that the two activities exceed the engineer's capability to drive safely.

A common approach to absolute workload and performance prediction is *time-line analysis*, which will enable the system designer to "profile" the workload of operators encountered during a typical mission, such as landing an aircraft or starting up a power-generating plant. In a simplified but readily usable version, it assumes that workload is proportional to the ratio of the time occupied performing tasks to total time available. If one is busy with some measurable task for 100 percent of a time interval, workload is 100 percent during that interval. Thus, the workload of a mission would be computed by drawing lines representing different activities, of length proportional to their duration. The total length of the lines would be summed and then divided by the total time (Parks & Boucek, 1989), as shown in Figure 9.8. In such a way the workload encountered by different members of a team (e.g., pilot, copilot, and flight engineer) may be compared and tasks reallocated if there is a great imbalance. Furthermore, epochs of peak workload or work overload, in which load is calculated as greater than 100 percent, can be identified as potential bottlenecks.

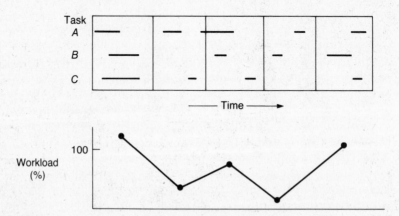

Figure 9.8 Time-line analysis. The percentage of workload at each point is computed as the average number of tasks per unit time within each window.

The basic research in multiple-task performance suggests at least four directions in which the time-line analysis described in Figure 9.8 should be extended. First, it is clear that operators may not necessarily choose to time-share two tasks (as dictated by a particular time line) if they have an opportunity to reschedule one or the other. Hence, effective time-line analysis should be coupled with models of task selection based on the kind of logic presented in Chapter 3. These kinds of advancement have been made in a number of predictive models reviewed by Wickens (1990b). Second, any time-line analysis must be sensitive to activities that are covert, such as planning or rehearsal, as well as the more overt actions that can be seen. Third, as discussed earlier, the analysis must be sensitive to the differences in resource demands of different tasks. Two time-shared tasks will not impose a 100 percent workload if they are easy but may very well exceed that value if they are difficult (Parks & Boucek, 1989). This resource demand feature has been incorporated into more recent versions of time-line analysis techniques (Aldrich, Szabo & Bierbaum 1989). Finally, an accurate time-line analysis model should incorporate the multiple-resources concept, recognizing that two tasks overlapping on the time-line could provide either very efficient performance or very disruptive performance, depending on their degree of resource conflict. This issue has been addressed in recent time-line models proposed by North and Riley (1989) and by Wickens, Larish, and Contorer (1989).

Assessing Mental Workload

Within the last two decades, the applied community has demonstrated considerable interest in the concept of *mental workload:* How busy is the operator? How complex are the tasks? Can any additional tasks be handled above and beyond those that are already performed? Will the operator be able to respond to uncertain stimuli? How does the operator feel about the tasks being performed? The growing number of articles, books, and symposia in the field (Eggemeier & O'Donnel, 1986; Gopher & Donchin, 1986; Hancock & Meshkati, 1955; Leplat & Welford, 1978; Moray, 1979, 1989; Smith, 1979; Wierwille & Williges, 1978, 1980; Williges & Wierwille, 1979) is testimony to the fact that system designers realize that workload is an important concern, which is, in fact, a major issue between management and labor in the airline industry. The Air Line Pilots Association has argued that the workload demanded at peak times in the class of narrow-body air transports such as the DC-9 or Boeing 737 is excessive for a two-person crew on the flight deck. They asserted that a three-person complement is required. Correspondingly, the airlines industry has argued that the workload can be adequately handled by the two-person crew (Lerner, 1983). The Federal Aviation Administration will soon require certification of aircraft in terms of workload metric, and the Air Force also imposes workload criteria on newly designed systems. All of these concerns lead to a very relevant question: What is mental workload and how is it measured?

Importance of Workload   Both the designers and the operators of systems realize that performance is not all that matters in the design of a good system. It is just as important to consider what demand a task imposes on the operator's limited resources. Demand may or may not correspond with performance. More specifically, the importance of research on mental work load may be viewed in three different contexts: workload prediction discussed previously, the assessment of workload imposed by equipment; and the assessment of workload experienced by the human operator. The difference between the second and third is their implications for action. When the workload of systems is assessed or compared, the purpose of such a comparison is to optimize the system. When the workload experienced by an operator is assessed, it is for the purpose of choosing between operators or providing an operator with further training. Workload in all three contexts may be initially represented by a simplified undifferentiated-capacity model of human processing resources, fundamentally similar to that shown in Figure 9.1. (We will consider the added complexities of multiple-resource theory later.) Figure 9.9 shows the relationship between the important variables in this model. The resources demanded by a task are shown on the horizontal axis. The resources supplied (or needed for adequate performance) are shown on the vertical axis. If adequate performance of a task demands more resources from the operator than are available, performance will break down. If, however, the available supply exceeds the demand, the amount of the excess expresses the amount of residual capacity.

We make the assumption in this chapter that the concept of workload is fundamentally defined by this relationship between resource supply and task demand. In the region to the left of the break point of Figure 9.9, workload is inversely related to reserve capacity. In the region to the right, it is inversely

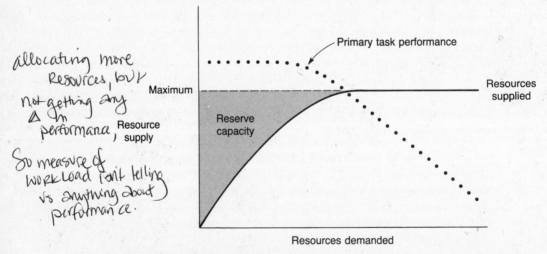

allocating more Resources, but not getting any Δ in performance,

So measure of workload isn't telling us anything about performance.

Figure 9.9 Schematic relationship among primary-task resource demand, resources supplied and performance.

related to the level of task performance. These two quantities are primarily what system designers should wish to predict. Note that changes in workload according to this conception may result either from fluctuations of *operator* capacity or from changes in *task resource* demands.

Equipment Assessment Although workload and performance prediction before a system is designed is desirable, it is often essential to measure the workload of a system already existing at some stage of production. This assessment may be made to identify those bottlenecks in system or mission performance in which resource demands momentarily exceed supply and performance breaks down. Alternatively, workload may be assessed to compare two alternate pieces of equipment that may achieve similar performance but differ in their resource demands because they possess differently shaped performance-resource functions (see, e.g., Figure 9.5). Sometimes the criterion of workload may offer the only satisfactory means of choosing between alternatives. An even greater challenge to workload assessment techniques is posed by the requirement to determine if the *absolute level* of workload imposed by a system is above or below a given absolute criterion level. The goal of developing workload certification criteria for complex systems has spawned the need for such absolute scales.

Assessing Operator Differences Work load measures may also assess differences in the residual resources available to a given operator and not that imposed by a given system. This may be done in one of two contexts: (1) The level of skill or automaticity achieved by different operators who may be equivalent in their primary-task performance may be compared. For example, Crosby and Parkinson (1979) and Damos (1978) showed that flight instructors differed from student pilots in their level of residual attention. Damos furthermore found that applied to students, this measure was a good predictor of success in pilot training. (2) Operators may be monitored on-line in real task performance. In this case, intelligent computer-based systems could decide to assume responsibility for the performance of certain tasks from the human operator when momentary demands were measured to exceed capacity (Enstrom & Rouse, 1977; Wickens & Gopher, 1977), although this form of on-line human-computer interaction requires a certain level of cooperation from the human operator (see Chapter 12).

Criteria for Workload Indexes O'Donnell and Eggemeier (1986) have proposed a number of criteria that should ideally be met by any technique to assess workload. Of course it is true that some of these criteria may trade off with one another, and so rarely if ever will one technique be found that satisfies all criteria. The following list of five criteria of a workload index is similar to the list proposed by O'Donnell and Eggemeier.

    Sensitivity. The index should be sensitive to changes in task difficulty or resource demand.

___*Diagnosticity.* An index should indicate not only when workload varies but also the cause of such variation. In multiple-resource theory, it should indicate *which* of the capacities or resources are varied by demand changes in the system. This information makes it possible to implement better solutions.

___*Selectivity.* The index should be selectively sensitive only to differences in capacity demand and not to changes in such factors as physical load or emotional stress, which may be unrelated to mental workload or information-processing ability.

___*Obtrusiveness.* The index should not interfere with, contaminate, or disrupt performance of the primary task whose workload is being assessed.

___*Bandwidth and reliability.* As with any measure of behavior, a workload index should be reliable. However, if workload is assessed in a time-varying environment (e.g., if it is necessary to track workload changes over the course of a mission), it is important that the index offer a reliable estimate of workload rapidly enough so that the transient changes may be estimated.

A myriad of workload assessment techniques have been proposed, some meeting many of these criteria but few satisfying all of them. These may be classified into four broad categories related to primary-task measures, secondary-task measures of spare capacity, physiological measures, and subjective rating techniques.

Primary-Task Measures   In evaluating any system or operator, one should always examine first the performance on the system of interest, like computer data-entry speed or driving deviations from the center of the lane. Because this is the target of evaluation, we refer to the task performed with this system as the *primary task*. Yet there are four important reasons why primary-task performance may be insufficient to reveal clearly the merits of the primary task. First, in Figure 9.9, two primary tasks may lie in the left-hand region of the supply-demand space (see the two tasks represented by the PRFs in Figure 9.5). Since both have sufficient reserve capacity to reach perfect performance, the latter measure cannot discriminate between them. Second, two primary tasks to be compared may differ in how they are measured or what those measures mean. A designer of prosthetic devices to enable the blind to read may find that the two kinds of devices produce qualitatively different forms of errors (e.g., semantic confusions versus letter confusions) or that they differ greatly in the speed-accuracy trade-off. As we saw in the previous chapter, comparisons of systems at different levels of speed and accuracy are possible but much less certain than if either accuracy or speed comparisons are identical.

Third, sometimes it is simply impossible to obtain good measures of the factors that go into primary-task performance. As noted in Chapter 7, decision making may impose tremendous cognitive demands on the operator, yet the performance outcome (right or wrong) is a very poor measure of all of the mental operations that were involved in reaching the final outcome. As another example, measuring primary-task performance in aviation is often quite difficult because of the expense in supplying instruments to detect deviations from a desired flight path.

Finally, two primary tasks may differ in their performance, not by the resources demanded to achieve that performance, but by differences in data limits. In decision making, for example, if a heuristic yields lower performance than a computational algorithm, it may be important information for the system or job designer, but this difference does not mean that the heuristic imposes greater workload. In fact, as suggested by the two PRFs in Figure 9.6, the difference may well be the result of the *lower* resource demands of the heuristic. Similarly, automated voice-recognition systems often yield poorer performance than manual entry systems for simple data strings. Yet this difference may be attributable to machine limits in the voice-recognition algorithm rather than to operator difference in the speed and resource demands of response production. In short, primary task performances may differ for a lot of reasons that are not related to workload, as the latter is defined in the context of Figure 9.9.

For these reasons, system designers have often turned to the three other workload assessment techniques—secondary-task performance, physiological measures, and subjective measures—which may assess more directly either the effort invested into primary-task performance or the level of residual capacity available during that performance.

The Secondary-Task Technique Imposing a secondary task as a measure of residual resources or capacity not utilized in the primary task (Ogden, Levine, & Eisner, 1979; Rolfe, 1973) is a technique that has a long history in the field of work load research. As described earlier, secondary-task performance is assumed to be inversely proportional to the primary-task resource demands. In this way, secondary tasks may reflect differences in task resource demand, automaticity, or practice that are not reflected in primary-task performance. The logic behind the secondary task in the PRF space is shown in Figure 9.10. The operator is requested to perform as well as possible on the primary task and then allocate any leftover resources to the secondary task. Thus, as we see in Figure 9.10*a*, three increasing levels of primary task difficulty will yield three successively smaller margins of available resources, and therefore three diminishing levels of secondary-task performance.

Garvey and Taylor (1959) found differences in tracking controls to be reflected only with the addition of a secondary-task measure. Bahrick and Shelly (1958) found that the secondary task was sensitive to differences in automaticity. Performance of their subjects on a serial RT task did not differ between a random and a predictable sequence of stimuli. However, performance on a secondary task did discriminate between them. With practice the repeated sequence required fewer resources.

A variant of the secondary-task technique is the use of a *loading task*

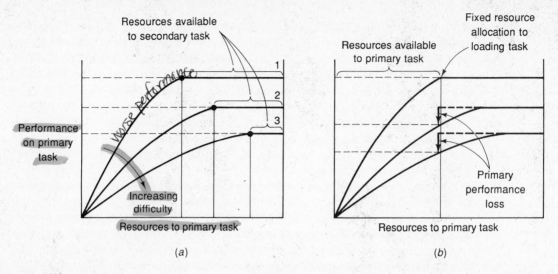

Figure 9.10 Relationships among the performance-resource function, resource allocation, and primary-task difficulty. (*a*) Secondary-task technique, (*b*) loading-task technique.

(Ogden, Levine, & Eisner, 1979; Rolfe, 1973). When using the secondary-task technique, the investigator is interested in variation in the secondary-task decrement (from a single secondary-task control condition) to infer differences in primary-task demand. The primary task is thus both the task of interest and the task whose priority is emphasized, as shown in Figure 9.10*a*. In the loading-task technique, shown in Figure 9.10*b*, different allocation instructions are provided. The subject is asked to devote all necessary resources to the loading task, and the degree of intrusion of this task on performance of the primary task is examined to compare differences between primary tasks.

Secondary Task Examples A multitude of secondary tasks have been proposed and employed at one time or another to assess the residual capacity of primary tasks. Although the reader is referred to reviews by Ogden, Levine, and Eisner (1979), O'Donnell and Eggemeier (1986), and Wierwille and Williges (1978, 1980), for a more exhaustive listing of these tasks, a few prominent candidates will be described here.

In the *rhythmic tapping* task, the operator must produce finger or foot taps at a constant rate (Michon, 1966; Michon & Van Doorne, 1967). Tapping variability increases as primary-task work load increases. *Random number generation* requires the operator to generate a series of random numbers (Baddeley, 1966; Logie et al., 1989; Wetherell, 1981). As work load increases, the degree of randomness declines and the operator begins to generate more repetitive sequences (e.g., 456, 456, 456). *Probe reaction time* tasks are commonly used as work load measurement techniques, as it is assumed that greater primary-task work load will prolong the reaction time to a secondary-task stimulus (Kantowitz, Bortolussi & Hart, 1987; Lansman & Hunt, 1982; Wetherell, 1981). A popular choice for a probe RT task has been the *Sternberg memory search task* (Crosby & Parkinson, 1979; Wetherell, 1981; Wickens, Hyman, Dellinger, Taylor, & Meador, 1986). The operator is presented with a verbal probe stimulus (e.g., a letter) and must respond yes or no regarding whether that letter is one of a previously memorized set of two to four letters. To do so, the operator must search through working memory (Sternberg, 1975). Performance on the Sternberg task thus provides a diagnostic index of the load on cognitive resources as well as the perceptual/motor ones. Wickens, Hyman, Dellinger, Taylor, and Meador discuss some of the specific limitations and constraints on the use of this technique.

Time estimation and *time production* are two related techniques with some-what different underlying assumptions. When the operator makes a *retrospective estimate* of the amount of time that has passed while performing a primary task whose demands may vary, a given interval of time is generally underesti-mated as work load increases. In other words, "time flies when you're keeping busy." If the operator is asked to produce time intervals of a constant duration (i.e., ten seconds), the intervals will tend to be overestimated when there are higher demands (Hart, 1975), as if the higher levels of workload interfere with (and postpone) whatever internal mechanism is responsible for mental time counting. (As we saw, high workload will also make these intervals more variable.)

Finally, the *critical instability tracking task* is a computer task that simulates the demands of balancing a stick on the end of your finger (Jex, 1967, 1979). The computer-simulated stick (viewed on a display and controlled by a joy-stick) gradually shortens over time, becoming less stable until control is lost (compare the ease of balancing a long pole versus a pencil). As primary-task demands become more difficult, the length of the stick at which stability is lost typically becomes longer and longer, and the length of the stick is found to be proportional to the visual time demands of the primary task. The commercially available task can be used either as a secondary task or a loading task (Jex, 1979).

Benefits and Costs of Secondary Tasks   The secondary-task technique has two very distinct benefits. First, it has a high degree of face validity. It is designed to predict the amount of residual attention an operator will have available if an unexpected failure or environmental event occurs. This validity places the technique in contrast with the physiological and subjective measures described below. Second, the same secondary task can be applied to two very different primary tasks and will give workload measures in the same units (which can therefore be compared). As we have seen, this is not the case with primary-task performance measures.

One difficulty is that the secondary-task technique must account for the fact that there are different kinds of resources (O'Donnell & Eggemeier, 1986). Workload differences that result from manipulating a primary-task variable can be greatly underestimated if the resource demands of the primary-task manipu-lation do not match those of most importance for secondary-task performance.

Thus the secondary-task index is not always *sensitive*. Wickens and Kessel (1980), for example, found that the response-loading critical-tracking task developed by Jex (1967) was not appropriate as a loading task when used with a perceptual monitoring task. Another example of such a mismatch would occur if an auditory word-comprehension or mental arithmetic task (auditory, verbal, perceptual/central demands) were used to assess the work load caused by manipulations of tracking-response load (visual, spatial, response demands). The previous discussions of time-sharing would suggest that these two tasks use very different resources in the model shown in Figure 9.7. Hence difficulty insensitivity would be likely to occur, a finding noted by Wickens, Sandry, and Vidulich (1983) when they used an auditory-voice Sternberg task to measure tracking workload.

Kahneman (1973) has proposed that the ideal secondary-task technique is one that employs a battery of secondary-task measures sensitive to different resources in the system. Schlegel, Gilliland, and Schlegel (1986) have proposed a structured set of tasks, known as the *criterion task set*, that are mapped onto different resource dimensions. When it is clear that one level of a dimension does not contribute to primary-task performance, the dimensionality of the battery may be reduced accordingly. For example, a verbal processing task with no spatial components need not be assessed by a spatial secondary task. However, in cases in which an activity is performed that potentially engages all "cells" of processing resources, as depicted in Figure 9.7, a secure workload measure should involve a battery that also incorporates those cells, or at least taps early and late processing of a verbal and spatial nature.

A second problem often encountered with the secondary-task technique is that it may interfere with and disrupt performance of the primary task; that is, the index suffers on the *obtrusiveness* criterion. On the one hand, this may be inconvenient or even dangerous if the primary task is one like flying or driving; a diversion of resources to the secondary task at the wrong time could lead to an accident. On the other hand, disruption of the primary task could present problems of interpretation if the amount of disruption suffered by the primary tasks to be compared is not the same.

Problems with obtrusiveness often affect the operator's attitude toward and willingness to perform the secondary task at all. In response, some researchers have advocated *embedded secondary tasks*. Here the "secondary task" is actually a legitimate component of the operator's total task responsibilities, but it is a component of lower priority in the task hierarchy than the primary task of interest (Raby & Wickens, 1990). For example, the latency of responding to a verbal request from air traffic control would be a good embedded task for assessing the pilot's workload demands of keeping the aircraft stable since the latter is of higher priority.

Physiological Measures *are used to measure* One solution to performance obtrusiveness is to record, unobtrusively, the manifestations of workload or increased resource mobilization through appropriately chosen physiological measures of autonomic or central nervous system activity (Kramer, 1987; see Figure 9.1). Three such techniques will be briefly described here, and a further review is provided by O'Donnell and Eggemeier (1986) and Wierwille (1979).

Evoked Brain Potential When the evoked brain potential (EP) described in Chapter 8 is used to assess workload, many of the same assumptions are made as those underlying secondary-task measures. That is, the EP is more like a measure of residual capacity than a measure of resource or effort mobilization. In using the EP, as a secondary task (Isreal, Chesney, Wickens, & Donchin, 1980; Isreal, Wickens, Chesney, & Donchin, 1980; Kramer, Sirivaag, & Braune, 1987), the subject sees or hears a Bernoulli series of stimuli (AABA-BABBB . . .) and is asked to count covertly one of the two classes of stimuli. The processing of the stimuli elicits a prominent late-positive or P300 component in the wave form of the evoked potential recorded from the scalp. Isreal and his colleagues found that introducing a concurrent primary task of a perceptual/cognitive nature, typical of the air traffic controller's task, will reduce P300 amplitude. Increasing the difficulty of the task by requiring more display elements to be monitored reduces P300 amplitude still further (see Figure 9.11). The EP measure is somewhat *diagnostic*, in that it reflects perceptual/cognitive load, but is relatively insensitive to variations in response load.

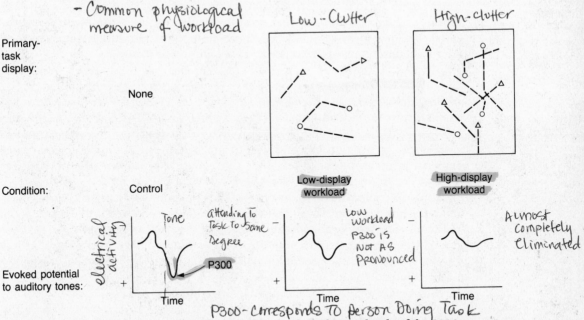

Figure 9.11 The evoked potential (EP) as a workload index. The amplitude of the P300 component of the EP to counted auditory tones declines systematically as the complexity of the simulated air traffic control display increases. Subjects must detect course changes of each circular element on the display. The dashed lines indicating the trajectory are not visible on the subjects' display. (*Source:* J. B. Israel, C. D. Wickens, G. L. Chesney, and E. Donchin, "The Event-Relation Brain Potential as an Index of Display Monitoring Workload," *Human Factors*, 22 (1980), pp. 214 and 217. Copyright 1980 by the Human Factors Society, Inc. Reproduced by permission.)

Therefore, its diagnosticity is obtained at the expense of *sensitivity*. The EP measure has two particular advantages. In contrast to the other physiological measures described below, it provides a graded measure of direct cognitive activity rather than an indirect measure of autonomic activity. In contrast to most secondary tasks, however, it does not require overt responses and therefore is less likely to be intrusive. Further discussions of the EP applications to workload and resource-allocation measures may be found in Parasuraman (1990) and Wickens (1990*a*).

② **Pupil Diameter**   Several investigators have observed that the diameter of the pupil correlates quite closely and accurately with the resource demands of a large number of diverse cognitive activities (Beatty, 1982). These include mental arithmetic (Kahneman, Beatty, & Pollack, 1967), short-term memory load (Beatty & Kahneman, 1966; Peavler, 1974), reaction time (Richer, Silverman, & Beatty, 1982), and logical problem solving (Bradshaw, 1968; see Beatty, 1982, for an integrative summary). This diversity of responsiveness suggests that the pupilometric measure may be highly sensitive, although as a result it is undiagnostic. It will reflect demands imposed anywhere within the system. Its disadvantage, of course, is that relevant pupil changes are in the order of tenths of a millimeter, which means that accurate measurement requires considerable head constraint and precise measuring equipment. Also, changes in ambient illumination must be monitored since these also affect the pupil. Because of its association with the autonomic nervous system, the measure will also be susceptible to variations in emotional arousal.

③ **Heart-Rate Variability**   A number of investigators have examined different measures associated with the variability or regularity of heart rate as a measure of mental load. Variability is generally found to decrease as the load increases (Mulder & Mulder, 1981). When this variability is associated specifically with the periodicities resulting from respiration, the measure is termed sinus arrhythmia (Kalsbeek & Sykes, 1967; Mulder & Mulder, 1981; Thornton, Vicente, & Moray, 1987). Like pupil diameter, the sinus arrhythmia measure is sensitive to a number of different difficulty manipulations and therefore appears to be more sensitive than diagnostic. Derrick (1988) investigated this measure with four quite different tasks performed in different combinations within the framework of the multiple-resource model. His data suggested that the variability measure reflected the total demand imposed on all resources within the processing system more than the amount of resource competition (and therefore dual-task decrement) between tasks.

④ **Costs and Benefits**   Physiological indexes have two great advantages: (1) They provide a relatively continuous record of data over time. (2) They are not obtrusive into primary-task performance. On the other hand, they do often require that electrodes be attached (EPs and heart measures) or some degree of physical constraints be imposed (pupilometric measures), and therefore they are not really unobtrusive in a physical sense. These constraints will influence user acceptance. They have a further potential cost in that they are, generally, one conceptual step removed from the inference that the system designers would like to make. That is, workload differences measured by physiological means must be used to *infer* that performance breakdowns would result or to *infer* how the operator would feel about the task. Secondary- or primary-task measures assess the former directly, whereas subjective measures assess the latter.

Subjective Measures A variety of techniques have been proposed to assess the subjective effort required to perform a task. Some of them use a structured rating scale to elicit a single dimensional rating (Wierwille & Casali, 1983), whereas others have adopted the view that subjective workload, like the resource concept itself, has several dimensions (Derrick, 1988; O'Donnell & Eggemeier, 1986). Two common multidimensional assessment techniques are the NASA TLX scale (Hart & Staveland, 1988), which assesses work load on each of five 7-point scales, and the SWAT technique (Reid & Nygren 1988), which measures work load on three 3-point scales (see Table 9.1). Each of these techniques has formal prescriptions for how the multiple scales may be combined to obtain a single measure. Although both scales tend to yield similar outcomes when they are applied to the same set of data (Vidulich & Tsang, 1986), the TLX technique, having a greater number of scales and greater resolution per scale, potentially allows it to convey more information.

Costs and Benefits The benefits of subjective techniques are apparent. They do not disrupt primary-task performance, and they are relatively easy to derive. Their costs relate to the uncertainty with which an operator's verbal statement truly reflects the availability of or demand for processing resources.

Relationship Between Workload Measures

If all measures of workload demonstrated high correlation with one another and the residual disagreement was due to random error, there would be little need for further validation research in the area. The practitioner could adopt whichever technique was methodologically simplest and most reliable for the work load measurement problem at hand. Generally, high correlations between measures will be found if the measures are assessed across tasks of similar structure and widely varying degrees of difficulty, for example, Jex and Clement's (1979) finding of the high correlation between subjective and secondary-task measures of flight-control difficulty. However, the correlations may not be high and may even be negative when quite different tasks are contrasted. For example, consider an experiment conducted by Herron (1980) in which an innovation designed to assist in a target-aiming task was subjectively preferred by users over the original prototype but generated reliably poorer performance than the original. Similar dissociations have been observed by Childress, Hart, and Bortalussi (1982) and Murphy et al. (1978), who measured pilot work load associated with cockpit-display innovations. It is important, then, to determine

Table 9.1   TWO MULTIDIMENSIONAL WORKLOAD RATING SCALES

SWAT Scale

Time load	Mental effort load	Stress load
1. Often have spare time. Interruptions or overlap among activities occur infrequently or not at all. 2. Occasionally have spare time. Interruptions or overlap among activities occur frequently. 3. Almost never have spare time. Interruptions or overlap among activities are very frequent, or occur all the time.	1. Very little conscious mental effort or concentration required. Activity is almost automatic, requiring little or no attention. 2. Moderate conscious mental effort or concentration required. Complexity of activity is moderately high due to uncertainty, unpredictability, or unfamiliarity. Considerable attention required. 3. Extensive mental effort and concentration are necessary. Very complex activity requiring total attention.	1. Little confusion, risk, frustration, or anxiety exists and can be easily accommodated. 2. Moderate stress due to confusion, frustration, or anxiety noticeably adds to workload. Significant compensation is required to maintain adequate performance. 3. High to very intense stress due to confusion, frustration, or anxiety. High to extreme determination and self-control required.

NASA TLX Scale:
Rating scale definitions

Title	Endpoints	Descriptions
MENTAL DEMAND	*Low/High*	How much mental and perceptual activity was required (e.g., thinking, deciding, calculating, remembering, looking, searching, etc.)? Was the task easy or demanding, simple or complex, exacting or forgiving?
PHYSICAL DEMAND	*Low/High*	How much physical activity was required (e.g., pushing, pulling, turning, controlling, activating, etc.)? Was the task easy or demanding, slow or brisk, slack or strenuous, restful or laborious?
TEMPORAL DEMAND	*Low/High*	How much time pressure did you feel due to the rate or pace at which the tasks or task elements occurred? Was the pace slow and leisurely or rapid and frantic?
PERFORMANCE	*Perfect/Failure*	How successful do you think you were in accomplishing the goals of the task set by the experimenter (or yourself)? How satisfied were you with your performance in accomplishing these goals?
EFFORT	*Low/High*	How hard did you have to work (mentally and physically) to accomplish your level of performance?
FRUSTRATION LEVEL	*Low/High*	How insecure, discouraged, irritated, stressed, and annoyed versus secure, gratified, content, relaxed, and complacent did you feel during the task?

which characteristics of a task most strongly influence one workload measure while leaving another measure unaffected.

The understanding of attention and resource theory can be quite useful in interpreting why these dissociations occur. Yeh and Wickens (1988) have paid particular attention to the dissociation between primary-task and subjective measures. Their assumption is that subjective measures directly reflect two factors: the effort that must be invested into performance of a task and the number of tasks that must be performed concurrently. These two factors, however, do not always influence performance. To illustrate, consider the following situations:

Reserve Capacity Figure 9.9.

1. *reserve Capacity*

1. If two different tasks are in the underload region on the left of Figure 9.9, more resources invested on the more difficult task (and therefore higher subjective workload) will not yield better performance.

2. *Loading Task*

2. If subjects perform the three tasks shown in Figure 9.10b while investing full resources, performance will differ, but the resources invested (and therefore the subjective workload experienced) will not differ. Subjective measures often fail to reflect differences due to data limits, particularly if the lower level of performance caused by the lower level of the data limit is not immediately evident to the performer who is giving the rating.

3. *inverse relationship*
4. *# of Tasks*

3. If two systems are compared, one of which induces a greater investment of effort, the latter will probably show higher subjective workload, even as its performance is improved (through the added effort investment). This dissociation is shown when effort investment is induced through monetary incentives (Vidulich & Wickens, 1986). However, it also appears that greater effort is invested when better display information is available to achieve better performance. Thus in tracking tasks, items like an amplified error signal (achieved through magnification or prediction—see Chapter 11), will increase tracking performance but at the expense of higher subjective ratings of workload (Yeh & Wickens, 1988).

no extra REZORCEZ

* Subjective level of workload is constant performance is changing *

* It can happen whereif something is reported Harder they do better *

4. Finally, Yeh and Wickens (1988) have concluded that a very strong influence on subjective workload is exerted by the number of tasks that must be performed at once. The subjective workload from time-sharing two (or more) tasks is almost always greater than that from a single task. We can see here the source of another dissociation with performance because a single task might be quite difficult (and result in poor performance as a result), whereas a dual-task combination, if the tasks are not difficult and use separate resources, may indeed produce a very good performance in spite of its higher level of subjective load.

*2 Factors effecting Subjective measures of workload
① effort
② # of Tasks*

The presence of dissociations often leave the system designer in a quandary. Which system should be chosen when performance and workload measures do not agree on the relative merits between them? The previous discussion, and the chapter as a whole, do not provide a firm answer to this question. However, the explanation for the causes of dissociation and its basis on a theory

, but these 2 measures don't always affect performance.

So there are Differences in what were measuring w/ performance Techniques.

of resources should at least help the designer to understand why the dissociation occurs, and thus why one measure or the other may offer a less reliable indicator of the true workload of the system in specific circumstances.

TRANSITION

In this chapter we have outlined the potential causes of multiple-task interference, discussing the role of switching, confusion, and cooperation but emphasizing most heavily the role of resource competition. The consideration of resources led to the discussion of the theory and measurement of mental workload and the fundamental importance of the resource concept to this theory.

Stress is often a consequence of high levels of mental workload, particularly if they are sustained for some time. Stress in turn will often produce changes in functioning of all of the information-processing components that we have discussed, and so will produce effects on performance. Of course stress may be experienced from other sources than high workload—sleep loss, noise, and anxiety, to name a few. In the following chapter we will examine these sources of stress and determine whether and how their effects might be predicted.

A second consequence of high workload is the occurrence of errors. As with stress, errors may be caused by other factors as well. In the following chapter we will see that one of the prominent effects of stress is a shift in the speed-accuracy trade-off to error-prone performance. Furthermore, errors may be caused by the characteristics of the task and by differences in the skill level of the operator. Thus, in the next chapter we will describe the causes and models of human error and show how human factors engineers have tried to use this information to predict human reliability and design error-tolerant systems. Because of the close linkage between stress and errors and because both may be related to all stages of processing, they are included together in a single chapter.

REFERENCES

Ackerman, P., & Wickens, C. D. (1982). Task methodology and the use of dual and complex task paradigm in human factors. In R. Edwards (ed.), *Proceedings of the 26th annual meeting of the Human Factors Society*. Santa Monica, CA: Human Factors Society.

Aldrich, T. B., Szabo, S. M., Bierbaum, C. R. (1989). The development and application of models to predict operator workload during system design. In G. R. McMillan, D. Beevis, E. Salas, M. H. Strub, R. Sutton, & L. Van Breda (eds.), *Applications of human performance models to system design* (pp. 65–80). New York: Plenum Press.

Allport, D. A. (1980). Attention. In G. L. Claxton (ed.), *New directions in cognitive psychology*. London: Routledge & Kegan Paul.

Allport, D. A., Antonis, B., & Reynolds, P. (1972). On the division of attention: A disproof of the single channel hypothesis. *Quarterly Journal of Experimental Psychology, 24,* 255–265.

Baddeley, A. (1966). The capacity for generating information by randomization. *Quarterly Journal of Experimental Psychology, 18,* 119–130.

Baddeley, A. D., & Hitch, G. (1974). Working memory. In G. Bower (ed.), *Recent advances in learning and motivation* (vol. 8). New York: Academic Press.

Bahrick, H. P., Noble, M., & Fitts, P. M. (1954). Extra task performance as a measure of learning a primary task. *Journal of Experimental Psychology, 48,* 298–302.

Bahrick, H. P., & Shelly, C. (1958). Time-sharing as an index of automization. *Journal of Experimental Psychology, 56,* 288–293.

Barnett, A., & Wickens, C. D. (1988). Display proximity in multicue information integration: The benefit of boxes. *Human Factors, 30,* 15–24.

Beatty, J. (1982). Task-evoked pupillary responses, processing load, and the structure of processing resources. *Psychological Bulletin, 91,* 276–292.

Beatty, J., & Kahneman, D. (1966). Pupillary changes in two memory tasks. *Psychonomic Science, 5,* 371–372.

Bettman, J. R., Johnson, E. J., & Payne, J. W. (1990). A componential analysis of cognitive effort and choice. *Organizational Behavior and Human Decision Processes, 45,* 111–139.

Bradshaw, J. L. (1968). Pupil size and problem-solving. *Quarterly Journal of Experimental Psychology, 20,* 116–122.

Braune, R., & Wickens, C. D. (1986). Time-sharing revisited: Test of a componential model for the assessment of individual differences. *Ergonomics, 29*(11), 1399–1414.

Brookings, J. B. (1987). A confirmatory factor analytic investigation of time sharing performance and cognitive abilities. *Proceedings of the 31st annual meeting of the Human Factors Society* (pp. 1062–1066). Santa Monica, CA: Human Factors Society.

Brooks, L. (1968). Spatial and verbal components of the act of recall. *Canadian Journal of Psychology, 22,* 349–368.

Carswell, M. (1990). Graphical information processing: The effects of proximity compatibility. *Proceedings of the 34th annual meeting of the Human Factors Society* (pp. 1494–1498). Santa Monica, CA: Human Factors Society.

Chernikoff, R., Duey, J. W., & Taylor, F. V. (1960). Effect of various display-control configurations on tracking with identical and different coordinate dynamics. *Journal of Experimental Psychology, 60,* 318–322.

Chernikoff, R., & Lemay, M. (1963). Effect of various display-control configurations on tracking with identical and different coordinate dynamics. *Journal of Experimental Psychology, 66,* 95–99.

Childress, M. E., Hart, S. G., & Bortalussi, M. R. (1982). The reliability and validity of flight task workload ratings. In R. Edwards (ed.), *Proceedings of the 26th annual meeting of the Human Factors Society.* Santa Monica, CA: Human Factors Society.

Crosby, J. V., & Parkinson, S. (1979). A dual task investigation of pilot's skill level. *Ergonomics, 22,* 1301–1313.

Damos, D. (1978). Residual attention as a predictor of pilot performance. *Human Factors, 20,* 435–440.

Damos, D., Smist, T., & Bittner, A. C. (1983). Individual differences in multiple task performance as a function of response strategies. *Human Factors, 25,* 215–226.

Damos, D., & Wickens, C. D. (1980). The acquisition and transfer of time-sharing skills. *Acta Psychologica, 6,* 569–577.

Derrick, W. L. (1988). Dimensions of operator workload. *Human Factors, 30*(1), 95–110.

Dornic, S. S. (1980). Language dominance, spare capacity, and perceived effort in bilinguals. *Ergonomics, 23,* 369–378.

Duncan, J. (1979). Divided attention: The whole is more than the sum of its parts. *Journal of Experimental Psychology: Human Perception and Performance, 5,* 216–228.

Enstrom K. D., & Rouse, W. B. (1977). Real-time determination of how a human has allocated his attention between control and monitoring tasks. *IEEE Transactions on Systems, Man, and Cybernetics, SMC–7,* 153–161.

Fogarty, G., & Stankov, L. (1982). Competing tasks as an index of intelligence. *Personality and Individual Differences, 3,* 407–422.

Fracker, M. L., & Wickens, C. D. (1989). Resources, confusions, and compatibility in dual axis tracking: Display, controls, and dynamics. *Journal of Experimental Psychology: Human Perception & Performance, 15,* 80–96.

Friedman, A., & Polson, M. C. (1981). Hemispheres as independent resources systems: Limited-capacity processing and cerebral specialization. *Journal of Experimental Psychology, 7,* 1031–1058.

Garner, W. R. (1974). *The processing of information and structure.* Hillsdale, NJ: Erlbaum.

Garvey, W. D., & Taylor, F. V. (1959). Interactions among operator variables, system dynamics, and task-induced stress. *Journal of Applied Psychology, 43,* 79–85.

Goettl, B. P. & Wickens, C. D. (1989). Multiple resources vs. information integration. *Proceedings of the 33rd annual meeting of the Human Factors Society.* Santa Monica, CA: Human Factors Society.

Gopher, D. (1982). A selective attention test as a prediction of success in flight training. *Human Factors, 24,* 173–184.

Gopher, D. (1991). The skill of attention control: Acquisition and execution of attention strategies. In D. Meyer & S. Kornblum (eds.), *Attention and performance IVX.* Hillsdale, NJ: Erlbaum.

Gopher, D., & Brickner, M. (1980). On the training of time-sharing skills: An attention viewpoint. In G. Corrick, M. Hazeltine, & R. Durst (eds.), *Proceedings of the 24th annual meeting of the Human Factors Society.* Santa Monica, CA: Human Factors Society.

Gopher, D., Brickner, M., & Navon, D. (1982). Different difficulty manipulations interact differently with task emphasis: Evidence for multiple resources. *Journal of Experimental Psychology: Human Perception and Performance, 8,* 146–158.

Gopher, D., & Donchin, E. (1986). Workload: An experimentation of the concept. In K. Boff, L. Kaufman, & J. Thomas (eds.), *Handbook of perception and performance* (vol. II). New York: Wiley.

Gopher, D., & Kahneman, D. (1971). Individual differences in attention and the prediction of flight criteria. *Perception and Motor Skills, 33,* 1335–1342.

Gopher, D., & Navon, D. (1980). How is performance limited? Testing the notion of central capacity. *Acta Psychologica, 46,* 161–180.

Gopher, D., & Sanders, A. F. (1985). S-Oh-R: Oh Stages! Oh Resources! In W. Prinz and A. F. Sanders (eds.), *Cognition and Motor Skills* (pp. 231–253). Amsterdam: North Holland.

Gopher, D., Weil, M., & Siegel, D. (1989). Practice under changing priorities: An approach to training of complex skills. *Acta Psychologica, 71,* 147–179.

Hancock, P. A., & Meshkati, N. (1988). *Human mental workload.* Amsterdam: North Holland.

Hart, S. G. (1975, May). Time estimation as a secondary task to measure workload. *Proceedings of the 11th Annual Conference on Manual Control* (NASA TMX–62, N75-33679, 53; pp. 64–77). Washington, DC: U.S. Government Printing Office.

Hart, S. G., & Staveland, L. E. (1988). Development of NASA-TLS (Task Load Index): Results of empirical and theoretical research. In P. A. Hancock & N. Meshkati (eds.), *Human mental workload.* Amsterdam: North Holland.

Hart, S. G., & Wickens, C. D. (1990). Workload assessment and prediction. In H. R. Booher (ed.), *MANPRINT: An emerging technology. Advanced concepts for integrating people, machines and organizations* (pp. 257–300). New York: Van Nostrand Reinhold.

Herron, S. (1980). A case for early objective evaluation of candidate displays. In G. Corrick, M. Hazeltine, & R. Durst (eds.), *Proceedings of the 24th annual meeting of the Human Factors Society.* Santa Monica, CA: Human Factors Society.

Hirst, W. (1986). Aspects of divided and selected attention. In J. LeDoux & W. Hirst (eds.), *Mind and brain* (pp. 105–141). New York: Cambridge University Press.

Hirst, W., & Kalmar, D. (1987). Characterizing attentional resources. *Journal of Experimental Psychology: General, 116*(1), 68–81.

Hunt, E., & Lansman, M. (1981). Individual differences in attention. In R. Sternberg (ed.), *Advances in the psychology of intelligence* (vol. 1). Hillsdale, NJ: Erlbaum.

Isreal, J., Chesney, G., Wickens, C. D., & Donchin, E. (1980). P300 and tracking difficulty: Evidence for a multiple capacity view of attention. *Psychophysiology, 17,* 259–273.

Isreal, J., Wickens, C. D., Chesney, G., & Donchin E. (1980). The event-related brain potential as a selective index of display monitoring load. *Human Factors, 22,* 211–224.

Jennings, A. E., & Chiles, W. D. (1977). An investigation of time-sharing ability as a factor in complex performance. *Human Factors, 19,* 535–547.

Jex, H. R. (1967). Two applications of the critical instability task to secondary task workload research. *IEEE Transactions on Human Factors in Electronics, HFE–8,* 279–282.

Jex, H. R. (1979). A proposed set of standardized subcritical tasks for tracking workload calibration. In N. Moray (ed.), *Mental workload: Its theory and measurement.* New York: Plenum Press.

Jex, H. R., & Clement, W. F. (1979). Defining and measuring perceptual-motor workload in manual control tasks. In N. Moray (ed.), *Mental workload: Its theory and measurement.* New York: Plenum Press.

Kahneman, D. (1973). *Attention and effort.* Englewood Cliffs, NJ: Prentice Hall.

Kahneman, D., Beatty, J., & Pollack, I. (1967). Perceptual deficits during a mental task. *Science, 157,* 218–219.

Kahneman, D., Ben-Ishai, R., & Lotan, M. (1973). Relation of a test of attention to road accidents. *Journal of Applied Psychology, 58,* 113–115.

Kahneman, D., & Chajczyk, D. (1983). Tests of the automaticity of reading: Dilution of Stroop effects by color-irrelevant stimuli. *Journal of Experimental Psychology: Human Perception and Performance, 9,* 497–501.

Kalsbeek, J. W., & Sykes, R. W. (1967). Objective measurement of mental load. *Acta Psychologica, 27,* 253–261.

Kantowitz, B. H., Bortolussi, M. R., & Hart, S. G. (1987). Measuring pilot workload in a motion base simulator: III. Synchronous secondary task. *Proceedings of the 31st Annual Meeting of the Human Factors Society* (pp. 834–837). Santa Monica, CA: Human Factors Society.

Kantowitz, B. H., & Knight, J. L. (1976). Testing tapping time-sharing: I. Auditory secondary task. *Acta Psychologica, 40,* 343–362.

Kantowitz, B. H., & Weldon, M. (1985). On scaling POC's: Caveat emptor. *Human Factors, 27,* 531–548.

Keele, S. W. (1973). *Attention and human performance.* Pacific Palisades, CA: Goodyear.

Keele, S. W., Neill, W. T., & DeLemos, S. M. (1977). *Individual differences in attentional flexibility* (Technical Report). Eugene: University of Oregon, Center for Cognitive and Perceptual Research.

Kinsbourne, M., & Hicks, R. (1978). Functional cerebral space. In J. Requin (ed.), *Attention and performance VII.* Hillsdale, NJ: Erlbaum.

Klapp, S. T. (1979). Doing two things at once: The role of temporal compatibility. *Memory & Cognition, 7,* 375–381.

Klapp, S. T. (1981). Temporal compatibility in dual motor tasks II: Simultaneous articulation and hand movements. *Memory & Cognition, 9,* 398–401.

Klein, G. S. (1964). Semantic power measured through the interference of words with color naming. *American Journal of Psychology, 77,* 576–588.

Kramer, A. F. (ed.). (1987). Special issue on cognitive psychophysiology. *Human Factors, 29,* whole issue #2.

Kramer, A. F., Sirevaag, E. J., & Braune, R. (1987). A psychophysiological assessment of operator workload during simulated flight missions. *Human Factors, 29(2),* 145–160.

LaBerge, D. (1973). Attention and the measurement of perceptual learning. *Memory & Cognition, 1,* 268–276.

Lansman, M., & Hunt, E. (1982). Individual differences in secondary task performance. *Memory and Cognition, 10,* 10–24.

Leplat, J., & Welford, A. T. (eds.). (1978). *Ergonomics, 21(3).*

Lerner, E. J. (1983). The automated cockpit. *IEEE Spectrum, 20,* 57–62.

Logie, R., Baddeley, A., Mane, A., Donchin, E., & Sheptak, R. (1989). Working memory in the acquisition of complex cognitive skills. *Acta Psychologica, 71,* 53–87.

Martin, G. (1989). The utility of speech input in user-computer interfaces. *International Journal of Man-Machine System Study, 18*, 355–376.

Martin, R. C., Wogalter, M. S., & Forlano, J. G. (1988). Reading comprehension in the presence of unattended speech and music. *Journal of Memory and Language, 27*, 382–398.

McLeod, P. (1977). A dual task response modality effect: Support for multiprocessor models of attention. *Quarterly Journal of Experimental Psychology, 29*, 651–667.

Michon, J. A. (1966). Tapping regularity as a measure of perceptual motor load. *Ergonomics, 9*, 401–412.

Michon, J. A., & Van Doorne, H. (1967). A semi-portable apparatus for measuring perceptual motor load. *Ergonomics, 10,* 67–72.

Moray, N. (1967). Where is attention limited? A survey and a model. *Acta Psychologica, 27*, 84–92.

Moray, N. (ed.). (1979). *Mental workload: Its theory and measurement.* New York: Plenum Press.

Moray, N. (1988). Mental workload since 1979. *International Reviews of Ergonomics, 2*, 123–150.

Mulder, G., & Mulder, L. J. (1981). Information processing and cardiovascular control. *Psychophysiology, 18*, 392–401.

Murphy, M. R., McGee, L. A., Palmer, E. A., Paulk, C. H., & Wempe, T. E. (1978). Simulator evaluation of three situation and guidance displays for V/STOL aircraft zero-zero landing approaches. *Proceedings of the IEEE International Conference on Cybernetics and Society* (pp. 563–471). New York: IEEE.

Navon, D. (1984). Resources: A theoretical soupstone. *Psychological Review, 91*, 216–334.

Navon, D., & Gopher, D. (1979). On the economy of the human processing systems. *Psychological Review, 86*, 254–255.

Navon, D., & Miller, J. (1987). The role of outcome conflict in dual-task interference. *Journal of Experimental Psychology: Human Perception and Performance, 13*, 435–448.

Nebes, R. D. (1977). Man's so-called minor hemisphere. In M. C. Wittrock (ed.), *The human brain.* Englewood Cliffs, NJ: Prentice Hall.

Norman, D., & Bobrow, D. (1975). On data-limited and resource-limited processing. *Journal of Cognitive Psychology, 7*, 44–60.

North, R. A., & Riley, V. A. (1989). A predictive model of operator workload. In G. R. McMillan, D. Beevis, E. Salas, M. H. Strub, R. Sutton, & L. Van Breda (eds.), *Applications of human performance models to system design* (pp. 81–90). New York: Plenum Press.

O'Donnell, R. D., & Eggemeier, F. T. (1986). Workload assessment methodology. In K. Boff, L. Kaufman, & J. Thomas (eds.), *Handbook of perception and performance* (vol. II). New York: Wiley.

Ogden, G. D., Levine, J. M., & Eisner, E. J. (1979). Measurement of workload by secondary tasks. *Human Factors, 21*, 529–548.

Parasuraman, R. (1990). Event-related brain potentials and human factors research. In J. Rohrbaugh, R. Parasuraman, & R. Johnson (eds.), *Event-related brain potentials: Basic issues and applications.* New York: Oxford University Press.

Parkes, A. M., & Coleman, N. (1990). Route guidance systems: A comparison of methods of presenting directional information to the driver. In E. J. Lovesey (ed.), *Contemporary ergonomics 1990* (pp. 480–485). London: Taylor & Francis.

Parks, D. L., & Boucek, G. P., Jr. (1989). Workload prediction, diagnosis, and continuing challenges. In G. R. McMillan, D. Beevis, E. Salas, M. H. Strub, R. Sutton, & L. Van Breda (eds.), *Applications of human performance models to system design* (pp. 47–64). New York: Plenum Press.

Payne, J. W., Bettman, J. R., & Johnson, E. J. (1988). Adaptive strategy selection in decision making. *Journal of Experimental Psychology: Learning, Memory, and Cognition, 14,* 534–552.

Peavler, W. S. (1974). Individual differences in pupil size and performance. In M. Janissee (ed.), Pupillary dynamics and behavior. New York: Plenum Press.

Peters, M. (1981). Attentional asymmetries during concurrent bimanual performance. Quarterly Journal of Experimental Psychology, 33A, 95–103.

Polson, M. C., & Friedman A. (1988). Task-sharing within and between hemispheres: A multiple-resources approach. *Human Factors, 30,* 633–643.

Raby, M., & Wickens, C. D. (1990). Planning and scheduling in flight workload management. *Proceedings of the 34th annual meeting of the Human Factors Society.* Santa Monica, CA: Human Factors Society.

Reid, G. B., & Nygren, T. E. (1988). The subjective workload assessment technique: A scaling procedure for measuring mental workload. In P. A. Hancock & N. Meshkati (eds.), *Human mental workload* (pp. 185–213). Amsterdam: North Holland.

Richer, F., Silverman, C., & Beatty, J. (1983). Response selection and initiation in speeded reactions: A pupillometric analysis. *Journal of Experimental Psychology: Human Perception and Performance, 9,* 360–370.

Rolfe, J. M. (1973). The secondary task as a measure of mental load. In W. T. Singleton, J. G. Fox, & D. Whitfield (eds.), *Measurement of man at work* (pp. 135–148). London: Taylor & Francis.

Rollins, R. A., & Hendricks, R. (1980). Processing of words presented simultaneously to eye and ear. *Journal of Experimental Psychology: Human Perception and Performance, 6,* 99–109.

Schlegel, R. E., Gilliland, K., & Schlegel, B. (1986). Development of the criterion task set performance data base. *Proceedings of the 30th Annual Meeting of the Human Factors Society* (pp. 58–62). Santa Monica, CA: Human Factors Society.

Schneider, W. (1985). Training high-performance skills: Fallacies and guidelines. *Human Factors, 27*(3), 285–300.

Schneider, W., & Detweiler, M. (1988). The role of practice in dual-task performance: Toward workload modeling in a connectionist/control architecture. *Human Factors, 30*(5), 539–566.

Schneider, W., & Fisk, A. D. (1982). Concurrent automatic and controlled visual search: Can processing occur without cost? *Journal of Experimental Psychology: Learning, Memory, and Cognition, 8,* 261–278.

Schneider, W., & Shiffrin, R. M., (1977). Controlled and automatic human information processing I: Detection, search, and attention. *Psychological Review, 84,* 1–66.

Senders, J. (1964). The human operator as a monitor and controller of multidegree freedom systems. *IEEE Transactions on Human Factors in Electronics, HFE–5,* 2–6.

Shaffer, L. H. (1975). Multiple attention in continuous verbal tasks. In S. Dornic (ed.), *Attention and performance V.* New York: Academic Press.

Shallice, T., McLeod, P., & Lewis, K. (1985). Isolating cognition modules with the dual-task paradigm: Are speech perception and production modules separate? *Quarterly Journal of Experimental Psychology, 37,* 507–532.

Sheridan, T. (1972). On how often the supervisor should sample. *IEEE Transactions on Systems, Science, and Cybernetics, SSC–6,* 140–145.

Shingledecker, C. A. (1989). Handicap and human skill. In D. H. Holding (ed.), *Human skills, 2nd ed.,* 249–279. New York: Wiley.

Smith, G. (ed.). (1979). *Human Factors, 21*(5).

Soede, M. (1980). *On the mental load of arm prosthesis control.* Leiden, Netherlands: TNO.

Sperling, G., & Dosher, B. A. (1986). Strategy and optimization in human information processing. In K. Boff, L. Kaufman, & J. Thomas (eds.), *Handbook of perception and performance* (vol. 1). New York: Wiley.

Sternberg, S. (1975). Memory scanning: New findings and current controversies. *Quarterly Journal of Experimental Psychology, 27,* 1–32.

Summers, J. J. (1989). Motor programs. In D. H. Holding (ed.), *Human skills, 2nd ed.,* 49–69. New York: Wiley.

Sverko, B. (1977). Individual differences in time-sharing performance. *Acta Instituti Psychologici, 79,* 17–30.

Tayyari, F., & Smith, J. L. (1987). Effect of music on performance in human-computer interface. *Proceedings of the 31st annual meeting of the Human Factors Society* (pp. 1321–1325). Santa Monica, CA: Human Factors Society.

Tsang, P. S., & Wickens, C. D. (1988). The structural constraints and strategic control of resource allocation. *Human Performance, 1,* 45–72.

Venturino, M. (1991). Automatic processing, code dissimilarity, and the efficiency of successive memory searches. *Journal of Experimental Psychology: Human Perception and Performance, 17,* 677–695.

Vicente, K. J., Thornton, D. C., & Moray, N. (1987). Spectral analysis of sinus arrhythmia: A measure of mental effort. *Human Factors, 29*(2), 171–182.

Vidulich, M. A. (1988). Speech responses and dual task performance: Better time-sharing or asymmetric transfer. *Human Factors, 30,* 517–534.

Vidulich, M. A., & Tsang, P. S. (1986). Techniques of subjective workload assessment: A comparison of SWAT and the NASA-bipolar methods. *Ergonomics, 29,* 1385–1398.

Vidulich, M. A., & Wickens, C. D. (1986). Causes of dissociation between subjective workload measures and performance. *Applied Ergonomics, 17,* 291–296.

Wetherell, A. (1981). The efficacy of some auditory-vocal subsidiary tasks as measures of the mental load on male and female drivers. *Ergonomics, 24,* 197–214.

Wickens, C. D. (1976). The effects of divided attention on information processing in tracking. *Journal of Experimental Psychology: Human Perception & Performance, 2,* 1–13.

Wickens, C. D. (1980). The structure of attentional resources. In R. Nickerson (ed.), *Attention and performance VIII* (pp. 239–257). Hillsdale, NJ: Erlbaum.

Wickens, C. D. (1984). Processing resources in attention. In R. Parasuraman & R. Davies (eds.), *Varieties of attention* (pp. 63–101). New York: Academic Press.

Wickens, C. D. (1986). Gain and energetics in information processing. In R. Hockey, A. Gaillard, & M. Coles (eds.), *Energetics and human information processing* (pp. 373–390). Dordrecht, Neth.: Martinus Nijhoff.

Wickens, C. D. (1989). Attention and skilled performance. In D. Holding (ed.), *Human skills* (2nd ed., pp. 71–105). New York: Wiley.

Wickens, C. D. (1990a). Applications of event related potential research to problems in human factors. In J. Rohrbaugh, R. Parasuraman, & R. Johnson (eds.), *Event-related brain potentials: Basic issues and applications.* New York: Oxford University Press.

Wickens, C. D. (1990b). Resource management and time-sharing. In J. I. Elkind, S. K. Card, J. Hochberg, & B. M. Huey (eds.), *Human performance models for computer-aided engineering* (pp. 181–202). Orlando, FL: Academic Press.

Wickens, C. D. (1991). Processing resources and attention. In D. Damos (ed.), *Multiple task performance.* London: Taylor & Francis.

Wickens, C. D., & Gopher, D. (1977). Control theory measures of tracking as indices of attention allocation strategies. *Human Factors, 19,* 249–366.

Wickens, C. D., Hyman, F., Dellinger, J., Taylor, H., & Meador, M. (1986). The Sternberg Memory Search task as an index of pilot workload. *Ergonomics, 29,* 1371–1383.

Wickens, C. D., & Kessel, C. (1980). The processing resource demands of failure detection in dynamic systems. *Journal of Experimental Psychology: Human Perception and Performance, 6,* 564–577.

Wickens, C. D., Larish, I. A., & Contorer, A. (1989). Predictive performance models and multiple task performance. *Proceedings of the 33rd annual meeting of the human factors society.* Santa Monica, CA: Human Factors Society.

Wickens, C. D., & Liu, Y. (1988). Codes and modalities in multiple resources: A success and a qualification. *Human Factors, 30,* 599–616.

Wickens, C. D., Mountford, S. J., & Schreiner, W. S. (1981). Multiple resources, task-hemispheric integrity, and individual differences in time-sharing efficiency. *Human Factors, 22,* 211–229.

Wickens, C. D., Sandry, D., & Vidulich, M. (1983). Compatibility and resource competition between modalities of input, output, and central processing. *Human Factors, 25,* 227–248.

Wickens, C. D., Vidulich, M., & Sandry-Garza, D. (1984). Principles of S-C-R compatibility with spatial and verbal tasks. *Human Factors, 26,* 533–543.

Wickens, C. D., & Yeh, Y. Y. (1985). POC's and performance decrements: A reply to Kantowitz and Weldon. *Human Factors, 27,* 549–554.

Wierwille, W. W., & Casali, J. G. (1983). A validated rating scale for global mental workload measurement applications. *Proceedings of the 27th Annual Meeting of the Human Factors Society.* Santa Monica, CA: Human Factors Society.

Wierwille, W. W., & Williges, R. C. (1978, September). *Survey and analysis of operator workload assessment techniques* (Report No. S-78-101). Blacksburg, VA: Systemetrics.

Wierwille, W. W., & Williges, B. H. (1980). *An annotated bibliography on operator mental workload assessment* (Report No. SY-27R-80). Patuxent River, MD: Naval Air Test Center.

Williges, R. C., & Wierwille, W. W. (1979). Behavioral measures of aircrew mental workload. *Human Factors, 21,* 549–574.

Yeh, Y-Y., & Wickens, C. D. (1988). The dissociation of subjective measures of mental workload and performance. *Human Factors, 30,* 111–120.

Chapter
10

Stress and Human Error

Integrative – Dual Causality.

Disasters such as the *Vincennes* incident (see Chapter 7) or the Three Mile Island accident (Chapter 1) have made people question the extent to which high levels of stress, either existing before the disaster or caused by the first few seconds of crisis, will degrade human information processing. This degradation may compound the effects of any initial error or failure that led to the crisis in the first place. Indeed, it often seems that stress and errors are tightly linked. When errors are made (and we become aware of them), they cause stress; and when high levels of stress exist, errors are more likely to occur. Hence, it is appropriate that we discuss in the same chapter these two important dimensions of human performance, each of which is relevant to all of the stages of processing and mental operations already examined. We begin by discussing stress, its definition, its measurement, and most important, its effects.

STRESS

The concept of stress is most easily understood in the context of Figure 10.1. On the left of the figure is a set of *stressors*, influences on information processing and cognition that are not inherent in the content of that information itself. Stressors may include such influences as noise, vibration, heat, dim lighting, and high acceleration, as well as such psychological factors as anxiety, fatigue, frustration, and anger. Such forces typically have three manifestations: (1) They are a phenomenological *experience*. For example, we are usually (but not always) able to report a feeling of frustration or arousal as a consequence of a stressor. (2) Closely linked, a change in physiology is often observable. This might be a short-term change—such as the increase in heart rate associated with flying an aircraft (Hart & Hauser, 1984) or the stress of air traffic controllers in high-load situations (Romhert, 1979)—or it might be a more sustained effect—such as the change in the output of catecholamines, measured

① subjective experience

② Δ physiology

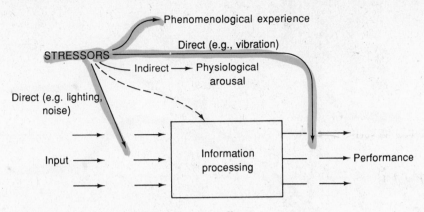

Figure 10.1 A representation of stress effects.

in the urine, after periods of flying simulated combat maneuvers in an F16 (Burton, Storm, Johnson, & Leverett, 1977) or actual battlefield events (Bourne, 1971). The phenomenological and physiological characteristics are often, but not invariantly, linked. (3) They affect the efficiency of information processing, although it need not always degrade performance.

As Figure 10.1 shows, these effects may be either direct or indirect. Direct effects influence the quality of information received by the receptors, or the precision of the response. For example, vibration will reduce the quality of visual input and motor input, and noise will do the same for auditory input. Time stress may simply curtail the amount of information that can be perceived in a way that will quite naturally degrade performance. Direct effects also include the effects of noise on working memory because noise interferes with the ability to use the phonetic loop in rehearsal (Poulton, 1976; see Chapter 6). They can also describe the "distraction" experienced by an operator with personal or family problems, who may redirect attention to thought about these problems rather than to the job at hand (Wine, 1971).

Some of these stressors, however—like noise or vibration—as well as others for which no direct effect can be observed—like anxiety, fear, or incentives—appear to influence the efficiency of information processing through mechanisms that have not yet been described. Many of the effects are mediated by *arousal*, and these will represent the major focus of the following discussion.

Arousal and the Yerkes-Dodson Law

One of the earliest laws in psychology was based on the performance of rats, in a discrimination learning task, under various levels of stress induced by an electric shock for incorrect performance. An inverted U-shaped function related the level of arousal to performance, as shown in Figure 10.2, and became known as the Yerkes-Dodson law after the researchers in the initial study

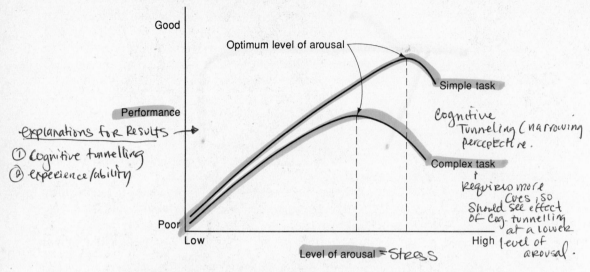

Figure 10.2 The Yerkes-Dodson law.

[Handwritten annotations:]

explanations for Results →
① cognitive tunnelling
② experience /ability

Cognitive Tunneling (narrowing perceptctive.

Complex task
↑
Requires more Cues, so Should see effect of Cog. tunnelling at a lower level of arousal.

more recent:
Simonov - Observed people Jumping out of a plane.

Level of arousal = Stress

(Yerkes & Dodson, 1908). Studying the role of stress, Simonov, Frolov, Evtu-shenko, and Suiridov (1977) measured the performance of parachute jumpers on a visual detection task as the time of their first jump approached. As time passed and their level of stress (assessed by physiological measures) contin-uously increased, performance first improved and then lessened until just before the jump.

Later analysis and synthesis of experimental results by Easterbrook (1959) suggested that the upward and downward "limbs" of the inverted U are the results of quite different factors. The upward limb may be thought of as the result of an "energizing," which simply expands the amount of resources available, in a manner described in the previous chapter (see Figure 9.1). Thus, for example, Kennedy and Coulter (1975) found that performance on a low-arousal vigilance task was actually improved by the stress caused by the threat of shock. In contrast, the downward limb is the consequence of a more specific effect of high arousal on the selectivity of attention or "tunneling" to different environmental or internal cues. Using the metaphor adopted in Chapter 3, high levels of stress narrow the spotlight of attention.

Extending this argument logically leads to the conclusion that high arousal should be more detrimental to tasks that are complex and require a diversity of cues than to those that are simple and depend on only a few cues (like a simple RT task). This is seen by the two curves in Figure 10.2: The optimum level of arousal for the simple task is higher than for the complex task.

For example, consider a classic experiment by Berkun (1964). Using Army soldiers engaged in simulated military tasks in the field, Berkun employed three different experimental manipulations to induce a very realistic experi-ence of stress. In one manipulation, the subjects were led to believe—as they attempted to fill out an insurance form—that the aircraft in which they were

flying was in danger of crashing. In a second manipulation, the subjects were
led to believe that artillery shells were exploding around them, the result of a
confusion in their location by the artillery, as they tried to follow procedures to
initiate a radio call to redirect the fire. In the third manipulation, the subjects
believed that a demolition had seriously injured one of their fellow soldiers,
and they needed to follow procedures in calling for help. Thus, in all cases, the
subjects believed that they or someone they felt responsible for was at serious
mortal risk. Berkun found that in all three cases, the high level of stress and
arousal induced by the anxiety or perceived danger led to a degradation in
following the necessary procedures. However, this degradation was less for
soldiers with greater experience.

Experience Decreased the cognitive tunneling

Berkun's (1964) finding that the negative effects of stress are reduced by
higher skill and ability, also observed by Lazarous and Erickson (1952), may be
placed into the context of Figure 10.2 if it is assumed that the more skilled
performer, described by the curve on the right, faces a task of reduced complexity and can therefore perform better at a similar stress level. The complexity may be reduced for skilled soldiers because their greater knowledge allows
chunking of the perceptual cues (see Chapter 6) or because their greater
knowledge allows a more direct and automatic retrieval of the appropriate
actions from long-term memory.

One of the problematic characteristics of the Yerkes-Dodson law is that it is
difficult if not impossible to know, a priori, where the optimum level of arousal
is for a particular task and, hence, whether the introduction of a stressor will
lead to an initial increase or decrease in task performance. This shortcoming
makes the law inadequate as a predictive model for stress effects. Nevertheless,
the inverted-U function has been useful in understanding stress effects and also
the interaction between them. (See Qualitative Pattern of Stress Effects.)

Berkun's (1964) study of Army soldiers has presented ethical concerns,
given the severe deception involved. Indeed, since the mid-1960s, government
and university regulations have been severely tightened. Whether the ethical
costs of deception are outweighed by the benefits gained in scientific knowledge is an issue that will always be of concern when efforts are made to induce
psychological stress experimentally in order to understand its effects. Yet in the
case of Berkun's research, it can be argued that the knowledge gained about
the *qualitative* manner in which performance is affected compensates for the
psychological risks imposed on the subjects.

Qualitative Pattern of Stress Effects

Hockey (1984, 1986) has expanded on the two factors (arousal and tunneling)
underling the Yerkes-Dodson law. Reviewing a large number of studies that
have examined the effects of stress on human performance, Hockey has dis-
tilled the pattern of stress effects into the form shown in Figure 10.3. Each
stressor appears to be identified by a profile or "signature" of effects across a
set of five critical information-processing components: general arousal, selectivity of attention, speed and accuracy of performance, and short-term memory

| | Performance indicators speeded responding | | | | | |
	GA	SEL	S	A	STM	Sources/Reviews
Noise	+	+	0	−	−	2, 3, 4, 5, 7, 8
Anxiety	+	+	0	−	−	4, 12
Incentive	+	+	+	+	+	2, 4, 5
Stimulant drugs	+	+	+	0	−	2, 4, 13
Later time of day	+	?	+	−	−	1, 2, 4, 5, 6, 8
Heat	+	+	0	−	0	2, 4, 11
Alcohol	−	+	−	−	−	2, 4, 7, 8, 13
Depressant drugs	−	−	−	−	−	2, 4, 10, 13
Fatigue	−	+	−	−	0	2, 4, 9
Sleep loss	−	−	−	−	0	2, 4, 5, 7, 8
Earlier time of day	−	?	−	+	+	1, 2, 4, 5, 6, 8

Figure 10.3 The patterning of stress effects across different performance indicators. The figure summarizes the typical outcome in various studies using these stress variables in terms of their effect on the five behavioral indicators shown: GA = general alertness/activation (subjective or physiological arousal); SEL = selectivity of attention; S and A refer to overall speed and accuracy measures in speeded responding tasks; STM = short-term memory. A plus (+) indicates a general increase in this measure, a zero either no change or no consistent trend across studies, and a minus (−) a general tendency for a reduction in the level of the indicator. A question mark is used to indicate cells where there is insufficient data. Sources of data: (1) Blake (1967a, 1971); (2) Broadbent (1971); (3) Broadbent (1978); (4) Davies & Parasuraman (1982); (5) M. W. Eysenck (1982); (6) Folkard (1983); (7) Hamilton, Hockey, & Rejman (1977); (8) Hockey (1979); (9) Holding (1983); (10) Johnson and Chernik (1982); (11) Ramsey (1983); (12) Wachtel (1967, 1968); (13) Wesnes and Warburton (1983). (*Source for this table:* G. R. J. Hockey, "Changes in Operator Efficiency as a Function of Stress, Fatigue, and Circadian Rhythms" in K. R. Boff, L. Kaufman, and J. P. Thomas (eds)., *Handbook of Perception and Human Performance.* Chichester, England: John Wiley & Sons, 1986. Copyright © 1986 by John Wiley & Sons. Reprinted by permission of John Wiley & Sons, Inc.)

(working memory) capacity. Hockey's synthesis is particularly important because it suggests that although certain stressors may be quite different from one another in how they are induced, they may nevertheless produce qualitatively similar patterns of effects. Consider noise and anxiety, for example, which both show an identical pattern of effects in spite of their different nature.

The information-processing characteristics outlined in Figure 10.3 have all been described in some fashion in the previous chapters. Our concern here is to focus on each one as it is influenced by a stressor that is an important characteristic of many operational environments: the anxiety, fear, or arousal associated with failures of task performance or dangerous, threatening environments. This fear is often coupled with the shortage of time to deal with the crisis (Svenson & Mahle, 1991). For example, a pilot following an engine stall in midair or an inexperienced bank teller facing a long line of impatient customers and an automated and uncooperative transaction system. We do not wish to deemphasize the importance of other stressors, such as sleep loss, alcohol, or vibration, whose effects are presented in Figure 10.3. The reader should refer to more detailed treatments by Broadbent (1971), Hamilton and Warburton (1984), and Hockey (1986). In the following review of research, we also capitalize on the psychological equivalence of signature patterns of noise and anxiety, allowing us to extrapolate from the effects of noise to those of danger-produced anxiety. We use the term *stress* to refer to these combined effects of danger, anxiety, and noise.

Attentional Narrowing Weltman, Smith, and Egstrom (1971) compared the performance of two groups of subjects on a central and peripheral detection task. One group was led to believe it was experiencing the conditions of a 60-foot dive in a pressure chamber, and the other was not. In fact, there was no change in pressure for either group. Both groups showed similar performance on the central task, but performance on the peripheral task was significantly degraded for the "pressure" group. This group also showed greater anxiety increases in heart rate, substantiating the increased level of stress. Similar perceptual-narrowing effects have also been found by other investigators (e.g., Bacon, 1974; Baddeley, 1972; Hockey, 1970) Although stress-produced perceptual tunneling will usually degrade performance, it is also possible to envision circumstances in which it may actually facilitate performance, that is, when focused attention is desired. Indeed, this positive effect was observed in a study by Houston (1969) in which the presence of noise facilitated rather than impaired performance on the Stroop task, described in Chapter 3. In this task, it will be recalled, one's ability to report the color ink in which a word is printed is disrupted if the word spells a color name. The failure to focus attention normally activates the competing but irrelevant color name in place of the ink color. However, the presence of noise apparently leads to greater focus on the relevant (ink color) aspect of the task and to improved Stroop performance.

As Houston's (1969) data would suggest, it appears that the stress effect on tunneling is not simply defined by a reduction of the spatial area of the attention spotlight, so that peripheral stimuli are automatically filtered. Rather the filtering effect seems to be defined by subjective importance, or *priority*. Performance of those tasks of greatest subjective importance remain unaffected—or perhaps enhanced (through arousal)—in their processing, whereas those of lower priority are filtered (Bacon, 1974; Broadbent, 1971; Hockey, 1970). In one sense this kind of tunneling is optimal, but it will provide undesirable effects if the *subjective* importance of the attended channel proves to be unwarranted. Such was the case, for example, in the Three Mile Island incident. Operators, under the high stress following the initial failure, appeared to fixate their attention on the one indicator supporting their belief that the water level was too high, thereby filtering attention from more reliable indicators that supported the opposite hypothesis. This narrowing effect can be directly related to biases in decision making.

Working Memory Loss   Davies and Parasuraman (1982) and Wachtel (1968) have directly identified the effects of anxiety on working memory. Many of the difficulties Berkun (1964) observed when his army subjects were placed under the stress of perceived danger can also be attributed to reduced working memory capacity. Noise, as well as danger and anxiety, has also produced consistent effects on working memory. Although it is intuitively evident that noise would disrupt the verbal phonetic working memory system (Poulton, 1976), it appears also that the combined effects of noise and anxiety may disrupt spatial working memory systems as well (Stokes & Raby, 1989). Indeed, in a simulation study of pilot decision making, Wickens, Stokes, Barnett, and Hyman (1991) observed that the effects of noise were greatest on problems that relied on spatial visualization for their successful resolution.

Given the important role of working memory in encoding new information into long-term memory, it would appear that stress would not lead to efficient learning (Keinan & Friedland, 1984). This reasoning is certainly one of the important factors behind the advocacy of *simulators* as useful training devices for dangerous activities such as flying or deep-sea diving (Flexman & Stark, 1987; see also Chapter 6).

Long-Term Memory   Although stress seems to hinder the encoding of new information into long-term memory, it appears not to disrupt the retrieval of information from long-term memory, to the extent that that information is well rehearsed and memorized. For example, in their study of pilot judgment, Wikens, Stokes, Barnett, and Hyman (1991) found that those judgments requiring direct retrieval of facts from long-term memory were relatively unimpaired by stress. Stokes, Belger, and Zhang (1990) found that expert pilots' decision-making ability was not impaired by a combination of noise and anxiety stress to the same extent as that of novices, and they assumed that the experts' judgments depended more on direct long-term memory retrieval (see Chapter 6). Similarly, in Berkun's (1964) study of Army soldiers, performance of those with greater experience was less degraded by stress. However, similar to its narrowing effect on perception and selective attention, stress also narrows the information retrieved from long-term memory specifically to those habits that are well learned or overlearned (Eysenck, 1976). Studies by Fitts and Seeger (1953) and Fuchs (1962), as well as a review of aircraft accidents by Allnut (1987), have all suggested that stress will lead to a regression to earlier and more compatible response patterns when they conflict with incompatible patterns. Collectively, these conclusions emphasize the importance of extensive training of those procedures and actions that may need to be taken in emergencies (when stress will probably be high). They further emphasize that such procedures should require only actions of high compatibility.

Strategic Shifts   Through the book we have discussed different styles of behavior, such as different sets for speed versus accuracy or conservative versus risky criterion placement in signal detection. These styles lie along a continuum, on which neither end is necessarily good or bad. Instead, the optimum point on the continuum will depend on external circumstances. There is some evidence that stress leads to consistent shifts in processing strategy. In their study of the anxiety brought on by the first parachute jump, for example, Simonov, Frolov, Evtushenko, and Suirdov (1977) observed a shift in signal detection performance to a "riskier" criterion setting with more hits and false alarms. Hockey (1986) concludes that there is a general effect of noise and anxiety stress on the speed-accuracy trade-off, a shift to less accurate but not slower performance (e.g., Lazarus & Erickson, 1952). In their study of pilot judgment, Wickens, Stokes, Barnett, and Hyman (1991) found that decisions were less accurate but not necessarily slower under the combined effects of noise, time pressure, and threat of loss of income.

The tendency of the stress of emergency to cause a shift in performance from accurate to fast (but error prone) responding has been cited as a concern in operator response to complex failure in nuclear power control rooms. As discussed in Chapter 8, the operator has a desire to "do something" rapidly, when in fact, this impatience is often counterproductive until the nature of the failure is well understood. The hasty action of the control room operators in response to the Three Mile Island incident was to shut down an automated device that had in fact been properly doing its job. To combat this tendency for a nonoptimal speed-accuracy shift in an emergency, nuclear power plant regulations in many countries explicitly require operators to perform no physical actions for a fixed time following an alarm while they gain an accurate mental picture of the nature of the malfunction.

Decision Making   The effects of stress on decision making has always been a topic of great interest and has been enhanced in its importance recently by the inquiry into the *Vincennes* incident (U.S. Navy, 1988; see Chapter 7). The concern that decisions degrade under stress is reinforced by anecdotes and case studies of poor pilot judgments that have occurred during bad weather, spatial disorientation, or aircraft failure (Jensen, 1982; Nagel, 1988; Simmel, Cerkonik, & McCarthy, 1987). Yet as noted in Chapter 7, without tight experimental control, it is often difficult to know if a real-world decision that failed was in fact a poor one in foresight as well as in hindsight. Furthermore, it is often difficult to tell whether the stress was itself a causal factor or whether the conditions that produced it also, for example, degraded the information available in such a way that the poorer decision became more likely.

To predict the effects of stress on decision making, it is possible to adopt a *componential* approach (Wickens & Flach, 1988). Since different decisions may involve varying dependence on such components as working memory, attention, and long-term memory retrieval, each decision will be affected differently by stress as a function of the components on which it depends (Wickens et al., 1991).

An alternative approach is through direct experimental manipulation of stress on decision making, diagnosis, and problem solving. Although few such studies exist, their results are consistent with the picture of stress-sensitive decision making. Such studies have shown not only that decisions of various sorts degrade under stress but also that this degradation takes specific forms. For example, Cowen (1952) found that subjects persevered longer with inappropriate or rigid problem solutions under the stress produced by threat of shock, a sort of action tunneling consistent with the idea of cognitive tunneling. Keinan, Friedland, and Ben-Porath (1987) found that the allocation of attention to a word problem became increasingly nonoptimal and unsystematic as stress was imposed by the threat of an electric shock. They also observed that this stressor produced a premature closure: The subjects terminated their decision process before all alternatives had been considered.

As discussed, Wickens, Stokes, Barnett, and Hyman (1991) observed that the combined stress of noise, time pressure, risk, and task loading produced a general degradation of pilot judgments on a computer-based flight simulation. The stress effect, however, was selectively observed only on problems that were difficult in spatial memory demand. Their data, indicating that decisions do not degrade when long-term memory retrieval is the primary mechanism, suggest that expert decision makers, who use a perceptual pattern-match technique, will be less likely than novices to suffer from the degrading effects of stress (Klein, 1989; see Chapter 7). As noted earlier, this hypothesis was confirmed by Stokes, Belger, and Zhang (1990).

Mediating Effects

The prediction of stress effects is complicated by the fact that the effects of combinations of stressors are often complex and follow neither the simple additive nor multiplicative form discussed in Chapter 8 (Broadbent, 1971). Thus, it is not possible to say that the effect of two stressors in combination is the sum of their individual effects (additive), nor even necessarily that it is greater than that sum (a positive interaction). Sometimes the effects of one stressor may compensate for another's degrading effects. Fortunately, however, these interactive effects may often be understood in the context of the inverted-U function of the Yerkes-Dodson law (Broadbent, 1971). For example, although both sleep loss and noise will typically degrade performance, their effect in combination is generally less than either effect alone (Wilkinson, 1963). This statement makes sense when it is realized that each effect is pulling arousal in a different direction on the function of Figure 10.2, and the net effect is to maintain arousal near the peak.

Individual differences between operator personality types describe a second mediating effect. These types are complex and not well understood, and their full treatment is well beyond the scope of this book. Two, however, will be briefly discussed. *Locus of control* describes the extent to which individuals believe that they, rather than other forces, have control over things that influence their lives. These two beliefs describe an *internal* versus *external* locus of control, respectively. There is some evidence that those with an internal locus of control are less stressed by an anxiety-provoking situation because of their belief that they can exert some control over it. The distinction between *introverted* and *extraverted* personality types has also been found to mediate the effects of stressors. In general, extroverts are more affected by all stressors (Hockey, 1986), although more specific patterns of differences between the two personality types, the time of day (A.M. versus P.M.), and stimulants that increase or decrease the level of arousal have also been reported (Revelle, Humphreys, Simon & Gilliland, 1980).

A third mediating effect on stressors is training, whose effects may be realized in at least two ways. First, as noted, well-trained habits will be less likely to degrade under stress than those that are less practiced. Second, successful experience in a stressful situation can greatly reduce the anticipated anxiety of repeated performance (Mandler, 1979). For example, Ursin, Baade, and Levine (1978) assessed the physiological measures of stress before the first and second parachute jumps of a group of trainees. The investigators observed a large drop in those measures between the two jumps, signaling the relief, as it were, that successful performance was possible. This finding emphasizes the importance of one's own perception of the situation in the experience of stress. Two people may be in an identical situation in terms of external stressors. But if they differ in how they believe they will be able to cope, their experienced stress will probably be quite different (Coyne & Lazarus, 1980).

Coping with Stress

A variety of techniques may be adopted in the effort to minimize the degrading effects of stress on human performance. Roughly these may be categorized as *design solutions*, which address the task, and *personal solutions*, which address the operator, either through training or through strategies.

Design Solutions Design solutions may focus on the human factors of displays. If perceptual narrowing among information sources or unsystematic scanning does occur, reducing the amount of unnecessary information (visual clutter) and increasing its organization will surely somewhat buffer the degrading effects of stress. For example, Zhang and Wickens (1990) found that integration of separate dimensions into a single object display (see Chapter 3) lessened the degrading effects of noise on performance in a multitask environment. Schwartz and Howell (1985) found that the degrading effects of time pressure on a simulated decision task were reduced by using a graphic rather than a digital display. Similarly, it is clear that any design efforts that minimize the need for operators to maintain or transform information in working memory should be effective. Emergency procedures that must be referred to on-line *must* be clear and simply phrased (see Chapter 5), as they will undoubtedly need to be followed under the very circumstances that make working memory for their contents extremely fragile.

It would seem important also that procedural instructions of what to do should be redundantly coded with speech as well as with print, should avoid arbitrary symbolic coding (abbreviations or tones, other than general alerting alarms), and should be phrased in direct statements of what action to take rather than as statements of what not to do (avoid negatives) and should augment any information that only describes the current state of the system. This is the policy inherent in voice alerts for aircraft in emergencies, in which commands are directed to the pilot of what to do to avoid collision ("Climb, climb, climb") (*Avionics*, 1990).

Training We have noted before the beneficial effects of training, in particular, *extensive* training of key emergency procedures so that they become the dominant and easily retrieved habits from long-term memory when stress imposes that bias. In fact, a case can possibly be made that training for emergency procedures should be given greater priority than training for routine operations, particularly when emergency procedures (or those to be followed in high-stress situations) are in some way *inconsistent* with normal operations. For example, the procedure to be followed in an automobile when losing control on ice is to turn in the direction toward the skid, precisely the opposite of our conventional turning habits in normal driving. Clearly, where possible, systems should be designed so that procedures followed under emergencies are as consistent as possible with those followed under normal operations. Adhering to fundamental principles of design compatibility, in all of its forms, is certainly one of the best techniques for minimizing the damaging effects of stress on performance.

There are also a number of strategies that can be adopted in designing a training program for performance under stress, although many of these lack clear empirical validation (Keinan & Friedland, 1984). Finally, certain strategies can be adopted by the performer to minimize the degrading effects of stress. Certainly one of the most effective is planning, anticipating, and rehearsing actions that may need to be taken under stress—either expected or produced by emergencies. Although advanced planning demands working memory capacity at the time it is being carried out, it can certainly compensate for the fact that that capacity will not be available under stressful conditions.

Naturally, planning is only as effective as the degree to which future possible events are envisioned, and it is always possible that the emergency that occurs imposes a condition and requires an action that was not planned for. In these circumstances, as well as in those emergencies for which plans have been made, psychologists have proposed many stress-coping techniques related to relaxation and lowering the level of arousal through self-control (Druckman & Swets, 1988). It is not clear, however, whether many of these techniques, based on biofeedback or cognitive restructuring, are as relevant to the immediate stress of emergencies as they are to prolonged stress of a difficult work setting or personal tension. Nevertheless, it is probably true that some degree of "calming" through attention to regular breathing can improve information-processing characteristics under situations of overly high arousal (Douglas, 1989).

Summary

It is clear that considerably less is known about stress effects and the appropriate techniques for their remediation than about many other aspects of performance. This shortcoming results in part from the great difficulty in conducting research in the area—that is, imposing realistic credible stressors in a controlled setting in a way that is also consistent with the ethics of research. It also results from the complex mediating effects of factors like personality and appraisal of the situation on the performance effects. Still, enough data are available from incident reports, the research described above, and extensive reviews by Broadbent (1971) and Hockey (1986) to construct a reasonably coherent picture of those effects and suggest some possible remediations.

As we have seen, one consequence of stress is a degradation of performance. Indeed, given the nature of the speed-accuracy trade-off, it appears that this degradation is more likely to appear as an increase in error rate rather than in processing time. Hence, we turn now to an in-depth discussion of errors—both those resulting from stress and those resulting from other causes.

HUMAN ERROR

In all phases of human performance, errors seem to be a frequent occurrence. It has been estimated in various surveys that human error is the primary cause of 60 to 90 percent of major accidents and incidents in complex systems such as nuclear power, process control, and aviation (Rouse & Rouse, 1983). Card, Moran, and Newell (1980) estimated that operators engaged in word processing make mistakes or choose inefficient commands on 30 percent of their choices. In one study of a well-run intensive care unit, doctors and nurses were estimated to make an average of 1.7 errors per patient per day (Gopher et al., 1989); and although the overall accident rate in commercial and business aviation is extremely low, the *proportion* of accidents attributable to human error is considerably greater than that due to machine failure (Nagel, 1988). In the face of these statistics, it is important to reiterate a point made in Chapter 1—that many of the errors people commit in operating systems are the result of bad system design or bad organizational structure rather than irresponsible action (Norman, 1988; Reason, 1990). Furthermore, although human error in accident analysis may be statistically defined as a contributing cause to an accident, usually the error was only one of a lengthy and complex chain of breakdowns—many of them mechanical—that affected the system and weakened its defenses (Perrow, 1984).

We have already discussed human error in various guises and forms, as we have discussed the different ways in which human performance can fall short. This failure may involve nonoptimal levels of beta and d', producing too many misses and false alarms, or errors in classification in an absolute judgment task. Indeed, many of the analyses of human error presented here point to concepts discussed elsewhere as contributing causes: poor discriminability and confusion of medical equipment in the intensive care unit (Gopher et al., 1989); failures of memory in the analyses of process control or ship accidents (Perrow, 1984; Reason, 1990); failures of data entry into the flight management computer of modern aircraft (Wiener, 1989); failures to interpret aviation displays correctly (Fitts & Jones, 1960a, b); breakdowns in aircraft communications from a lack of redundancy (Nagel, 1988); biases in diagnosis and decision making in a variety of cognitive tasks in the nuclear power industry (Reason, 1989; Woods, 1984); or selection of compatible (but incorrect) responses in using simple household devices (Norman, 1988).

We have also considered human error directly in the context of the speed-accuracy trade-off and reaction time. Cognitive sets for fast performance produce more errors; some design or environmental factors (e.g., stress or voice display) lead to faster but more error-prone responding. When processing is easy, faster responses are more likely to be in error; but when processing is more difficult, because of low perceptual quality or high memory load, fast responses are more likely to be correct.

Why focus, then, explicitly on the *product* of human error, having covered the *processes* by which that error is generated? One reason is that the study of human error has itself recently emerged as an important and well-defined discipline (Hurst & Hurst, 1982; Norman, 1981a; Reason, 1990; Senders & Moray, 1991). Many human factors practitioners have realized that errors made in operating systems are far more important and costly than delays of the 10 to 500 msec magnitude typically observed in RT studies. This realization has forced human performance theorists to consider the extent to which design guidelines based on RT generalize to error prediction; it has also led researchers to consider classes of errors that do not necessarily result from the speed stress typical of the RT paradigm—for example, forgetting to change a mode switch on a computer or pouring orange juice rather than syrup on your waffles (Norman, 1981a, 1988; Reason, 1984, 1990). A second reason is that a number of important human factors concerns in the treatment of human error cannot be isolated only in errors of a certain kind (e.g., nonoptimal beta settings or fast guesses in reaction time). Rather, these concerns, related to the statistics of error description and prediction and the approaches to error remediation, address errors of all kinds, no matter from which stage of processing they originate. Hence, it is appropriate to treat them together in a single chapter.

We first chart recent developments in categorizing human error within a framework that is consistent with the information-processing model presented in this book. We next turn to efforts to predict human errors statistically through the applications of reliability analyses. We then place human error in the context of larger organizational factors, and finally, we consider its remediation. How should training, system design, or task design be brought to bear to reduce its likelihood or impact.

Categories of Human Error: An Information-Processing Approach

Although various forms of error taxonomies have been proposed through the years, two roughly parallel developments by Donald Norman (1981, 1988) and James Reason (1984, 1990) have revealed an important dichotomy: mistakes and slips. The context for understanding this dichotomy is presented in Figure 10.4. The human operator, confronting a state of the world represented by stimulus evidence, may or may not interpret that evidence correctly and, given an interpretation, may or may not intend to carry out the right action to deal

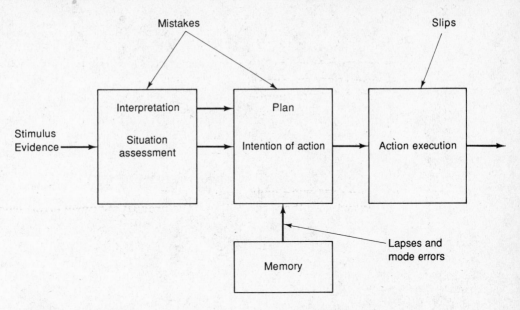

Figure 10.4 Information processing context for representing human error.

with the situation and may or may not execute that intention correctly. Errors of interpretation or of the choice of intentions are called *mistakes*. Thus, the misdiagnosis of the status of the nuclear power plant at Three Mile Island is a clear example of a mistake. So, too, would be the misunderstanding of the meaning of a button on a push-button phone — a misunderstanding that would lead to its incorrect use.

Quite different from mistakes are *slips, in which* the understanding of the situation is correct and the correct intention is formulated, but the wrong action is accidentally triggered. One common example is the typist who presses the wrong key or the pilot who grabs the control of the flaps instead of the landing gear.

As shown in Figure 10.4, it is possible for either or both kinds of errors to occur in a given operation. In fact, it may even be possible for both kinds of errors to cancel one another out in various ways. For example, you mistakenly formulate the intention to push the wrong button, but you slip and push what actually turns out to be the right one. Mistakes and slips have a number of interesting characteristics and discriminating features, but a full understanding of error in human information processing requires the categorization of error types to be somewhat expanded to include those related to memory failures. Following are five categories that have been synthesized from the more detailed schemes (and excellent readings) of Norman (1981*a*, 1988) and Reason (1984, 1990).

Mistakes　Mistakes — failing to formulate the right intentions — can actually result from the shortcomings of perception, memory, and cognition. Reason

(1990), using Rasmussen's (1983) terminology, has discriminated between knowledge-based and rule-based mistakes (discussed in Chapter 8). *Knowledge-based-mistakes* are very much like the kinds of errors made in decision making, in which incorrect plans of actions are arrived at because of a failure to understand the situation. Such failures result from the influences of many of the biases described in Chapters 5, 6, and 7. Operators misinterpret communications, their working memory limits are overloaded, they fail to consider all the alternatives, they succumb to a confirmation bias, and so forth. In short, the human operator is often overwhelmed by the complexity of evidence and lacks the knowledge or clear display of information to interpret it correctly.

Rule-based mistakes, in contrast, occur when operators are somewhat more sure of their ground. They know (or believe they know) the situation, and they invoke a rule or plan of action to deal with it. The choice of a rule follows an "if-then" logic. When understanding the environmental conditions (diagnosis) matches the "if" part of the rule or when the rule has been used successfully in the past, the "then" part is activated. The latter may be an action—"If my computer fails to read the disk, I'll reload and try again"—or simply a diagnosis—"If the patient shows a set of symptoms, the patient has a certain disease."

Why might rules fail? Reason (1990) proposes three possibilities. First, a good rule might be misapplied when the "if" conditions that trigger it are not actually met by the environment. This mistake often occurs as exceptions to rules are learned. The rule has worked well in most cases, but subtle distinctions in the environment or context now indicate that it is not always appropriate. These distinctions or qualifications might be overlooked, or their importance might not be realized. For example, although it is usually appropriate to turn a vehicle in the direction in which you wish to go, an exception occurs when skidding on ice. The correct rule then is to turn first toward the direction of the skid to regain control of the vehicle. Second, rule-based mistakes can result when a "bad rule" is applied. The cause might be an encoding deficiency in which the "if" part of the environment is simply misinterpreted. Third, the "then" part of the rule may be incorrect or poorly chosen.

Reason (1990) argues that the choice of a rule is guided very much by frequency and reinforcement. Rules that have frequently been employed in the past under certain circumstances, particularly if they have been successful and therefore reinforced, will be chosen. But whatever the cause, rule-based mistakes tend to be made with a relative degree of certainty, as the operator believes that the triggering conditions are right and that the rule is appropriate and correct. Thus, Reason describes rule-based mistakes as "strong but wrong."

There are some important differences between rule-based and knowledge-based mistakes. Rule-based mistakes will be performed with confidence, whereas in a situation in which rules do not apply and where knowledge-based mistakes are more likely, the operator will be less certain. The latter situation will also involve far more conscious effort, and the likelihood of making a mistake while functioning at a knowledge-based level is higher than at a rule-based level (Reason, 1990) because there are so many more ways in which information processing can fail—through shortcomings of attention, working memory, logical reasoning, and decision making.

Slips   In contrast to mistakes, in which the intended action is wrong (either because the diagnosis is wrong or the rule for action selection is incorrect), slips are errors in which the right intention is incorrectly carried out. A common class of slips are *capture errors*, which result when the intended stream of behavior is "captured" by a similar, well-practiced behavior pattern. Such a capture is allowed to take place for three reasons: (1) The *intended* action (or action sequence) involves a slight departure from the routine, frequently performed action; (2) some characteristics of either the stimulus environment or the action sequence itself is closely related to the now inappropriate (but more frequent) action; and (3) the action sequence is relatively automated and therefore not monitored closely by attention. As Reason (1990) says, "When an attentional check is omitted, the reins of action or perception are likely to be snatched by some contextually appropriate strong habit (action schema), or expected pattern (recognition schema)."

Pouring orange juice, rather than syrup, on the waffles while reading the newspaper is a perfect example of a slip. Clearly the act was not *in*tended, nor was it *at*tended since attention was focused on the paper. Finally, both the stimulus (the tactile feel of the pitcher) and the response (pouring) of the intended and the committed action were sufficiently similar that capture was likely to occur. A more serious type of slip—related to the same underlying cause—occurs when the incorrect one of two similarly configured and closely placed controls is activated, for example, flaps and landing gear on some classes of small aircraft. Both controls have similar appearance, feel, and direction; they are located close together; both are relevant during the same phases of flight (takeoff and landing); and both are to be operated when there are often large attention demands in a different direction (outside the cockpit). One might also imagine slips occurring in a lengthy procedure of checks and switch setting that is operated in one particular way when a system is in its usual state but involves a change, midway through the sequence when the system is in a different state. In the absence of close attention, the standard action sequence could easily capture the stream of behavior.

Lapses   Whereas slips represent the commission of an incorrect action, lapses represent the failure to carry out an action. As such they can be directly tied to failures of memory, but they are quite distinct from the knowledge-based mistakes associated with working memory overload typical of poor decision making. Instead the typical lapse is what is colloquially referred to as *forgetfulness*. One walks into a room, is interrupted, and then forgets why one went there in the first place. Important lapses may also involve the omission of steps in a procedural sequence. Here again, an interruption is what often causes the sequence to be stopped, then started again a step or two later than it should have been, with the preceding step now missing. The analysis of the crash of a

Northwest Airlines MD–80 outside of the Detroit airport revealed that the pilots, progressing through a routine taxi checklist, were interrupted midway by air traffic control, who requested a runway change. When the checklist was resumed, a critical step of setting the flaps had been skipped, a major cause of the subsequent failure of the aircraft to take off safely (NTSB, 1988).

Mode Errors   Mode errors are a different facet of memory failure than lapses. They result when a particular action that is highly appropriate in one mode of operation is performed in a different, inappropriate mode because the operator has not correctly remembered the appropriate context. An example would be an attempt to raise the landing gear while the aircraft is still on the runway, the pilot having perceived incorrectly the plane is in the air. Mode errors are becoming of increasing concern in more automated cockpits, which have various modes of autopilot control (Wiener, 1988, 1989). Mode errors are also of major concern in human-computer interactions if the operator must deal with keys that serve very different functions, depending on the setting of another part of the system. Even on the simple typewriter, a typist who intends to type a string of digits (e.g., 1965) may mistakenly leave the case setting in the uppercase mode and so produce !(^ %. Mode errors may occur in computer text editing, in which a command that is intended to delete a line of text may instead delete an entire page (or data file) because the command was executed in the wrong mode. Certain computer text-editing systems are particularly unforgiving in this respect because mode errors are quite likely to occur, and their consequences can be drastic (Norman, 1981*b*). Also the design of many modern military aircraft is replete with multimode switches and displays.

Mode errors are a joint consequence of relatively automated performance or of high workload—when the operator fails to be aware of which mode is in operation—and of improperly conceived system design, in which such mode confusions can have major consequences. The reason, of course, that mode errors can occur is that a single action may be made in both appropriate and inappropriate circumstances.

Distinctions Between Error Categories   These categories of error can be distinguished in a number of respects. For example, as already noted, knowledge-based mistakes tend to be characteristic of a relatively low level of experience with the situation and a high attention demand focused on the task, whereas rule-based mistakes, and particularly slips, are associated with higher skill levels. Slips also typify attention directed away from rather than toward the problem in question.

One of the most important contrast between slips, on the one hand, and mistakes and lapses, on the other, is in the ease of detectability. The detection of slips appears to be relatively easy because people typically monitor, consciously or unconsciously, their motor output, and when the feedback of this output fails to match the expected feedback (based on the correctly formulated intentions), the discrepancy is often detected. Thus, as noted in Chapter 8, typing errors (usually slips) are very easily detected (Rabbitt & Vyas, 1970). In contrast, when the intentions themselves are wrong (mistakes), or a step is omitted (lapse) any feedback about the error arrives much later, and errors cannot easily be detected on-line. This distinction is clearly backed up with data. In an analysis of simulated nuclear power plant incidents, Woods (1984) categorized operator errors as slips and mistakes; half of the slips were detected by the operators themselves, whereas none of the mistakes were noted. Reason (1990) summarized data from other empirical studies to conclude that the ease of error correction, as well as error detection, also favors slips over mistakes. This factor is in part related to the easier cognitive process of revising an action rather than reformulating an intention, rule, or diagnosis. However, system design principles, related to the *visibility* of feedback and the *reversibility* of action to be discussed below, can have a large impact on how easy it is to recover from a slip.

Given the many differences between slips and mistakes, it is logical that the two major categories should have somewhat different prescriptions for remediation. Although this issue will be dealt with in more detail later, it is easy to conclude that the heaviest emphasis on preventing slips should focus on system and task design, addressing issues like S-R compatibility and stimulus and control similarity. For the prevention of mistakes, in contrast, it is necessary to focus relatively more on design features related to effective displays (supporting accurate updating of a mental model) and on training (Rouse & Morris, 1987).

Human Reliability Analysis

In the wake of the public's concern for the possibility of a nuclear accident and, therefore, the reliability of nuclear power systems, within which the human operator represents a vital link, a number of efforts have been directed toward applying engineering reliability analysis to the human operator. The objective of this analysis is to predict human error.

A fairly precise analytic technique can predict the reliability (probability of failure or mean time between failures) of a complex mechanical or electrical system consisting of components of known reliabilities that are configured in series or in parallel (Figure 10.5). For example, consider a system consisting of two components, each with a reliability of 0.9 (i.e., a 10 percent chance of failure during a specified time period). Suppose the components are arranged in series, so that if either fails the total system fails (Figure 10.5*a*). This describes "the chain is only as strong as its weakest link" situation. The probability that the system will *not* fail (the probability that both components will work successfully) is $0.9 \times 0.9 = 0.81$. This is the system reliability. Therefore, the probability of system failure is precisely $1 - (0.9 \times 0.9) = 1 - 0.81 = 0.19$. In contrast, if the two components are arranged in parallel (redundantly), as in Figure 10.5*b*, so that the system will fail only if *both* fail, the probability of system failure is $0.1 \times 0.1 = 0.1$. Its reliability is 0.99.

The work of Swain and his colleagues on the *technique for human error rate prediction* (THERP) has attempted to bridge the gap between machine and

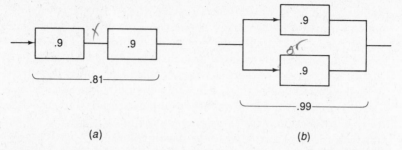

Figure 10.5 (*a*) Two components in series. (*b*) Two components in parallel. The numbers in the boxes indicate the component reliabilities. The numbers below indicate the system reliabilities. Probability of error = 1 reliability.

human reliability in the prediction of human error (Miller & Swain, 1987; Swain, 1990; Swain & Weston, 1988). THERP has three important components.

1. *Human error probability* (HEP) is expressed as the ratio of the number of errors made on a particular task to the number of opportunities for errors. For example, for the task of routine keyboard data entry, a HEP = 1/100. These values are obtained, where possible, from data bases of actual human performance. Given that such data are often lacking, they are instead estimated by experts, although such estimates can be heavily biased and are not always terribly reliable (Reason, 1990). HEPs are typically estimated within a ± 95 percent confidence bracket.

2. When a task analysis is performed on a series of procedures, it is possible to work forward through an *event tree*, or *fault tree*, such as that shown in Figure 10.6. In the figure, the two events (or actions) performed are A and B, and each can be performed either correctly (lower case) or in error (capital). An example might be an operator who must read a value from a table (event A) and then enter it into a keyboard (event B). Following the logic of parallel and serial components, and if the reliability of the components can be accurately determined, it is possible to deduce the probability of successfully completing the combined procedure or, alternatively, the probability that the procedure will be in error, as shown at the bottom of the figure.

3. The HEPs that make up the event tree can be modified by *performance-shaping factors*, multipliers that predict how a given HEP will increase or decrease as a function of expertise or the stress of an emergency. Table 10.1 is an example of the predicted effects of these two variables.

Further advances of human reliability analysis have focused specifically on speed stress, by modeling the probability that specific events will be diagnosed and acted on correctly within a given period of time, subsequent to a system failure (Wreathall, 1982). This and other techniques for estimating human reliability are described in some detail by Reason (1990).

Human reliability analysis represents an admirable beginning to the devel-

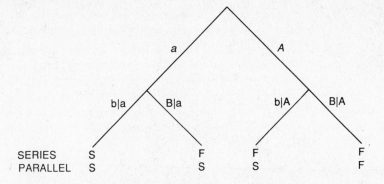

SERIES   S   F   F   F
PARALLEL S   S   S   F

TASK "A" = The first task
TASK "B" = The second task
a = Probability of successful performance of task "A"
A = Probability of unsuccessful performance of task "A"
$b|a$ = Probability of successful performance of task "B" given a
$B|a$ = Probability of unsuccessful performance of task "B" given a
$b|A$ = Probability of successful performance of task "B" given A
$B|A$ = Probability of unsuccessful performance of task "B" given A

For the series system:
$$Pr[S] = a(b|a)$$
$$Pr[F] = 1 - a(b|a) = a(B|a) + A(b|A) + A(B|A)$$

For the parallel system:
$$Pr[S] = 1 - A(B|A) = a(b|a) + a(B|a) + A(b|A)$$
$$Pr[F] = A(B|A)$$

Figure 10.6 Fault tree. (Source: D. Miller and A. Swain, "Human Reliability Analysis," in G. Salvendy (ed.), *Handbook of Human Factors* (New York: John Wiley & Sons, 1987). Copyright © 1987 by John Wiley & Sons, Inc. Reprinted by permission.)

Table 10.1 MODEL ACCOUNTING FOR STRESS AND EXPERIENCE IN PERFORMING ROUTINE TASKS

	Increase in error probability	
Stress Level	Skilled	Novice
Very Low	×2	×2
Optimum	×1	×1
Moderately high	×2	×4
Extremely high	×5	×10

Source: D. Miller & A. Swain, "Human Reliability Analysis," in G. Salvendy (ed.), *Handbook of Human Factors*. New York: John Wiley & Sons, 1987. Copyright © 1987 by John Wiley & Sons, Inc. Reprinted by permission.

opment of predictive models of human error. Its advocates have argued that it can be a useful tool for identifying critical human factors deficiencies. Furthermore, as noted in Chapter 1, providing hard HEP numbers, the output from the model, which document poor human factors in the form of increased predicted errors, can be an effective tool for lobbying designers to incorporate human factors concerns (Swain, 1990). In spite of its potential value, however, human reliability analysis has a number of major shortcomings, which have been carefully articulated by Adams (1982), Reason (1990), and Dougherty (1990; see also Apostolakis, 1990).

Lack of Data Base   There are empirically based data for simple acts, such as dial reading or keyboard entry under nonstressed conditions. However, the data on which estimates of the important cognitive factors, related to diagnosis and problem solving, along with those concerning stress effects, are scanty at best and rely much more heavily on expert opinion. These opinions may be faulty, and so it may sometimes be dangerously misleading to assign precise numerical values to HEPs or performance-shaping factors.

Error Monitoring   When machine components fail, they require outside repair or replacement. Yet as we have seen, humans normally have the capability to monitor their own performance, even when operating at a relatively automated level. As a result, they often correct errors before those errors ultimately affect system performance, particularly capture errors or action slips (Rabbitt, 1978). The operator who accidently activates the wrong switch may be able to shut it off quickly and activate the right one before any damage is done. Thus it is quite difficult to associate the probability of a *human error* with the probability that it will induce a *system error*.

Nonindependence of Human Errors   The assumption is sometimes made in analyzing machine errors that the probability of the failure of one component is independent of that of another. Although this assumption is questionable when dealing with equipment (Perrow, 1984), with humans it is particularly untenable. If we make one mistake, our resulting frustrations may sometimes increase the likelihood of a subsequent bungle. At other times, the first mistake may increase our care and caution in subsequent operations and make future errors *less* likely. Whichever the case, it is impossible to claim that the probability of making an error at time T is independent of whether an error was made at time $T - N$, a critical assumption normally made in reliability analysis. The actuarial data base on human error probability, which is used to predict reliability, will not easily capture these dependencies because they are determined by mood, caution, personality, and other uniquely human properties (Adams, 1982).

A similar lack of independence can characterize the parallel operation of two human "components." When machine reliability is analyzed, the operation of two parallel (or redundant) components is assumed to be independent. For example, three redundant autopilots are often used on an aircraft so that if one fails, the two remaining in agreement will still give the true guidance input. None of the autopilots will influence the others' operation (unless they are all affected by a superordinate factor such as a total loss of power). This independence, however, cannot be said to hold true of human behavior. In the control room of a power station, for example, it is unlikely that the diagnosis made by one operator in the face of a malfunction will be independent of that made by another. Thus, it is unlikely that there will be independent probabilities of error shown by the two operators. Social factors may make the two operators relatively more likely to agree than had they been processing independently, particularly if one is in a position of greater authority (see also Chapter 5). Their overall effect may be to make correct performance either more or less likely, depending on a host of influences that are beyond the scope of this book.

Integrating Human and Machine Reliabilities Adams (1982) argues that it is difficult to justify mathematically combining human-error data with machine-reliability data, derived independently, to come up with joint reliability measures of the total system. Here again a nonindependence issue is encountered. When a machine component fails (or is perceived as being more likely to fail), it will probably alter the probability of human failure in ways that cannot be precisely specified. It is likely, for example, that the operator will become far more cautious, trustworthy, and reliable when interacting with a system that has a higher likelihood of failure or with a component that itself has just failed than when interacting with a system that is assumed to be infallible. This is a point that will be considered again in the discussion of automation in Chapter 12.

The important message here, as stated succinctly by both Reason (1990) and Adams (1982), is that a considerable challenge is required to integrate actuarial data of human error with machine data to estimate system reliability. Unlike some other domains of human performance (see particularly manual control in Chapter 11), even if the precise mathematical modeling of human performance were achieved, it would not appear to allow accurate prediction of *total system* performance. Although the potential benefits of accurate human reliability analysis and error prediction are great, it seems likely that the most immediate human factors benefits will be realized if effort is focused on case studies of individual errors in performance (Pew, Miller, & Feehrer, 1981). These case studies can be used to diagnose the resulting causes of errors and to recommend the corrective system modification.

Errors in the Organizational Context

The greatest focus of public awareness on human error has resulted from major accidents and disasters, such as the nuclear meltdown at Chernobyl; the explosion of *Challenger IV*; or the chemical plant accident at Bhopal, India, in which over 1000 lives were lost. In all of these cases, human error has been singled out as a contributing cause. But Reason (1988, 1990) has carefully analyzed these and two other major accidents—the sinking of the ferry boat *Herald of*

Free Enterprise in 1987 (188 lives lost) and the Three Mile Island incident—and has identified human operator error as only one small component in a set of more serious organizational deficiencies.

Reason (1988, 1990) has drawn the important distinction between local triggers, or active failures, and resident pathogens, or latent failures. He argues that the triggering human error that caused each of the five accidents, along with numerous others, is only the final event in a series of poor design, management, and maintenance decisions, many of which existed long before the local trigger. The *local trigger* is, we might imagine, the tip of the iceberg, whereas the faulty design and management decisions—the base of the iceberg—are a collection of factors that represent an accident waiting to happen. Like a silent virus in the human body, Reason refers to these latter factors as *resident pathogens.* The Three Mile Island incident is certainly consistent with this analysis. The poor panel design, the poor valve design, and poor procedures (the fact that plant status had been changed because of maintenance, a change of which the operators on duty were unaware) contributed to the operators' mistake in dealing with the local trigger.

Figure 10.7 presents Reason's (1990) representation of the relationship between decision errors, made at various points in a complex system, and the final unsafe acts, which can create the conditions for a local trigger. All of the factors in the boxes may be thought of as the different dimensions of resident pathogens. Together they promote the conduct of unsafe acts and also lower the defenses of the system against the potentially disastrous consequences. Full descriptions of each box are well beyond the scope of this book, and the reader is referred to Reason and Perrow (1984) for highly readable details. However, the top and bottom boxes are worth some further discussion. Hardware defects refer to all of the human factors problems of displays and controls. But also important here is the design of the system itself, rather than the human interface. Perrow has argued that very complex, multicomponent systems, because of their very complexity, are guaranteed (1) to fail and (2) to fail in ways that are beyond the capacity of human cognition to readily understand, a point that will be expanded in Chapter 12.

The bottom box in Figure 10.7 draws the important distinction between errors and violations. In preceding discussions, we have assumed that errors are unintentional. Reason (1990) has distinguished errors from *violations*, which are intentional departures from specified operating procedures but are not intended to create accidents or to cause harm. Driving 60 in a 55 mph speed zone is a violation, and so is riding without a seat belt. The incident at Chernobyl was the direct result of a violation, in which operators intentionally "experimented" with the plant at unsafe operating conditions (*Nature*, vol. 323, 1986). Furthermore, violations of procedures appear to make up a majority of pilot errors responsible for major aircraft accidents (Nagel, 1988). Violations, then, are caused by operating conditions in which safety is not stressed or by management goals that run contrary to safety. Usually the latter involve an emphasis on production and profit.

In addressing the conditions in many large organizations that lead to errors

Mediating Factors

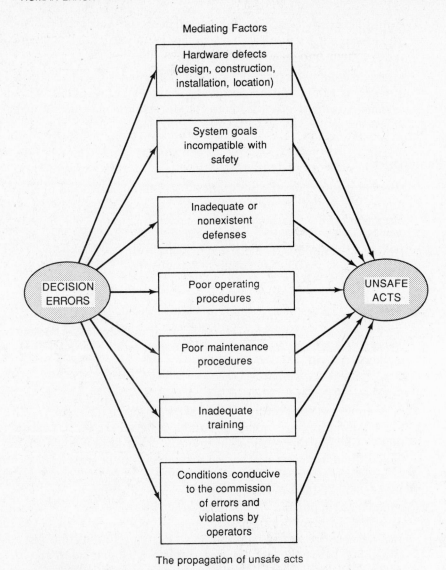

The propagation of unsafe acts

Figure 10.7 Framework for understanding resident pathograms. (*Source:* J. Reason, *Human Error.* New York: Cambridge University Press, 1990. This draft version reproduced with permission of the author.)

and violations, Reason (1990) points to the choices that are often faced by industrial managers who must allocate scarce resources either to production (a sure gain) or to safety (avoiding a risky loss). Using the analysis of *decision framing* described in Chapter 7, it is easy to see why the choice is so often biased toward the more certain gain (enhance production) over the risky alternative (avoid the low probability accident by implementing safety programs and procedures). Reason notes also that the reinforcement to managers from emphasizing production is generally frequent, immediate, and represented by the presence of positive evidence (money). The reinforcement for emphasizing safety, in contrast, is less salient and tangible and is usually characterized by the *absence* of evidence (the accident that does not happen). This unfortunate asymmetry all too often leads to the operating conditions in which violations will occur and resident pathogens will thrive.

Error Remediation

We will now discuss the solutions offered to minimize the likelihood of errors or the potential damage that they might cause. Clearly many of the ways to eliminate resident pathogens, as suggested by Figure 10.7, lie in organizational climate and effectiveness—topics that lie beyond the domain of this book. However, implicit in our discussions and in those of Norman (1988), Senders and Moray (1991), and Reason (1990) has been a number of specific remediations.

Task Design   Designers should try to minimize operator requirements to perform tasks that impose heavy working memory load under conditions of stress or other tasks for which human cognitive mechanisms are poorly suited. Such efforts will generally decrease the frequency of mistakes.

Equipment Design   There are a number of equipment design remedies that in Norman's (1988) terms, reduce the *affordance* for errors:

1. Minimize perceptual confusions. Norman (1988) has described the care that is taken in the automobile to ensure that fluid containers and apertures look distinct from one another, so that oil will not be poured into the antifreeze opening nor antifreeze into the battery and so forth. Such a design stands in stark contrast to the identical appearance of different fluid tubes and fluid containers supporting the patient in an intensive care unit, a situation that describes an error waiting to happen (Gopher et al., 1989; see also Figure 10.7). There are, of course, a series of design solutions that can ensure discriminability between controls and displays: distinct color and shape, spatial separation, distinct feel, and different control motions (see Chapter 8).

2. Make the execution of action and the response of the system *visible* to the operator (Norman, 1988). When slips occur, they cannot easily be detected (and hence corrected) if the consequences of actions cannot be seen. Hence, feedback from switches and controls that change a state should be clearly visible. If it is not too complex, the way a system carries out its operations should be revealed. Unfortunately, extreme simplicity, economy, and aesthetics in engineering design can often mask the visibility of response feedback and system operation, a visibility useful in preventing errors.

3. Use constraints to "lock out" the possibility of errors (Norman, 1988). Sometimes these can be cumbersome and cause more trouble than they are worth. For example, interlock systems that prevent a car from starting before

the seat belts are fastened have proven to be so frustrating that people disconnect the systems. On the other hand, an effective constraint is that seen in the car doors that cannot be locked unless the key is inserted and turned on the outside. This slight inconvenience will prevent the key from being locked in the car. Other constraints may force a sequence of actions in the computer that will prevent the commission of major errors—like erasing important files.

4. Avoiding multimode systems. Systems, like the multimode digital watch, in which identical actions accomplish different functions in different contexts, are a sure invitation for mode errors. When they cannot be avoided, the designer should make the discrimination of modes as *visible* as possible by employing salient visual cues. A continuous flashing light on a computer system, for example, is a salient visual reminder that an unusual mode is in effect.

Training   Because lack of knowledge is an important source of mistakes, it is not surprising that increased training will reduce their frequency (although, as we have seen, training may have little effect on slips). But Reason (1990) has also argued that error-free training should not be required. If operators are not practiced at correcting errors that occur during training, they will not know how to deal with the errors that might occur in real system operation (see also Chapter 6).

Assists and Rules   Both assists and rules can represent designer solutions to error-likely situations, and some of these make very obvious sense. For example, such assists as memory aids for procedures checklists can be extremely valuable (Rouse, Rouse, & Hammer, 1982), whether for operators of equipment following a start-up procedure or for maintenance personnel carrying out a complex sequence of lapse-prone steps. If rules are properly explained, are logical, and are enforced, they can reduce the likelihood of safety violations. However, if the implications of rules adopted for complex systems, like nuclear and chemical process control plants, are not thought through, they can create unforeseen problems of their own. As Reason (1990) describes it, the "band-aid" approach to human error may only make the situation worse. Rules may unexpectedly prohibit necessary behavior in times of crisis, in a way that the rule designer had not anticipated, and automated assists may themselves lead to failures and errors (see Chapter 12).

Error-Tolerant Systems   Although human error is typically thought of as undesirable, it is possible to see its positive side (Senders & Moray, 1991). In discussing both signal detection theory (Chapter 2) and decision theory (Chapter 7), we saw that in a probabilistic world, certain kinds of errors will be inevitable, and engineering psychologists are concerned as much with controlling the different *kinds* of errors (e.g., misses versus false alarms) as with eliminating them. In Chapter 8, we saw that the optimal setting of the speed-accuracy trade-off was usually at some intermediate level, where at least a small number of errors was better than none at all. In Chapter 6, we saw that error is often necessary for learning to occur (so long as the error is not repeated).

Finally, as discussed in Chapter 1, error may be viewed as the inevitable downside of the valuable flexibility and creativity of the human operator. Rasmussen (1989) addresses this issue explicitly in the context of many large, complex systems, such as the nuclear power plants we will be discussing in Chapter 12. There are so many possible strategies for accomplishing plant goals that specification of a single "correct," precise sequence of step-by-step procedures is not possible. The human operator must be opportunistic, responding differently according to the conditions of the moment. Under such circumstances, it becomes nearly essential for the operator to be able to explore the limits of the system, particularly when the operator is learning the system characteristics and developing a good mental model. This makes a certain amount of error inevitable, if not desirable.

Understanding the inevitable and sometimes even desirable properties of human error has forced a rethinking of conventional design philosophies, in which all errors were to be irradicated (Rasmussen, 1989). Instead, researchers such as Rouse and Morris (1987) advocate the design of *error-tolerant* systems. An error-tolerant design, for example, would attempt to avoid irreversible actions. A file-delete command on a computer might not irreversibly delete the file but simply remove it and "hold" it in another place for some period of time (e.g., until the computer is turned off). Then the operator would have the chance to recover from the slip, which in this case would be an incorrect deletion command (Norman, 1988).

In describing their conceptual approach to error-tolerant design, Rouse and Morris (1987) propose a fairly sophisticated intelligent monitoring system that continuously makes inferences about human intentions. If such a system then infers, on the basis of human output and system status, that those intentions are in danger of violating safety, or if human actions have been committed that are inconsistent with the inferred intentions, a graded series of interventions can be implemented. These run from increased vigilance of human performance monitoring by the system to feedback of the nature of the error (which allows the operator to disregard it if he or she chooses) to the final level of direct intervention and control. Here the system will step in and take over if it infers that the consequences of error are severe.

The philosophy of error-tolerant systems has a great deal of appeal and has been echoed in Wiener's (1989) recommendation of a "glass cocoon" surrounding the pilot of advanced aircraft. In such a concept, computer-based recommendations and controls will intervene only if the pilot "flies" the aircraft beyond this cocoon in a way that is deemed unsafe. The design of modern commercial airplanes such as the McDonnel Douglas MD–11 are adapting this philosophy in many respects (Hughes, 1990). It is apparent that effective error-tolerant systems of the sort envisioned by Rouse and Morris (1987) impose greatly increased responsibility on the intelligence of the computer-based monitoring system (Wiener, 1989). The extent to which computer intelligence is ready for this challenge and the way in which the human operator will respond to imperfect levels of machine intelligence are issues that will be addressed in Chapter 12.

TRANSITION

This chapter has focused on operator stress and human error. Both, but particularly the latter, are relevant to the nuclear and energy process control environment discussed in Chapter 12. Human error is also critically linked to issues of automation—considered by some to be the safeguard against human error. We saw the role of automation in supporting the error-tolerant system proposed by Rouse and Morris (1987), and it will be discussed again in Chapter 12. However, it is first necessary to discuss the elements of continuous *manual* control because how people manually control dynamic systems, like automobiles, ships, and aircraft, is directly relevant to the way they control more complex chemical and nuclear processes.

REFERENCES

Adams, J. J. (1982). *Simulator study of a pictorial display for instrument flight* (NASA Technical Paper no. 1963). Hampton, VA: NASA Langley Research Center.

Allnut, M. F. (1987). Human factors in accidents. *British Journal of Anaesthesia, 59,* 856–864.

Anderson, K. J., & Revelle, W. (1982). Impulsivity, caffeine, and proofreading: A test of the Easterbrook hypothesis. *Journal of Experimental Psychology: Perception & Performance, 8,* 614–624.

Apostolakis, G. E. (1990). *Reliability engineering and system safety.* Eng.: Elsevier Science Publishers.

Avionics (December, 1990). TCAS for Transports: Part III (pp. 22–45).

Bacon, S. J. (1974). Arousal and the range of cue utilization. *Journal of Experimental Psychology, 102,* 81–87.

Baddeley, A. D. (1972). Selective attention and performance in dangerous environments. *British Journal of Psychology, 63,* 537–546.

Berkun, M. M. (1964). Performance decrement under psychological stress. *Human Factors, 6,* 21–30.

Blake, M. J. F. (1967). Time of day effects in performance on a range of tasks. *Psychonomic Science, 9,* 349–350.

Blake, M. J. F. (1971). Temperament and time of day. In W. P. Colquhoun (ed.), *Biological rhythms and human behavior.* London: Academic Press.

Bourne, P. G. (1971). Altered adrenal function in two combat situations in Vietnam. In B. E. Elefheriou and J. P. Scott (eds.), *The physiology of aggression and defeat.* New York: Plenum.

Broadbent, D. E. (1971). *Decision and stress.* New York: Academic Press.

Broadbent, D. E. (1978). The current state of noise research: Reply to Poulton. *Psychological Bulletin, 85,* 1052–1067.

Burton, R. R., Storm, W. F., Johnson, L. W., & Leverett, S. D., Jr. (1977). Stress responses of pilots flying high-performance aircraft during aerial combat maneuvers. *Aviation Space & Environmental Medicine, 48*(4), 301–307.

Card, S., Moran, T., & Newell, A. (1983). *The psychology of human-computer interaction*. Hillsdale, NJ: Erlbaum.

Cowen, E. L. (1952). The influence of varying degrees of psychosocial stress on problem-solving rigidity. *Journal of Abnormal and Social Psychology, 47*, 512–519.

Coyne, J. C., & Lazarus, R. S. (1980). Cognitive style, stress perception, and coping. In I. L.Kutash, L. B. Schlesinger, & Associates (eds.), *Handbook on stress and anxiety* pp. 144–158. San Francisco: Jossey-Bass.

Davies, D. R., & Parasuraman, R. (1982). *The psychology of vigilance*. London: Academic Press.

Dougherty, E. M. (1990). Human reliability analysis—Where shouldst thou turn? *Reliability Engineering and System Safety, 29*, 283–299.

Douglas A. (1989). Fear, panic can wreck pilot ability. *Aviation Safety, 9*(15) 1–6.

Druckman, D., & Swets, J. A. (eds.). (1988). *Enhancing human performance*. Washington, DC: National Academic Press.

Easterbrook, J. A. (1959). The effect of emotion of cure utilization and the organization of behavior. *Psychological Review, 66*, 183–201.

Eysenck, M. W. (1976). Arousal, learning, and memory. *Psychological Bulletin, 83*, 389–404.

Eysenck, M. W. (1982). *Attention and arousal: Cognition and performance*. Berlin: Springer-Verlag.

Fitts, P. M., & Jones, R. E. (1960a). Analysis of factors contributing to 460 "pilot-error" experiences in operating aircraft controls. In H. Wallace Sinaiko (ed.), *Selected Papers on Human Factors in the Design and Use of Control Systems* (pp. 332–358). New York: Dover Publications.

Fitts, P. M., & Jones, R. E. (1960b). Psychological aspects of instrument display. I: Analysis of 270 "pilot-error" experiences in reading and interpreting aircraft instruments. *Selected Papers on Human Factors in the Design and Use of Control Systems* (pp. 359–396). New York: Dover Publications.

Fitts, P. M., & Seeger, C. M. (1953). S-R compatibility: Spatial characteristics of stimulus and response codes. *Journal of Experimental Psychology, 46*, 199–210.

Flexman, R., & Stark, E. (1987). Training simulators. In G. Salvendy (ed.), *Handbook of human factors*. New York: Wiley.

Folkard, S. (1983). Diurnal variation. In G. R. J. Hockey (ed.), *Stress and fatigue in human performance*. Chichester, Eng.: Wiley.

Fuchs, A. (1962). The progressive-regressive hypothesis in perceptual-motor skill learning. *Journal of Experimental Psychology, 63*, 177–181.

Gopher, D., Olin, M., Badhih, Y., Cohen, G., Donchin, Y., Bieski, M., & Cotev, S. (1989). The nature and causes of human errors in a medical intensive care unit. *Proceedings of the 32nd annual meeting of the Human Factors Society*. Santa Monica, CA: Human Factors Society.

Hamilton, P., Hockey, G. R. J., & Rejman, M. (1977). The place of the concept of activation in human information processing theory: An integrative approach. In S. Dornic (ed.), *Attention and performance* (vol. 6). New York: Academic Press.

Hamilton, V., & Warburton, D. M. (eds.). (1984). *Human stress and cognition: An information processing approach*. Chichester, Eng.: Wiley.

Hart, S. G., & Hauser, J. R. (1987). In-flight application of three pilot workload measurement techniques. *Aviation, Space, and Environmental Medicine, 58*, 402–410.

Hockey, G. R. J. (1970). Effect of loud noise on attentional selectivity. *Quarterly Journal of Experimental Psychology, 22*, 28–36.

Hockey, G. R. J. (1979). Stress and the cognitive components of skilled performance. In V. Hamilton & D. M. Warburton (eds.), *Human stress and cognition: An information processing approach*. Chichester, Eng.: Wiley.

Hockey, G. R. J. (1984). Varieties of attentional state: The effect of the environment. In R. S. Parasuraman & D. R. Davies (eds.), *Varieties of attention*. Orlando, FL: Academic Press.

Hockey, G. R. J. (1986). Changes in operator efficiency as a function of environmental stress, fatigue, and circadian rhythms. In K. R. Boff, L. Kaufman, & J. P. Thomas (eds.), *Handbook of perception and human performance* (vol. II). New York: Wiley.

Holding, D. H. (1983). Fatigue. In G. R. J. Hockey (ed.), *Stress and fatigue in human performance*. Chichester, Eng.: Wiley.

Houston, B. K. (1969). Noise, task difficulty, and Stroop color-word performance. *Journal of Experimental Psychology, 82*, 403–404.

Hughes, D. (1990, October 22). Extensive MD–11 automation assists pilots, cuts workload. *Aviation Week & Space Technology*, pp. 34–45.

Hurst, R., & Hurst, L. R. (eds.). (1982). *Pilot error: The human factors*. New York: Jason Aronson.

Jensen, R. S. (1982). Pilot judgment: Training and evaluation. *Human Factors, 24*, 61–74.

Johnson, L. C., & Chernik, D. A. (1982). Sedative-hypnotics and human performance. *Psychopharmacology, 76*, 101–113.

Keinan, G., & Freidland, N. (1984). Dilemmas concerning the training of individuals for task performance under stress. *Journal of Human Stress, 10*, 185–190.

Keinan, G., & Friedland, N. (1987). Decision making under stress: Scanning of alternatives under physical threat. *Acta Psychologica, 64*, 219–228.

Kennedy, R. S., & Coulter, X. B. (1975). Research note: The interactions among stress, vigilance, and task complexity. *Human Factors, 17*, 106–109.

Klein, G. A. (1989). Recognition-primed decisions. In W. Rouse (ed.), *Advances in man-machine systems research* (vol. 5, pp. 47–92). Greenwich, CT: JAI Press.

Lazarus, R. S., & Ericksen, C. W. (1952). Effects of failure stress on skilled performance. *Journal of Experimental Psychology, 43*, 100–105.

Mandler, G. (1984). Thought processes, consciousness, and stress. In V. Hamilton & D. M. Warburton (eds.), *Human stress and cognition: An information processing approach*. Chichester, Eng.: Wiley.

Miller, D., & Swain, A. (1987). Human reliability analysis. In G. Salvendy (ed.), *Handbook of human factors*. New York: Wiley.

Nagel, D. C. (1988). Human error in aviation operations. In E. Wiener & D. Nagel (eds.), *Human factors in aviation* (pp. 263–303). New York: Academic Press.

National Transportation Safety Board. (1988). *Northwest Airlines, Inc. McDonnell Doug-*

las DC-9-82 N312RC. *Detroit Metropolitan Wayne County Airport, Romulus, Michigan, August 16, 1987* (Report No. NTSB-AAR-88-05). Washington, DC: Author.

Norman, D. A. (1981a). Categorization of action slips. *Psychological Review, 88,* 1–15.

Norman, D. A. (1981b). The trouble with Unix. *Datamation, 27,* 139–150.

Norman, D. (1988). *The psychology of everyday things.* New York: Harper & Row.

Perrow, C. (1984). *Normal accidents: Living with high-risk technology.* New York: Basic Books.

Pew, R. W., Miller, D. C., & Freehrer, C. E. (1981). Evaluating nuclear control room improvements through analysis of critical operator decisions. In R. C. Sugarman (ed.), *Proceedings of the 25th annual meeting of the Human Factors Society.* Santa Monica, CA: Human Factors Society.

Poulton, E. C. (1976). Continuous noise interferes with work by masking auditory feedback and inner speech. *Applied Ergonomics, 7,* 79–84.

Rabbitt, P. M. A. (1978). Detection of errors by skilled typists. *Ergonomics, 21,* 945–958.

Ramsey, J. D. (1983). Heat and cold. In G. R. J. Hockey (ed.), *Stress and fatigue in human performance.* Chichester, Eng.: Wiley.

Rasmussen, J. (1983). Skills, rules, and knowledge: Signals, signs, and symbols, and other distinction in human performance models. *IEEE Transactions on System, Man, & Cybernetics, 13,* 257–266.

Rasmussen, J. (1989). Human error and the problem of causality in analysis of accidents. *Invited paper for Royal Society meeting on human factors in high risk situations,* London.

Reason, J. T. (1984). Lapses of attention. In R. Parasuraman & R. Davies (eds.), *Varieties of attention.* New York: Academic Press.

Reason, J. (1988). Resident pathogens and risk management. World Bank workshop on safety control and risk management, Washington, DC, October 18–20.

Reason, J. (1990). *Human error.* New York: Cambridge University Press.

Revelle, W., Humphreys, M. S., Simon, L., & Gilliland, K. (1980). The interactive effect of personality, time of day, and caffeine: A test of the arousal model. *Journal of Experimental Psychology: General, 109,* 1–31.

Romhert, W. (1979). Determination of stress and strain at real work places: Methods and results of field studies with air traffic control officers. In N. Moray (ed.), *Mental workload.* New York: Plunum Press.

Rouse, S. H., Rouse, W. B., & Hamner, J. M. (1982). Design and evaluation of an onboard computer-based information system for aircraft. *IEEE Transactions on Systems, Man, and Cybernetics, SMC–12,* 451–463.

Rouse, W. B., & Morris, N. M. (1987). Conceptual design of a human error tolerant interface for complex engineering systems. *Automatica, 23*(2), 231–235.

Rouse, W. B., & Rouse, S. H. (1983). Analysis and classification of human error. *IEEE Transactions on Systems, Man, and Cybernetics, SMC–13,* 539–549.

Schwartz, D. R., & Howell, W. C. (1985). Optional stopping performance under graphic and numeric CRT formatting. *Human Factors, 27,* 433–444.

Senders, J., & Moray, N. (1991). *Human error: Cause, prediction and reduction.* Hillsdale, NJ: Erlbaum.

Simmel, E. C., Cerkovnik, M., & McCarthy, J. E. (1987). Sources of stress affecting pilot judgment. *Proceedings of the fourth international Symposium on Aviation Psychology* (pp. 190–194). Ohio State University, Department of Aviation, Columbus, OH.

Simonov, P. V., Frolov, M. V., Evtushenko, V. F., & Suiridov, E. P. (1977). *Aviation, Space, and Environmental Medicine, 48*, 856–858.

Stokes, A. F., & Raby, M. (1989). Stress and cognitive performance in trainee pilots. *Proceedings of the 33rd annual meeting of the Human Factors Society.* Santa Monica, CA: Human Factors Society.

Stokes, A. F., Belger, A., & Zhang, K. (1990). Investigating factors comprising a model of pilot decision making II: Anxiety and cognitive strategies in expert and novice aviators. Savoy, IL: University of Illinois aviation research lab technical report no. ARL 90-8/SCEEE 90-2.

Svenson, O., & Maule, J. (eds.). (1991). *Time pressure and stress in human judgment and decision making.* Cambridge, Eng.: Cambridge University Press.

Swain, A. D. (1990). Human reliability analysis: Need, status, trends and limitations. *Reliability engineering and system safety, 29*, 301–313.

Swain, A. D., & Weston, L. M. (1988). An approach to the diagnosis and misdiagnosis of abnormal conditions in post-accident sequences in complex man-machine systems. In L. Goodstein, H. Anderson, & S. Olsen (eds.). *Tasks, errors, and mental models.* London: Taylor & Francis.

Ursin, H., Baade, E., & Levine, S. (eds.). (1978). *Psychobiology of stress: A study of coping men.* New York: Academic Press.

U.S. Navy. (1988). *Investigation report: Formal investigation into the circumstances surrounding the downing of Iran air flight 655 on 3 July 1988.* Washington, DC: Department of Defense Investigation Report.

Wachtel, P. L. (1967). Conceptions of broad and narrow attention. *Psychological Bulletin, 68*, 417–429.

Wachtel, P. L. (1968). Anxiety, attention and coping with threat. *Journal of Abnormal Psychology, 73*, 137–143.

Weltman, G., Smith, J. E., & Egstrom, G. H. (1971). Perceptual narrowing during simulated pressure-chamber exposure. *Human Factors, 13*(2), 99–107.

Wesnes, K., & Warburton, D. M. (1983). Stress and drugs. In G. R. J. Hockey (ed.), *Stress and fatigue in human performance.* Chichester, Eng.: Wiley.

Wickens, C. D., & Flach, J. (1988). Human information processing. In E. Wiener & D. Nagel (eds.), *Human factors in aviation* (pp. 111–155). New York: Academic Press.

Wickens, C. D., Stokes, A. F., Barnett, B., & Hyman, F. (1991). The effects of stress on pilot judgment in a MIDIS simulator. In O. Svenson & J. Maule (eds.), *Time pressure and stress in human judgment and decision making.* Cambridge, Eng.: Cambridge University Press.

Wiener, E. L. (1988). Cockpit automation. In E. L. Wiener & D. C. Nagel (eds.), *Human factors in aviation* (pp. 433–461). San Diego: Academic Press.

Wiener, E. L. (1989). Reflections on human error: Matters of life and death. *Proceedings of the 33rd annual meeting of the Human Factors Society* (pp. 1–7). Santa Monica, CA: Human Factors Society.

Wilkinson, R. T. (1963). Interaction of noise with knowledge of results and sleep deprivation. *Journal of Experimental Psychology, 66,* 332–337.

Wine, J. (1971). Test anxiety and direction of attention. *Psychological Bulletin, 76,* 92–104.

Woods, D. D. (1984). Some results on operator performance in emergency events, *Institute of Chemical Engineers Symposium Series, 90,* 21–31.

Wreathall, J. (1982). *Operator action trees: An approach to quantifying operator error probability during accident sequences* (Report No. NUS-4159). Gaithersburg, MD: NUS Corporation.

Yerkes, R. M., & Dodson, J. D. (1908). The relation of strength of stimulus to rapidity of habit formation. *Journal of Comparative Neurological Psychology, 18,* 459–482.

Zhang, K., & Wickens, C. D. (1990). Effects of noise and workload on performance with two object displays vs. a separated display. *Proceedings of the 34th annual meeting of the Human Factors Society.* Santa Monica, CA: Human Factors Society.

Chapter
11

Manual Control

OVERVIEW

In the performance of most tasks, the information that is encoded and the decisions that are made must be translated into action. The preceding chapters of this book have assumed for convenience that the form of this action is relatively simple compared to perceptual or central-processing activities. Little attention was given to the analog form or the time-space trajectory of the response. In this chapter we will consider the class of tasks in which this trajectory is critical—the domain of continuous control. In most applications, this control is exerted manually. Hence we focus our discussion on *manual* control but also briefly consider voice control.

Human performance in manual control has been considered from two quite different perspectives: skills and dynamic systems. Each has used different paradigms and different analytical tasks, and each has been generalized to different applied environments. The skills approach primarily involves analog motor behavior, in which the operator must produce or reproduce a movement pattern from memory when there is little environmental uncertainty. The gymnast performs such a skill when executing a complex maneuver; so too does the assembly-line worker who coordinates a smooth, integrated series of actions around a set of environmental stimuli—the products to be assembled—that are highly predictable from one instance to the next. Because there is little environmental uncertainty, such skills in theory may be performed perfectly and identically from trial to trial. In Figure 1.3 the behavior is described as open loop since there is little need to process the visual feedback from the response. The emphasis of experiments on skills has focused heavily on the time course of skill acquisition and the optimal conditions of practice, whether addressed from the traditional learning point of view (Bilodeau & Bilodeau, 1969) or from a more recent information-processing perspective (Holding, 1989; Stelmach, 1978).

In contrast to the skills approach, the dynamic systems approach examines human abilities in controlling or *tracking* dynamic systems to make them conform with certain time-space trajectories in the face of environmental uncertainty (Kelley, 1968; Poulton, 1974, Wickens, 1986). Most forms of vehicle control fall into this category, and so increasingly do computer-based cursor positioning tasks. The research on tracking has been oriented primarily toward engineering, focusing on mathematical representations of the human's analog response when processing uncertainty. Unlike the skills approach, which focuses on learning and practice, the tracking approach generally addresses the behavior of the well-trained operator.

The general treatment of manual control that follows reflects this dichotomy but acknowledges that as is so often the case in psychology, the dichotomy is really more of a continuum. Basketball players who execute a skilled, highly practiced maneuver by themselves may be engaging in a pure open-loop skill, but when they do so in the middle of a game, with a defender providing some degree of environmental uncertainty, the response becomes more of a compromise between open-loop skills and tracking.

We will begin by considering at an atomistic level the simplest form of analog response—the minimum time taken to move from a starting point to a target with constrained accuracy. The data describing this skill, it turns out, are quite well captured by a basic "law" of motor control, whose principles seem to underlie both open-loop skills and the tracking of dynamic systems. More complex forms of open-loop skills will then be discussed before turning to an extensive treatment of manual control in tracking and vehicle control. We will then address the strengths and weaknesses of different devices for discrete control in computer-based tasks. Finally, the chapter supplement will consider in some detail the efforts that have been made to model the human operator in tracking and will provide an introduction to the mathematical language of frequency-domain analysis, which has been used in these modeling efforts.

OPEN-LOOP MOTOR SKILLS

Discrete Movement Time

Pioneering investigations by Woodworth (1899) and Brown and Slater-Hammel (1949) found that the time required to move the hand or stylus from a starting point to a target obeys the basic principles of the speed-accuracy trade-off. Quite intuitively, faster movements terminate less accurately in a target, whereas targets of small area, requiring increased accuracy, are reached with slower responses. The amplitude of a movement also influences this speed-accuracy relationship. It takes a longer time to move a greater distance into a target of fixed area. However, if precision is allowed to decline with longer movements, movement time is essentially unchanged with length.

Fitts (1954) investigated the relationship among the three variables of time, accuracy, and distance in the paradigm shown in Figure 11.1. Here the subject is to move the stylus as rapidly as possible from the start to the target area. Fitts found that when movement amplitude (A) and target width (W) were manipulated, their joint effects were summarized by a simple equation that has subsequently become known as Fitts's law:

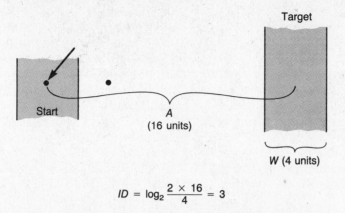

$$ID = \log_2 \frac{2 \times 16}{4} = 3$$

Figure 11.1 The Fitts movement-time paradigm. The movement may be either a single discrete movement from start to target or a series of alternating taps between the two targets.

$$\text{Movement time } (MT) = a + b \log_2 \left(\frac{2A}{W} \right)$$

where a and b are constants. This equation describes formally the speed-accuracy trade-off in movement: Movement time and accuracy (target width W) are reciprocally related. Longer movements can be made (increasing A), but if their time is to be kept constant, accuracy must suffer proportionately. That is, the target into which the movement will terminate, W, must be widened.

Fitts described the specific quantity $\log_2 (2A/W)$ as the *index of difficulty* (*ID*) of the movement. In Figure 11.1, *ID* = 3. Movements of the same index of difficulty can be created from different combinations of A and W but will require the same time to complete. Figure 11.2 shows the linear relationship between *MT* and *ID* obtained by Fitts when different combinations of amplitude and target width were manipulated. Each *ID* value shows the similar movement time created by two or three different amplitude/width combinations. The high degree of linearity is evident for all but the lowest condition, in which the linear relationship slightly underpredicts *MT*.

Several investigations have demonstrated the generality of Fitts's law (Jagacinski, 1989). For example, Fitts and Peterson (1964) found that the law is equally accurate for describing single, discrete movements or reciprocal tapping between two targets. Langolf, Chaffin, and Foulke (1976) observed that the relationship accurately describes data for manipulating parts under a microscope. Drury (1975) found that the law accurately describes movement of the foot to pedals of varying diameter and distance, and Card (1981) employed the law as a basic predictive element of key-reaching time in keyboard tasks (Card, Newell, & Moran, 1986). Jagacinski, Repperger, Ward, and Moran (1980) extended the model to predict performance in a dynamic target-acquisition task.

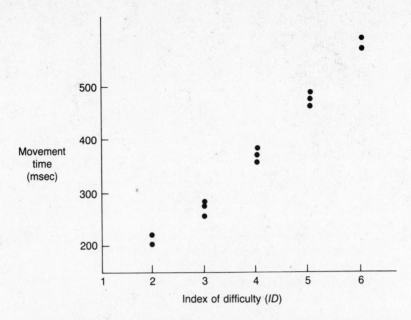

Figure 11.2 Data on movement time as a function of the index of difficulty. (*Source:* P. M. Fitts, "The Information Capacity of the Human Motor System in Controlling the Amplitude of Movement," *Journal of Experimental Psychology, 47* (1954), 385.)

Other investigators have examined more theoretical properties of the basic relationship. For example, Fitts and Peterson (1964) studied the relationship between movement time and the time to initiate the movement and found that they were relatively independent of one another. Increasing the index of difficulty, which made movement time longer, did not affect the reaction time to initiate the movement. Conditions that varied reaction time (i.e., single versus choice) had no effect on subsequent movement time. Kelso, Southard, and Goodman (1979) examined movement times for simultaneous two-handed movements to two targets of varying index of difficulty. They observed that the two movements were not independent of one another but that movement time of the easier (lower *ID*) hand was slowed down to be in synchrony with the time taken for the more difficult and therefore slower movement.

Models of Discrete Movement Figure 11.3*a* shows a typical trajectory or time history recorded as the stylus approaches the target in the paradigm of Figure 11.1. Two important characteristics of this pattern are apparent: (1) The general form of the movement is that of an exponential approach to the target, with an initial high-velocity approach followed by a smooth, final, "homing" phase. In the earliest research in this area, Woodworth (1899) distinguished between these two phases, labeling the first the *initial ballistic* and the second

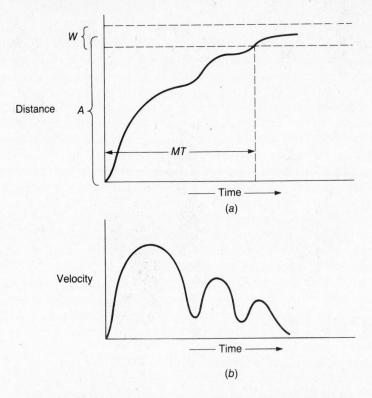

Figure 11.3 Typical position (*a*) and velocity (*b*) profile of Fitts's law movement.

current control. (2) The velocity profile of the movement shown in Figure 11.3*b* reveals that control is not continuous but appears to consist of a number of discrete corrections, each involving an acceleration and a deceleration.

Although numerous investigators have attempted to provide precise models of the processes that generate the typical time histories shown in Figure 11.3 (see Glencross & Barrett, 1989, for a review), two characteristics typify most of these modeling and experimental approaches: (1) All agree that the basic form provides a reasonably close and parsimonious approximation to the data even if it is not entirely accurate. (2) Most are based on a feedback processing assumption. As operators approach the target, the remaining error to the target is sampled, either continuously or intermittently, and proportional (to the error) corrections are implemented to nullify the error. This sample-and-correct process is continued until the target boundary is crossed (Pew, 1974). Such behavior will produce the generally exponential approach shown in Figure 11.3, in which stylus velocity is roughly proportional to momentary error. This characteristic of target-aiming responses is important in describing continuous tracking skill, as well as discrete control.

Motor Schema

Visually guided responses such as those described by Fitts's law are, of course, critical components in a wide variety of real-world skills such as those required in assembly-line work, target acquisition, or performance on complex or unfamiliar keyboards. Yet with highly learned skills performed under conditions of minimal environmental uncertainty, it is evident that visual feedback is not necessary. We say that these skills may be performed in *open-loop* fashion. In fact, sometimes this visual feedback may actually be harmful. Shoe tying, touch typing, or the performance of a skilled pianist are good examples of skilled performance that does not require visual feedback.

Psychologists and motor-learning theorists have identified two general characteristics of such well-learned motor skills: (1) They may well be dependent on feedback, but the feedback is *proprioceptive*. Information from the joints and muscles is relayed back to central movement-control centers to guide the execution of the movement in accordance with centrally stored goals, or "templates" of the ideal time-space trajectory (Adams, 1971, 1976). (2) The pattern of desired muscular innervation may be stored centrally in long-term memory and executed as an open-loop *motor program* without benefit of visual feedback correction and guidance (Schmidt, 1975; Summers, 1989).

The terms *motor program* and *motor schema* have been used to label highly overlearned skills that, as a consequence of this learning, are not dependent on guidance from visual feedback (Keele, 1968; Schmidt, 1975; Shapiro & Schmidt, 1982; Summers, 1989). The concept of the motor program or schema may be best defined in terms of four correlated attributes: high levels of practice, low attention demand, single-response selection, and consistency of outcome.

High Levels of Practice   Extensive practice is perhaps the most critical defining attribute of the motor program. Skills that are not highly practiced are unlikely to possess the following attributes.

Low Attention Demand   The motor program tends to be *automated* in the terms described in Chapter 9. In that discussion, task practice was assumed to have a major influence on resource demand. Here the limited resource demand is a major criterion for defining the motor program. A well-learned complex sequence of responses may be executed while disrupting only slightly the performance of a concurrent task (Bahrick & Shelley, 1958).

Single-Response Selection   Within the framework of the information-processing model in Chapter 1, it is assumed that a single-response selection is required to activate or "load" a single motor program, even though the program itself may contain a number of separate, discrete responses. Thus resources are demanded only once, at the point of initiation, when the program is selected.

Programs may vary in their complexity. Investigators such as Klapp and Erwin (1976); Martenuik and MacKenzie (1980); and Shulman, Jagacinski, and Burke (1978) have argued that motor programs of greater complexity will take longer to "load." Therefore, choice reaction times will be longer when responses of greater complexity are chosen. For example, Shulman, Jagacinski, and Burke found that reaction time to initiate a double key press was longer than for a single press. The relationship between program complexity and reaction time only appears to hold, however, as long as the program cannot be loaded in advance. In a simple RT task, for example, it is possible to load or activate the entire response sequence in advance of the imperative signal since that response is the only one possible. In this case, the relationship between program complexity and response latency is no longer observed (Klapp and Erwin, 1976; Martenuik & MacKenzie, 1980).

Consistency of Outcome: Programs versus Schemata A motor program is assumed to generate very consistent space-time trajectories from one replication to another. A number of investigators have pointed out that what is consistent is not the *process* of muscular innervation (and therefore the specific pattern of neural commands) by the *product* of the response (Pew, 1974; Schmidt, 1975, 1988; Shapiro & Schmidt, 1982; Summers, 1989). Thus the signature of one's name meets the criteria of a motor program. Yet the signature may vary drastically in the actual muscular commands used (or even total muscle groups), depending on the context in which one's name is signed — whether, for example, on a small horizontal piece of paper or on a large vertical blackboard. In his original discussion, Bartlett (1932) pointed out that the specific muscular patterns involved in a tennis player's swing were quite different in the different circumstances. MacNeilage (1970) argued that the articulation of familiar words is an example of motor programs. The product of an articulation is roughly the same whether the speaker speaks normally or through clenched teeth. Yet the process of muscular innervation is totally changed between these two conditions.

In both of these examples, drastic changes in motor patterns have occurred. Yet certain characteristics of the time-space trajectory have remained invariant across the modification. Thus whatever is learned and stored in long-term memory cannot be a specific set of muscle commands but must represent a more generic or general set of specifications of how to reach the desired goal. These specifications were labeled by Bartlett (1932) and Schmidt (1975) as a *motor schema*. Once a schema is selected, the process of loading requires the specific instance parameters to be specified to meet the immediate goals at hand (Pew, 1974). Two major attributes appear to be preserved in the final output: (1) the *relative* timing of highlights (directional changes) in the movement, which may be slowed down or speeded up in this absolute value, and (2) the *relative* positioning of these highlights in x, y, and z coordinates of space, even as the absolute extent of the movement may be expanded or shrunk along any of these three dimensions.

TRACKING OF DYNAMIC SYSTEMS

In performing manual skills we often guide our hands through a coordinated time-space trajectory. Yet at other times we use our hands to guide the position of some other analog system or quantity. At the simplest level, the hand may merely guide a pointer on a blackboard or a light pen on a video display. The hand may also be used to control the steering wheel and thereby guide a vehicle on the highway, or it may be used to adjust the temperature of a heater or the closure of a valve to move the parameters of a chemical process through a predefined trajectory of values over time. When describing human operator control of physical systems, research moves from the domain of perceptual motor skills and motor behavior to the more engineering domain of tracking. This shift in domain results primarily from the great influence of three inanimate elements on the performance of the operator who must make a system state correspond to a desired goal: (1) the *dynamics* of the system itself: how it responds in time to the forces applied; (2) the *input*, or desired trajectory of the system; and (3) the *display*, the means whereby the operator views or hears the information concerning the desired and actual state of the system. These three elements interact with many of the human operator's limitations to present difficulties to tracking in the real world.

Real-world tracking is demonstrated in almost all aspects of vehicle control, ranging from bicycles to aircraft, ships, and space vehicles. It also characterizes many of the tasks performed in complex chemical and energy process control industries, when flow, pressure, and temperature must be controlled and regulated. It is increasingly common in direct manipulation computer system design (see Chapter 4), when continuous analog movement is used to position cursors on display screens. In the experimental laboratory, the tracking paradigm is typically one in which the subject controls a system whose dynamics are generated by a computer, by manipulating a control stick and observing the response as a moving symbol on a visual display.

The Tracking Loop: Basic Elements

Figure 11.4 presents the basic elements of a tracking task. These elements will be described within the context of automobile driving, although the reader should realize that they may generalize to any number of different tracking tasks. Each element produces a time-varying output, which is expressed as a function of time.

When driving an automobile, the *human operator* perceives a discrepancy or error between the desired state of the vehicle and its actual state. The car may have deviated from the center of the lane or may be pointing in a direction away from the road. The driver wishes to reduce this error function of time, $e(t)$. To do so, a force (actually a torque), $f(t)$, is applied to the steering wheel, or control. This force in turn produces a rotation, $u(t)$, of the wheel itself. The relationship between the force applied and the steering wheel movement is defined as the *control dynamics*. Movement of the wheel or control by a given

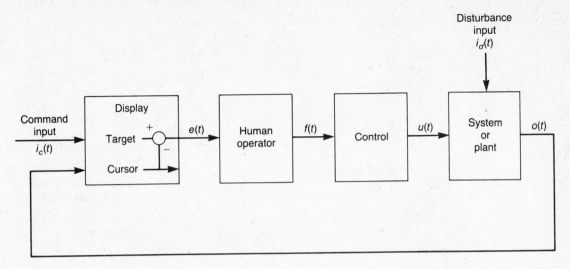

Figure 11.4 The tracking loop.

time function, $u(t)$, in turn causes the vehicle's actual position to move laterally on the highway. This movement is the *system output*, $o(t)$. The relationship between control position, $u(t)$, and system response, $o(t)$, is defined as the *plant dynamics*. When presented on a display, the representation of this output position is called the *cursor*. If the operator is successful in the correction, it will reduce the discrepancy between vehicle position on the highway, $o(t)$, and the desired, or "commanded," position at the center of the lane, $i(t)$. On a display, the symbol representing the input is called the *target*. The difference between the output and input signals is the error, $e(t)$, the starting point of our discussion. The good driver will respond in such a way as to keep $o(t) = i(t)$ or $e(t) = 0$. It should be clear from the course of this discussion, which has taken us around the loop, and from the form of Figure 11.4 why tracking is often called closed-loop behavior.

Because errors in tracking stimulate the need for corrective responses, the operator need never respond at all as long as there is no error. However, errors typically arise from one of two sources. *Command inputs*, $i_c(t)$, are changes in the *target* that must be tracked. For example, if the road curves, it will generate an error for a vehicle traveling in a straight line and so will necessitate a response. *Disturbance inputs*, $i_d(t)$, are those applied directly to the system. For example, a wind gust that buffets the car off the highway is a disturbance input. So also is an accidental movement of the steering wheel by the driver. Either kind of input may be *transient*, such as a step displacement or a gradual shift. In the first case, called a *step*, imagine that the crosswind on a highway suddenly shifts. In the second case, called a *ramp*, imagine that the crosswind gradually increases as a car goes around a curve. Alternatively, the input may be *continuous*, in which case it may be described as either predictable and periodic or random. Examples of these four different kinds of inputs are shown in Figure

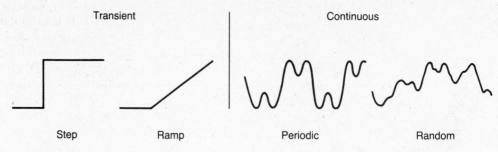

Figure 11.5 Tracking inputs.

11.5. As described in Chapter 5, either random or periodic inputs may be represented in the frequency domain as spectra. The representation of tracking signals in the frequency domain will be discussed more fully in the supplement to this chapter.

The source of all information necessary to implement the corrective response is the *display*. For the automobile driver, the display is simply the field of view through the windshield, but for the aircraft pilot making an instrument landing, the display is represented by the instruments depicting pitch, roll, altitude, and course information. An important distinction may be drawn between *pursuit* and *compensatory* displays. A pursuit display presents independent movement of both the target and the cursor. Thus the driver of a vehicle views a pursuit display since movement of the automobile can be distinguished and viewed independently from the curvature of the road (the command input). A compensatory display presents only movement of the error relative to a fixed reference on the display. The display provides no indication of whether this error arose from a change in system output or command input. Flight navigation instruments are typically compensatory displays. Most compensatory displays are artificial means of depicting real-world conditions, and these will be discussed in some detail later in the chapter.

Finally, tracking performance is typically measured in terms of error. It is calculated at each point in time and then cumulated and averaged over the duration of the tracking trial. Kelley (1968) discusses different means of calculating tracking performance.

Transfer Functions

Figure 11.4 presents three examples in which a time-varying input to a system produces a time-varying response: The operator's force, $f(t)$, applied to the control-produced control displacement, $u(t)$. The displacement, $u(t)$, produced a change in system position, $o(t)$. Finally, in the same terms, we may think of the error, $e(t)$, "applied to," or viewed by, the human as generating the force, $f(t)$. In all cases, the *transfer function* represents the mathematical relationship between the input and output of a system. (When describing human behavior in terms of transfer functions, the output of the human is usually considered to be the control position $u(t)$, rather than the force, $f(t)$. It is assumed that the human "intends" to produce a given position. The force, $f(t)$, used to achieve this position is usually achieved fairly automatically.) The transfer function may be expressed either by a mathematical equation or graphically by showing the time-varying output produced by a given time-varying input. When systems are said to be *linear*, their transfer functions may be thought of as built from the combinations of a number of fundamental, atomistic dynamic elements. Because the limits of human tracking performance depend in important ways on the transfer function of the system being controlled, and because many of the models of tracking behavior have used transfer functions to describe human performance, it is important to describe these fundamental dynamic elements.

Figure 11.6 shows the dynamic response of six of these basic elements to the step input in Figure 11.5; this response is sometimes called the *step re-*

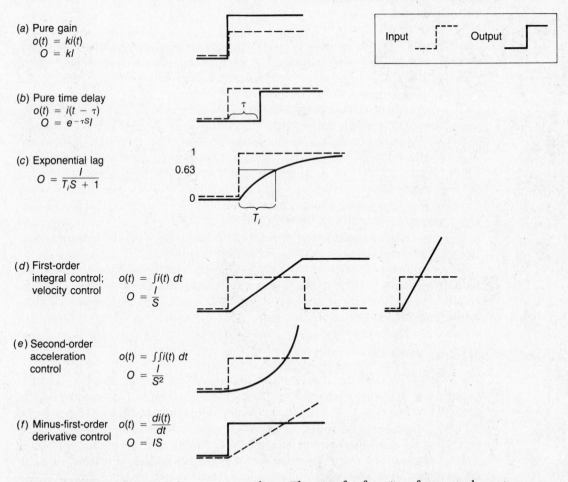

Figure 11.6 Basic dynamic elements in tracking. The transfer functions for most elements are presented in both the time domain (lowercase letters) and the Laplace domain (capital letters).

sponse. In addition, the mathematical equation that reflects output to input is presented in two different formats: differential equations in the time domain and the Laplace transform in the frequency domain. The reader should not be concerned with the Laplace function at this point, as it will be covered in more detail in the chapter supplement.

Pure Gain A pure-gain element describes the ratio of the amplitude of the output to that of the input. The element in Figure 11.6a has a gain of 2 since the output is twice the size of the input. High-gain systems, like the steering mechanism of a sports car, are highly responsive to inputs. However, they may sometimes lead to instability. Low-gain systems tend to be described as sluggish since large inputs produce only small outputs. Output and input do not need to be measured in the same units to describe gain. Thus one can speak of the gain of a radio volume control as the ratio of loudness change to angle of knob rotation. This description has meaning when the relative gains of two different systems are compared.

Pure Time Delay The pure time delay, or transmission lag, delays the input but reproduces it in identical form T seconds later (Figure 11.6b). This would describe the response of a remotely located robot equipped with a television camera on the surface of the moon, controlled from a master on earth. Feedback from control signals sent to the robot would be reproduced precisely at the operator's terminal, but only after the four-second delay necessary for the signal to be relayed to the robot and for the picture to be relayed back to earth. A pure time delay has no effect on gain, nor does gain have any effect on time delay.

Exponential Lag Some lags do not reproduce the input identically but instead gradually "home in," or stabilize, on the target input. The *exponential lag*, shown in Figure 11.6c, is defined by its *time constant*, T_I, which is the time that the output takes to reach 63 percent of its final value. The response bears considerable resemblance to the human target-acquisition response described by Fitts's law in Figure 11.3a. In a more general sense, it describes the response of many systems with a built-in negative feedback loop to ensure that an output is reached. A similar response would be shown by the tires on a car with a hydraulic power-steering system following the command indicated by steering-wheel position change or by the hydraulic response of many airplane control surfaces to inputs applied by the pilot.

Integrator, Velocity-Control, or First-Order System The step response shown in Figure 11.6d is a constant velocity (change in position per unit time) with a magnitude that is proportional to the step size. In calculus this response is defined by the *time integral* of the input. Notice that if the input is withdrawn, the velocity returns to zero but the output is at a new location. Such systems are frequently encountered in manual control. An example is the relationship between the angle of steering wheel deflection and the heading of a vehicle. A constantly held wheel position will produce a constant rate of change of heading, or rate of turn. In an aircraft, a constant bank will also lead approximately to a constant rate of turn. Any first-order or velocity-control system must also be defined by its gain. A low-gain system is shown on the left of Figure 11.6*d*; a high-gain system is on the right. The term *order* refers to the number of time integrations in the transfer function. Therefore, since it contains one integration, the system shown in Figure 11.6*d* is a first-order system. (The systems in Figure 11.6*a* and Figure 11.6*b* are zero order.) The first-order dynamic response is closely related to the exponential lag shown in Figures 11.3 and 11.6*c*. If a system makes a first-order response to the *error* rather than to the command input, the result will be an exponential lag. Since the response velocity to correct the error is proportional to the size of the error, as the error is reduced response velocity is reduced proportionately. In Figure 11.6*c* the velocity approaches zero, on the right side, as the error approaches zero.

Double-Integrator, Acceleration-Control, or Second-Order System A second-order system combines two integrators in series. The step response shown in Figure 11.6*c* is therefore the constant acceleration that would be obtained if the velocity response in Figure 11.6*d* were integrated a second time. The pure second-order system is typical of any physical system with large mass and therefore great inertia when a force is applied. It is *sluggish*, in that it will not respond immediately, particularly if the gain is low. When tracked, second-order systems also tend to be *unstable*, or difficult to control, because once the system does begin to respond, its high inertia will tend to keep it going in the same direction and cause it to overshoot its destination. The operator will have to make a series of reverse corrections, which often produces oscillatory behavior. As an intuitive example of a pure second-order system, try to imagine rolling an orange or bowling ball from one end of a flat board to the other by tilting the board. The relationship between the board angle and the position of the orange on the board is a second-order one. Second-order systems are very prevalent in aviation, seagoing vehicles, and chemical processes.

Differentiator The minus-first-order, or differentiator, control system, shown in Figure 11.6*f*, will produce an output of a value equal to the rate of change of the input. The step response of the differentiator is theoretically a spike of infinite height and zero width since the "step " is an instantaneous change in position (and so has infinite velocity). As a result, the step response is not shown in Figure 11.6*f*. Instead, the *ramp response* is depicted. The system response is therefore just the opposite of the first-order system shown in Figure 11.6*d*. In the calculus representation, these two are also opposite. If a time function is differentiated and then integrated, the original function will be recovered. In isolation, differential control systems are not frequently observed. An example might be an electrical generator in which the output (current) is proportional to the rate of turn of the input coils. However, differential control systems are of critical importance when they are placed in series with systems of higher order. They can reduce the "effective" order of the system by "canceling" one of the integrators and so make it easier to control. For example, a differentiator placed in series with the second-order system of Figure 11.6*e* will produce a first-order system. This point will be considered later, when we see how humans should track second-order systems.

Frequency-Domain Response   The transient dynamic response of the elements shown in Figure 11.6 is in the time domain. Yet engineers are often more concerned about the response of these elements to periodic or random inputs. Indeed, most tracking studies involve continuous inputs. For some of the dynamic elements of Figure 11.6, the response to random or periodic inputs is intuitively, as well as formally, quite predictable from the step response. A pure gain, for example, will reproduce a periodic signal perfectly but at a higher or lower amplitude, given by the value of the gain. For other elements, however, the response to continuous inputs is considerably more complex. In the chapter supplement we will consider the frequency-domain response of different elements as they are used in human operator modeling.

Human Operator Limits in Tracking

The previous chapters of this book have identified a number of limitations in human information processing. Five of these limits in particular influence the operator's ability to track: processing time, information transmission rate, predictive capabilities, processing resources, and compatibility. Each of these will be described briefly in the context of manual control and then considered in more detail as they influence specific aspects of the manual-control task.

Processing Time   The discussion of reaction time in Chapter 8 suggested that humans do not process information instantaneously. In tracking, a perceived error will be translated to a control response only after a lag, referred to as the effective time delay (McRuer & Jex, 1967). Its absolute magnitude seems to depend somewhat on the order of the system being controlled. Zero and first-order systems are tracked with time delays from 150 to 300 msec. For a second-order system, the delay is longer, about 400 to 500 msec, reflecting the more complex decisions that need to be made (McRuer & Jex, 1967).

Time delays, whether the result of human processing or system lag, are harmful to tracking for two reasons: (1) Obviously, any lag will cause output to line up no longer with input. The error thus resulting, shown as the shaded region in Figure 11.7, will grow with the magnitude of the delay. (2) Often more seriously, when periodic or random inputs are tracked, delays will induce problems of *instability*, producing oscillatory behavior. These will be disucssed later in the chapter.

Bandwidth   Tracking involves the transmission of information, whether displayed as a command or as a disturbance-induced error. As discussed in Chapter 2, time-varying input and output signals may be quantified through continuous information theory, and it is not surprising that the same limitations of

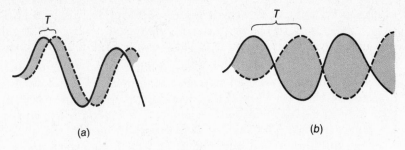

Figure 11.7 Error resulting from time delay: (*a*) small, (*b*) large.

information transmission in discrete tasks are evident in continuous tracking as well. Thus, Baty (1971), Crossman (1960), and Elkind and Sprague (1961), found that the limit of information transmission in tracking is between 4 and 10 bits per second, depending on the particular conditions of display. Elkind observed that the transmission rate increases if a preview of the input is available before it is tracked, much as automobile drivers preview segments of the road ahead before the vehicle actually reaches them.

In Chapter 8 it was argued that the limits of serial reaction time were defined by the frequency of decisions, not by their complexity. In tracking, too, there appears to be an upper limit in the frequency with which corrective decisions can be made that is more restrictive than the limit imposed by their complexity. This frequency limit in turn determines the maximum *bandwidth* of random inputs that can be tracked successfully; it is normally found to be between 0.5 and 1.0 Hz (Elkind & Sprague, 1961). This value corresponds quite closely with estimates that the maximum frequency with which corrections are exerted in tracking is roughly two times per second (Craik, 1948; Fitts & Posner, 1967). Since two corrections are required for each cycle, this limit corresponds to a bandwidth of one cycle per second. This limit appears to be a central one, related to processing uncertainty in the tracking signal, rather than a motor one, because operators have no difficulty in tracking *predictable* signals as high as 2 to 3 Hz (Pew, 1974; Pew, Duffenbach, & Fensch, 1967). The limit of two corrections per second in continuous tracking is close to the maximum decision-making speed in the serial RT paradigm of 2.5 decisions per second (Debecker & Desmedt, 1970; see Chapter 8).

Prediction and Anticipation Fortunately, human operators are rarely placed in real-world environments in which they must track inputs at bandwidths so high that the limits on processing rate become restrictive. The more serious limits instead appear to be imposed when operators track systems like ships and aircraft that have lags. Here the operator must *anticipate* future errors on the basis of present values to make control corrections that will be realized by the system output only after a considerable lag. Consider the pilot of a supertanker in a channel who realizes the vessel is off course and wishes to correct this error. Because of the high inertia of the ship and its higher-order control characteristics, a correction delivered to the rudder will not substantially alter the ship's course for a matter of tens of seconds. Therefore, to stay within the limits of the channel effectively, the operator must base corrections on future error and not present error. Corrections based on present error will be too late.

Future error, of course, equals the difference between future input and future output. In the case of ship control, future command input can easily be *previewed* (this is the view of the channel or path to be negotiated). But future output must be derived and anticipated, a function, as we have noted in Chapter 3 and again in Chapter 7, that humans do not perform effectively. In tracking, this limitation occurs in part because higher derivatives (velocity and acceleration) of the error signal must be perceived. Where a signal *is* at the moment is best indicated, of course, by where it is. But where a signal *will be* in the future is best indicated by its present velocity and acceleration. Ample data are available to suggest that humans perceive position changes more precisely than velocity changes, and both velocity and position changes more precisely than acceleration (Fuchs, 1962; Gottsdanker, 1952; Kelley, 1968; McRuer et al., 1968; Runeson, 1975). Thus, when tracking slow, sluggish systems, the operator's perceptual mechanisms are called on to perform functions for which they are relatively ill equipped.

Processing Resources   Another source of difficulty in anticipation relates to the resource demands of spatial working memory, as discussed in Chapters 6 and 9. When anticipating where a sluggish, higher-order system like a super-tanker will be in the future, it helps to be able to perceive its acceleration, but it is also important to be able to perform calculations and estimations of where that system will be in the future, given an *internal model* of the system's dynamics (Eberts & Schneider, 1986; Gill, Wickens, Donchin, & Reid, 1982; Pew & Baron, 1978). For the operator who is not highly trained, the operations based on this internal model demand the processing of working memory. Tracking thus is readily disrupted by concurrent tasks. The limits of human resources also account for tracking limitations when the operator must perform more than one tracking task at once, that is, in dual-axis tracking, to be discussed later in this chapter.

Compatibility   The discussion of S-R compatibility in Chapter 8 emphasized that certain spatial compatibility relationships were relatively "natural." Because tracking is primarily a spatial task, it is apparent that these relationships should affect tracking performance. The research on control and display relationships in tracking suggests that indeed they do. Incompatibility is seen in compensatory displays, as when a left-moving error cursor requires a right-moving response. It is also seen whenever the axis of control is not aligned with the axis of display. The disruption of tracking performance becomes particularly bad when the misalignment is greater than 45 degrees (Kim et al., 1987).

Effect of System Dynamics on Tracking Performance

The interaction between human limitations and the dynamic properties of the system to be controlled determines the level of tracking performance. We will consider the effects on performance of three important characteristics of those system dynamics: gain, time delay, and order. In certain combinations these variables produce problems of *stability*, a factor that will be considered separately.

Gain Both tracking performance and subjective ratings of effort appear to follow an inverted U-shaped function of system gain (Gibbs, 1962; Hess, 1973; Wickens, 1986). Whether tracking steps or compensating for random disturbances, systems with intermediate levels of gain are easiest to track. The advantage of middle-gain systems results from the trade-off between the costs and benefits of more extreme gains. When gain is high, minimal control effort is required to produce large corrections. For example, the steering wheel on a sports car has to be turned only slightly to round a curve. Thus, in a sense, high gain is economical of effort, and this economy is quite valuable when continuous corrections are required to track random input. On the other hand, gain that is too high can lead to overcorrections, oscillations, and instability if there are lags in the system. This result also is undesirable and can be eliminated by reducing gain. The crossover point of the two functions describing instability at high gain and effort at low gain, determines the optimal level of gain, which cannot be precisely specified in a general sense because it is determined by the extent to which effort at too low a level of gain and instability at high gain is the more important concern to be avoided.

Time Delay Pure time delays are universally harmful in tracking, and tracking performance gets progressively worse with greater delays. The reason is apparent from the discussion of processing time limits. If a control input will not be realized until some point in the future (the consequences of the delay), the corrective input generated by the human operator must be based on the future value of error rather than on its present value. Such anticipation, as noted, is imperfectly done and demands resources.

The effects of exponential lags, however, are often less harmful. An exponential lag is, in a sense, a combination of a zero-order, or position, control and a first-order, or velocity, control. In Figure 11.6c note that immediately following the step input, the response of the exponential lag looks very much like that of the velocity control system in Figure 11.6d. Only later does it look like the response of the time delay in Figure 11.6b. In fact, when controlled with high-frequency corrections, a zero-order system with an exponential lag behaves very much like the first-order system. We will see that first-order systems have some substantial advantages over systems of zero order. These advantages prevent exponential lags controlled at higher frequencies from causing the kinds of harmful effects that the pure time delay does.

System Order The effects of system order on all aspects of performance may be best described in the following terms: Zero-order and first-order systems are roughly equivalent, each having its costs and benefits. Both are also equivalent to exponential lags, which as we have seen are a sort of combination of zero and first order. However, with orders above first, both error and subjective work load increase dramatically (Wickens, 1986). The reason that zero- and first-order systems are nearly equivalent may be appreciated by realizing that successful tracking requires both position and velocity to be matched. Under some circumstances, matching one of these quantities might be more important than matching the other. Compare the two functions in Figure 11.8. In Figure 11.8*a* position error is reduced quite frequently to zero, but the velocities of input and output are rarely matched. In Figure 11.8*b*, although velocity is closely matched, the positions of input and output rarely agree. Which form of tracking is superior? Clearly the answer to this question depends on the circumstances. If one were a passenger in an aircraft, the response in Figure 11.8*a* would not suggest a comfortable ride compared to that of *b*, but if the aircraft were flying at a low level, following a precise course with a minimum margin for error, the performance in *b* might be disastrous.

If we view performance as a mixture of position matching and velocity matching, the fact that the input to a zero-order control system directly accomplishes the former and that input to a first-order velocity system accomplishes the latter indicates why neither is unequivocally preferable to the other. The intermediate level between zero- and first-order control can be created either by linearly combining the outputs of the two pure orders (called a *rate-aided* system) or by varying the time constant of an exponential lag (Wickens, 1986). These systems generate performance that is also equivalent to either pure first- or pure zero-order control, depending on the relative importance of making a position or velocity match (Chernikoff & Taylor, 1957).

Another contrast between system orders to note is economy of movement and space. In a velocity control system, any change in output position can eventually be accomplished by displacing the control only a small amount (how rapidly, of course, depends on the gain). In a position control system, larger position changes must be accomplished by moving the control through a larger physical space. When the amount of area for control space is limited, as in the cockpit of a high-performance aircraft, constraints are imposed on the use of position controls, unless these are rotary controls.

Control systems of second order and higher are unequivocally worse than either zero- or first-order systems (Kelley, 1968). The problems with second-

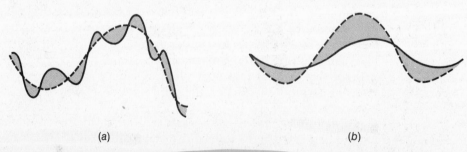

(a)   (b)

Figure 11.8 Styles of control: (*a*) minimized position error but high velocity error, (*b*) minimized velocity error but high position error.

order control are many. As noted, to control any higher-order system effectively, one must anticipate its future state from its present. To do so requires that higher-error derivatives be perceived as a basis for correction, a process known as *generating lead*, and humans, as we have discussed, do not perform this function well. As McRuer and Jex (1967) observed, the operator's effective time delay is also longer when higher derivative must be processed under second-order control. This increased lag contributes an additional penalty to performance.

Second-order systems may be controlled by two strategies. The one described above requires the operator to perceive the higher-error derivatives continuously and respond smoothly on the basis of this information. An alternative strategy of second-order control is sometimes referred to as "bang-bang," double-impulse, or time-optimal control (Hess, 1979; Wickens & Goettl, 1985; Young & Miery, 1965). The operator perceives an error and reduces it in the minimum time possible with an open-loop "bang-bang" correction. As shown in Figure 11.9, this is accomplished by throwing the stick "hard over" in one direction to generate maximum acceleration for half the interval and then reversing the stick to produce maximum deceleration for the other half, monitoring and checking the result at the end. Because the double-impulse strategy reduces large errors in the shortest possible time, it is referred to as a form of optimal control (Young, 1969). Imagine using a bang-bang control strategy to roll an orange from one end of a board to the other.

Although the double-impulse control eliminates the need for continuous perception of error derivatives of smooth analog control, it does not necessarily reduce the total processing burden (Wickens & Goettl, 1985). More precise timing of the responses is now required, and an accurate internal model of the state of the system must be maintained in working memory, in order to apply the midcourse reversal at the appropriate moment (Jagacinski & Miller, 1978). Also, as suggested by Figure 11.9, the bang-bang strategy will produce high velocities. As in the response shown in Figure 11.8*a*, there are conditions when a lower-velocity "smooth ride" is preferable. The *optimal control model* of

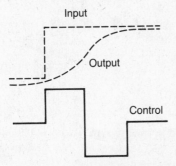

Figure 11.9 Required bang-bang response to reduce an error with a second-order system in minimum time.

tracking behavior discussed in the chapter supplement dictates the appropriate strategy, given the importance of a smooth ride versus low error.

Instability A major concern in the control of real-world dynamic systems is whether or not control will be *stable*—that is, whether the output will follow the input and eventually stabilize without diverging or producing excessive oscillations. Oscillatory and unstable behavior can result from two quite different causes: positive and negative feedback.

To illustrate *positive feedback systems,* imagine two people sleeping under an electric blanket with dual temperature controls, one for each side. Suppose that the controls inadvertently become switched. If A now feels cold, A will turn the heat up. This action will lead to an increase in heat on the other side, causing B to turn the heat down, and thereby leading A to feel still colder. Person A will then adjust the heat still higher, and the resulting chain of events is evident. This is an example of a positive feedback system: An error once in existence is magnified.

Whereas this example is unlikely to occur in real-world systems, unless switches are inadvertently misconnected, there is a second kind of positive feedback that is not unusual in aviation systems. An intuitive example occurs when one must balance a stick on the end of one's finger. A computer analog of this task was built as the *critical instability tracking task*, described as a workload assessment technique in Chapter 9 (Allen & Jex, 1968; Jex, McDonnel, & Phatak, 1966). The feedback loop of the critical task, as it is commonly called, is shown in Figure 11.10. An error, *e*, once detected by the system, will generate a proportional output velocity, *o*, whose value is determined by the gain, λ. However, unlike "purposeful" human control, in which negative feedback subtracts this velocity from the existing error, the positive feedback loop in the critical task *adds* the velocity to the error, thereby increasing the rate of error movement away from the center. This response is similar to the dynamics of a balanced stick. If there is a small error (from the vertical) it will begin to

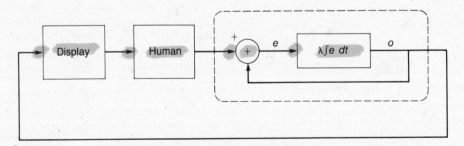

Figure 11.10 Dynamics of critical instability tracking task. (*Source:* H. R. Jex, J. P. McDonnel, and A. V. Phatak, "A Critical Tracking Task for Manual Control Research," *IEEE Transactions on Human Factors in Electronics, HFE*–8 (1966), p. 139. Copyright 1966 by the Institute of Electrical and Electronic Engineers, Inc. Reprinted by permission.)

fall, and its rate of falling (increase in error) will increase as it falls farther. The overall rate of falling will furthermore be proportional to the shortness of the stick, analogous to the value of λ in the critical task.

Positive feedback loops characterize a number of complex dynamic vehicles, for example, a booster rocket, controlled by swiveling, tail-mounted engines. Another example is the control task of wheeling a bicycle backward or backing up a trailer. Positive feedback loops may also be found in certain aspects of the control of helicopters and other complex aircraft. Like second-order systems, those with positive feedback are universally harmful for the obvious reason that they cannot be left unattended. If control is not exercised they will eventually diverge, just as the stick balanced on the fingertip must eventually fall unless the finger is moved back under the top of the stick.

Negative feedback systems are more typical. Humans and most well-designed systems function in such a way as to reduce rather than increase detected errors. This is the property of a negative feedback system, described clearly by Jagacinski (1977), Kelley (1968), and Toates (1975). Good stability normally results from such "purposeful" control action. However, there are certain occasions when even a negative feedback loop with the best error-correcting intentions produces oscillatory or even unstable behavior. A potentially disastrous example in aircraft control occurs when *pilot-induced oscillations* are produced. The vertical path of the aircraft swings violently up and down with growing amplitude as a consequence of the pilot's inappropriate corrections (Hess, 1981).

Instability caused by negative feedback results from high gain coupled with large phase lags or any delay around the closed loop, as shown in Figure 11.4. For example, because the second-order system is sluggish with a long lag in its response, control of the system tends to be unstable. However, instability may also occur with lower-order systems when there are long delays in system response.

The reason why high gain and long phase lag collectively produce instability may be appreciated by the following example taken from Jagacinski (1977). Imagine that you are adjusting the temperature of shower water to your ideal comfort value (a command input). You are controlling to reduce the error and so acting as a negative feedback system. However, because of the plumbing, there is a lag between your adjustment of the faucet and the change in water temperature—the source of the perceived error used to guide correction. If your gain is high, then when you feel initially cold, you increase the hot water by a large amount and will continue to increase it as long as you feel cold. As a consequence, you will probably overshoot and scald yourself, and the error will now be on the "hot" side. If your gain remains high, the compensatory cooling correction will also be overapplied, and the water will, after a lag, become too cold. The eventual temperature-time history will be a series of growing oscillations (and discomfort). Clearly in these circumstances you must reduce your gain to avoid the unstable behavior resulting from the time lag. A gain reduction involves applying a smaller corrective turn of the faucet in response to the detected error—tolerating a mild discomfort now in anticipation of an eventual stable response.

The difference between high- and low-gain systems and their association with unstable and sluggish behavior, respectively, is an important one in human performance because gain may be thought of as a "bias parameter," like the response criterion in signal detection or the speed-accuracy set in reaction time. It is a parameter, then, that can be strategically adjusted to different values according to different environmental conditions or strategic goals. The difference in the tracking performance of the two systems shown in Figure 11.8, for example, could be attributed in part to a difference in the gain of the feedback system that is tracking the error: high in Figure 11.8a, low in b.

The role of stability and its dependence on gain and lag is, of course, critically important in the design and testing of piloted vehicles and represents a major application of manual-control research. However, engineering-oriented research on stability normally uses the language of the *frequency* domain, described briefly in Chapter 5. The supplement to the present chapter covers the frequency domain analysis of tracking in more detail, with the goal of making the sometimes mystical language of this area more understandable to the psychologist.

There is an alternative control strategy for making corrections when the lag is long: to base control correction on the trend of the error rather than its absolute level. Thus, in the shower, if you feel that the water is getting warmer even if you are still too cold in an absolute sense, this trend can serve as a signal to stop increasing the heat.

This control strategy, based on error rate rather than error value, should by now be familiar. Earlier in this chapter we suggested that it formed the basis of anticipation when controlling systems with long lags. Here we see that this strategy is essential because it is often necessary to avoid instability. Also when describing the control elements in Figure 11.6, we suggested that a differentiator could cancel an integration in controlling higher-order systems. When humans respond predictively on the basis of trends, they are effectively becoming differentiators and so are canceling one of the integrators of second-order dynamics. We will see how certain tracking displays have been modified to make this prediction easier and induce humans to control as differentiators do.

Displays

Input Prediction and Preview   The problems associated with prediction and anticipation in tracking can be divided into those of predicting the command input and those of predicting the future trajectory of the system output. The future of the output, in turn, is determined by operator corrections or by disturbance inputs. When there are lags in the tracking loop of T seconds caused either by the human operator's time delay or by system lags, it is essential to know what the error will be T seconds into the future so that corrections can be formulated on the basis of this *predicted* error, not on the present information. Thus, if a future error is perceived and corrected now, the system will realize its appropriate correction the *T* seconds later that will be appropriate.

Clearly the future input will be most accurately available when there is *preview*. The automobile driver, for example, has preview of the course of the road ahead, except when driving in the fog. Figure 11.11 shows preview as it might be presented on a typical tracking display, the future course shown on the top. The large benefits of preview (Crossman, 1960; Elkind & Sprague, 1961; Grunwald, Robertson, & Hatfield, 1981; Reid & Drewell, 1972) occur primarily because it enables the operator to compensate for processing lags in the tracking loop. The fact that the operator's time delay is only in the order of 200 to 500 msec suggests that half a second of preview should be all that is needed when one is tracking systems that have no lags of their own. Reid and Drewell (1972) varied the amount of preview available while subjects tracked a first-order system. The results shown for one subject in Figure 11.12 suggested that performance did indeed benefit greatly for the first half-second of preview offered but improved minimally with greater amounts. However, when there are long system lags, preview is used further into the future. The supertanker pilot is tracking the channel several hundred yards ahead of the bow of the ship. This is a command input signal that will not be traversed by the ship until minutes later. For aircraft pilots, flight path preview that is tens of seconds into the future will be useful.

In the absence of preview, the human operator must use whatever information and computational facilities are available to *predict* the future course of

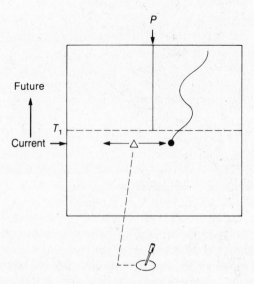

Figure 11.11 Tracking with preview.

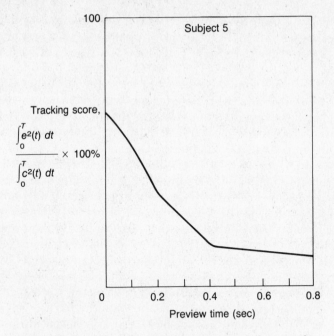

100

Subject 5

Tracking score,

$$\frac{\int_0^T e^2(t)\ dt}{\int_0^T c^2(t)\ dt} \times 100\%$$

0   0.2   0.4   0.6   0.8

Preview time (sec)

Figure 11.12 Effect of varied preview interval on tracking performance. (*Source:* D. Reid and N. Drewell, "A Pilot Model for Tracking with Preview," *Proceedings, 8th Annual Conference on Manual Control* (Wright Patterson AFB Ohio Flight Dynamics Laboratory Technical Report AFFDL–TR-72, 92) (1972), Washington DC: U.S. Government Printing Office, p. 200.)

the input. To some extent this prediction may be based on the statistical properties of the input. For example, if the bandwidth of a random input is low, the present position and velocity provide some constraints on future position. Even if we had no preview of the input function shown in Figure 11.11, we would consider it unlikely that the input would be to the left of point P at time T_1. Our past experience with this input tells us that it just doesn't change that rapidly.

To the extent that the input is nonrandom or contains periodicities, the knowledge of future input is increased considerably and prediction becomes easier. When this occurs, the operator can track by using what Krendel and McRuer (1968) described as a *precognitive mode*.

Output Prediction and Quickening As described previously, the future trajectory of the input may be predicted with some confidence, given its current position, velocity, and acceleration. Correspondingly, the best estimate of where a higher-order system with some mass and inertia will be in the future is provided by a combination of its present position and its higher derivatives. This kind of prediction, as we have seen, is not easily done and extracts a heavy toll on operator resources. In trying to reduce this burden, engineering psychologists have developed displays in which a computer estimates error (or output) derivatives and explicitly presents them as predicted symbols of future position (Gallagher, Hunt, & Williges, 1977; Kelley, 1968). This format is called a *predictive display*. A typical one-dimensional predictive display for aircraft control is that used in a modern commercial aircraft, shown in Figure 11.13. Figure 11.14 shows a three-dimensional flight path predictor (Wickens, Haskell, & Harte, 1989). Note that the future aircraft predictor symbol (the black aircraft) is accompanied by *preview*, represented as a desired flight path tunnel.

Any number of different computation techniques can be used to estimate where a vehicle will be likely to be in the future (Grunwald, 1985). For example, the predictive elements may be driven by directly computing the position, velocity, and acceleration of the present system state and adding these values together with appropriate weights. Alternatively, these values may be inferred by directly measuring different internal states of the system (Gallagher, Hunt, & Williges, 1977). To provide an example of these two computational procedures, a predictive display of a car's future lateral position on the highway could be driven by computing the present values of its lateral position, velocity, and acceleration. Alternatively, since these three values are roughly equivalent to the current position of the car, the heading of the car, and the deflection of the steering wheel, respectively, the three variables could be

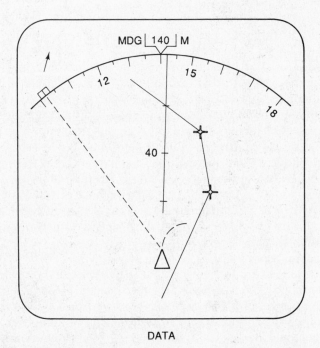

Figure 11.13 Predictor element on map display for Boeing 757. The curved arc from the triangular aircraft symbol is the predicted flight path.

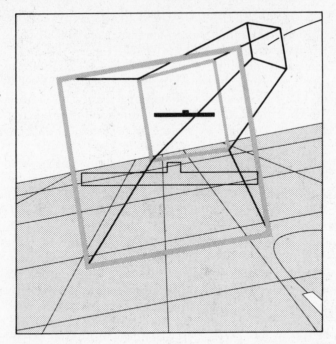

Figure 11.14 Three-dimensional predictor of flight path information.

directly measured, weighted, and summed to provide an accurate predictive display.

Every predictive display must make some assumptions about the future forces acting on the system (either from operator control inputs or from disturbances). For example, is it assumed in predicting the future trajectory of the system that the operator will apply an optimal correction, a suboptimal correction, or no correction at all? The nature of these assumptions may differ somewhat from display to display and will determine the accuracy and therefore the effectiveness of the display. It is also apparent that the accuracy of any prediction will decline into the future. Just how far into the future a prediction is valid depends on such factors as the sluggishness of the system and the frequency of control or disturbance inputs. Prediction will be more accurate longer into the future if systems are more sluggish and disturbance inputs are of lower frequency. Thus, a supertanker (which is sluggish) can have a longer predictive accuracy than an aircraft, and so can a spacecraft traveling in a wind-free environment (no turbulence). For a light aircraft traveling in turbulence, however, the predictive interval will be short.

No matter how the predictive information is derived, predictive displays have proven to be of great assistance in the tracking of higher-order systems. Wickens, Haskell, and Harte (1989) demonstrated the value of prediction (see also Grunwald et al. 1987) in the cockpit display shown in Figure 11.13. In an excellent discussion of the topic, Kelley (1968) described the tremendous benefit of predictive displays for submarine depth control, an example of very sluggish third-order dynamics.

In 1954, Birmingham and Taylor proposed a technique known as *quickening* that was closely related to the predictive display. A quickened display presents only a single indicator of quickened tracking error, which is calculated by combining the present error position, velocity, and acceleration. Like the prediction element in a predictive display, the quickened element indicates where the system will be likely to be in the future if it is not controlled. Unlike the predictive display, a quickened display has no indication of the current error. The justification for this absence is that the current error provides no information that is useful for correction. This lack, of course, has a disadvantage: There are certainly times when you want to know where you are and not just where you will be. Quickened displays are used in the flight director of many modern commercial airliners.

Pursuit versus Compensatory Displays   The goal of tracking is to match the output to the input, or to minimize the error. These two seemingly equivalent statements define the *pursuit* and *compensatory* display formats, respectively. The operator with a pursuit display views the command input and the system output moving separately with respect to the display frame. The flight path display shown in Figure 11.14 is a pursuit display since the pilot can view changes in the command input (the tunnel) independently of changes in the aircraft's position. In the compensatory display, only the difference between these two — the error — is portrayed. Pursuit displays generally provide superior performance to compensatory displays (Poulton, 1974) for two major reasons: the ambiguity of compensatory information and the compatibility of pursuit displays.

On the compensatory display, the operator is unable to distinguish among the three potential causes of error: command input, disturbance input, and the operator's own incorrect control actions. As a consequence, control is more difficult than in the pursuit display, in which command and disturbance inputs can be distinguished. If, however, there is only one source of input, the advantage of the pursuit display decreases since errors on the compensatory display are now less ambiguous. Flying an aircraft toward a fixed runway, for example, is a tracking task in which there is information only in the disturbance input. The runway, representing the command input, does not move. In contrast, flying an aircraft toward the runway on an aircraft carrier now has a command input added if the carrier is moving or rolling in the sea swells.

Whenever there is a changing command input, the pursuit display will also provide some advantage because its stimulus-response compatibility is greater (see Chapter 8). If the command input is suddenly displaced to the left, it will require a leftward correction on the pursuit display. A left-moving stimulus thereby is corrected with a leftward response. This is an inherent motion compatibility that is consistent with the operator's tendency to move toward the source of stimulation (Roscoe, 1968; Roscoe, Corl, & Jensen, 1981; Simon, 1969). In contrast, in the compensatory display, the left-moving command input will be displayed as a right-moving error. In this case, a right-moving stimulus requires an incompatible leftward response. Whenever some portion of tracking input is command, this compatibility factor will benefit the pursuit display.

Pursuit versus Compensatory Behavior   Previous discussion has emphasized the importance of the closed, negative feedback loop in tracking. We assume that operators continuously process the difference between where they are and where they would like to be and respond appropriately. That is, they *compensate* for an error. This style of tracking is referred to as closed-loop, or *compensatory*, behavior. The operator need not, however, process the error directly. When tracking with a pursuit display, the operator may attend only to the input. If there are no disturbances acting on the plant and the operator knows the plant dynamics, the operator may ignore the output and assume that it will follow the commands that he or she has provided to track the input. If the output is ignored, the operator *cannot* be processing the error since the error is by definition the difference between input and output. This strategy is referred to as open-loop, or *pursuit*, behavior. It leads to more efficient tracking because, unlike compensatory behavior, pursuit behavior does not require an error to be present in the first place to generate a corrective response.

The contrast between pursuit and compensatory behavior is illustrated by the two mechanisms the human eyeball uses to track moving targets to keep them in foveal vision (Young & Stark, 1963). The *saccadic* mode is compensatory. Based on a discrepancy between where the target is located and where the eye is fixated, a discrete jump, or saccade, is programmed to reduce the error. (This is essentially a zero-order system. The size of the movement is proportional to the size of the error.) There is, however, a *pursuit* mode that generates a constant velocity of eyeball rotation whenever the target shows constant velocity motion. The two mechanisms work fundamentally independently and in parallel.

Although it appears logical that pursuit behavior will occur with pursuit displays and compensatory behavior with compensatory displays, this association is not necessary in a pursuit display because the operator may focus attention either on error or input. Pew (1974) has pointed out that much of human tracking behavior is a combination of the two behaviors. The operator will track in the pursuit mode when possible, but will supplement it with compensatory corrections when errors build up. Conversely, pursuit behavior is possible even with a compensatory display when there are no disturbances, although this is more difficult. An operator who knows precisely the effects of control manipulations on the plant response can mentally subtract this contribution from the perceived error. The difference is the command input, which can be tracked directly with pursuit behavior.

Krendel and McRuer (1968) have contrasted compensatory and pursuit behavior to represent the progression of skill acquisition in tracking. In their *successive organization of perception* (SOP) model of tracking, the authors propose that three modes of tracking, as shown in Figure 11.15, describe the

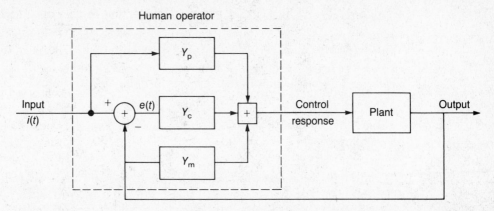

Figure 11.15 Successive operation of perception (SOP) model of Krendel and McRuer (1968) shows compensatory loop (Y_c), pursuit loop (Y_p), and memory or precognitive loop (Y_m). (*Source:* D. T. McRuer and H. R. Jex, "A Review of Quasi-linear Pilot Models," *IEEE Transactions on Human Factors in Electronics, 8* (1967), p. 240. Copyright © 1967 by the Institute of Electrical and Electronics Engineers, Inc. Reprinted by permission.)

progression of tracking behavior with practice. In the compensatory mode (Y_c), the operator acts as an error corrector with a corresponding time lag, which leads to poorer performance. With greater skill, the operator becomes progressively more able to track in the pursuit mode (Y_p), thereby responding directly to the input and nullifying some of the existing lag. Finally, a third mode of tracking, known as the *precognitive* mode (Y_m), is possible only when the input is nonrandom. In this case, the operator can store the input patterns in long-term memory (the box labeled Y_m in the figure) and respond on the basis of this stored information. Since the inputs can be predicted in advance, tracking can be carried out with no lag whatsoever. This behavior is truly open-loop since it may take place in the total absence of feedback. For example, as we traverse the curves of a familiar driveway in a familiar car, we may do so if we choose with our eyes closed, an example of precognitive tracking.

Precognitive, pursuit, and compensatory behavior may be experimentally separated from one another by the following means: If the total display can be blanked for periods of time, and the operator continues to reproduce the input successfully, behavior must be precognitive. If just the output symbol is eliminated and tracking is successful, behavior is pursuit. If not, behavior must have been compensatory. Unfortunately these manipulations somewhat distort the tracking task in their effort to determine how the operator can track. A more ingenious way of naturally distinguishing the characteristics of pursuit and compensatory behavior without artificially disrupting portions of the display is to exploit the fact that disturbance inputs can be corrected only with compensatory behavior. Since these are not shown as command inputs, they cannot be pursued. If the operator tracks an input function that is made up of several independent components and if these are distributed between command and disturbance, two separate transfer functions of the human operator simultaneously tracking the two sets of inputs can be compared (Allen & Jex, 1968). If the two transfer functions are identical, all behavior must be compensatory as this is the lowest common denominator of tracking behavior. If they differ, the differences will reflect the better pursuit behavior used to track the command input.

MULTIAXIS CONTROL

People must often perform more than one tracking task simultaneously. The aircraft pilot, for example, controls both the pitch and roll of the aircraft; when we drive we track lateral position, heading, and velocity. Even riding a bicycle involves tracking lateral position while also stabilizing the vertical orientation of the bike. In general, as discussed in Chapter 9, there is a cost to multiaxis control that results from the division of processing resources between tasks. However, the severity of this cost may be influenced by the nature of the relationship between the two (or more) variables that are controlled and the way in which they are physically configured.

Cross-Coupled and Hierarchical Systems

A major distinction can be drawn between multiaxis systems in which the two variables to be controlled as well as their inputs are essentially independent of one another and those in which there is cross-coupling, so that the state of the system or variable on one axis partially constrains or determines the state of the other. An example of two basically independent axes is provided by the control of attitude and elevation in a gunnery task. In contrast, there is a small degree of cross-coupling between the pitch and roll axes of aircraft control. What the pilot does to control the roll (rotation around the long axis of the aircraft fuselage) has a small effect on the pitch (nose up or nose down). At the far extreme of cross-coupling, control of the heading and lateral position of an automobile on the highway are highly cross-coupled axes. Control of vehicle heading directly affects lateral position. In this case, the two cross-coupled tasks are considered to be *hierarchical*. That is, lateral position cannot be changed independently of a control of heading. The steering wheel, which directly controls headings, is *used* to obtain a change in lateral position.

Many higher-order control systems in fact possess similar hierarchical relationships. Lower-order variables must be controlled to regulate or track higher-order variables. Figure 11.16 shows analogous representations of three such hierarchically organized control systems: automobile control, aircraft heading control, and submarine depth control. In each case the operator controls the variable on the far left, with the final goal of tracking the variable on the far right. For driving, this is a second-order task since there are two integrals in the control loop between the change in steering wheel position and the lateral position on the highway. For flying and submarine control, it is a

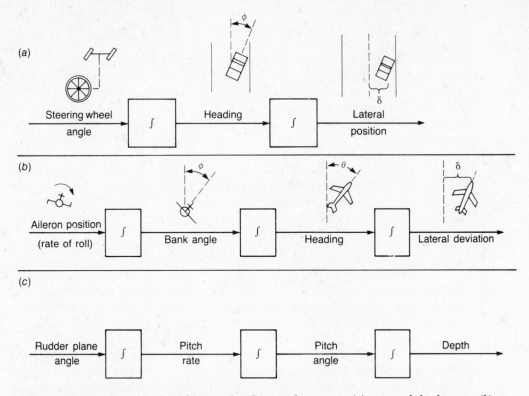

Figure 11.16 Three examples of hierarchical control systems: (*a*) automobile driving, (*b*) aircraft heading control, (*c*) submarine depth control.

third-order task. Variables on the left are said to be inner-loop, and those on the right define outer-loop variables. Kelley (1968) and Wickens (1986) explain hierarchically organized control systems.

Whereas hierarchical systems often have a number of *displayed* elements (e.g., steering wheel angle, vehicle heading, and lateral position), there is normally only one control element that gives control over the inner-loop variable (e.g., the steering wheel). Kelley (1968) and Roscoe (1968) have argued that it is important to organize the displayed elements coherently in a manner that conforms to the operator's internal model of the system being controlled. Thus, if three variables have an ordered causal relation to one another by virtue of three increases in system order, as in Figure 11.16c, displays of the three variables should be presented in such a format that the coherent ordering is preserved.

The strategy of hierarchical loop control is one in which the operator typically sets goals for the highest-order, or outer-loop, variable (e.g., a change in the aircraft lateral position from a desired path). To accomplish this, the operator must, in turn, control the variable of the next lower level (a change in aircraft heading, which is equal to lateral velocity). This, in turn, places constraints on the variables of the next inner loop (bank angle equal to rate of change of heading), whose rate of change must then be controlled by the innermost loop, aileron control. Hierarchical control thus involves the parallel efforts to control outer loops through the regulation of inner loops. At any given time the operator may be focusing attention on errors of inner-loop variables, outer-loop variables, or both. Because the inner-loop control is often rapid and is somewhat "mindless" in slavish pursuit of the more purposeful, cognitive outer-loop goals, systems are being designed with automation of the inner-loop control so that the operator has a direct means of controlling outer-loop variables. For example, the automated cockpit allows the pilot to dial in a desired heading on the autopilot, and the automated control system will accomplish the necessary tracking of inner-loop variables (ailerons or bank angle) to attain the goal. Automated tracking control will be discussed further in Chapter 12.

Factors That Influence the Efficiency of Multiaxis Control

Display Separation   Whether hierarchical, cross-coupled, or independent, multiaxis control will obviously be harmed if the error or output indicators are more separated across the visual field. When the separation is so great that the indicators are not simultaneously in foveal vision, then as discussed in Chapter 3, operator scan patterns may provide a useful index of the sequence of information extraction from the display. In some of the earliest classic work on formatting of aviation displays, Fitts, Jones, and Milton (1950) provided fundamental data on the importance of various sources of information in flight control derived from instrument scan patterns. The principles of display formatting derived from scan information were summarized in Chapter 3.

The problems associated with visual scanning and sampling strategies were also discussed in some detail in Chapter 3. The resulting loss of performance was attributed to peripheral interference. When the eye is fixated on one display, the other will be in peripheral vision and therefore generate data of lower quality. In tracking this does not mean, however, that a display can be tracked only if it is fixated on. There is good evidence that a considerable amount of peripheral information concerning both position and velocity may be used to attain effective control (Allen, Clement, & Jex; 1970; Levison, Elkind, & Ward, 1971; Wickens, 1986), although tracking is still poorer in the periphery than in the fovea (Weinstein, 1990). Wickens, Sandry, and Hightower (1982) have found, furthermore, that a peripheral axis in a multiaxis display is tracked considerably better when it falls in the left visual field than in the right. This finding makes sense in terms of the structure of the visual system and cerebral hemispheres. The left visual field has direct access to the right hemisphere. The right hemisphere processes spatial information and so would play a dominant role in tracking.

The obvious solution to the problems of degraded performance with the separation of tracking displays is the same as with the discrete tasks discussed in Chapter 3: Minimize display separation by bringing the displayed axes closer together. In the extreme, two control dimensions may be represented by the motion of a single variable in the x and y axes of space—an integrated object display such as the aircraft attitude display indicator (see Chapter 3). In this case, peripheral interference no longer contributes a cost to multiaxis tracking, and other sources of diminished efficiency may be identified. Three such sources, related to resource demand, control similarity, and proximity compatibility, will now be considered.

Resource Demand   Navon, Gopher, Chillag, and Spitz (1982) note that the cost to multiaxis tracking with a single display and control is surprisingly small. It seems that the integration of the two axes as two integral dimensions of a single object fosters a certain amount of cooperation between the tasks. Nevertheless, to the extent that some interference between tracking tasks results from resource competition, as discussed in Chapter 9, it is not surprising to find that the cost of dual-axis control increases as the resource demands of a single axis are increased. For example, Baty (1971) had subjects time-share the tracking of two zero-order, two first-order, and two second-order systems. He found little evidence for a difference in interference between zero- and first-order control—the two were described earlier as substantially similar in their demands when performed singly. However, the magnitude of interference imposed by time-shared second-order control was considerably greater. Similar results have been obtained by Fracker and Wickens (1989).

Similarity of Control Dynamics   An experiment performed by Chernikoff, Duey, and Taylor (1960) indicates that the increasing resource demand of higher-order control may under certain circumstances be lessened by making the control dynamics more similar between the two axes. Different orders of control require the operator to adopt different control strategies. It is apparently more difficult for the operator to time-share control with two different strategies than to maintain a single strategy for both axes.

The subjects were asked to perform dual-axis tracking with all three control orders (zero, first, and second) in all pairwise combinations. It was therefore possible to evaluate error on a given axis as a function of the order of control on the time-shared axis. The data are shown in Figure 11.17. For zero-order tracking, error increases when there is higher order on the paired axis. However, for first-order tracking, error is lower when the time-shared axis is also of first order than it is when time-shared with the lower but different zero-order system. Likewise, for second-order tracking, performance is no worse when the shared axis is also second-order than when shared with the lower but different first- and zero-order controls. In fact, it is a good bit higher when shared with zero order. The advantage of the lower resource demands of zero- and first-order tracking is nullified by the greater interference resulting from the fact that separate dynamics must be controlled. Wickens, Tsang, and Benel (1979) have found that the requirement to share different dynamics also contributes to increased subjective work load, as well as reduced time-sharing efficiency.

These findings can be placed in the context of the causes of dual-task

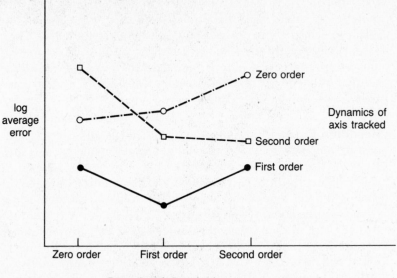

log average error

Zero order

Dynamics of axis tracked

Second order

First order

Zero order   First order   Second order

Dynamics of shared axis

Figure 11.17 Effect of time-shared control order on tracking performance at different orders. Note that the absolute height of the three curves is arbitrary. What is important is the change in error within each curve. (*Source:* R. Chernikoff, J. W. Duey, and F. V. Taylor, "Two-Dimensional Tracking with Identical and Different Control Dynamics in Each Coordinate," *Journal of Experimental Psychology, 60* (1960), p. 320. Copyright 1960 by the American Psychological Association. Adapted by permission of the authors.)

interference, described in Chapter 9. Task interference was generally assumed to be an increasing function of task similarity. An exception to this principle was noted, however, when similarity also allows some degree of *cooperation* between tasks. In Chapter 3, this cooperation was seen in the phenomenon of redundancy gain. In the present example, the cooperation is seen when a single control can be activated in working memory and applied to both axes simultaneously.

Display and Control Integration When two axes are tracked, the degree of display or control integration may be varied independently. The four quadrants in Figure 11.18 show the four different display-control combinations that can be generated by integrating or separating the axes on both displays and controls. Further options are available when the separate-separate display is employed (quadrant IV), for here it is possible to present the two axes either in parallel or at right angles to one another. Comparing the latter two configurations, Levison, Elkind, and Ward (1971) found that the right-angle placement is superior. In this case the dissimilarity (between the two axes of motion) is

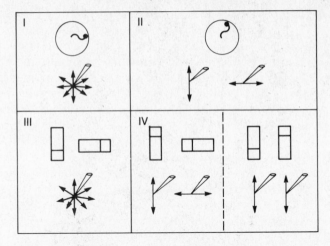

Figure 11.18 Different configurations of display and control integration.

helpful because the control responses on the two axes are independent. To the extent that the two hands are moving in a common plane, it will be more difficult to maintain independence of their separate responses. Kelso, Southard, and Goodman (1979) reported a similar difficulty in maintaining independent timing of responses by two hands moving in similar trajectories.

Investigations by Baty (1971), Chernikoff and Lemay (1963), and Fracker and Wickens (1989) compared the four quadrants of Figure 11.18. These studies concluded that when two axes with like dynamics are shared, there is an advantage to display integration and control integration (DI and CI), over separation (DS and CS) (quadrant I is better than quadrant IV). Chernikoff and Lemay also compared conditions of quadrant II (DI and CS) with those of quadrant III (DS and CI) and found performance in quadrant II to be superior. That is, the effect of integrating displays was generally more beneficial than that of integrating controls. This finding is to be expected. Integrating displays produces a clear reduction in visual scanning and allows a holistic object perception (see Chapter 3). Integrating controls reduces total motor activity but increases the possibility of response interference since motor control for both tasks must now be generated by the common hemisphere (see Chapter 9).

The most interesting results of Chernikoff and Lemay's (1963) study occurred when the same set of display-control configurations were employed with two axes of different dynamics (zero- and second-order). Under these conditions, as previously noted, there is an overall cost to performance. Chernikoff and Lemay found, however, that this cost of dissimilarity was least when the display and control were both separate, as in quadrant IV (the condition that had generated greatest cost with similar dynamics), and was greatest when the control was integrated, as in quadrants I and II.

The dual-axis tracking study by Fracker and Wickens (1989), with a similar design, yielded analogous although not identical results. They observed that the benefits of an integrated display were realized only when axes with like dynamics were tracked and not when tracking was carried out with different (first- and second-order) dynamics. Fracker and Wickens also found that conditions with both integrated controls and displays, or both separated controls and displays, gave rise to more efficient tracking, with a shorter time delay and less effort, than did conditions in which there was a mismatch between the integration of display and control.

The results of both of these studies are in keeping with the proximity compatibility principle discussed in Chapters 3 and 9. When cooperation is possible between some aspects of two tasks, there is a benefit to be gained by maintaining high proximity in their configuration; in this case the proximity is achieved by display and control integration. When there is competition — different control strategies required — proximity should be minimized by separating the tasks.

The standard design of the helicopter separates two controls between the right and left hand and feet. Because these three dimensions must often be integrated in flight control, investigators have proposed and successfully tested a more integrated control in which all axes are combined on a single 4-axis control stick that can move forward and sideways, and be twisted and lifted (Aiken & Merrill, 1980; Sinclair & Morgan, 1981).

Auditory Displays

A major difficulty with multiaxis control occurs when the operator must scan between displays. As noted, this difficulty may be reduced by display integration, but this procedure can also impose some cost if the dynamics differ. Furthermore, there are clearly limits to how tightly information can be condensed in the visual field without encountering problems of display clutter (see Chapter 3). One potential solution is to present tracking-error information through the auditory modality. The discussion of multiple-resource theory in Chapter 9 suggested that the auditory modality was indeed capable of processing spatial information. After all, we do this whenever we turn our head to the source of sound. It is important to realize, however, that audition is probably less intrinsically compatible with spatial processing than is vision (see Chapter 8). These properties suggest that single-task auditory tracking will never be superior and will probably be inferior to single-axis visual tracking, but it may provide benefits in environments with a heavy visual load.

Generally speaking, all of these hypotheses have been confirmed by the experimental data (Wickens, 1986). The earliest investigation of auditory tracking, carried out in World War II, concerned whether the auditory modality could be used to convey turn, bank angle, and airspeed information in an aircraft simulator (Forbes, 1946). The display, known as "flybar," indicated turn by a sweeping tone, from one ear to the other, whose sweep rate was proportional to the turn rate. Bank angle was indicated by pitch changes in one ear or the other, and airspeed by the frequency of interruption of a single pitch.

With sufficient training on this display, Forbes found that both pilots and nonpilots could fly the simulator as well as with the visual display.

More recently, with more precise performance measurement, both Isreal (1980) and Vinge (1971) have found that auditory tracking is nearly but not quite equivalent to visual. In the single-axis display used by these investigators, error was represented by the apparent spatial location of a tone, which was adjusted by playing tones of different relative intensity through stereo headphones. In addition, the absolute value of the error was represented redundantly by tone pitch. Low error was indicated by a low pitch. Vinge (1971) found that control over two independent tracking axes was superior when one was presented auditorily and the other visually, as compared to a visual-visual condition. These results replicate the findings of within-modality resource competition discussed in the dual-task literature in Chapter 9. Although Isreal (1980) found that single-task auditory and visual tracking was nearly equivalent, he also found that auditory tracking was more disrupted by a secondary task displayed in *either* modality. These results suggest that in terms of Figure 9.3, auditory tracking is more resource-limited than visual tracking, a condition that probably reflects the fact that we rarely use our sense of hearing to make fine manual adjustments in space. As discussed in Chapter 8, the auditory modality is less spatially compatible than the visual modality.

In spite of its early success in the flybar experiment and subsequent studies showing near equivalence with vision, auditory displays in tracking have received only minimal investigation over the past four decades. This neglect results in part from the fact that the auditory channel is more intrinsically tuned to the processing of verbal (speech) information. Hence the feeling is that the auditory channel should be dedicated to speech processing. Also, the auditory modality is hampered somewhat because it does not have spatial reference points that are precisely defined, as vision does. Nevertheless, it does appear that under certain conditions auditory spatial displays could provide valuable supplementary and redundant information, particularly if this information were presented along channels that do not peripherally mask the comprehension of speech input. Such redundancy can be of considerable use in environments that are visually loading.

DISCRETE CONTROL DEVICES

The discussion of manual control has focused mostly on continuous vehicle control, and for this task the physical implement of control has usually taken the form of a continuously moving device, like a joystick, yoke, or steering wheel. Yet with the emergence of the computer screen for human-machine dialogue, considerable emphasis has recently been placed on evaluating the tools for discrete control involved in moving a cursor to a fixed position (target) on the display. Should the designer use a mouse, a trackball, spatially oriented cursor keys, touch-panel overlays, joysticks, or voice control? Evaluation of the merits of these different control devices can proceed from three different perspectives: identifying certain physical constraints associated with each, comparing the movement capabilities supported by each, and directly comparing their performance in different cursor manipulation tasks.

Discrete Manual Control

Physical Constraints   Both light pens and touch panels, in which the hand (or an extension thereof) must physically touch a vertically mounted display, can be fatiguing for long periods of use because the arm is not supported. Additional anthropometric disadvantages are that the user must be close to the display and the hand may obscure parts of the display. Both of these problems can be remedied if the pressure-sensitive panel is removed from the display screen and positioned on a horizontal surface. In this case, however, additional structural constraints are imposed by the added physical space required to accommodate the control surface. Space needs will be severely constraining in environments like the automobile dashboard or aircraft cockpit. Similar space constraints occur when a mouse is used since it must be moved across a flat surface. Finally, keyboard cursor devices (up, down, left, right) have a structural advantage if cursor positioning is to be required in the midst of a typing task, as is typically the case during word-processing screen editing. Control devices that require the user to remove one hand from the keyboard have an obvious disadvantage (Shneiderman, 1987).

Movement Effects   Most analog devices for discrete movement control will obey the movement time function specified by Fitts's law (Card, English, & Burr, 1978; Flach, Hagan, O'Brien, & Olson, 1990). For a given target width (e.g., the constant precision required to designate a point on the computer screen), the implications of Fitts's law are that movement time will increase logarithmically with movement distance, as shown by the solid line in Figure 11.19. However, for keyboard control, in which depressing the key will move the cursor at a constant rate, movement time is a linear function of distance — the dashed line in the figure. This difference suggests that analog control devices will generally be superior to key presses when longer moves are required (e.g., when different screen-displayed items must be tagged and then moved to distant parts of the screen, or words and letters must be captured at unpredictable points on a page).

We have discussed already the effects of control order on continuous manual control, and similar effects are observed with discrete control. In particular, the goal of most discrete control devices is to acquire a particular *position* in space. This goal is compatible with zero-order (position) control devices, which means that the touch panel, light pen, mouse, and trackball are ideally suited because all four can allow rapid movement to a new position. The trackball and mouse also make poor candidates for velocity controllers since neither has a "natural" resting (zero) point at which the cursor will not drift. The joystick is less well suited for position control since it typically has a

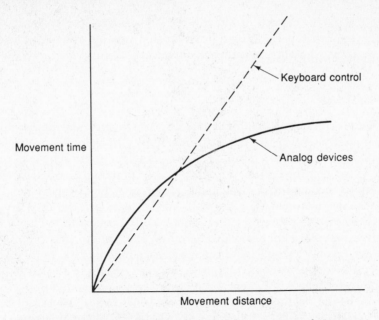

Figure 11.19 Movement time function for analog and discrete control devices. The slope of both functions depends on the gain of the controller. For very short distances, a single key press may almost instantly accomplish the movement and hence be more rapid than the analog device. But the constant movement rate penalizes longer movements for keyboard control relative to analog devices, which follow the logarithmic increase predicted by Fitts's law.

smaller range of movement and therefore enables less precise positioning. Hence when joysticks are used, they are more compatible with first-order velocity control.

Performance Effects Epps (1987) has compared a variety of discrete control devices across a number of different tasks, such as line drawing and positioning, and found that both the mouse and trackball yielded generally better performance and lower subjective difficulty ratings than did touch panels, light pens, and joysticks with first-order (velocity) control. Given the physical constraints of touch panel and light pens, and the fact that the tasks called for position rather than velocity matching, these findings are not altogether surprising. Shneiderman (1987) also notes that touch panels and light pens often provide performance that is more rapid but less accurate than performance with other control devices.

Card, English, and Burr (1978) compared the mouse with a first-order joystick; spatial keyboard control (up, down, left, right); and symbolic command keys that could designate paragraphs, lines, and words. The mouse and joystick were both found to conform to Fitts's law, represented in Figure

11.19, with the mouse producing the overall best performance. The mouse furthermore had a Fitts's law function that was indistinguishable from direct eye-hand movement, indicating its extremely natural characteristics.

Voice Control

As noted in Chapter 9, the availability of cheaper and more reliable voice-recognition systems has made the possibility of voice control more feasible. Furthermore, it is both intuitively obvious and supported by experimental data that voice control will be of considerable benefit (1) when both hands are otherwise engaged in a physical task, (such as airline baggage handlers, calling out city destinations (Shneiderman, 1987); (2) when the eyes are closely engrossed in visual processing and cannot easily process the visual feedback that is often necessary for manual data entry (e.g., photo interpretation); and (3) when even only a single hand is engaged in a manual task. In the last case there is good evidence—consistent with multiple-resource theory (Chapter 9)—that interference will be less if the other task is responded to by voice and not also by hand (Martin, 1989; Tsang & Wickens, 1988; Vidulich, 1988; Wickens & Liu, 1988; Wickens, Sandry, & Vidulich, 1983; Wickens, Vidulich, & Sandry-Garza, 1984). In spite of its intuitive attractiveness, however, voice control is not without its drawbacks. It is sometimes slow, for example; the rate of entering a series of voice commands is considerably slower than the rate of key presses (Leggett & Williams, 1984). Furthermore, the technology of accurate machine-based voice-recognition systems is still far from fully developed, and feedback from a failure to correctly recognize an utterance may very well disrupt other concurrent activities, a disruption that voice control was intended to eliminate (Frankish & Noyes, 1990).

Finally, the goal of off-loading manual control to voice must be exercised in the context of S-C-R compatibility (Wickens, Sandry, & Vidulich, 1983; see Chapter 8). According to this principle, the manual modality is ideally suited for conveying spatial and continuous information, and the voice modality for conveying discrete categorical information, but not the converse. It goes without saying that the voice cannot move continuously through space nor identify precise locations in near space to the same degree as the hand; but it also appears that the voice is less proficient at conveying categorical spatial information. This limitation was demonstrated in an experiment by Wickens, Zenyuh, Culp, and Marshak (1985). The subjects time-shared a verbal categorical task with a discrete tracking task, in which a tracking cursor could be made to follow a target with discrete commands of up, down, left, right, and stop. These commands could be issued by depressing spatially compatible keys or by voice articulation. The investigators found a dramatic difference favoring performance when tracking was manual and the verbal task was vocal (S-C-R compatible) over the reverse condition (vocal tracking and manual verbal task). The S-C-R compatible configuration produced substantially better dual-task performance than did a two-handed control of condition. The incompatible condition produced a performance that was substantially worse.

Conclusion

In conclusion, it appears that there is no single best control device for discrete control, although the mouse generally does quite well. Intelligent selection depends on the environment in which it is to be used, the need for position versus velocity control, the nature of the task, and a host of other factors.

MODELING THE HUMAN OPERATOR IN MANUAL CONTROL

There are many circumstances in vehicle control in which the need to keep error low but also to maintain stability greatly constrains the controller's freedom of action to engage in different strategies of control. These constraints have one great disadvantage. They allow human performance in manual control to be modeled and predicted with a far greater degree of precision than is possible in many other tasks. In fact, the mathematical models of tracking performance that have been derived have been some of the most accurate, successful, and useful of any of the models of human performance that we have examined. As discussed in Chapter 1, if systems designers know before it is built whether an aircraft with a given set of dynamics is flyable, by combining the model of vehicle dynamics with the transfer function of a pilot, they can realize a tremendous savings in engineering cost. The supplement to this chapter will describe two such models—the crossover model (McRuer & Krendel, 1959) and the optimal control model (Baron, 1988; Kleinman, Baron, & Levison, 1971). Both of these share certain characteristics with previous models discussed in the book. They specify optimal behavior and allow for some trade-off of operator strategies. Before presenting these models in the supplement, however, we will describe in some detail the engineering language of the frequency domain, used primarily to discuss and interpret these models.

TRANSITION

Motor and manual control is often difficult, time-consuming, and heavily loading. At the same time, as noted previously, many aspects of such control reflect lower-level, noncognitive processes that might readily be assigned to machines in a systems analysis. This approach is clearly being adopted to some extent. Robots in industrial assembly tasks are performing tracking, and so also are autopilots and stability augmentation devices in aircraft control. Unfortunately, however, the trend toward automation is not without a number of problems, and this fact has led some to conclude that automation in certain contexts may have proceeded far enough already (Wiener, 1988; Wiener & Curry, 1980). Chapter 12 will examine process control as a broader extension of manual control. It will then examine other aspects of control related to supervisory control and air traffic control. Finally it will consider automation in the broader context of other decision-making and cognitive tasks that have been discussed in previous chapters.

SUPPLEMENT: ENGINEERING MODELS OF MANUAL CONTROL

As noted previously, it is possible to think of the human operator as perceiving an error and translating it to a response in the same conceptual terms used for any other dynamic element responding to an input to produce an output. Figure 11.6 shows examples of several such elements. The objective of human operator models in tracking is to describe the human in terms similar to these dynamic elements.

FREQUENCY-DOMAIN REPRESENTATION

Figure 11.6 shows the step response of the different dynamic elements as functions of time—that is, the response in the *time domain*. The mathematical expression of these elements—a differential or integral equation—is also a function of time. Each different element is *uniquely* described by its step response and its time-domain transfer function.

In manual-control research in engineering psychology, it is sometimes preferable to represent transfer functions in the *frequency domain* in terms of spectra. This procedure was described briefly in our discussion of the speech signal in Chapter 5. In tracking, we assume that the subject tracks an error, which is a continuous signal varying in time (see Figure 11.5), to produce a response, also a continuous time-varying signal. Spectral analysis breaks down each of the signals into its component frequencies of oscillation, as shown in Figure 5.9. Then the transfer function between the two signals, input and output, is specified by two fundamental relationships that exist between two signals specified at each frequency of oscillation: the *gain*, or *amplitude ratio*, and the *lag*. The gain, as we have seen, is the ratio of output to input amplitude. The lag is the amount by which the output trails the input. Lag is normally expressed in degrees of a cycle rather than in units of time. Thus, a one-second time delay will be a one-cycle lag at a frequency of one cycle per second (1 Hz) but will be a two-cycle lag at a 2-Hz frequency.

These two properties are shown at the top of Figure 11.20. On the left, at a frequency of oscillation of 1 Hz, the output amplitude is twice the magnitude of the input. Hence the amplitude ratio, or gain, is 2/1, or 2. The output also lags behind the input by one-fourth of a cycle: The output reaches its peak when the input is already halfway to its trough. Hence, the phase lag is said to be 90 degrees. To the right is a higher frequency of 10 Hz. Here the gain is less than 1 since the input amplitude is larger than the output, and the phase lag is 180 degrees: Peaks of the inputs occur at troughs of the output and vice versa. The two signals are out of phase.

Typically the frequency-domain representation of a transfer function is depicted in a *Bode plot* (pronounced *Bodey*), an example of which is shown in the middle of Figure 11.20. Each Bode plot actually consists of two functions, plotting gain and phase on the same frequency axis. Therefore, at the top of the Bode plot, the amplitude ratio expressed in *decibels* is shown as a function of the logarithm (base 10) of the frequency. Across the bottom, phase lag in degrees is expressed again as a function of log frequency. The particular gain and phase relationships of the 1 Hz and 10 Hz frequencies at the top of the figure are shown as four points in the Bode plot. In fact, these points have been connected with solid lines to show the gain and phase values that would have been observed if the dynamic system whose input and output we were measuring had the time-domain transfer function of the form

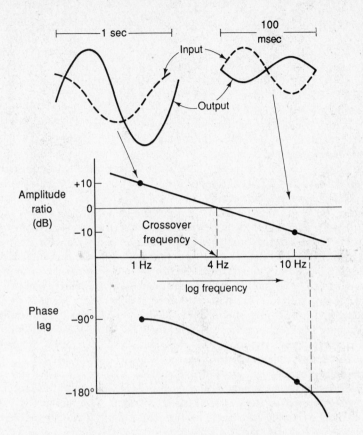

Figure 11.20 Bode plot of system $O(t) = k \int i(t - \tau) dt$.

$$Y = K \int e(t - \tau_e) \, dt$$

This function is a first-order system with gain K and time delay τ_e, a combination of the dynamic elements shown in Figure 11.6a, b, and d. The importance of this particular function will be seen below.

In the Bode plot, the value of K determines the intercept of the gain function. Changes in K will shift the curve up and down but will leave its slope unchanged. The constant time delay causes a greater phase lag at higher frequencies. Hence there is an exponential drop-off with the increase in log frequency. The phase lag would increase linearly if frequency were plotted on a linear rather than a logarithm scale. Finally, on a Bode plot the single integral in the transfer function contributes an amplitude ratio that becomes progressively smaller at high frequencies and is reduced at a rate of -20 dB for each decade increase of frequency. The phase lag produced by a first-order system is always a constant 90 degrees at all frequencies. In Figure 11.20 the lag caused by the integrator and the lag caused by the time delay are simply added together. At very low frequencies the time delay lag is negligible, and so the constant integrator lag of 90 degrees is the only one seen.

Often two transfer functions are placed in series. For example, in the tracking loop shown in Figure 11.4 the transfer function of the human operator, the control, and the plant are all in series. In this case, the components of the combined Bode plot are simply added. Thus, since a second-order system (Figure. 11.6e) is just two first-order systems in series (Figure 11.6d), the second-order Bode plot would correspond to the added components of two first-order Bode plots. The result would be a plot with an amplitude ratio slope of −40 dB/decade (gains are multiplied, so their logarithmic values in decibels are added) and phase lag of 180 degrees.

At this point it is possible to see why higher-order systems lead to instability in some closed-loop negative-feedback systems. Recall from our earlier discussion that instability was caused by a combination of high phase lag and high gain. The long phase lag in responding to periodic signals with higher-order systems (180 degrees for second order) is present at every frequency. The other element that leads to instability—high gain—is present only at lower frequencies. These two characteristics jointly lead to a critical principle in the analysis of system stability: *If the gain is greater than 1 (0 dB) at frequencies at which the phase lag is also greater than 180 degrees, the system will be unstable when responding to frequencies of that value.* The reason is that when the phase lag is 180 degrees, a correction intended to reduce an error at that frequency will, by the time it is realized by the system response (one-half cycle, or 180 degrees, later), be *added to* rather than subtracted from the error since the error will now have reversed in polarity. If the gain is greater than 1, this counterproductive correction will increase the error, leading to the kinds of oscillations described earlier.

According to this principle of closed-loop stability, the system shown in Figure 11.20 is stable. The frequency at which the phase lag becomes greater than 180 degrees (about 12 Hz) is higher than the frequency at which the gain curve dips below 1 (above 4 Hz). This latter frequency measure—critical for stability analysis—is called the *crossover frequency*. Figure 11.21 shows a Bode plot of the same system in Figure 11.20, responding with higher gain, thereby raising the amplitude ratio curve. This change moves crossover frequency to a higher value, which is now greater than the critical frequency at which the phase lag becomes greater than 180 degrees. The system in Figure 11.21 will therefore be unstable.

First- and Second-Order Lags in the Frequency Domain

Figure 11.22 shows Bode plots for pure zero-, first-, and second-order systems. Each order adds a phase lag of 90 degrees and increases the slope of the gain function by −20 dB/decade. Earlier we suggested that the first-order, or exponential, lag shown in Figure 11.6c represents something of a compromise between a zero- and first-order system. Examination of the Bode plot of the first-order lag, in Figure 11.23, shows how this compromise is realized in the frequency domain. When a first-order lag is driven by low-frequency inputs, the system responds as a zero-order system. There is no phase lag, and the gain is constant at all frequencies. When driven by high frequencies, in contrast, the system responds as the first-order system of Figure 11.22: a 90-degree phase lag and decreasing amplitude ratio. The break-point transitioning from zero to first order is given by the frequency $1/T_I$. T_I is referred to as the *time constant* of the lag.

Operations in the Frequency and Laplace Domain

In Figure 11.6, the dynamic elements are described by differential equations in the time domain. The functions are of the form $f(t)$. When dealing with systems in the frequency domain, dynamic elements are expressed in the form $F(j\omega)$, where $j\omega$ represents a

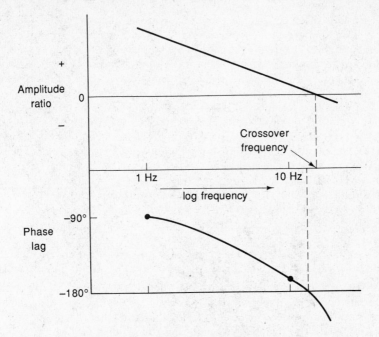

Figure 11.21 Bode plot of the system in Figure 11.19, now unstable because of a higher value of K.

particular frequency characteristic. Alternatively, they may be represented in the *Laplace domain* in the form $F(S)$. The Laplace operator S is one that accounts for both frequency-domain and time-domain characteristics, and so it is the most general way of describing dynamic systems and their associated inputs and outputs. A description of the mathematical nature of the frequency-domain and Laplace-domain operations is beyond the scope of this chapter, and the reader is referred to treatments by Licklider (1960), Moray (1981), Toates (1975), and Wickens (1986) for intuitive discussions of their derivations.

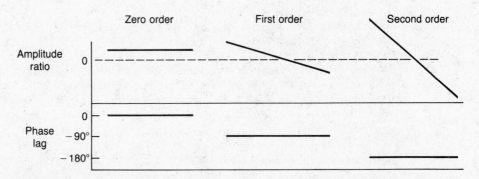

Figure 11.22 Bode plot of zero-, first-, and second-order system.

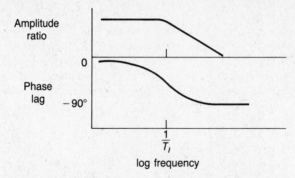

Figure 11.23 Bode plot of a first-order lag, showing zero-order behavior at low frequency and first-order behavior at high frequency.

We will focus on one major benefit of the Laplace representation, the reason for its attractiveness to engineering analysis—both signals and dynamic elements can be represented by Laplace transforms. To determine the output of a dynamic element from its input, all that needs to be done is to multiply the Laplace transform of the input by the transform of the dynamic element, and the product will be the transform of the output. If we know the Laplace transform of the input and the output, we can compute the transfer function of the dynamic element simply by dividing output by input. This calculation turns out to be a lot simpler than performing differential and integral calculus in the time domain. The Laplace-domain transfer functions of some of the dynamic elements are shown in Figure 11.6. Here the simplicity is evident. The Laplace representation of an integrator is K/S, and that of a differentiator is KS. Hence placing the two transfer functions in series produces $K/S \times KS = K^2$, that is, a pure gain. In the discussion of models of human-operator tracking, both the time-domain and the Laplace-domain transfer functions will be considered—the time domain because it is perhaps slightly more intuitive, the frequency domain or Laplace domain because it provides a bridge to the engineering literature, in which these models are often used.

MODELS OF HUMAN OPERATOR TRACKING

The initial efforts to model human tracking behavior in the late 1940s and the 1950s were described as *quasi-linear* models (Licklider, 1960; McRuer, 1980; McRuer & Jex, 1967; McRuer & Krendel, 1959). The term *quasi-linear* derives from the engineer's assumption that the operator's control behavior in perceiving an error and translating it to a response can be modeled as a linear transfer function such as those dynamic elements shown in Figure 11.6. However, they acknowledge that this representation is indeed only an approximation to linear behavior, which is why the modifier *quasi* is attached. Because the human response is not truly linear, it is referred to as a *describing function* rather than a transfer function. Quasi-linear models have been applied with greatest success to describing tracking behavior in the frequency domain.

The Crossover Model

Early efforts to discover the invariant characteristics of the human operator as a transfer function relating perceived error, $e(t)$, to control response, $u(t)$, encountered considerable frustration (see Licklider, 1960, for an excellent discussion of these models). The most successful of these early approaches, the crossover model developed by McRuer and Krendel (1959; McRuer & Jex, 1967), was successful because it departed from previous efforts in one important respect, which may be appreciated by reviewing Figure 11.4. Rather than looking for an invariant relationship between error and operator control, $u(t)$, McRuer and Jex examined that between error and system response, $o(t)$. In this form, their model allows the operator's describing function to be flexible and to change with the plant transfer function in order to achieve the characteristics of a "good" control system. And humans seem to behave in this way.

As described earlier, the two primary characteristics of good control are low error and a high degree of system stability. To meet these criteria, the crossover model asserts that the human responds in such a way as to make the total open-loop transfer function —the function that relates perceived error to system output—behave as a first-order system with gain and effective time delay. That is,

$$o(t) = K\int e(t - \tau_e)dt$$

or, in the Laplace domain,

$$O(S) = (KE^{-\tau_e S})/S$$

$$(KE - \tau_e S)/S$$

This transfer function is the simple crossover model, and its frequency-domain representation was in fact that shown in Figure 11.20. The way in which the crossover model describes the human and the plant together is shown in Figure 11.24. As noted, in the Laplace domain the transfer function of two components in series can simply be multiplied together, which is why the Laplace domain function is equal to *HG*.

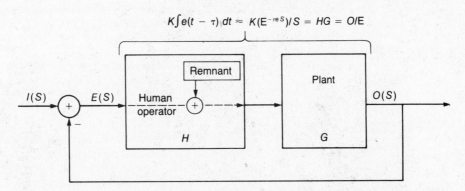

Figure 11.24 The crossover model of McRuer and Krendel (1959). (*Source:* D. T. McRuer and H. R. Jex, "A Review of Quasi-linear Pilot Models," *IEEE Transactions on Human Factors in Electronics,* 8 (1967), p. 240. Copyright 1967 by the Institute of Electrical and Electronics Engineers, Inc. Adapted by permission.)

The crossover model is described by two parameters: the gain and the effective time delay. The gain is of course the ratio of output velocity to perceived error. (Output is expressed as velocity rather than amplitude because the transfer function is an integrator that produces velocity output from position input.) Humans adjust their own gain to compensate for increases or decreases in plant gain, to maintain the total open-loop gain, the ratio O/E, at a constant value (McRuer & Jex, 1967). The variable τ_e is the effective time delay, described as the continuous analog of the human operator's discrete reaction time. Unlike gain, for which there are advantages and disadvantages at both high and low levels, long time delays are invariably harmful. There is not much that the human operator can do to shorten this parameter unless prediction and preview are available.

The third element of the model is its first-order characteristics—the single-time integration. It is important in understanding the crossover model to consider why the operator chooses to behave in a way that makes the operator-plant team respond as a first-order system. The answer may be given in terms of two compromises—one expressed intuitively, one formally. At an intuitive level it is possible to view a first-order response as a compromise between the costs and benefits of zero- and higher-order controls. Zero-order controls are tight and, as shown in Figure 11.6a, instantly correct detected errors. However, if the errors are sudden steps, as in Figure 11.6, an instant response of this kind might be quite unpleasant if, for example, one were riding in a vehicle corrected in this manner. Greater smoothness of correction is thereby obtained by control at higher order. However, second-order control, as we have noted, is so sluggish that it becomes unstable. Hence, first-order control makes an appropriate compromise between jerkiness of control and sluggish instability.

The formal compromise expresses the rationale for adopting a first-order control system in terms of an attempt to meet the two criteria of a good control system: low error and stability (McRuer & Jex, 1967). Considering the system depicted in Figure 11.24, it is evident that low error (at least as spatially defined) will be accomplished by a system in which the gain of the entire *closed*-loop transfer function (relating command input, I, to system output, O) is equal to 1. In this case, inputs will be directly matched in amplitude by outputs, and barring any phase lags, the error will be zero. However, to accomplish a unity gain of the closed-loop transfer function, the *open-loop* function, describing the relationship of error to output, must have an infinite (or very high) gain (i.e., small errors should be corrected with large corrective responses in the opposite direction). This property can be shown formally (see Jagacinski, 1977, and Wickens, 1986).

Whereas "tight" closed-loop control and low error can therefore be obtained by making the open-loop gain very high, as we have seen, a high-gain response strategy will generate instability problems whenever system phase lags are greater than 180 degrees of the frequencies being corrected. These problems could occur any time an operator with effective time delay enters the control loop or whenever the plant itself has phase lags. As a consequence, at high frequencies good control *must* keep the open-loop gain to a value of less than 1 at a frequency, the crossover frequency, below that where the phase lag is greater than 180 degrees. The first-order system, with its downsloping amplitude ratio in the frequency domain, shown in Figures 11.21 and 11.22, nicely accomplishes this function.

The human operator, the "flexible" element of the open-loop function, HG, meets these two criteria by responding in such a way as to make HG behave like a first-order system, with the unavoidable time delay. Thus the form of the crossover model, $o(t) = K \int e(t - \tau_e) dt$, is adopted. Referring to Figure 11.20, we see that such a function will produce high gain at low frequencies and low gain at high frequencies, thereby jointly meeting the criteria of minimizing low-frequency error and maintaining high-frequency stability.

To achieve this form of the open-loop transfer function, $O/E = GH$, the ideal human operator must adapt to changes in the system transfer function. This adaptation is accomplished by changing the form of H (the operator's control response to an error) to first-order when the system is zero-order, to zero-order when the system is first-order, and to a minus-first-order, or derivative control, system when the system is second-order. In the Laplace domain, the reason for this adaptation is very easy to see. The total transfer function, HG, must be first-order, K/S. Thus when the plant is zero-order (K), the human is $1/S$ and $K \times 1/S = K/S$. When the plant is first-order ($1/S$), the human is zero-order (K) and $1/S \times K = K/S$. When the plant is second-order ($1/S^2$), the human is a derivative controller, KS, and $KS \times 1/S^2 = K/S$. With the second-order control dynamics, the fact that the human becomes a derivative controller agrees with our earlier discussion of second-order control. We said that perceiving the derivative of the error signal (behaving as a KS controller) is an aid to anticipation and prediction. Here we see that it is not only desirable but also may be essential to maintain stability. If third-order dynamics are to be controlled, the crossover model is less applicable because it requires the human operator to adopt a minus-second-order, KS^2, or double-derivative control function to maintain stability, responding directly to error acceleration. As noted previously, human abilities to perceive acceleration are limited (Fuchs, 1962; McRuer, 1980).

In an extensive series of validation studies, McRuer and his colleagues (McRuer, 1980; McRuer & Jex, 1967; McRuer & Krendel, 1959; McRuer et al., 1968) found that the human behaves similarly to the crossover model when performing compensatory tracking tasks with random input. When Bode plots of the human transfer function between error and output are constructed in a compensatory task, more than 90 percent of the variance of well-trained operators can be accounted for by the simple two-parameter model. The model, therefore, compares favorably with other models of human behavior that we have discussed. Figure 11.25, taken from McRuer and Jex, shows the unchanging form of the crossover model Bode plot HG as humans track zero-, first-, and second-order systems. The figure then indicates how the human operator adapts to compensate for the changing dynamics of the system. Only when the system becomes second-order does the fit of data to the model begin to deteriorate somewhat. This result occurs because, as noted previously, with second-order systems operators are likely to engage in some nonlinear bang-bang behavior that cannot be captured by the linear transfer function.

The mathematical equation of the crossover model describing functions cannot predict all the output that will be observed when the human tracks a given error signal. The remaining variance in system response that is not accounted for by the linear describing function is referred to as the *remnant*. Some of the remnant results from nonlinear forms of behavior such as the bang-bang impulse control shown in second-order tracking. In addition, a remnant is also caused by such factors as time variations in the describing function parameters or random "noise" in human behavior. It is often depicted as a quantity injected into the human processing signal, as in Figure 11.24 (Levison, Baron, & Kleinman, 1969; McRuer & Jex, 1967).

The crossover model has proven to be quite successful in accounting for human behavior in dynamic systems. It has allowed design engineers to predict the closed-loop stability of piloted aircraft by combining the transfer function of the aircraft provided by aeronautical engineers with the crossover model of its pilot. It has also provided a useful means of predicting the mental work load encountered by aircraft pilots from the amount of lead or derivative control that the pilot must generate to compensate for higher-order control lags (Hess, 1977). Finally, the time delay, gain, and remnant measures of the model provide a convenient means of capturing the changes in the frequency domain that occur as a result of such factors as stress, fatigue, dual-task loading, practice, or supplemental display cues (Fracker & Wickens, 1989; Wickens & Gopher, 1977). To the extent that these three parameters capture fundamental changes in human processing mechanisms, their expression forms a more economical means of representation than does the raw Bode plot.

The Optimal Control Model

Despite the great degree of success that the crossover model has had in accounting for tracking behavior, it is not without some limitations (McRuer, 1980; Pew & Baron, 1978). (1) It is essentially a frequency-domain model and so does not readily account for time-domain behavior. (2) The form of the model and its parameters are based purely on fits of the equations to the input-output relations of tracking; they are not derived from consideration of the processing mechanisms actually used by the human operator. (3) Unlike models of reaction time, signal detection, or dual-task performance, the cross-over model does not readily account for different operator strategies of performance. Although the operator is assumed to be flexible — adapting the order of the describing function or compensatorily adjusting gain to the form of the plant — these adjustments are dictated by characteristics of the system. They are not chosen according to "styles" of tracking or instructional sets in the same way that beta was adjusted in signal detection theory. On the other hand, the *optimal control model*, in contrast to the crossover model, incorporates an explicit mechanism to account for this sort of strategic adjustment (Baron, 1988; Levison, 1989; Pew & Baron, 1978).

The basic components of the optimal control model are shown in Figure 11.26. The human operator perceives a set of displayed quantities and must exercise control to minimize a quantity, J, known as the quadratic *cost functional* and shown in the middle of the figure. This quantity J, a critical component of the optimal control model, is often expressed in the form

$$J = \int (A\dot{u}^2 + Be^2) \, dt$$

The integrated quantity within parentheses that is to be minimized is a weighted combination of squared control velocity and squared error. The relative constants of the two terms will depend on the relative importance of control precision, e, as opposed to control effort, u (the square of control velocity turns out to be a very good measure of control effort). These may vary from occasion to occasion or from operator to operator. As shown in the comparison of the two signals in Figure 11.8, attempts to minimize error at all costs may require fairly rapid control, but attempts to make a ride comfortable or to save fuel if controlling a space vehicle require less control action and therefore increase error. Thus as the operator trades off between tracking with low error and tracking with smooth control, the quantities $\dot{u}^2$ and e^2 will also trade off accordingly.

The assumption that the operator attempts to minimize the cost function represents one aspect of the optimal characteristic of the optimal control model. To minimize this quantity, the operator adjusts the gains that are used to translate the various perceived quantities (i.e., error and error velocity) into control. These gains are represented by the *optimal gain matrix* in Figure 11.26, a set of rules whose formulations are beyond the scope of the present treatment (see Baron, Kleinman, & Levison, 1970).

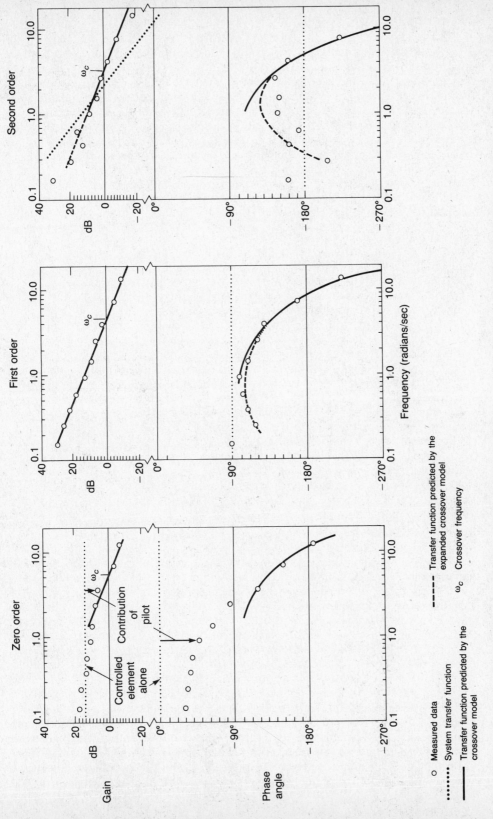

Figure 11.25 Human operator adaptation to plant dynamics with the crossover model. (*Source:* D. T. McRuer and H. R. Jex, ''A Review of Quasi-linear Pilot Models,'' *IEEE Transactions on Human Factors*, 8 (1967), p. 235. Copyright 1967 by the Institute of Electrical and Electronics Engineers, Inc. Adapted by permission.)

Optimal control is not perfect control. The human operator suffers two kinds of limitations: time delay and disturbance. Typically all of the time delays in operator response are lumped into one delay, shown in Figure 11.26. The sources of disturbance are attributable either to external forces or to the remnant of the operator. In Figure 11.26, these are shown as a noise in perceiving (observation noise) and a noise in responding (motor noise). To provide the most precise and current estimate of the current state of the system on which to base the optimal control, the operator must engage in two further processing operations: (1) *Optimal prediction* is done to compensate for the time delay, taking into account the manner in which future states may be predicted from past states. (2) *Estimation* of the true state of the system from the noisy state perturbed by the remnant is accomplished by applying a *Kalman filter*. The Kalman filter is an optimization technique similar in some respects to the Bayesian decision-making procedures described in Chapter 7. It combines estimates of system noise, operator remnant, and an internal model of the system to make a "best guess" of true system state. Application of the Kalman filter represents a second way in which the human is assumed to behave optimally since the estimation of system state is based on an optimal weighting of system noise and operator remnant. The final, current, best guess of the system state, $x(t)$, is then used as the input to the control gains that were selected by the cost function, J, and from this process the appropriate motor response is generated, although perturbed by motor noise.

Like the crossover model, the optimal control model has also proven quite successful in accounting for much of the variance of human performance in tracking. A key feature is the derived reciprocal relationship between the observation noise and the *fraction of attention* that is allocated to a tracking axis. This relationship allows the designer to predict in quantitative terms the degree of degradation of the perceived input signal and hence, by applying the model, the amount of loss of control accuracy when attention is diverted to other aspects of information processing (Baron, 1988; Levison, 1989; Wewerinki, 1989). However, the computational complexities of the optimal control model, as well as the greater number of parameters in the model that must be specified to fit the data, make it somewhat more difficult to apply. Nevertheless, the ability of the model to account for shifts in operator strategies gives it a desirable degree of flexibility that the crossover model does not have. Discussions by Baron, (1988); Baron and Levison (1977); Curry, Kleinman, and Hoffman (1977); Hess (1977); Levison (1989); and Levison, Elkind, and Ward (1971) provide examples of the manner in which the model has been applied to optimize the design of aviation systems and to assess operator work load and attention allocation in a quantitative model of attention.

SUMMARY

Models of human tracking have come about as close as any to providing useful predictive information that can assist in system design. As we have pointed out, however, these models are accurate precisely because they describe behavior that is fairly tightly constrained. The pilot of certain aircraft simply cannot depart from the error and stability prescriptions of the crossover model or else the plane will crash. Yet, ironically, it is these very constraints on flexibility that have made manual controls some of the easiest functions to replace by automation, thereby potentially making the models obsolete. However, the extent to which humans will be replaced by automatic controllers is determined by a host

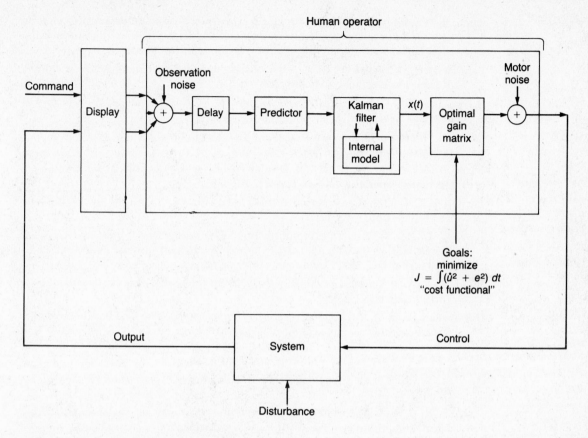

Figure 11.26 The optimal control model.

of other factors above and beyond the simple ability of automatic controllers to perform as well as humans. These issues will be addressed in Chapter 12.

REFERENCES

Adams, J. A. (1971). A closed loop theory of motor learning. *Journal of Motor Behavior*, 3, 111–150.

Adams, J. A. (1976). Issues for a closed loop theory of motor learning. In G. E. Stelmach (ed.), *Motor control: Issues and trends*. New York: Academic Press.

Aiken, E. W., & Merrill, R. K. (1980, May). Results of a simulator investigation of control system and display variations for an attack helicopter mission. *Presented at the 36th Annual National Forum of the American Helicopter Society*. Washington, DC.

Allen, R. W., Clement, W. F., & Jex, H. R. (1970, July). *Research on display scanning, sampling, and reconstruction using separate main and secondary tracking tasks* (NASA CR–1569). Washington, DC: NASA.

Allen, R. W., & Jex, H. R. (1968, June). *An experimental investigation of compensatory and pursuit tracking displays with rate and acceleration control dynamics and a disturbance input* (NASA CR–1082). Washington, DC: NASA.

Bahrick, H. P., & Shelly, C. (1958). Time-sharing as an index of automatization. *Journal of Experimental Psychology, 56,* 288–293.

Baron, S. (1988). Pilot control. In E. Wiener & D. Nagel (eds.), *Human factors in aviation.* New York: Academic Press.

Baron, S., Kleinman, D., & Levison, W. (1970). An optimal control model of human response. *Automatica, 5,* 337–369.

Baron, S., & Levison, W. H. (1977). Display analysis with the optimal control model of the human operator. *Human Factors, 19,* 437–457.

Bartlett, F. C. (1932). *Remembering: A study in experimental social psychology.* London: Cambridge University Press.

Baty, D. L. (1971). Human transinformation rates during one-to-four axis tracking. *Proceedings of the 7th Annual Conference on Manual Control* (NASA SP-281). Washingon, DC: U.S. Government Printing Office.

Bilodeau, E. A., & Bilodeau, I. McD. (eds.). (1969). *Principles of skill acquisition.* New York: Academic Press.

Birmingham, H. P., & Taylor, F. V. (1954). *A human engineering approach to the design of man-operated continuous control systems.* Washington, DC: U.S. Naval Research Lab, report no. 433.

Brown, J. S., & Slater-Hammel, A. T. (1949). Discrete movements in a horizontal plane as a function of their length and direction. *Journal of Experimental Psychology, 10,* 12–21.

Card, S. K. (1981). The model human processor: A model for making engineering calculations of human performance. In R. Sugarman (ed.), *Proceedings of the 25th annual meeting of the Human Factors Society.* Santa Monica, CA: Human Factors Society.

Card, S. K., English, W. K., & Burr, B. J. (1978). Evaluation of mouse, rate-controlled isometric joystick, step keys, and task keys for text selection on a CRT. *Ergonomics, 21*(8), 601–613.

Card, S., Newell, A., & Moran, T. (1986). The model human processor. In K. Boff, L. Kaufman, & J. Thomas (eds.), *Handbook of perception and performance* (vol. II, chpt. 45). New York: Wiley.

Chernikoff, R., Duey, J. W., & Taylor, F. V. (1960). Two-dimensional tracking with identical and different control dynamics in each coordinate. *Journal of Experimental Psychology, 60,* 318–322.

Chernikoff, R., & Lemay, M. (1963). Effect on various display-control configurations on tracking with identical and different coordinate dynamics. *Journal of Experimental Psychology, 66,* 95–99.

Chernikoff, R., & Taylor, F. V. (1957). Effects of course frequency and aided time constant on pursuit and compensatory tracking. *Journal of Experimental Psychology, 53,* 285–292.

Craik, K. W. J. (1948). Theory of the human operator in control systems II: Man as an element in a control system. *British Journal of Psychology, 38,* 142–148.

Crossman, E. R. F. W. (1960). The information capacity of the human motor system in pursuit tracking. *Quarterly Journal of Experimental Psychology, 12,* 1–16.

Curry, R. E., Kleinman, D. L., & Hoffman, W. C. (1977). A design procedure for control/display systems. *Human Factors, 19*, 421–436.

Debecker, J., & Desmedt, R. (1970). Maximum capacity for sequential one-bit auditory decisions. *Journal of Experimental Psychology, 83*, 366–373.

Drury, C. (1975). Application to Fitts' law to foot pedal design. *Human Factors, 17*, 368–373.

Eberts, R., & Schneider, W. (1985). Internalizing the system dynamics for a second-order system. *Human Factors, 27*(4), 371–393.

Elkind, J. I., & Sprague, L. T. (1961). Transmission of information in simple manual control systems. *IRE Transactions on Human Factors in Electronics, HFE-2*, 58–60.

Epps, B. W. (1987). A comparison of cursor control devices on a graphic editing task. *Proceedings of the 31st annual meeting of the Human Factors Society* (pp. 442–446). Santa Monica, CA: Human Factors Society.

Fitts, P. M. (1954). The information capacity of the human motor system in controlling the amplitude of movement. *Journal of Experimental Psychology, 47*, 381–391.

Fitts, P. M., Jones, R. E., & Milton, J. L. (1950). Eye movements of aircraft pilots during instrument-landing approaches. *Aeronautical Engineering Review, 9*(2), 24–29.

Fitts, P. M., & Peterson, J. R. (1964). Information capacity of discrete motor responses. *Journal of Experimental Psychology, 67*, 103–112.

Fitts, P. M., & Posner, M. A. (1967). *Human performance.* Pacific Palisades, CA: Brooks/Cole.

Flach, J., Hagen, B., O'Brien, D., & Olson, W. A. (1990). Alternative displays for discrete movement control. *Human Factors, 32*, 685–695.

Forbes, T. W. (1946). Auditory signals for instrument flying. *Journal of Aeronautical Science, 13*, 255–258.

Fracker, M. L., & Wickens, C. D. (1989). Resources, confusions, and compatibility in dual axis tracking: Displays, controls, and dynamics. *Journal of Experimental Psychology: Human Perception and Performance, 15*, 80–96.

Frankish, C., & Noyes, J. (1990). Sources of human error in data entry tasks using speech input. *Human Factors, 32*, 697–716.

Fuchs, A. (1962). The progression regression hypothesis in perceptual-motor skill learning. *Journal of Experimental Psychology, 63*, 177–192.

Gallagher, P. D., Hunt, R. A., & Williges, R. C. (1977). A regression approach to generate aircraft predictive information. *Human Factors, 19*, 549–566.

Gibbs, C. B. (1962). Controller design: Interactions of controlling limbs, time-lags, and gains in positional and velocity systems. *Ergonomics, 5*, 385–402.

Gill, R., Wickens, C. D., Donchin, E., & Reid, R. (1982). The internal model: A means of analyzing manual control in dynamic systems. *Proceedings of the IEEE Conference on Systems, Man, and Cybernetics.* New York: Institute of Electrical and Electronics Engineers.

Glencross, D. J., & Barrett, N. (1989). Discrete movements. In D. H. Holding (ed.), *Human skills* (2nd ed., pp. 107–1146). New York: Wiley.

Gottsdanker, R. M. (1952). Prediction-motion with and without vision. *American Journal of Psychology, 65*, 533–543.

Grunwald A. J. (1985, September/October). Predictor laws for pictorial flight displays. *Journal of Guidance, Control, and Dynamics, 8*(5).

Grunwald, A. J., Robertson, J. B., & Hatfield, J. J. (1981). Experimental evaluation of a perspective tunnel display for three-dimensional helicopter approaches. *Journal of Guidance and Control, 4*(6), 623–631.

Hess, R. A. (1973). Nonadjectival rating scales in human response experiments. *Human Factors, 15,* 275–280.

Hess, R. A. (1977). Prediction of pilot opinion ratings using an optimal pilot model. *Human Factors, 19,* 459–475.

Hess, R. A. (1979). A rationale for human operator pulsive control behavior. *Journal of Guidance and Control, 2,* 221–227.

Hess, R. A. (1981). An analytical approach to predicting pilot-induced oscillations. *Proceedings of the 17th Annual Conference on Manual Control.* Pasadena, CA: Jet Propulsion Laboratory.

Holding, D. H. (ed.). (1989). *Human skills* (2nd ed.). New York: Wiley.

Isreal, J. (1980). Structural interference in dual task performance: Behavioral and electrophysiological data. Unpublished Ph.D. dissertation, University of Illinois, Champaign.

Jagacinski, R. J. (1977). A qualitative look at feedback control theory as a style of describing behavior. *Human Factors, 19,* 331–347.

Jagacinski, R. J. (1989). Target acquisition: Performance measures, process models, and design implications. In G. R. McMillan, D. Beevis, E. Salas, M. H. Strub, R. Sutton, & L. Van Breda (eds.), *Applications of human performance models to system design* (pp. 135–150). New York: Plenum Press.

Jagacinski, R. J., & Miller, D. (1978). Describing the human operator's internal model of a dynamic system. *Human Factors, 20,* 425–434.

Jagacinski, R. J., Repperger, D. W., Ward, S. L., & Moran, M. S. (1980). A test of Fitt's law with moving targets. *Human Factors, 22,* 225–233.

Jex, H. R., McDonnel, J. P., & Phatak, A. V. (1966). A "critical" tracking task for manual control research. *IEEE Transactions on Human Factors in Electronics, HFE–7,* 138–144.

Keele, S. W. (1968). Movement control in skilled motor performance. *Psychological Bulletin, 70,* 387–403.

Kelley, C. R. (1968). *Manual and automatic control.* New York: Wiley,

Kelso, J. A., Southard, D. L., & Goodman, D. (1979). On the coordination of two-handed movements. *Journal of Experimental Psychology: Human Perception and Performance, 5,* 229–259.

Kim, W. S., Ellis, S. R., Tyler, M., Hannaford, B., & Stark, L. (1987). A quantitative evaluation of perspective and stereoscopic displays in three-axis manual tracking tasks. *IEEE Transactions on Systems, Man, and Cybernetics, SMC-17,* 61–71.

Klapp, S. T., & Erwin, C. I. (1976). Relation between programming time and duration of response being programmed. *Journal of Experimental Psychology: Human Perception and Performance, 2,* 591–598.

Kleinman, D. L., Baron, S., & Levison, W. H. (1971). A control theoretic approach to manned-vehicle systems analysis. *IEEE Transactions in Automatic Control, AC–16,* 824–832.

Krendel, E. S., & McRuer, D. T. (1968). Psychological and physiological skill development. *Proceedings of the 4th Annual NASA Conference on Manual Control* (NASA SP–182). Washington, DC: U.S. Government Printing Office.

Langolf, C. D., Chaffin, D. B., & Foulke, S. A. (1976). An investigation of Fitts' law using a wide range of movement amplitudes. *Journal of Motor Behavior, 8,* 113–128.

Leggett, J., & Williams, G. (1984). An empirical investigation of voice as an input modality for computer programming. *International Journal of Man-Machine Studies, 21,* 493–520.

Levison, W. H. (1989). The optimal control model for manually controlled systems. In G. R. McMillan, D. Beevis, E. Salas, M. H. Strub, R. Sutton, & L. Van Breda (eds.), *Application of human performance models to systems design* (pp. 185–198). New York: Plenum Press.

Levison, W. H., Baron, S., & Kleinman, D. L. (1969). A model for human controller remnant. *IEEE Transactions in Man-Machine Systems, MMS–10,* 101–108.

Levison, W. H., Elkind, J. I., & Ward, J. L. (1971, May). *Studies of multivariable manual control systems: A model for task interference* (NASA CR–1746). Washington, DC: NASA.

Licklider, J. C.R. (1960). Quasilinear operator models in the study of manual tracking. In R. D. Luce (ed.), *Mathematical psychology.* Glencoe, IL: Free Press.

MacNielage, P. F. (1970). Motor control of serial ordering of speech. *Psychological Review, 77,* 182–196.

McRuer, D. T. (1980). Human dynamics in man-machine systems. *Automatica, 16,* 237–253.

McRuer, D. T., Hoffmann, L. G., Jex, H. R., Moore, G. P., Phatak, A. V., Weir, D. H., & Wolkovitch, J. (1968). *New approaches to human-pilot/vehicle dynamic analysis* (AFFDL–TR-67-150). Dayton, OH: Wright Patterson AFB.

McRuer, D. T., & Jex, H. R. (1967). A review of quasi-linear pilot models. *IEEE Transactions on Human Factors in Electronics, 8,* 231.

McRuer, D. T., & Krendel, E. S. (1959). The human operator as a servo system element. *Journal of the Franklin Institute, 267,* 381–403, 511–536.

Martenuik, R. G., & MacKenzie, C. L. (1980). Information processing in movement organization and execution. In R. S. Nickerson (ed.), *Attention and performance VIII.* Hillsdale, NJ: Erlbaum.

Martin, G. (1989). The utility of speech input in user-computer interfaces. *International Journal of Man-Machine System Study, 18,* 355–376.

Moray, N. (1981). The role of attention in the detection of errors and the diagnosis of errors in man-machine systems. In J. Rasmussen & W. Rouse (eds.), *Human detection and diagnosis of system failures.* New York: Plenum Press.

Navon, D., Gopher, D., Chillag, W., & Spitz, G. (1982, August). *On separability and interference between tracking dimensions in dual axis tracking* (Technical Report). Haifa: Technion Israeli Institute of Technology.

Pew, R. W. (1974). Human perceptual-motor performance. In B. Kantowitz (ed.), *Human information processing.* Hillsdale, NJ: Erlbaum.

Pew, R. W., & Baron, S. (1978). The components of an information processing theory of skilled performance based on an optimal control perspective. In G. E. Stelmach (ed.), *Information processing in motor control and learning.* New York: Academic Press.

Pew, R. W., Duffenback, J. C., & Fensch, L. K.(1967). Sine-wave tracking revisited. *IEEE Transactions on Human Factors in Electronics, HFE–8,* 130–134.

Poulton, E. C. (1974). *Tracking skills and manual control.* New York: Academic Press.

Reid, D., & Drewell, N. (1972). A pilot model for tracking with preview. *Proceedings of the 8th Annual Conference on Manual Control* (Wright Patterson AFB, Ohio Flight Dynamics Laboratory Technical Report AFFDL–TR-72, 92). Washington, DC: U.S. Government Printing Office.

Roscoe, S. N. (1968). Airborne displays for flight and navigation. *Human Factors,* 10, 321–322.

Roscoe, S. N., Corl, L., & Jensen, R. S. (1981). Flight display dynamics revisited. *Human Factors,* 23, 341–353.

Runeson, D. (1975). Visual predictions of collisions with natural and nonnatural motion functions. *Perception & Psychophysics, 18,* 261–266.

Schmidt, R. A. (1975). A schema theory of discrete motor skill learning. *Psychological Review,* 82, 225–260.

Schmidt, R. A. (1988). *Motor control and learning: A behavioral emphasis* (2nd ed.). Champaign, IL: Human Kinetics.

Shapiro, D. C., & Schmidt, R. A. (1982). The schema theory: Recent evidence and development of implications. In J. A. S. Kelso & J. E. Clark (eds.). The development of movement control and coordination. Norwich, Eng.: Wiley.

Shneiderman, B. (1987). *Designing the user interface: Strategies for effective human-computer interaction.* Reading, MA: Addison-Wesley.

Shulman, H. G., Jagacinski, R. J., & Burke, M. W. (1978). *The time course of motor preparation.* Paper presented at the 19th annual meeting of the Psychonomics Society, San Antonio, TX.

Simon, J. R. (1969). Reactions toward the source of stimulation. *Journal of Experimental Psychology, 81,* 174–176.

Sinclair, M., & Morgan, M. (1981). *An investigation of multi-axis isometric side-arm controllers in a variable stability helicopter.* Canada: National Research Council.

Stelmach, G. E. (ed.). (1978). *Information processing in motor control and learning.* New York: Academic Press.

Summers, J. J. (1989). Motor programs. In D. H. Holding (ed.), *Human skills* (2nd ed.) New York: Wiley.

Toats, F. (1975). *Control theory in biology and experimental psychology.* London: Hutchinson Education.

Tsang, P. S., & Wickens, C. D. (1988). The structural constraints and strategic control of resource allocation. *Human Performance, 1,* 45–72.

Vidulich, M. A. (1988). Speech responses and dual-task performance: Better time-sharing or asymmetric transfer? *Human Factors,* 30, 517–534.

Vinge, E. (1971). Human operator for aural compensatory tracking. *Proceedings of the 7th Annual Conference on Manual Control* (NASA SP–281). Washington, DC: U.S. Government Printing Office.

Weinstein, L. F. (1990). The reduction of central-visual overload in the cockpit. *Pro-*

ceedings of the 12th symposium on psychology in the Department of Defense. Colorado Springs, CO: U.S.A.F. Academy.

Wewerinke, P. H. (1989). *Models of the human observer and controller of a dynamic system.* Twerte, Neth.: University of Twente.

Wickens, C. D. (1986). The effects of control dynamics on performance. In K. Boff, L. Kaufman, & J. Thomas (eds.), *Handbook of perception and performance* (vol. II, pp. 39–1/39–60). New York: Wiley.

Wickens, C. D., & Goettl, B. (1985). The effect of strategy on the resource demands of second order manual control. In R. Eberts & C. Eberts (eds.), *Trends in ergonomics and human factors.* North Holland, Neth.: North Holland.

Wickens, C. D., & Gopher, D. (1977). Control theory measures of tracking as indices of attention allocation strategies. *Human Factors, 19,* 349–365.

Wickens, C. D., Haskell, I., & Harte, K. (1989). Ergonomic design for perspective flight-path displays. *IEEE Control Systems Magazine, 9*(4), 3–8.

Wickens, C. D., & Liu, Y. (1988). Codes and modalities in multiple resources: A success and a qualification. *Human Factors, 30,* 599–616.

Wickens, C. D., Sandry, D. C., & Hightower, R. (1982, October). *Display location of verbal and spatial material: The joint effects of task-hemispheric integrity and processing strategy* (Technical Report EPL–82-2/ONR-82-2). Champaign: University of Illinois, Engineering Psychology Research Laboratory.

Wickens, C. D., Sandry, D., & Vidulich, M. (1983). Compatibility and resource competition between modalities of input, output, and central processing. *Human Factors, 25,* 227–248.

Wickens, C. D., Tsang, P., & Benel, R. (1979). The dynamics of resource allocation. In C. Bensel (ed.), *Proceedings of the 23rd annual meeting of the Human Factors Society.* Santa Monica, CA: Human Factors Society.

Wickens, C. D., Vidulich, M., & Sandry-Garza, D. (1984). Principles of S-C-R compatibility with spatial and verbal tasks: The role of display-control location and voice-interactive display-control interfacing. *Human Factors, 26,* 533–543.

Wickens, C. D., Zenyuh, J., Culp, V., & Marshak, W. (1985). Voice and manual control in dual task situations. *Proceedings of the 29th annual meeting of the Human Factors Society.* Santa Monica, CA: Human Factors Society.

Wiener, E. (1988). Cockpit automation. In E. Wiener & D. Nagel (eds.), *Human factors in aviation.* New York: Academic Press.

Wiener, E. L., & Curry, E. R. (1980). Flight deck automation: Problems and promises. *Ergonomics, 23,* 995–1012.

Woodworth, R. S. (1899). The accuracy of voluntary movement. *Psychological Review, 3* (Monograph Supplement No. 2).

Young, L. R. (1969). On adaptive manual control. *Ergonomics, 12,* 635–675.

Young, L. R., & Meiry, J. L. (1965). Bang-bang aspects of manual control in higher-order systems. *IEEE Transactions on Automatic Control, 6,* 336–340.

Young, L. R., & Stark, L. (1963). Variable feedback experiments testing a sampled data model for eye tracking movements. *IEEE Transactions on Human Factors in Electronics, HFE–4,* 38–51.

Chapter 12

Process Control
and Automation

PROCESS CONTROL

In Chapter 11 we discussed manual control and tracking of dynamic systems. We assumed either implicity or explicity that the system was a vehicle, but this need not be the case. The dynamics of many industrial, chemical, and energy conversion processes must also be tracked, by manipulating controls to compensate for disturbances and to follow prescribed inputs. However, we treat process control in a separate chapter for four reasons: (1) These processes are generally more complex than vehicle dynamics, with a greater number of interacting variables. (2) The system responses are often so slow that human manual control is more discrete than analog. (3) As a consequence of this slower control it is an area that is less constrained by motor limitations but that makes issues of decision making, perception, and memory, discussed in Chapters 2 through 7, much more relevant. (4) Finally, the process control task is closely tied to concepts of automation, discussed at the end of this chapter.

Overview

Process control is certainly not synonymous with automation, as much of the control is carried out manually. However, because of the tremendous complexities of the processes involved and because hazardous environments and toxic materials are often employed, the process control environment is one in which automation is an inevitable companion. Regulating and controlling large chemical, energy, or thermal processes impose many demands that are simply beyond the capabilities of the human operator. Some of these limitations result from obvious physical constraints. Humans, for example, cannot readily manipulate the chemicals that are involved in many industrial processes, they cannot handle the radioactive fuel in nuclear processes, nor can they come in physical contact with elements at the extreme temperatures of many energy-conversion systems. Other constraints are directly related to the complexity of the process variables involved, which impose limits on human cognitive processes. Thus many components of process control have been automated, and increasing computer technology makes it inevitable that automation will probably proceed a good bit further. It is essential, therefore, that the nature of human involvement in process control be clearly specified so that the automation that is implemented will be appropriate and functional (Bainbridge, 1983).

We will first focus on process control during normal operations, describing its cognitive aspects and the human factors design implications. We will then consider issues of problem solving and fault diagnosis, under the critical and dangerous conditions of abnormal operation, and some of the solutions to these issues provided by the application of *cognitive ergonomics* (Woods, 1988). Because of its high risk, complexity, and public visibility, much of this discussion will revolve around nuclear process control. However, we will also integrate some examples from other process industries. We will then briefly discuss research in other related domains in which operators must monitor and supervise complex multivariate systems: air traffic control, automated manufacturing, and robot supervision. Finally, we will turn directly to automation itself, as implemented in a wide variety of environments but focusing most directly on two: aviation and, again, process control.

Our discussion of process control will provide a number of links to treatments of basic human processes and limitations covered in earlier chapters. Indeed, process control is a domain that at one time or another can impose on all of the limitations of human performance that have been discussed, and so it makes a fitting conclusion to the book. The reader is referred to excellent treatments by Edwards and Lees (1974); Rasmussen and Rouse (1981); Sheridan and Johannsen (1976); Umbers (1979); and Woods, O'Brien, and Hanes (1987) for detailed discussions of the process control field.

Characteristics of Process Control

The specific kinds of processes of concern to the human operator in this domain are diverse. They include the management of cake-baking ovens (Beishon, 1969) and the control of blast furnaces (Bainbridge, 1974), paper mills (Beishon, 1966), distribution of electric power (Umbers, 1976; Williams, Seidenstein, & Goddard, 1980), distillation of gas (Queinnec, DeTerssac, & Thon, 1981), and of course nuclear power (Hollnagel, Mancini, & Woods, 1988; Moray & Huey, 1988; Rasmussen, 1983; Sheridan, 1981; Woods, O'Brien, & Hanes, 1987). These diverse examples have four general characteristics that allow them to be classified together.

First, the process variables that are controlled and <u>regulated are slow</u>. They have long time constants, relative to the controlled or tracked systems discussed in Chapter 11. Thus the control delivered by an operator in process control may not produce a visible system response for seconds and sometimes even minutes. Using the terminology introduced in Chapter 11, we say that the operator is more often controlling *outer-loop variables*, whereas automated adjustment and feedback loops handle the inner-loop control. Thus the operator of a blast furnace may choose a set point of desired temperature, and automated inner-loop control will provide the amount of fuel and energy to the furnace necessary to achieve that temperature some minutes later.

Second, although controls are often adjusted in discrete fashion, the variables that are being controlled are essentially analog, continuous processes. Thus, in terms of the concepts discussed in Chapters 4 and 6, the operator's ideal internal model of the processes should also be analog and continuous rather than discrete and symbolic.

Third, the processes typically consist of a large number of interrelated variables. Some are hierarchically organized, as in Figure 11.16; others are cross-coupled, so that changes in one variable will influence several other variables simultaneously. The staggering magnitude of this complexity is demonstrated by the display confronting the typical supervisor of a nuclear power plant control room (Figure 12.1). Grimm (1976), for example, has noted that the complexity of power station control rooms, as measured by the number of controls and displays, grew geometrically from under 500 in 1950 to around 1500 in 1970 to more than 3000 in 1975. This complexity of interactions can severely tax the operator's mental model of the status of the plant, a model whose level of accuracy is important for normal control and critical for the response to malfunctions or failures.

Fourth, many of these processes, particularly those in nuclear power, are characterized by high risk. As incidents at Chernobyl, Three Mile Island, and Bhaupal have demonstrated (Reason, 1990), the consequences of human error in terms of costs to society and to human life can be severe. Yet operators must often function under conflicting goals, balancing productivity and profit against safety. This difficult trade-off, affected by management attitudes (see Chapter 10), will influence many aspects of the operator's decisions and actions (Roth & Woods, 1988).

A conceptual overview of the nuclear power plant environment is seen in Figure 12.2. At the left is a reactor, and when the control rods (C_R) are withdrawn, the fission reaction of the uranium atoms produces heat. This heat is removed from the reactor by a primary steam loop (bottom right), in which the water is somewhat radioactive. In a pressurized water reactor, the water circulates through a heat exchanger (center of the figure). Here the heat is transferred to a secondary cooling loop that contains pure water and makes steam, which passes at high pressure to the turbines to drive them to generate electricity (top right). The steam in the primary loop is condensed back to water, preheated, and returned to the core by powerful pumps. An automatic shutdown device (top left) shuts down, or *scrams*, the reactor if the delicate balance between heat and mass is disrupted so that pressure becomes either too high or too low within the reactor. The reactor scram is a major event that has substantial financial costs within the plant. Even when no damage occurs, it can

Figure 12.1 Typical nuclear power plant control room. Controls are placed on the benchboard or just above, quantitative displays are placed on the vertical segment of the boards, and qualitative annunciator displays are placed in the uppermost segments of the boards. (*Source:* Electrical Power Research Institute, *Human Factors Review of Nuclear Power Plant Design; Final Report* (Project 501, NP-309-SY, March 1977). Sunnyvale, CA: Lockhead Missiles and Space Co.)

cost up to $250,000 a day to buy electricity from other utilities to replace the lost power.

In the following pages, two contrasts are emphasized. One is the contrast between normal routine operation and abnormal operation, which may characterize either start-up operations or fault management. The other is the contrast between the efforts to model the cognitive and information-processing requirements of the operator's task and the actual human factors efforts to improve control room design. The latter efforts, in turn, belong to two categories: the standard human factors principles of control-display design and anthropometry (Seminara, Pack, Seidenstein, & Eckert, 1980) and more sophisticated efforts to optimize design on the basis of cognitive models of the human operator — the field of *cognitive engineering* (Woods, 1988).

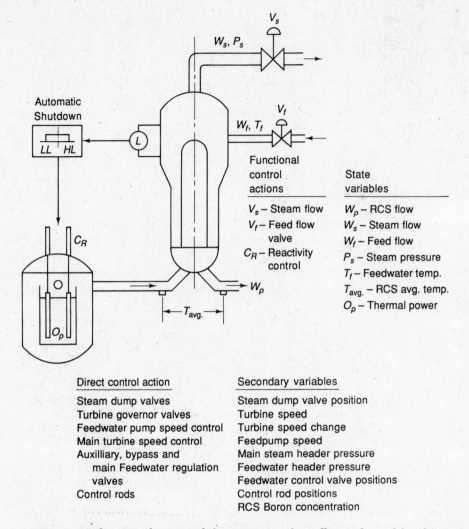

Functional control actions

V_s – Steam flow
V_f – Feed flow valve
C_R – Reactivity control

State variables

W_p – RCS flow
W_s – Steam flow
W_f – Feed flow
P_s – Steam pressure
T_f – Feedwater temp.
$T_{avg.}$ – RCS avg. temp.
O_p – Thermal power

Direct control action	Secondary variables
Steam dump valves	Steam dump valve position
Turbine governor valves	Turbine speed
Feedwater pump speed control	Turbine speed change
Main turbine speed control	Feedpump speed
Auxilliary, bypass and	Main steam header pressure
main Feedwater regulation	Feedwater header pressure
valves	Feedwater control valve positions
Control rods	Control rod positions
	RCS Boron concentration

Figure 12.2 Schematic diagram of the processes that affect indicated level in a steam generator of a steam supply system in a nuclear reactor. (*Source:* E. M. Roth and D. D. Woods, "Aiding Human Performance: I. Cognitive Analysis," *Le Travail Humain,* *51* (1988), p. 46. Copyright © 1986 Westinghouse Electric Corporation.)

Control versus Diagnosis

The process control operator's task has typically been described as hours of intolerable boredom punctuated by a few minutes of pure hell. This dichotomy, although perhaps somewhat overstated, serves nicely to discriminate between the two major functions of the process controller: the normal control and regulation of the process and timely detection, diagnosis, and corrective action in the face of the very infrequent malfunctions that may occur. As the incident at Three Mile Island indicates, the low frequency at which these failures occur must not diminish concern for their accurate detection and diagnosis.

The duties of the process controller normally begin with the start-up of the process—for example, bringing a nuclear reactor on-line. Although a fairly standardized set of procedures is followed during start-up, the cognitive demands of this phase of operation are heavy and the task is complex. As described later, a large number of warning signals, designed to be appropriate for steady-state operation, may be flashing during the transient conditions of start-up.

During this period, a very delicate balance must be maintained between energy and mass flow among the various reactor elements, and in spite of the standardization of procedures, there is a high degree of uncertainty in the precise display readings an operator will perceive and in the actions that must be taken (Roth & Woods, 1988). Once the system is in a steady state, the operator typically engages in periodic adjustments, or "trimming," of process variables, to keep certain critical process parameters within designated bounds and meet changing production criteria. In the framework of the tracking paradigm, these adjustments may be made because of *disturbances* (e.g., a reduction in water quality will require more purification) or because of *command inputs*. For example, the need for electric power will increase at peak times, requiring the production of more high-pressure, high-temperature steam to drive the turbines.

Although the control aspects of process control are thus somewhat allied with the discussion of manual control in Chapter 11, the detection and diagnosis aspects (along with many of those related to start-up) are more closely related to the discussion of those topics in Chapters 2 and 7, as well as to the treatment of mental representations in Chapters 4 and 6. In addition to the obvious differences in the quantitative work load associated with the two phases, there are a number of other ways in which routine control and abnormal diagnosis and problem solving differ from one another. Landeweerd (1979) notes that the controller in normal circumstances must focus on the forward flow of events: what *causes* what. The diagnostician in times of failure, in contrast, must often reverse the entire pattern and think about what was *caused by* what. The difficulties people have in dealing with diagnosis rather than prediction—in reversing the normal causal sequence of events—was discussed in some detail in Chapter 7.

Moray (1981) suggests that visual scan patterns and information-seeking strategies will also differ between these phases. Scanning strategies appropriate for normal operation may be inappropriate if there is a failure; and if these normal strategies are continued in the presence of a failure, they may prevent the operator from realizing that the failure is there.

If the two phases are truly independent, then those who are good controllers may not necessarily be effective at detection and diagnosis and, conversely, good diagnosticians may not be good controllers. A study by Landeweerd (1979) suggested that individual differences in ability of these two components were indeed independent. Landeweerd assessed the accuracy of operators' visual-spatial images of a chemical distillation process (drawing a schematic diagram) and the accuracy of their verbal-causal models (statements of what caused what in the process). He then assessed the operators' abilities both to control the process under normal operating conditions and to diagnose failures. He found that the accuracy of the visual-spatial image correlated reliably with detection and diagnosis performance but not with control; causal-model accuracy correlated with control ability but not with diagnosis. Drawing a similar conclusion in the realm of training, rather than individual differences, Kessel and Wickens (1982) found that operators who had prior training in the detection of dynamic system failures were not helped by this training in the ability to control these systems.

We have emphasized that control versus detection and diagnosis, both integral aspects of the overall process control task, may be independent in terms of operator abilities. Because this distinction is seen in a number of different ways, the two aspects of the process controller's task will be considered separately.

Control

Two characteristics of routine process control have received special attention by engineering psychologists and human factors engineers. These relate, first, to some of the fundamental principles of good human factors applied to control-room design and, second, to a careful analysis and modeling of the human control process itself.

Human Factors Design Process   Around the time of the Three Mile Island incident, a number of human factors specialists had taken a critical look at several features of current control-room design and found substantial violations of several design principles (Electrical Power Research Institute, 1977; Hopkins et al., 1982; Seminara, Pack, Seidenstein, & Eckert, 1980).

1. Lack of consistency. Two side-by-side control panels had been designed with mirror image symmetry, so that identical controls to the right side as the operator faced one panel were to the left as he or she faced the other panel. Or as shown in Figure 12.3a and b, two identical (and important) functional switches for a reactor shutdown had different characteristics, or two dials rotated in opposite directions to achieve the same goal (12.3c).
2. Violations of basic anthropometry. Many controls were hard to reach and displays hard to see.
3. Violations of S-R compatibility. Displays were not necessarily positioned in close proximity to the controls that they reflected, violating the colocation principle discussed in Chapter 8.
4. Absence of organization and functional grouping. Sets of controls and displays, like those in Figure 12.4a, were not clearly delineated into meaningful clusters, a feature of importance discussed in Chapter 3.

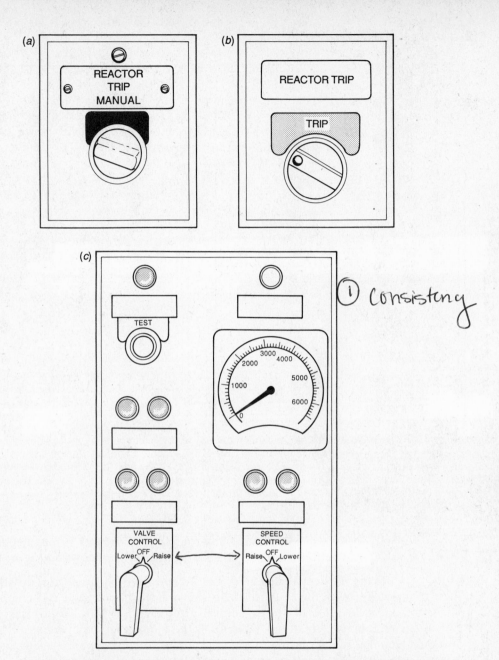

(1) consistency

Figure 12.3 Inconsistent labeling and coding on critical controls (the control on the left is coded with a black background; the one on the right with a lightly shaded background). (*a*) Control located in reactor control area. (*b*) Control located on turbine panel. (*c*) Inconsistent direction-of-movement relationships. (*Source:* D. D. Woods, J. F. O'Brien, and L. F. Hanes, "Human Factors Challenges in Process Control: The Case of Nuclear Power Plants," in G. Salvendy (ed.), *Handbook of Human Factors* (New York: John Wiley & Sons, 1987). Figures copyrighted 1979 by Electric Power Research Institute, *EPRI NP-1118, Volume I: Human Factors Method for Nuclear Power Control Room Design.* Reproduced by permission.)

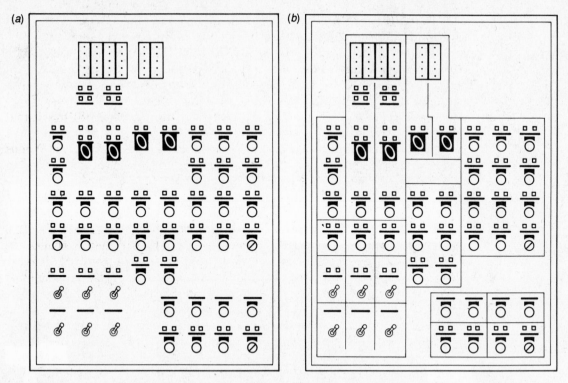

Figure 12.4 (*a*) Control panel before enhancements. (*b*) Same control panel as (*a*) after enhancements. (*Source:* D. D. Woods, J. F. O'Brien, and L. F. Hanes, "Human Factor Challenges in Process Control: The Case of Nuclear Power Plants," in G. Salvendy (ed.), *Handbook of Human Factors* (New York: John Wiley & Sons, 1987). Figures copyrighted 1979 by Electric Power Research Institute, *EPRI NP-1118, Volume I: Human Factors Method for Nuclear Power Control Room Design.* Reproduced by permission.)

     5. Response confusion. Numerous control devices, often affecting different actions, were identically structured, as shown in Figure 12.1.

    Many of these deficiencies can be and have been addressed with relative ease. For example, Figure 12.4*b* shows how simple paint can emphasize important features of the structural grouping. Ivergard (1989) has written a comprehensive textbook describing the appropriate application of many fundamental human factors design principles to control-room layout and anthropometry. Naturally, many of these design principles are as important to abnormal operations as they are to the normal ones involved in control.

Performance Strategies The operator's primary responsibilities during normal process control are to monitor system instruments and periodically adjust control settings to maintain production quantities within certain bounds. Although the task bears some similarity to tracking, the difference is that tracking normally involves continuous closed-loop behavior, whereas much of the responding in process control uses a discrete open-loop strategy (see Chapter

11). In a simulated process control task, Crossman and Cooke (1962) and McLeod (1976) found that subjects employed both modes of control but performed better on the open-loop mode. Beishon (1966) reported that skilled paper mill operators would initiate their adjustments with open-loop control action, followed later by discrete closed-loop adjustment. The open-loop strategy, however, becomes available only after the process is well learned. Kragt and Landeweerd (1974) and Moray, Lootsteen and Pajak (1988) found that operators' performance evolved from closed- to open-loop strategies when they received training in a simulated process control task. Roth and Woods (1988) observed that the expert controller of a feed-water reactor tended to use more open-loop control than did the nonexpert.

Open-loop is preferable to closed-loop behavior in process control because the long time constants typically involved cause a closed-loop strategy to be inefficient and potentially unstable (see Chapter 11). Process variables often change so slowly that operators cannot readily perceive their velocity and acceleration in a manner that makes for stable closed-loop control (McLeod, 1976). In many reactor tasks, the difficulties imposed by the lag are amplified by the presence of counterintuitive "shrink-swell" dynamics, as shown in Figure 12.5. The initial response of a variable produced by a step control input designed to increase its value is a *decrease* in value (shrink), followed only later by an appropriate increase (swell) (Roth & Woods, 1988).

When the task of the process controller is carefully analyzed, successful control, like successful tracking, requires three important components: (1) a clear specification and understanding of the future goals of production, that is, a *command input*; (2) an accurate *mental representation* of the current state of the process, which because of the sluggish nature of the variables involved, must be used to predict future state; and (3) an accurate *mental model* of the dynamics of the process. This third element is essential if open-loop control is to be employed since it represents the means by which a plan of control action is formulated to bring the process to a future state (Moray, Lootsteen, & Pajak, 1986). If there is no mental model, the operator must respond, wait to see what happens, and then respond again. That is, the operator must engage in slow and inefficient closed-loop control. As discussed in more detail in Chapter 3, another function of the mental model in normal process control is to guide visual scanning and information sampling (Moray & Rotenberg, 1988).

These three components — goals, a mental representation of the current and future state, and a mental model of the process dynamics — along with the inherently sluggish nature of these processes, highlight the great importance of the future in process control. In Chapter 11 we showed how the planning of future activities in control may be partitioned into two components: planning and anticipating future *goals* (these may be command inputs or internally specified goals) and anticipating or predicting the future *system responses*. Since the system in process control responds slowly and sometimes, initially, in a backward direction, actions implemented now must be based on future, not present, states. Further evidence of the importance of both future goals and future states comes from verbal protocols taken from process and industrial

Shrink-Swell
dynamics

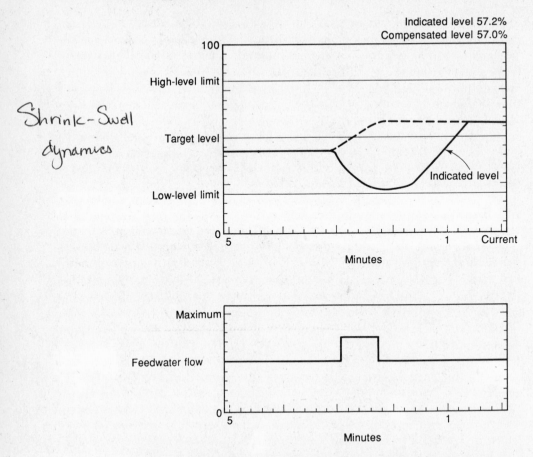

Figure 12.5 Illustration of nonminimum phase (shrink/swell) dynamics. Following an increase in feedwater flow (bottom graph), the indicated water level first decreases, then approaches its longer-term asymptotic value. The dotted line illustrates what level the behavior would be, given an increase in feedwater flow, if the system dynamics were minimum phase. (*Source:* E. M. Roth and D. D. Woods, "Aiding Human Performance: I. Cognitive Analysis," *Le Travail Humain, 51* (1988), p. 52. Figure copyrighted 1986 by Westinghouse Electric Corporation.)

control operators, which indicate that a significant portion of their time is spent in planning operations (Bainbridge, 1974), anticipating future demands or slacks in the process. Roth and Woods (1988) found substantial improvements from novice to expert controllers in the extent to which operators anticipated future states in their control activity and mentally kept track of controls they had issued that were "in the pipe line," that is, controls whose effects had not yet been realized by display changes.

Display Implications   Given the importance of future information, the use of computer aiding can be of considerable value when a prediction of the system response is displayed, having the same advantages as those described in Chapter 11; however, these advantages are typically even greater in process control because of the longer time constants involved. Long time constants usually mean that little motion is observable in display indicators (McLeod, 1976). As a consequence, status indicators that are based only on present time do not provide adequate trend information. Trend information may be derived from one of three sources.

1. *Historical displays*, typically presented by strip charts, give some estimate of future trends to the extent that the future may be mentally extrapolated from the past. An example of a historical display used in nuclear power plants is shown in Figure 12.6.

2. *Predictive displays*, as described in Chapter 11, generate predicted outputs based on computer models of the plant. West and Clark (1974) found that considerable assistance was provided by the predictive display but little by the historical strip chart. Woods and Roth (1988) have developed a two-element predictor display to aid in controlling the flow of feedwater to a steam generator, as shown in Figure 12.7. Recall that this is a cognitively difficult task because of the shrink-swell phenomenon: Corrections to increase level produce an initial response in the opposite direction. To address these difficulties, the display portrays both the actual level of steam generation and a predictor of

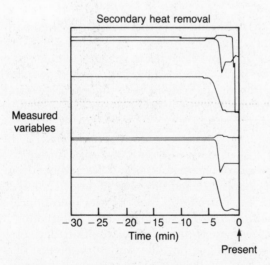

Figure 12.6 Historical strip-chart display. (*Source:* D. D. Woods, J. Wise, and L. Hanes, "An Evaluation of Nuclear Power Plant Safety Parameter Display Systems," *Proceedings of the 25th Annual Meeting of the Human Factors Society* (1981), p. 111. Santa Monica, CA: Human Factors. Copyright © 1981 by the Human Factors Society, Inc. Reproduced by permission.)

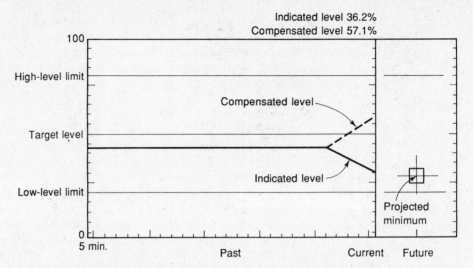

Figure 12.7 Predictive display for assisting in setting the water level in a nuclear reactor. (*Source:* D. D. Woods and E. M. Roth, "Aiding Human Performance: II. From Cognitive Analysis to Support Systems," *Le Travail Humain,* 51 (1988), p. 149. Figure copyrighted 1986 by Westinghouse Electric Corporation.)

the future minimum (or maximum) level that will be reached with the present control input. This display gives the operator an estimate of how a given control will affect future values and how close those values will come to critical limits that may shut down the plant. The *compensated level,* shown in the figure, is another form of prediction and indicates where the steam-generation level will eventually stabilize after the cognitively difficult shrink-swell effects have been computationally removed.

3. The third source of predictive information, found to be of use by Brigham and Liaos (1975), is derived from the display of *intervening process variables* between the control manipulations and the actual output changes. In the terms described in Chapter 11 and shown in Figure 11.16, this is analogous to the information that the controller of higher-order hierarchical systems may derive from a display of lower-order variables—the information, for example, that the automobile driver controlling lateral position derives from vehicle heading and steering-wheel angle, both derivative predictors of future lateral position. In process control, the heat applied (e.g., by a furnace), is an intervening variable that predicts the actual temperature.

Abnormal Operations

When discussing abnormal operation, fault detection, and fault diagnosis in process control, our treatment comes full circle, touching base again with the concepts discussed in Chapter 2 (vigilance and detection) and Chapter 7 (decision making and diagnosis). However, we now address these concepts distinctly from the point of view of the process monitor's task. In spite of the formal equivalence between failure detection in the process control environment and the concepts of detection and vigilance discussed in Chapter 2, there are three characteristics of process control that emphasize its differences from the simulated environment that is most often employed in laboratory vigilance studies.

1. In process control the operator is not typically waiting passively between failures but is intermittently engaged in moderate levels of control adjustments, parameter checking, and log keeping necessary to maintain a current mental representation of the system. These are the sort of intervening activities that maintain at least a modest level of arousal and reduce this source of vigilance decrement.

2. When a failure does occur it is normally indicated by a visual and/or auditory alarm. The latter will be sufficiently salient to call attention to itself so that the probability of a miss (in the vigilance sense) is, in fact, quite low. Nevertheless, the consequences of a miss may still be drastic.

3. More than in most vigilance situations, the process control environment is characterized by an excessive number of machine false alarms, that is, alarms that sound (or warning lights that illuminate) when there is in fact nothing abnormal with the plant. These machine false alarms occur either because the set points of the alarms are too sensitive or because an abnormal level of a variable in one context (appropriately triggering an alarm) may in fact be quite normal in a different context.

These three qualifications do not eliminate the concern about failure detection in process control. In the design of any discrete warning device, it is necessary to assume a threshold for the indicated variable that is sufficiently tolerant that random variations in the process variables do not trigger false alarms. To attain this level of tolerance, it is likely that information regarding true failures that occur gradually rather than catastrophically will be available in trends in the variables that are visible before the activation of the warning. A sensitive, alert operator with a well-formulated mental representation will be able to use this advance information to prepare for the upcoming event and possibly take corrective action before the alarm sounds. Moray (1981) has suggested that the detection process can be facilitated by providing operators with displays that are scaled in terms of their *probability* of normal value, rather than absolute physical units, and by providing a more integrated configural representation of variables.

Alarms may be triggered because what is normal in one context is abnormal in another. This state is best seen in Figure 12.8, which illustrates the complexity of power plant design (Woods, O'Brien, & Hanes, 1987). The indicators in the middle represent a series of parameters set to trigger alarms when their value exceeds or falls below a certain level. The circles above represent various root-cause failures. Each failure may affect different (but overlapping) sets of indicators, and each indicator may be triggered by a number of possible failures. Pictured below are a series of nonfailure states, such as those that might occur during maintenance or start-up. These states may also produce some of

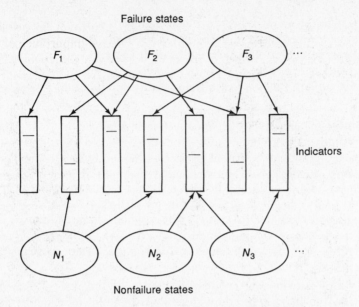

Figure 12.8 Complexity of nuclear power plants operations.

the same parameter values and alarms associated with failure states. Hence, many alarms are ambiguous both with regard to what a failure might be and whether a failure exists at all. Woods, O'Brien, and Hanes note that some operators relate to certain alarms as old friends because their appearance is assumed to signal no danger or simply because, if there are several hundred or more displays, some of the displays themselves will probably be faulty.

In summary, either a vigilant operator with a current mental representation or one equipped with well-configured displays will be likely to suspect or detect a malfunction prior to the activation of annunciator systems (De Keyser, 1988). However, even if this is not the case, detection will nevertheless be fairly well guaranteed by the alarms and lights of the annunciation system. Then *diagnosis* becomes the next phase of operator response. However, there is a certain paradoxical relationship between detection and diagnosis: The very characteristics of annunciators that may guarantee timely detection — salience and prominence — can severely inhibit their effective use in diagnosis.

Fault Diagnosis

Once a failure or abnormality is detected, the operator is faced with a choice of what actions to take in order to meet three criteria. These criteria, in decreasing order of importance, are (1) actions that will ensure plant safety, (2) actions that will not jeopardize plant economy and efficiency, and (3) diagnostic actions that will localize and correct the fault. Unfortunately, these criteria are not always compatible with one another, and operators who must function in a probabilistic world, in which outcomes of actions cannot be predicted with certainty, should at least be certain about the relative importance of each. This importance must be emphasized, for example, when a choice must be made to take a turbine or plant off the line, thereby ensuring safety (criterion 1) but potentially sacrificing economy (criterion 2). The status of the third criterion, diagnosis, is also somewhat ambivalent. In the nuclear industry, written procedures emphasize that control actions that will restore critical system variables to their normal range must take priority over diagnosis of a fault (Zach, 1980). The intent of this prioritization is to emphasize that the operator need not worry about what caused the failure until the system has been stabilized. However, diagnosis is often a necessary precursor to restoring or maintaining plant safety. As the incident at Three Mile Island revealed (see Chapter 1), misdiagnosis can lead to disastrous circumstances. Hence the overall ordering of priorities remains somewhat ambiguous for the operator.

Concerns for problem solving and diagnosis under abnormal conditions have generally taken two complementary approaches. One is to try to model and understand the cognitive difficulties that operators have in this problem-solving environment (Roth & Woods, 1988), and the other is to examine possible remedial solutions (Woods & Roth, 1988). The nature of the problems encountered in both fault diagnosis and in control under abnormal states, such as plant start-up, have been revealed by four different methods.

Case Studies The case study of Three Mile Island provides a clear example of how cognitive performance can break down. In this case, the analysis highlights the role of *cognitive tunnel vision*, discussed more fully in Chapter 7. Following each nuclear power incident, a fairly detailed failure analysis is carried out (by the Nuclear Regulatory Commission in the United States) that focuses both on human as well as equipment limitations.

Error Analysis Woods (1984; Woods, O'Brien, & Hanes, 1987) has made a more systematic effort to collect and categorize sources of error committed by power plant operators. A critical source of difficulty was found in the operator's ability to modify the adherence to prespecified procedures if this modification was necessary in changing and uncertain environmental conditions. Reactor designers have carefully documented numerous procedures that should be followed in abnormal circumstances. But these procedures should not be followed in a rigidly open-loop fashion, without carefully monitoring the effect of each action on the state of the plant and adjusting the procedures if necessary when that effect is not what was intended or does not occur precisely when it was intended.

Novice-Expert Differences Roth and Woods (1988) have carried out a systematic analysis of the differences between novices and experts performing a steam generator start-up procedure. Although this is not truly a failure or fault situation, it is a highly unstable, dynamic, and abnormal condition, taxing operator problem-solving capabilities extensively. Characteristics of expert performance that distinguished them from those of novices were found in (1) the ability to *anticipate* the future; (2) a superior *mental model* of the process, its time constants and interconnections (as we saw, this mental model provides a partial basis for anticipation); (3) a conscious setting of the *speed-accuracy trade-off* to *go-slow* (see Chapter 8), indicating that experts know that rapid actions carried out with sluggish and complex systems can be an invitation to error and instability; (4) a broader spotlight of attention, guarding against *cognitive tunneling*; and (5) a better ability to *communicate* and *coordinate* with other operators in the complex, multitask environment of the control room.

Cognitive Task Analysis   Major research efforts by Rasmussen (e.g., Goodstein & Rasmussen, 1988; Rasmussen, 1983, 1986; Vicente & Rasmussen, 1990) and Woods (1988; Roth & Woods, 1988) have focused on analysis of the task requirements in problem solving and diagnoses, revealing two key concepts. First, Woods has focused extensively on the *brittleness* of preprogrammed procedures designed to deal with unusual events. As noted, such lists of procedures do not easily take account of the dynamic and uncertain characteristics of the plant environment. These characteristics dictate that procedures must often be bent or modified on the fly to deal with an unexpected plant response (or with the consequences of operator error in carrying out the procedures). Such procedures are often written as a series of actions or rules, and each action may be conditional on a particular state of the plant ("if x do y"). But too often the "if x" part is just assumed, and y is done. A careful evaluation of whether condition x actually exists is not performed. Another way of looking at this problem is in terms of Rasmussen's trichotomy among skill-based behavior, rule-based behavior, and knowledge-based behavior (discussed in Chapter 8). Operators must be sensitive to when skill- and rule-based behavior is no longer appropriate and they must, therefore, move into a knowledge-based problem-solving domain.

Second, Rasmussen and his colleagues have focused attention on the need for operators to problem-solve flexibly by moving up and down to different *levels of abstraction*. Sometimes operators must think at very concrete levels in terms of variables like steam or water flow, heat measurement, and valve settings; at other times they must conceptualize at more abstract levels related to the thermodynamics of the energy conversion process under their supervision — thinking, for example, about the appropriate balance between mass and energy. At still other times the required level of thinking may be still more abstract, defined in terms of concepts like plant safety, human risk, and company profits. These different levels all interact, such as the multiloop control systems in Figure 11.16, and a major hurdle for effective problem solving is to provide the operator with the necessary tools to think conceptually at these different levels and rapidly make transitions among them (Vicente & Rasmussen, 1990).

Collectively, the problems confronted by the operator in the fault diagnosis problem-solving phase are formidable; they have been summarized in Table 12.1, along with some fairly abstract requirements necessary to remediate the

Table 12.1 HUMAN OPERATOR PROBLEMS AND SYSTEM REQUIREMENTS IN FAULT
DIAGNOSIS

Problem	Requirements
Cognitive tunnel vision *Diagnoses*	Need for fresh perspective
Sluggish process	Need for anticipation, good predictive displays
Complexity of system connections	Need for good mental model, communications, good displays
Brittleness of rigid procedures _words word_	Need for flexible adaptation of procedures (pay attention to "if" part of "if-then" rule) *(Often ignored If part)*
Pressure to go fast	Need to slow down
Multilevel problem solving	Need to move up and down in level of abstraction *Integrate concrete → abstract*

problems. The goal of *cognitive ergonomics* is to use this problem analysis to define a series of more concrete remediation recommendations.

Remedies to Address Abnormality Problems

It is apparent that human factors remedies to support problem solving and fault management in process control are not straightforward. Nevertheless, researchers have identified a number of approaches that can be taken to improve the accuracy and efficiency of such diagnosis. Globally these can be categorized into system design issues (including error tolerance and several display issues), training issues, and organizational factors.

Error Tolerance and Prediction   The concept of error-tolerant systems, introduced in Chapter 10 (Rouse & Morris, 1986), is nowhere of greater potential importance than in the complex process control environment, where reversible operator actions should be considered. A related option is to allow the operator the opportunity to experiment with the control of variables and predict their influence in a computer-simulated off-line mode. Ideally this mode should have a fast-time model of the plant, so that potential solutions can be tried out to establish their effectiveness (or reveal their disastrous consequences) before they are finally put into the system.

Display Issues   Researchers in the nuclear power control field have continually called attention to the difficulties operators have in maintaining a mental representation of the evolving situation (Bainbridge, 1983; De Keyser, 1988; Goodstein & Rasmussen, 1988; Woods, 1988; Vincente & Rasmussen, 1990). This problem can be most directly remediated by considering improvements in display design over the simple "one variable, one indicator" (1V-1I) mapping shown in Figure 12.1. As we have seen in Figure 12.8, such displays are problematic because the meaning of a given reading on any single indicator depends on the *context* of the other indicators. Also problematic is a centralized CRT display to replace the wide-panel display in Figure 12.1, in which information must be called up to the CRT screen by keyboard data entry. This means of information retrieval denies the operator the continuously available readings that can be accessed by a mere scan of the eyes or a turn of the head (Moray, 1981). Instead of 1V-1I designs or centralized CRT displays, a number of more sophisticated display innovations are available, nine of which follow.

1. *Predictive displays.* The value of predictive displays has been discussed in several contexts. The information they provide is as valuable in diagnosis as it is in control.

2. *Feedback.* Woods (1988) has noted that one of the greatest problems in the process control environment is the absence (or poor quality) of feedback on the effects and implications of a control action, illustrating the principle of *visibility* (discussed in Chapter 10).

3. *Annunciator and alarm information.* A good deal of preliminary information concerning the nature of a failure is *potentially* available in the alarms and annunciators that first indicate its existence. The word *potentially* is emphasized, however, because from the operator's point of view, the information is often essentially uninterpretable. This unfortunate state of affairs occurs because the vast interconnectedness of the modern process control or nuclear power plant often means that one primal failure will drive conditions at other parts of the plant out of their normal operating range so rapidly that within minutes or even seconds scores of annunciator lights or buzzers will create a buzzing-flashing confusion. At one loss-of-coolant incident at a nuclear reactor, more than 500 annunciators changed status within the first minute and more than 800 within the first two minutes (Sheridan, 1981). Operators on the scene at Three Mile Island complained that this rapid growth prevented them from obtaining good information concerning the initial, primal conditions that led to the other, secondary failures (Goodstein, 1981). The problem can be exacerbated by a central master alarm (usually auditory), which will sound whenever a fault appears anywhere in the system. Difficulties can occur when this alarm, often very annoying, is turned off. When this happens, some systems are configured to turn off the causal alarms, or to freeze their state, so that even if the variables return to normal conditions this information is not registered (Goodstein, 1981; Sheridan, 1981). Thus the very annoyance of the master alarm that guarantees failure *detection* induces a form of behavior that may be counterproductive to effective diagnosis.

Several investigators have concluded that the confusing nature of annunciator systems could be greatly reduced by implementing certain basic principles, which would also increase the diagnostic value of the annunciators. The next four categories of innovations appear to be particularly useful: sequencing, grouping and prioritizing, color, and informativeness.

4. *Sequencing.* This procedure allows the operator to recover an accurate picture of a progressive series of annunciators and therefore be better able to deduce the primal fault. A simple variant of sequencing is the concept of a "first-out panel," in which the first alarm to appear in a series is somehow distinctively identified for a prolonged period, for example, by a flashing light (Benel, McCafferty, Neal, & Mallory, 1981). The more complex form of se-

quencing, dependent on computer services, requires a buffer memory that could replay on a schematized display — at a speed chosen by the operator — the sequence of annunciator appearances. Lees (1981) discusses the role of the computer in analyzing the alarm sequence and in using its own intelligent logic to help identify the primal fault. Sequencing is thus another example of how smart displays can assist the fallable records of human memory. This intelligence could also alleviate the problem (illustrated in Figure 12.8) in which the deeper meaning of any given alarm indicator is influenced by the context of other variables.

5. *Grouping and prioritizing.* Of course any computer-driven replay of an event sequence should ideally be organized spatially in such a manner that displays of systems that are closely related functionally are also closely spaced physically (see Figure 12.9b). This recommendation is equally important for the placement of annunciators themselves. Annunciators for functionally related systems should be physically close. In this way, it is likely that a given fault will trigger annunciation of a primal cluster of indicators and lead to a pattern far easier to interpret than a random appearance of lights distributed all across the control panel.

Of course trade-offs may have to be made. The concept of functional proximity, for example, may be defined in several dimensions. The indicators of the same pressure valve on two different reactors are, in one sense, highly similar. Yet in another sense, their similarity is low when compared to that between the valve indicator and the indicator of pressure on its input side. The latter two, belonging to a single system, are more likely to be causally related in a failure and therefore belong to the same fault cluster. The similarity of *priority* is yet another dimension that, as some have argued, should guide annunciator proximity. Certain warnings are of high priority (e.g., potential exposure of the radioactive core in a nuclear reactor or excessive steam pressure approaching explosion point), whereas others are less so. Analysts have argued that one physical dimension of annunciator displays should define priority. Benel, McCafferty, Neal, and Mallory (1981) suggest a matrix array in which the vertical dimension defines priority and the horizontal defines some functional proximity. Alternatively, priority may be defined by color. In summary, in agreement with the general concept of stimulus-central-processing compatibility put forth in Chapter 4, the principles of grouping here assert that the conceptual, cognitive dimensions that define annunciators (functional proximity and urgency) should correspond with the physical dimensions that define their placement and format.

6. *Color.* As discussed in Chapters 2 and 3, color, if properly and consistently used, can aid greatly in interpreting the diagnostic information available from alarms. Yet, as we also discussed in Chapter 3, it is extremely important for display colors to be consonant with conceptual population stereotypes (e.g., red = danger; yellow = caution; green = normal) and for these stereotypes to be consistently adhered to (Osborne, Barsam, & Burgy, 1981). A classic example of the lack of consistency has been pointed out by Osborne, Barsam, and Burgy in process plants, where green indicates the status of an open, flowing

valve and red the status of a closed one. On the one hand, this coding runs contrary to stereotypes in the electronics industry, in which green means "circuit not live" and red means "circuit hot, do not touch." This convention would place alternate coding interpretations side by side where electrical and hydraulic systems are mixed. Furthermore, under normal plant operations some valves will be open and some will be closed. The resulting "Christmas tree" depiction of normal operation may be indistinguishable at a glance from the appearance of a failure. The writers contrast this coding with a higher conceptual level of coding in which green indicates the normal valve position (whether open or closed) and red the abnormal. In this "greenboard" display, abnormalities will thereby show up with great salience as any red indicator in a sea of green.

The complex issues involved in good display design are revealed, however, when it is seen that even this concept has certain limitations: The same valve positions that are normal during one phase of operation — the steady running of a nuclear plant — may be abnormal during alternate phases such as start-up and shutdown. Thus the Christmas-tree pattern will reappear during these transient phases and provide an uninterpretable array that cannot readily distinguish normal from abnormal conditions. This situation is particularly dangerous because start-up and shutdown are periods of the highest cognitive load. The only solution is to incorporate computer logic to make the assignment of color code to valve position not only by the criterion of normality but also by that of the operational phase. With complex systems such as those found in the nuclear industry, this is an extremely challenging undertaking.

7. *Informativeness.* Thompson (1981) suggests that annunciators should be made to supply a fair degree of information, in the formal sense of the word discussed in Chapter 2. We concluded in Chapter 8 that humans can process a small number of information-rich stimuli more efficiently than a large number of stimuli of small information content. This characteristic reinforces the value of more informative, higher-level annunciator systems that can provide a single piece of evidence for a set of symptoms that covary, rather than one annunciator for each. Further economy is gained if the single physical annunciator may be configured to provide more than one bit of information (i.e., present information on the likelihood that a variable is abnormal) (Sorkin & Woods, 1985; Sorkin, Kantowitz, & Kantowitz, 1988).

8. *Display integration.* In Chapter 3 we discussed object display, a method for integrating a large number of separate variables as a single contoured object. This approach has two distinct advantages: (a) It can reduce visual clutter if there are a large number of variables to be displayed (certainly the case within the typical process control room), and (b) it can be configured to form *emergent features*, which are properties of the object's size and shape. Cognitive demands will be reduced if such features can be mapped onto sets of display variables that should be mentally integrated (Jones, Wickens, & Deutsch, 1989).

Figure 12.9 presents an example of one such object display — the safety parameter display developed by Woods, Wise, and Hanes (1981; Woods,

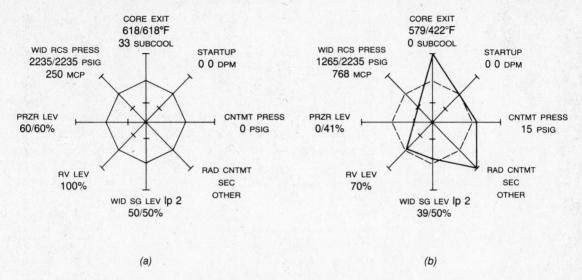

Figure 12.9 Integrated spoke or polar display for monitoring critical safety parameters in nuclear power. (*a*) Normal operation. (*b*) Wide-range iconic display during loss-of-coolant accident. (*Source:* D. D. Woods, J. Wise, and L. Hanes, "An Evaluation of Nuclear Power Plant Safety Parameter Display Systems," *Proceedings of the 25th Annual Meeting of the Human Factors Society* (1981), p. 111. Santa Monica, CA: Human Factors. Copyright © 1981 by the Human Factors Society, Inc. Reproduced by permission.)

O'Brien, & Hanes, 1987). Critical parameters are configured as a polygon. When certain parameters fall beyond their expected tolerance, the systematic octagon shape is distorted, thus facilitating failure *detection*. Furthermore, different failure states are characterized by specific *shapes*—emergent features—thereby facilitating diagnosis. This concept is illustrated by the distinct shape created by a loss-of-coolant accident portrayed in Figure 12.9*b*.

Some indirect empirical support for the concept of the polygon display in system monitoring was provided by Jones, Wickens, and Deutsch (1990) in a simulated process control experiment in which subjects had to integrate five dynamic variables displayed either as a pentagon or as separate bar graphs. As predicted by the proximity compatibility principle (Chapter 3), integration performance was better with the pentagon display. In a different process monitoring study, results also supporting the principle indicated better performance with a separated display when the task called for focused attention on each variable (Casey, 1987).

Figure 12.10*a* presents another example of an object display—the integrated representation of pressure and temperature (Goodstein & Rasmussen, 1988), which describes the trajectory formed as these variables change state within the plant. Here again, the integrated representation produces a shape, an *emergent feature*, that can help the operator understand the momentary system state. Figure 12.10*b* presents a third example of a set of displays whose emergent features are designed to alert the operator of imbalances in energy and temperature (Vicente, Flach, & Sanderson, 1989; Vicente & Rasmussen, 1990).

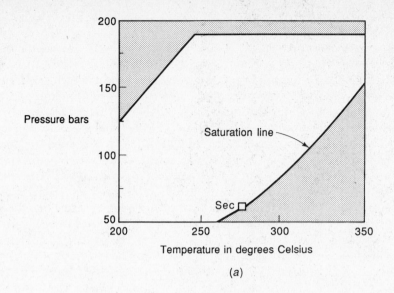

(a)

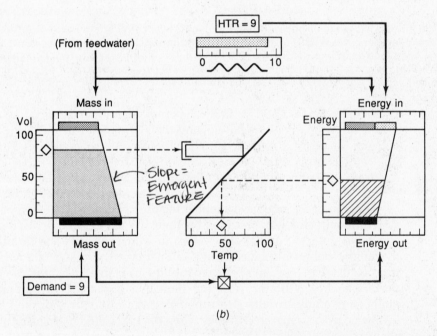

(b)

Figure 12.10 (a) Pressure temperature display. (*Source:* L. P. Goodstein and J. Rasmussen, "Representation of Process State, Structure, and Control," *Le Travail Humain, 51* (1988), pp. 19–37. (b) Integrated display of mass-energy balance designed by Vicente and Rasmussen. The shapes of the shaded boxes provide emergent features, which describe this balance. (*Source:* K. Vincente and J. Rasmussen, "On Applying the Skills, Rules, Knowledge Framework to Interface Design," *Proceedings of the 32nd Annual Meeting of the Human Factors Society* (1988), p. 258. Copyright 1988 by the Human Factors Society, Inc. Reproduced by permission.)

9. *Visual momentum.* A potential danger with any multielement display system is that the operator might become cognitively lost in the wealth of information presented, forgetting what he or she is viewing or how it relates to a display element seen previously. As we discussed in Chapter 4 this state has a direct analogy to the sense of spatial "lostness" that occurs, say, in a city where the position, direction, and shape of the buildings simply don't make sense. Woods (1984; Woods, O'Brien, & Hanes, 1986) described the display concept of *visual momentum* as one that tries to support and preserve the cognitive orientation within a set of displays, so that the relationship between all displays, as well as that between each display and the "big picture," is made clear; that is, a "momentum" between them is preserved. The visual momentum concepts of graceful transitions, anchors, world maps, and consistency were discussed in Chapter 4. Here we reemphasize that the importance of consistency may be illustrated in the design of displays that support diagnostic thinking at different levels of abstraction. We noted that operators should be able to view the state of the plant at different levels (Goodstein & Rasmussen, 1988), sometimes considering abstract variables, like energy flow, and sometimes fairly concrete ones, like water and steam pressure. Goodstein and Rasmussen, along with Vicente and Rasmussen (1990), have pointed out that there are *consistent* elements among displays at all levels of abstraction: *what* a process accomplishes (e.g., a pump moves cold water), *why* that process is carried out in support of purposes at higher levels of abstraction (e.g., to remove heat from a hot source), and *how* that process is accomplished in more concrete levels (e.g., by rotary blades). Visual momentum in making transitions between these displays will be preserved if the what, why, and how information is spatially arranged in a consistent manner across displays (Goodstein, 1981).

Training Successful performance in a control room depends on knowledge of procedures and on a good mental model of the system—both a model of its normally operating state and a model of its failure states, although these two will not be identical (Kragt & Landeweerd, 1980). In the case of normal operations, the model is a set of *expectancies* of what control actions lead to what changes in variables and when. In the case of abnormal operations, the necessary model may be more complex, involving a heavy degree of visualization. At issue, then, is how knowledge of system relations should be trained.

De Keyser (1988) has argued that this knowledge is best acquired through the actual experience of incidents (either real or simulated) which forms kernels of knowledge that eventually constitute knowledge of the plant. In contrast to this on-the-job training, research in this area has often focused on whether explicit training should involve fundamental theory or more specific control heuristics.

Both Crossman and Cooke (1962) and Kragt and Landeweerd (1974) report that an understanding of the general scientific principles underlying the process provides no assistance in the operator's ability to control it. This finding also suggests that an internal model is not a highly general concept. Other studies support the conclusion that instruction in theory does not help performance in these operational process control contexts (Patrick, et al., 1989; Duncan, 1981). In contrast, understanding the *specific* principles underlying the given system to be controlled does assist the operator in process control (Attwood, 1970; Brigham & Liaos, 1975). Landeweerd, Seegers, and Praageman (1981), for example, found that operators with two hours of instruction on the dynamics of a complex chemical process learned to control the process more rapidly than those who received information on the input/output relations for the same duration of study. These findings reinforce the validity of the mental model. They suggest that effective control not only is learned by acquiring a series of stimulus-response mappings but also is based on an understanding of the structure of causal sequencing in the plant.

The importance of learning a causal sequence, based on an internal model for a particular system, suggests that training should *not* be focused only on following rote procedures. As discussed, Woods (1988) points out the brittleness of such procedures and the dangers inherent in following them rigidly without paying attention to whether or not each step is effective. And if a given step is not effective, a well-trained troubleshooter should know why the procedure was intended to be carried out in that context, allowing for the necessary flexibility (Roth & Woods, 1988; Woods, 1988).

Organizational Structure   In Chapter 10 we described how the nuclear power plant typifies the complex, multioperator system that serves as a breeding ground for *resident pathogens* (Reason, 1990). These pathogens are latent error opportunities, which are embedded in system complexity, procedures, training inadequacies, organizational structure or management policies, and attitudes (e.g., attitudes toward safety versus economy and production). A recent review of human factors in nuclear process control rooms highlighted the importance of organizational factors over specific human factors in influencing overall plant safety (Moray & Huey, 1988). Two specific examples will be briefly mentioned.

1. *Communications.* In Chapter 5, we discussed briefly the human factors of communications, focusing heavily on aviation systems. The coordination of process control plants is also very much of a team operation, and smooth team performance and communications can be critical for success. Roth and Woods (1988) have highlighted this feature in their analysis of expert performance in the steam generator start-up phase. Similarly, it is easy to understand how the confirmation bias, identified as a contributing cause to the Three Mile Island disaster, might have been reduced by better communications between operators. A pair of "fresh eyes" might have interpreted the symptoms differently (and correctly), thereby alleviating the chain of incorrect actions, which was based on the mistaken belief that reactor pressure was too high.

2. *Conflicting goals.* As noted, most process industries operate under somewhat conflicting goals of production and profits versus safety. The relative priorities between these two objectives can be clearly stated by the management of a company (or the policy of a regulatory agency), or they may remain ambiguous. But their impact, whether established from above or chosen by the

control-room operators, will directly influence performance. Heavy emphasis on production in a steam-generating plant, for example, may lead to strategies that will proceed through the costly and time-consuming process of bringing a reactor on-line in the most rapid fashion possible, thereby ignoring the go-slow strategy and shortcutting safety (Roth & Woods, 1988). Within the framework of signal detection theory and risky decision making, these attitudes may also influence the detection criterion an operator adopts in categorizing a warning signal as a failure rather than as a machine false alarm. General attitudes toward safety will of course dictate the extent to which allowable procedures are followed or violated, the Chernobyl accident in the Soviet Union illustrating a case in which such procedures were intentionally violated (Reason, 1990).

Finally, a critical organization issue — and one that is sometimes advocated as a solution to many of the human performance problems in process control — is the continued development and introduction of higher and higher levels of control room automation (Bainbridge, 1983). However, because the issues of automation transcend the process control environment, we shall address them in a broader context later.

Conclusion

Human factors research on complex processes, both their control and their fault management and diagnosis, is a relatively young domain. Many of the recommendations concerning display layout, anthropometry, control location, lighting and legibility, and clearly written procedures translate directly from the general practice of good human factors (Bailey, 1989; McCormick & Sanders, 1988; Salvendy, 1987), and this approach has been implemented with some degree of success. Yet it is equally clear that great improvements are possible by combining theories of cognitive psychology with increasingly available computer and display technology. This area of cognitive ergonomics is one in which relatively little engineering psychology research has been conducted to assess the utility of the recommendations made by such researchers as Rasmussen and Woods. The potential payoffs of this research appear to be quite high, in light of the fact that systems are evolving toward greater complexity with increasing levels of automation.

OTHER COMPLEX SYSTEMS

Although we have focused on the process control industry, many other complex systems, requiring human supervision, have received considerably less attention from human factors researchers. There is, for example, ship control. Most ships share the long time constants and sluggish responses typical of many process variables, and as discussed in Chapter 11, they too can greatly benefit from predictive displays. The analysis of many ship accidents shows a great deal of complexity between interacting systems that may characterize both the root cause of the accident and the efforts of supervisory personnel to deal with its immediate consequences—fire, loss of buoyancy, loss of power, and so forth (Wagenaar & Groeneweg, 1988).

Another critical control task is that exerted by the air traffic controller, who must manage the flight paths of numerous aircraft to keep them spatially separated (Hopkin, 1988; Lenorovitz & Phillips, 1987). The task of the controller, in times of high work load, imposes a tremendous cognitive load because of the complexity of the spatial interactions involved among the multiple elements. Furthermore, it also imposes a need for prediction, planning, and anticipation so that impending collision courses can be dealt with before they have become crises. As one might guess then, these demands make relevant the concerns about display design. Whitefield, Ball, and Ord (1980) have advocated predictive displays in which the controller may "try out" different solutions to a developing conflict before issuing the actual command for implementation. Automation issues are relevant as well. On the one hand, expert systems may be available as "consultants" to recommend optimal conflict resolution. On the other hand, the development of the automated digital data link between the air traffic controller and the aircraft may remove many of the voice communication responsibilities (Lee, 1989).

Finally, a host of human factors issues are of importance in the general domain of *supervisory control* (Sheridan, 1987; Sheridan & Hennessy, 1984) in which the human operator may be responsible for the control of automated robots engaged in tasks such as remote exploration in hazardous environments or manufacturing (Sanderson, 1989; Sharit, 1985). The relatively small amount of human factors research in the important domain of manufacturing supervision has focused primarily on two general thrusts (Sanderson, 1989). First, the task of supervisory manufacturing (whether by robots or people) involves a very heavy burden on planning and scheduling. Will machine X be available for task A before machine Y? How long will machine Z take to perform task B? These scheduling, planning, and prediction questions become particularly taxing when machines are *flexible* (capable of performing more than one different task) and when task completion times themselves are variable and uncertain. Naturally an important human factors response to these challenges has been the development of various forms of predictive displays (Liaos, 1978; Sharit, 1985; Smith & Crabtree, 1975).

A second thrust has focused on the "John Henry" question: What is the relative superiority of the human versus the automated scheduler at optimizing production through scheduling? (See Sanderson, 1989, for a good review.) It is not altogether clear that such contests by themselves are meaningful since the "winner" is so dependent on the skills of the automation programmer and on the evolution of expert systems technology. However, the studies are important to the extent that they illustrate the cognitive strengths and weaknesses of both the human and the automated scheduler. Such qualitative comparisons then provide important data about the general issues of automation: When is it useful? When is it not? And what problems may it create?

AUTOMATION

The call for increased automation of complex systems has been a natural response to instances in which human error has contributed to disasters or near disasters. One philosophy of automation was articulated in the early 1950s by Birmingham and Taylor (1954), who proposed automation in manual control and stated that "speaking mathematically he [man] is best when doing least." The trend toward automation has been encouraged both by the increasing availability of sophisticated technology and by the fact that in many situations, particularly in aviation, increasing automation leads to more economical, fuel-efficient operation (Lerner, 1983; Wiener, 1988; Wiener & Curry, 1981) and may reduce workload in some situations (Hughes, 1990). It is also a trend that may ultimately reduce personnel costs, as operators—often expensive to train and employ—are replaced by automated components. Despite the apparent attractions of automation, however, there are a number of less obvious shortcomings.

Examples and Purposes of Automation

Automation varies from that which totally replaces the human operator by computer or machine to computer-driven aids that help an overloaded operator. The different purposes of automation may be assigned to three general categories.

1. *Performing functions that the human operator cannot perform because of inherent limitations.* This category describes many of the complex mathematical operations performed by computers (e.g., those involved in statistical analysis). In the realm of dynamic systems, examples include control guidance in a manned booster rocket, in which the time delay of a human operator would cause instability (see Chapter 11); aspects of control in complex nuclear reactions, in which the dynamic processes are too complex for the human operator to respond to on-line; or robots for the manipulation of materials in hazardous or toxic environments. In these and similar circumstances, automation appears to be essential and unavoidable, whatever its costs.

2. *Performing functions that the human operator can do but performs poorly or at the cost of high workload.* Examples include the autopilots that control many aspects of flight on commercial aircraft (Lerner, 1983) and the automation of certain complex monitoring functions, such as the ground proximity warning systems (GPWS) (Danaher, 1980; Wiener & Curry, 1981). For pilots, the GPWS integrates information about altitude, location, sink rate, and other factors and decides whether the aircraft is approaching dangerously close to the ground in the wrong circumstances. If it is, an auditory alert is sounded.

The Traffic Alert and Collision Avoidance System (TCAS) is an automated function that commands pilots to make particular maneuvers to avoid a threatened midair collision (Chappell, 1989, 1990). Many recent efforts in automation have been directed toward automating diagnosis and decision processes in such areas as medicine (Shortliffe, 1983), nuclear process control (Woods & Roth, 1988), and air combat (Rouse, Geddes, & Curry, 1987). These approaches generally involve the implementation of computer-based artificial intelligence or expert systems (Madni, 1988).

3. *Augmenting or assisting performance in areas in which humans show limitations.* This category is similar to the preceding one, but automation is intended not as a replacement for integral aspects of the task but as an aid to peripheral tasks necessary to accomplish the main task. As we have seen, there are major bottlenecks in performance, because of limitations in human working memory and prediction/anticipation, for which automation would be useful. An automated display, or visual echo, of auditory messages is one such example, as discussed in Chapter 6. In aviation this might involve a speech recognition device that interprets a command received from air traffic control and prints the command on a visual display. This procedure would eliminate the need for the pilot to rehearse the information in working memory until it is entered. It also would relieve the operator of the requirement to jot down such information manually. In the development of the new air traffic control system, the notion of an automated, digital data link between traffic control and the aircraft makes such a concept quite feasible (Lee, 1989).

A second example is the identification or tagging of aircraft symbols with their flight number, altitude, and other factors on computer-generated air traffic control displays (Danaher, 1980). A reduction of working memory load and/or manual writing is accomplished by this procedure. Another example is a computer-displayed "scratch pad" of the output of diagnostic tests in fault diagnosis of the chemical, nuclear, or process control industries. As suggested in Chapter 7, this procedure would greatly reduce memory load. Computer-displayed checklists of required procedures, coupled with a voice-recognition system, would allow operators to indicate orally when each procedure had been completed. The automated display could then check off each step accordingly.

As noted several times throughout this book, any sort of predictive display that would off-load the human's cognitive burden of making predictions would be of great use. Yet another example of an automated aid is the display "decluttering" option, which can remove unnecessary detail from an electronic display when it is not needed, thereby facilitating the process of focused and selective attention (Stokes, Wickens, & Kite, 1990).

Recent advances in computer technology have led to a state in which all three levels of automation are often dependent on some degree of computer-based *artificial intelligence*. Thus, it is appropriate to address briefly the nature of artificial intelligence (**AI**) and one particular domain of AI, *expert systems*.

Artificial Intelligence and Expert Systems

The study of **AI** is devoted to developing computer programs that will mimic the product of intelligent human problem solving, perception, and thought.

Historically, two approaches have been taken to achieve this objective. One has been based on the symbolic, logical reasoning capabilities of the computer, programming large numbers of rules, which typically take the form of if-then statements. We might imagine an **AI** aid for failure troubleshooting that would proceed with this logical reasoning by mapping symptoms (if) to failure states (then). An alternative approach, somewhat more recent, has tried to employ computers to process information more in the fashion of the human nervous system, in which large numbers of neurons operate essentially in parallel and whose collective activations are more analog than digital. This approach has been alternatively described by the label *neural networks, parallel distributed processing*, or *connectionism*, the last term because of the importance of vast numbers of connections among the various computer-simulated "neuronal units" (Rumelhart & McClelland, 1986). The reader is referred to an excellent volume of readings by Graubard (1988) about the debate between the two approaches.

In spite of their recent successes, both approaches also encounter a number of limitations when compared with the product of human thought. Foremost among these is the overwhelming *size* of the knowledge base (the store of facts and procedures in long-term memory) that humans have available to solve problems. For this reason, AI techniques have enjoyed relatively greater success as they attempt to tackle problems in a very restricted domain, such as petroleum drilling, recommending bank loans, or diagnosing infectious diseases (Shortliffe, 1983). Systems of this sort, imitating the behavior of a domain expert, are called *expert systems* and typically operate in one of three fashions: They may replace human performance entirely; they may act as consultants to a human operator, providing advice as needed or requested; or they may be used to train the novice in a domain, providing instructions based on the system's stored knowledge (Madni, 1988).

In his review of expert systems, Madni (1988) identifies the human factors concerns during four critical stages of development and use. The first of these is *knowledge acquisition*. Since the expert system is designed to imitate expert behavior, a challenge is how to extract that knowledge from the human expert so that it may be programmed into computer-based rules (Chignell & Peterson, 1988). Two human factors problems are encountered here: first, experts cannot always verbalize all the knowledge that guides their thinking (Bainbridge, 1979; Hayes & Broadbent, 1988), and so alternative ways to knowledge elicitation must be explored. Second, experts' knowledge, when it is verbally expressed, may be subject to many of the sorts of biases and heuristics discussed in Chapter 7. Therefore, engineering psychology applications must focus on techniques for "debiasing" this information (Cleaves, 1987).

The second stage concerns the representation of expert knowledge in the data base of the computer. How is this knowledge organized in a way that is compatible with the mental model of the system user, so that the latter may retrieve it and understand it? How do human users then relate the stored concepts and facts to each other? This stage obviously demands an appreciation of the particular needs and capabilities of the user (e.g., Is the user an expert or novice? For what purpose is the system required?). This stage of *knowledge representation* links very closely with the third stage, *knowledge utilization* by the user. Human factors principles must be applied to the computer interface design, an issue that we have touched on repeatedly in this book. What displays and controls should be used? Should the interaction be carried out with natural language or restricted vocabulary? A related issue concerns that of trust. How should the expert system convey to the user its varying degree of confidence in the advice that it provides? The final stage concerns *knowledge maintenance*. Is the system designed so that new knowledge can be easily added or old knowledge modified by program designers or human experts?

At all four stages, there are important design questions, but unfortunately the data base from which their answers can be provided remains sparse. There is no doubt that expert systems will provide, with increasing frequency, useful information in domains as diverse as air combat (Rouse, Geddes, & Hammer, 1990) and bank loans as long as the domain of expertise is restricted and well structured. At the same time the difficulties in creating good systems in even narrowly defined areas highlight the major limitations that will be encountered in the foreseeable future (White, 1990). These difficulties ensure the presence of human operators in complex systems for a long time to come and, hence, forecast the continuing importance of the engineering psychology field. We now turn to a discussion of the specific issues confronted by the introduction of automation into the workplace: its benefits and its costs.

Costs and Benefits of Automation

The potential benefits of automation can be easily demonstrated. In almost every incident, such as Three Mile Island, in which human error has led to a disaster, subsequent analyses conclude, "If only such and such a function had been automated, then the human operator would not have had the opportunity to make the blunder that produced the event." On the other hand, a number of incidents and accidents may be attributed to the direct or indirect effects of automation itself. The problem is that neither examples in which human error is obvious nor those in which automation is a problem can by themselves provide objective evidence of whether automation is, on the whole, a good or a bad thing. This uncertainty is reflected in Figure 12.11, which may be interpreted within the framework of the discussion of program evaluation by Einhorn and Hogarth (1978), presented in Chapter 7. The visible examples of failures fall in the right cells of the figure. However, as described in Chapter 7, an objective assessment of the overall merits of automation must take into account data in the left cells — data that are so routine in nature that they are rarely tabulated. How many times have automated systems or human operators worked successfully? Furthermore, to provide more reliable data in these two cells, one would have to know how automated and manual systems would have performed doing equivalent kinds of jobs. Although it is possible to establish tight control of job equivalency in laboratory investigations, these data are not readily available when assessing real-world performance.

A specific example of the lack of complete data is discussed by Wiener and

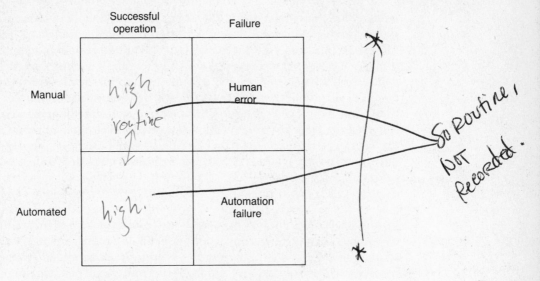

Figure 12.11 Failure in manual and automated systems.

Curry (1981) and pertains to the ground proximity warning system in aviation. The GPWS, it will be recalled, is an alarm system that warns the pilot if the plane is flying too close to the ground or is approaching the ground too rapidly. Pilots have complained about false alarms and other difficulties with the system, thereby in highlighting the failures of automation (Lerner, 1983). Yet it is impossible to document how many potential crashes have been avoided because of the presence of the GPWS in the modern cockpit. Statistics on the reduced accident rate since its mandatory incorporation in 1974, however, suggest that these advantages exist and that the failures of the system should be tolerated. Occasional failures are not sufficient grounds for eliminating the system.

Given the absence of data concerning normal operation of both manual and automated systems, it is therefore somewhat difficult to make empirically based statements that automation is better or worse than manual performance in certain environments, that is, to answer the "John Henry" question. What is possible, however, is to provide a more accurate subjective evaluation of the kinds of problems that will be encountered by automation, thereby emphasizing the guidelines for implementing corrective solutions. Such evaluations still would not allow an explicit statement of whether manual performance could have done better in the equivalent task.

Problems with Automation

Increased Monitoring Load As described in Chapter 2, the automation of functions once carried out by humans moves the human operator to the higher level of a supervisory controller (Sheridan, 1987). The operator now has less to do but, as a consequence of the increased number of system components in the automated system, has many more indicators of component status to monitor. Furthermore, humans do not make terribly effective monitors (Parasuraman, 1987). An intuitive solution that could be adopted is to automate the monitoring function as well. This was the solution implemented when "smart" or centralized alarm systems were incorporated in process control systems (Lees, 1981) or aviation (Hansen et al., 1982). Yet automated monitoring functions also have their problems: They add still more components, which are also subject to failure, and they leave the operator even further removed from the ultimate process under his or her responsibility. This increased distance can be quite harmful to successful control and diagnosis in case a failure occurs (Bainbridge, 1983; Lerner, 1983).

Component Proliferation and System Complexity   Automation has created two related problems here. First, because automated systems, unlike humans, are not limited in their information-processing bandwidth, system designers have capitalized on them to build systems of ever-increasing complexity, with more components (Wohl, 1983). Second, when any single function is automated, it usually increases by at least three the number of "things" that must be monitored and could fail—the function itself, the health of the automated device designed to accomplish that function, and the indicator of that health (e.g., a light bulb that could fail). The proliferation of components means that there is a greater likelihood that something somewhere in the system will fail. This is a consequence of the basic reliability equation discussed in Chapter 10: The probability of any one thing failing is equal to one minus the product of the reliability of all components. Since component reliabilities are less than one, the product will decrease (and one minus the product will increase) as more components are included (Adams, 1982). The proliferation of components requires more displayed elements—some 886 separate annunciators on the Lockheed L–1011 aircraft, for example. This increase in display complexity, particularly if the elements are not carefully configured, will magnify the problems associated with fault diagnosis.

Monitoring performance is also complicated by the increase in complexity brought on by automation: There may be failures of the initial system (like the aircraft engine) or of the automated device that controls the system (like the autopilot) or of the supervisory monitor that monitors the health of both of these systems. To complicate matters further, the monitor could fail in one of two ways: It could fail to indicate a malfunction in the systems that it monitors (a miss), or it could incorrectly signal a failure in a healthy system (a false alarm). As we have noted, these false alarms can lead to a general mistrust of the warning signals, and therefore to the possibility of ignoring some that are genuine.

Two examples illustrate this added complexity. As discussed in Chapter 10, a factor responsible for the fatal crash of a Boeing 737 just taking off from the Detroit airport was the failure of pilots to lower the flaps (NTSB, 1988). Furthermore, an automated aid to remind the pilots to lower the flaps failed to function; and to further complicate matters, the indicator of the health of this system was itself nonfunctioning. Another accident illustrates a different form of monitoring failure: the Eastern Airlines L–1011 crash in the Everglades in 1972 (described in Chapter 3). In this case, a warning light incorrectly indicated the status of the plane's landing gear. That problem diverted the crew's attention from guidance of the aircraft and ultimately led to the disaster. To help address these problems, Wiener and Curry (1981) have argued that annunciator systems should be designed so that the operator is able to check the validity of an alarm.

Trust (Overtrust and Mistrust)   The human operator's trust in an automated device is critical to the smooth functioning of the human-machine system. Muir (1988) speaks of the *calibration* of trust. Automated systems that never fail should be trusted perfectly, totally unreliable devices should be ignored, and those with intermediate levels of reliability should have intermediate levels of trust. Yet there is some evidence that humans do not show this optimally calibrated sense of trust toward automated devices (this shortcoming was also discussed in Chapter 2 in the context of eyewitness testimony). Instead they may flip too rapidly from complete trust to complete distrust once an automated device has shown itself to be fallible.

The sense of overtrust, or the false sense of security that an operator may feel when dealing with an automated device can be illustrated by the tragic crash of the Eastern Airlines L–1011, discussed under Component Proliferation (Danaher, 1980; Wiener, 1977). The flight deck personnel, while in holding pattern, were attempting to diagnose the cause of the malfunctioning landing-gear indicator. However, a major contributing cause of disaster was their unwarranted faith that the autopilot would hold them at altitude while their full attention was diverted to diagnosis. The autopilot did not hold, and the plane gradually descended into the Everglades, with 99 fatalities. Clearly with no autopilot on board, the pilot would have had to continue to fly and hold altitude, delegating the fault diagnosis to other members of the flight deck.

If automated devices are sometimes overtrusted, what then causes the rapid drop in trust when they are seen to be fallible? As Muir (1988) notes, "trust, once betrayed, is hard to recover." With many automated devices, it is quite likely that there is a confirmation bias of the sort discussed in Chapter 7. When the use of automation is optional (as it usually is), the occurrence of an error will lead to sufficient distrust that the operator may cease to use it, thereby depriving the device of the opportunity to demonstrate instances of healthy performance. Both the availability of the failure in memory and the recency of the failure will leave in the operator's mind the representation of a faulty device. This distrust will be likely to exist even if the failure were just as likely to have occurred had a human operator been performing the same task.

Surprisingly, there is little empirical research on human trust in machine automation, although Muir's (1988) model of automation trust is founded on research or trust between people. Her discussion emphasizes the importance of calibrating trust to the level of system reliability and possible ways to do so. A

① increased
monitoring lead

② complexity

③ Trust
overtrust
mistrust

④ out of the
loop —
Long Team

⑤ Higher level
errors —
Set-uperrors
in design.
↑complicated
error is with
the programmer

⑥ Cooperation

study by Moray and Lee (1990) examined operator trust in a simulated auto-mated pasteurizing plant, which was guilty of occasional failures of operation. Validating Muir's assertion, they found that once trust was lost, because of a system failure, it recovered only slowly with repeated failure-free operation. The authors also found, however, that the decision to use an automated system was not based only on the employees' trust of the system. Equally important was their level of *self-confidence* in their own abilities.

As discussed in Chapter 7 in the context of decision aids, this level is often overestimated (Kleinmunt, 1990). A study by Fuld, Liu, and Wickens (1987) illustrates this point. Subjects performed a scheduling task on a computer display in which they were to assign incoming "customers" (boxes of different sizes) to one of three queues that had the shortest wait. The time to process customers through a queue was proportional to the box size, and so the line with the shortest wait was the one with the least total area of waiting boxes. (Imagine choosing the shortest line in a supermarket. Box sizes are analogous to the size of the load in the different grocery carts.) Because the task was difficult—when the lines were very long, it wasn't always easy to judge which line would be shortest—sometimes subjects would make an error. In the manual condition, the subjects' task was both the assignment of customers and the detection of their own errors. In the automation condition, subjects watched an automated scheduler perform the task, again detecting errors. Unknown to the subjects, however, the performance of the "automated de-vice" was actually the subjects' own performance, recorded on an earlier manual trial and then redisplayed. Although the subjects were neither more nor less *sensitive* in detecting errors in the automated condition, they were found to be more apt to detect errors in the "automated scheduler." That is, they detected its errors more often than they detected their own, but they also falsely classified its correct performance as an error more often than their own—an increased false alarm rate.

④ **Out-of-the-Loop Unfamiliarity** When the operator is replaced by an auto-matic controller, the level of interaction or familiarity with the state of the system is reduced. There is some evidence that when a malfunction of the system does occur (e.g., an engine failure), the operator will be slower to detect it and will require a longer time to jump into the loop and exert the appropriate corrective action if he or she is not integrally part of the loop. For example, an aircraft mishap at New York's LaGuardia Airport in 1984, in which a Scandina-vian Airlines jet overshot the runway on landing, was attributed to this prob-lem. The National Transportation and Safety Board accident report blamed the incident on the pilot's lack of awareness of his airspeed, a variable that was under automatic control at the time (NTSB, 1984). In an analogous incident, an Air China Boeing 747 flying over the Pacific suffered a gradual engine failure. The autopilot did such a successful job of compensating for the failure that the pilot was unaware of anything wrong, until catastrophically the autopilot was unable to compensate any more, the plane went into a stall, fell several thou-sand feet, and was only recovered a few seconds above the ocean (*U.S. News &*

World Report, 1989). Bainbridge (1983) discusses similar phenomena when process controllers are monitoring automated processes they had once controlled. The increased latency and reduced accuracy of failure detection when out of the loop was also demonstrated in the laboratory, using both laboratory tracking tasks and more realistic flight simulation, by Bortolusse and Vidulich (1989), Ephrath and Young (1981), Kessel and Wickens (1982), Wickens and Kessel (1979, 1980), and Young (1969). In all of these experiments operators were required to detect changes in the dynamics of a tracking system that was either controlled manually by the operator or by an autopilot. Operators were consistently slower and less accurate in failure detection when they were out of the loop.

The role of system familiarity applies to the immediate, real-time loss of information regarding the momentary state of the system. There also appear to be long-term consequences of being removed from the control loop, evident to the extent that pilots or controllers may lose proficiency as they receive less and less hands-on experience (Bainbridge, 1983). This skill remains one of critical importance as long as the potential remains for them to intervene, for example, in cases of autopilot failure. Loss of familiarity is particularly evident in civil aviation when copilots gain experience flying heavily automated wide-bodied jets (e.g., Boeing 747, DC-10), and then transfer to the less automated narrow-body jets, where they must revive their proficient manual control skills (Wiener, 1988; Wiener & Curry, 1981). A recent survey indicated that half of the pilots of highly automated aircraft, like the Boeing 757 and 767, expressed this concern over degradation of their skills (Hughes, 1989). The fact that familiarity may be gradually lost from out-of-the-loop experience argues strongly for the importance of frequent retraining periods. During these sessions, the operator of complex systems may enter the loop in simulators and experience the dynamic relationship between system variables.

Ⓢ **Higher-Level Operator Error** Automation does not eliminate the possibility of human error; it merely relocates its sources to a different level, as may be seen in several different situations. For example, there are set-up errors in which an automated system is programmed incorrectly prior to its use. Automatic inertial navigation systems that make corrective changes in midflight require pilots to enter the coordinates of these changes into the system prior to takeoff. A mistake in this entry is a potential source of error that may go undiagnosed even at the critical turn point (Wiener & Curry, 1980). It is likely that this sort of error was responsible for the flight path deviation of the Korean Airlines flight KAL–007, a deviation that caused it to stray into Soviet airspace in 1983. The aircraft was shot down by Soviet fighters, with the tragic loss of all passengers (Wiener, 1988).

There are also sources of human error in the manufacture and maintenance of automated equipment. Errors at any of these levels can be disastrous. For example, in the Three Mile Island incident, a pressure relief valve indicator was designed to display to the operator what the valve was commanded to do and not what it actually did. When the two did not correspond, the operators formed the wrong impression of the state of the system and initiated actions that worsened the situation.

As automated systems become more and more complex, greater responsibility often shifts to the computer programmer, who must design programs to foresee possible combinations of events that might require automated responses. Yet as noted in Chapter 1, it is just this sort of creative problem solving in response to low-probability events that benefits from the unique contributions the human operator at the scene has to offer. Programming cannot take into account all possible eventualities. If an automated fault-diagnosis system is not equipped to interpret a certain event that actually does occur (involving, e.g., difficult-to-foresee compound failures), what should be its appropriate response? If it makes a best (but wrong) guess, this choice might be far worse than making no response at all because of the operator's tendency to trust the system. If the operator does believe the system and so maintains an incorrect hypothesis, this trust would thereby induce all of the consequences of cognitive tunnel vision in hypothesis testing described in Chapter 7 (Sheridan, 1981).

A final difficulty is that computer programmers, like human operators of other systems, are themselves fallible. Yet exhaustively checking for bugs in computer programs is incredibly time-consuming, and often it may not be feasible. For example, one computer module containing 100 lines of assembly code was found to contain 28 million possible paths, of which 1/2 million could be followed with valid input data (Beatson, 1989).

The solutions to the fallibility of automated diagnostic systems are not known, although human factors engineers have acknowledged the heavy responsibility that is placed on programmers and system designers as a result (Goodstein, 1981; Lees, 1981). It would seem advisable for computerized systems to err on the side of caution in attaching confidence ratings to their own decision-making or diagnostic performance, particularly when confronting ill-structured problems, where the rigidity of computer-programmed solutions is likely to lead to nonoptimal performance (Buck & Hancock, 1978; Laughery & Drury, 1979).

⑥ Loss of Cooperation Between Human Operators   Automation by definition removes some functions from human operator control. In his excellent treatment of issues in automation, Danaher (1980) points out the sometimes intangible benefits that result from the interaction of two thinking, cooperative human beings in a complex system, in contrast to the impersonal communications between human and computer. In the air transport system, in particular, this level of flexibility and cooperation between air traffic controller and pilot may help to resolve ambiguities that could otherwise prove difficult. This is an important factor that must be considered if an automated data link is set up between pilots and controllers (Lee, 1989).

Conclusion   Given that we have dwelt at length on the problems of automation, it is appropriate to reemphasize that automation does have many tangible benefits. If it is carefully crafted and not overapplied, it can lead to lower work load, safer system operation, and favorable operator opinion. Indeed the majority of pilots questioned in a recent survey said that flight deck automation on the Boeing 757 and 767 had improved the quality of their piloting jobs (Hughes, 1989). Balancing costs and benefits, then, we can safely conclude that automation should be carefully crafted to minimize its problems and should also probably be flexible.

Flexible Automation

It is safe to say that automation is not effective for all tasks for all people at all times. Some tasks do not lend themselves well to being automated, in particular, those decision tasks with fuzzy, ill-defined rules (Harris & Owens, 1984). Different people also differ in their appreciation of automation or their desire to use it. For example, pilots of automated aircraft range widely from those who use autopilots or computer-based flight management systems at every available opportunity to those who use these devices rarely (Hughes, 1989; Wiener & Curry, 1980). Both consumers and office workers also vary widely in their acceptance and reliance on microcomputers. In short, the conditions and level of automation should probably be flexible and not fixed.

There are undoubtedly several factors that influence the preference to use or to avoid automated systems. One is the user's mental model of the automated device (Aretz, Hickox, & Kesler, 1987; Rouse & Morris, 1985). If the device functions in a manner similar to that of the human operator, it is more likely to be used. Closely related are differences in trust and the tolerance for automation error. Some may view an error of an automated device as merely a statistical data point, which is not qualitatively different from a human error. For these users, the calibrated use of the automation (or trust in its reliability) would be lowered just proportionally to its error rate. Others may view the error as a sign of a fundamental flaw and choose then to ignore the automated device entirely. Beyond differences between people, both Fuld, Liu, and Wickens (1987) and Morris and Rouse (1986) have noted that users in general tend to be more confident of their own ability to do tasks than of the device's ability. Hence, an operator will decide to use automation only if the perceived level of automated performance is significantly better than the self-perceived level of the operator's own performance (Kleinmuntz, 1990; Moray & Lee, 1990). This confidence (or overconfidence) in one's own ability may vary widely between people. One way of allowing flexible automation is simply to allow the operator to turn the automation on or off. A more sophisticated approach is through the use of closed-loop adaptive systems.

Closed-Loop Adaptive Systems Adaptive systems are those in which some characteristic of the system changes or adapts, usually in response to measured or inferred characteristics of the human user. The nature of what adapts may be broad and varied. For example, in Chapter 6 we spoke of adaptive training, in which the difficulty of a learning task was changed in response to the inferred skill of the learner. However, we will now focus on systems in which the level of automation is adapted (enhanced or reduced) as a function of the inferred *workload* of the human operator (Rouse, 1988; Wickens & Kramer, 1985). Automation levels for a given operator may vary over time, being introduced to lighten work load when demands are high and withdrawn when demands are lowered. The aircraft pilot, for example, may depend on autopilot control while he or she is trying to diagnose a potential system failure, but not during routine flight.

Because the level of automation changes over time, and this change is normally instigated by a measured or inferred characteristic of the performance that the automation is designed to influence, such systems are often called *closed-loop adaptive systems* (Wickens & Kramer, 1985). In the small number of empirical investigations that have been carried out, such systems have been found to perform at least as well as fully unaided systems, and therefore show considerable promise (Rouse, 1988). A typical example might be found in aviation. We might imagine an intelligent system that senses that the pilot is having difficulty with flight control and infers that it is the result of attention directed to a potential malfunction (of which its sensors are also aware). In response to these inferences, the system immediately assumes autopilot control and takes responsibility for heightened monitoring for (and alerting of) outside traffic. Figure 12.12 provides a framework for discussing several of the important features of such a system (Rouse, 1988).

1. *Who's in charge?* The issue of responsibility for assuming and relinquishing control of a task, the "switch" in the middle of Figure 12.12, is a critical one. One alternative is for human operators to turn the automation on or off as they deem necessary. A second is to leave this decision in the hands of the intelligence of the automated system itself. This option then raises the question of the appropriate decision rule that the automated switch should use

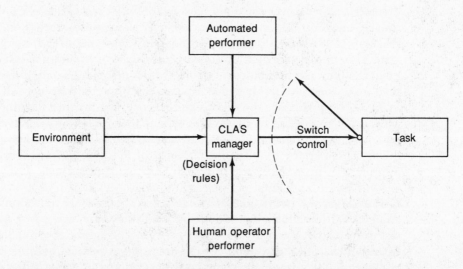

Figure 12.12 Closed-loop adaptive system.

to assume tasks or hand them off to the operator. A third approach is to allow human operators to choose to implement an automated decision rule and also give them the opportunity to set or adjust the decision criterion for implementing automation. Thus, at one time, the human may want to relinquish control only when help is desperately needed. At other times, perhaps when fatigue sets in, a more liberal decision rule will be chosen. A final option is to create a *hysteresis* in the authority: Enable machines to assume responsibility automatically when it infers that the human is overloaded (because humans are often poor estimators of their own performance) and allow humans to reassume control when they feel that their workload has declined (Rouse, 1988).

There are no substantial data to indicate which algorithm is best, although concern has been expressed by researchers that it is important to give the human some level of authority (Rouse, 1988). In this regard, a study by Morris and Rouse (1986) found that performance in an adaptive aiding system was improved if the human operator was in charge of the decision process. This improvement was actually realized in those segments of a task when the aid was *not* on, as if the human's knowledge of his or her own responsibility to delegate performance was itself a source of performance improvement.

2. *Form of communications.* It seems fairly critical to have clear, unambiguous, and explicit communications between human and automation about who is in charge of what task(s), and in particular, about warning of any change in task responsibility. A second aspect relates to the degree of authority with which automation assumes or relinquishes tasks. Rouse and Morris (1986) argue that this should be in the form of recommendations and advice ("suggest that autopilot control be implemented"), rather than assertions ("autopilot control is taken"). Still, the cost of making recommendations rather than assertions is that the former impose an added decision requirement on the operator, who may already be overburdened.

3. *Decision rules.* As shown in Figure 12.12, these rules constitute the heart of any closed-loop adaptive system. In essence, there are three conditions from which a computer can infer the momentary level of the operator's work load. The simplest approach is to base the inference on *external task conditions*. For example, automation can assume that landing an aircraft is more difficult than level flight; flying through turbulence in the clouds is more difficult than flying in clear, still air; and night flight is more difficult than day flight. In each case, detection of the more difficult environmental condition would automatically trigger implementation of some automated aid.

A second approach is to base aid on what Rouse (1988) refers to as *leading indicators* — direct measures of system performance that suggest that performance either has declined or is about to do so. An example is the detection of serious departures from the altitude assigned to an aircraft or of deviations from an approach path. Leading indicators tend to be very specific to a particular task and, furthermore, may not always reveal the causes of the departure. For example, the pilot who departs from an assigned altitude because of a lack of vigilance in low-arousal conditions might need a very different form of aid (an alert) than one who does so because of excessive workload.

The third approach is based on a model of the operator's *attentional state* (Walden & Rouse, 1978, Wickens et al., 1976), which will discriminate overload from underload and whether performance has failed because a primary task has become too difficult or because another activity has diverted attention. Such a model may be based on a carefully crafted integration of measures of system performance (e.g., flight path deviations), operator behavior (e.g., delay in responding to environmental events or flight control activity), and possibly physiological indicators of workload such as heart rate or evoked potentials (see Chapter 9).

4. *Closed-loop instability.* Any closed-loop system, in which a response is made to correct a perceived imbalance or error, may become unstable if the gain (size of correction) is too high, the time delay is too long, or the frequency of required corrections is too high (see Chapter 11). Closed-loop adaptive systems also suffer from these constraining limitations. Imagine that a model-based system infers pilot workload to be high because of, say, a momentary burst of turbulence. To collect sufficient data from the pilot to make a statistically reliable inference that workload really is high, some amount of time will be required, perhaps ten seconds. The system then initiates automation just at the very time that the turbulence has eased, and the operator is no longer burdened. Concerns about instability therefore dictate careful consideration of the kind of dynamic changes in workload that may be encountered and the reliability with which operator parameters can be measured, so that model inferences can be rapidly updated.

Conclusion   The concept of closed-loop adaptive systems is a conceptually attractive approach to human-machine system design, capitalizing on the strengths of human and machine in a dynamic and cooperative fashion. The concept certainly remains in the forefront of the thinking of designers of many highly automated complex systems (Hancock & Chignell, 1989). Yet as we have discussed, there are a large number of issues that must be addressed before viable systems can become effective or even feasible. Most important, these will depend on a continued and better understanding of the fundamentals of human attention, along with fascinating areas of human performance theory that have only recently received interest in the human factors domain—communication, cooperation, and trust.

SUMMARY

The discussion of process control and automation in this final chapter has alluded to every previous chapter in the book. This global characteristic emphasizes an important general point: The best design of the system must take into account human limitations and strengths, but these are not isolated components, as the somewhat arbitrary division of chapters might suggest. The human brain is complex. Perception, memory, attention, and action all interact. A well-designed system must consider the way in which these processes interrelate as well as their limitations. A system that is so designed will never eliminate the possibility of human error. But it can at least remove the system designer as a contributing factor, and that will be a major step.

REFERENCES

Adams, J. A. (1982). Issues in human reliability. *Human Factors, 24,* 1–10.

Aretz, A. J., Hickox, J. C., & Kesler, S. R. (1987). Dynamic function allocation in fighter cockpits. In *Proceedings of the 31st annual meeting of the Human Factors Society* (pp. 414–418). Santa Monica, CA: Human Factors Society.

Attwood, D. D. (1970). The interaction between human and automatic control. In F. Bolam (ed.), *Paper making systems and their control.* London: British Paper and Board Makers Association.

Bailey, R. W. (1989). *Human performance engineering* (2nd ed.). Englewood Cliffs, NJ: Prentice Hall.

Bainbridge, L. (1974). Analysis of verbal protocols from a process control task. In E. Edwards & F. P. Lees (eds.), *The human operator in process control.* London: Taylor & Francis.

Bainbridge, L. (1979). Verbal reports as evidence of the process operator's knowledge. *International Journal of Man-Machine Studies, 11,* 411–436.

Bainbridge, L. (1983). Ironies of automation. *Automatica, 19*(6), 775–779.

Beatson, J. (1989, April 2). Is America ready to "fly by wire"? *Washington Post,* p. C3.

Beishon, R. J. (1966). *A study of some aspects of laboratory and industrial tasks.* Unpublished D.Phil. thesis, University of Oxford, Oxford, Eng.

Beishon, R. J. (1969). An analysis and simulation of an operator's behavior in controlling continuous baking ovens. In F. Bressen & M. deMontmollen (eds.), *The simulation of human behavior.* Paris: Durod.

Benel, D. C. R., McCafferty, D. B., Neal, V., & Mallory, K. M. (1981). Issues in the design of annunciator systems. In R. C. Sugarman (ed.), *Proceedings of the 25th annual meeting of the Human Factors Society.* Santa Monica, CA: Human Factors Society.

Birmingham, H. P., & Taylor, F. V. (1954). *A human engineering approach to the design of man-operated continuous control systems* (Report no. 433). Washington, DC: U.S. Naval Research Laboratory.

Bortolussi, M. R., & Vidulich, M. A. (1989). The benefits and costs of automation in advanced helicopters: An empirical study. In R. S. Jensen (ed.), *Proceedings of the 5th international symposium on aviation psychology* (pp. 594–599). Columbus: Ohio State University, Department of Aviation.

Brigham, F. R., & Liaos, L. (1975). Operator performance in the control of a laboratory process plant. *Ergonomics, 29,* 181–201.

Buck, J. R., & Hancock, W. M. (1978). Manual optimization of ill-structured problems. *International Journal of Man-Machine Studies, 10,* 95–111.

Chappell, S. L. (1989). Avoiding a maneuvering aircraft with TCAS. *Proceedings of the Fifth Symposium on Aviation Psychology.* Columbus: Ohio State University.

Chappell, S. L. (1990). Pilot performance research for TCAS. *Managing the Modern Cockpit: Third Human Error Avoidance Techniques Conference Proceedings* (pp. 51–68). Warrendale, PA: Society of Automotive Engineers.

Chignell, M. H., & Peterson, J. G. (1988). Strategic issues in knowledge engineering. *Human Factors, 30,* 381–394.

Cleaves, D. A. (1987). Cognitive biases and corrective techniques: Proposals for improving elicitation procedures for knowledge-based systems. *International Journal of Man-Machine Systems, 27,* 155–166.

Crossman, E. R. F. W., & Cooke, J. E. (1962). *Manual control of slow response systems.* Paper presented at the International Congress on Human Factors in Electronics, Long Beach, CA.

Danaher, J. W. (1980). Human error in ATC systems operations. *Human Factors, 22,* 535–545.

De Keyser, V. (1988). How can computer-based visual displays aid operators? In E. Hollnagel, G. Mancini, & D. D. Woods (eds.), *Cognitive engineering in complex dynamic worlds* (pp. 15–22). London: Academic Press.

Edwards, E., & Lees, F. P. (1974). *The human operator in process control.* London: Taylor & Francis.

Einhorn, J. H., & Hogarth, R. M. (1978). Confidence in judgment: Persistence of the illusion of validity. *Psychological Review, 85,* 396–416.

Electrical Power Research Institute (1977, March). *Human factors review of nuclear plant design* (Project 501, NP–309-SY). Sunnyvale, CA: Lockheed Missiles and Space Co.

Ephrath, A. R., & Young, L. R. (1981). Monitoring vs. man-in-the-loop detection of aircraft control failures. In J. Rasmussen & W. B. Rouse (eds.), *Human detection and diagnosis of system failures.* New York: Plenum Press.

Fuld, R., Liu, Y., & Wickens, C. D. (1987). *Computer monitoring vs. self monitoring: The impact of automation on error detection.* (Technical Report ARL–87-3/NASA–87-4). Savoy: University of Illinois, Aviation Research Laboratory.

Goldstein, L. P. (1981). Discriminative display support for process operators. In J. Rasmussen & W. B. Rouse (eds.), *Human detection and diagnosis of system failures.* New York: Plenum Press.

Goodstein, L. P., & Rasmussen, J. (1988). Representation of process state, structure and control. *Le Travail Humain, 51,* 19–37.

Graubard S. P. (ed.) (1988). *The artificial intelligence debate.* Cambridge, MA: MIT Press.

Grimm, R. (1976). Autonomous I/O-colour-screen-system for process-control with virtual keyboards adapted to the actual task. In T. B. Sheridan & G. Johannsen (eds.), *Monitoring behavior and supervisory control.* New York: Plenum Press.

Hancock, P. A., & Chignell, M. H. (eds.). (1989). *Intelligent interfaces: Theory, research and design.* North-Holland: Elsevier Science Publishers B.V.

Hanson, D., Boucek, G., Smith, W., Hendrickson, J., Chickos, S., Howison, W., & Berson, B. (1982). Flight operations safety monitoring effects on the crew alerting system. In R. Edwards (ed.), *Proceedings of the 26th annual meeting of the Human Factors Society.* Santa Monica, CA: Human Factors Society.

Harris, S. D., & Owens, J. M. (1984). *Some critical factors that limit the effectiveness of machine intelligence technology in military systems applications* (Special Report 84–2). Pensacola, FL: Naval Aerospace Medical Research Laboratory.

Hayes, N. A., & Broadbent, D. E. (1988). Two modes of learning for interactive tasks. *Cognition, 28,* 249–276.

Hollnagel, E., Mancini, G., & Woods, D. D. (eds.). (1988). *Cognitive engineering in complex dynamic worlds.* London: Academic Press.

Hopkin, V. D. (1988). Air traffic control. In E. L. Wiener & D. C. Nagel (eds.), *Human factors in aviation* (pp. 639–663). San Diego, CA: Academic Press.

Hopkins, C. D., Snyder, H., Price, H. E., Hornick, R., Mackie, R., Smillie, R., & Sugarman, R. C. (1982). *Critical human factor issues in nuclear power regulation and a recommended comprehensive human factors long-range plan* (vols. 1–3; NUREG/CR–2833). Washington, DC: U.S. Nuclear Regulatory Commission.

Hughes, D. (1989, August 7). Glass cockpit study reveals human factors problems. *Aviation Week & Space Technology,* pp. 32–36.

Hughes, D. (1990, October 22). Extensive MD-II automation assists pilots, cuts workload. *Aviation Week & Space Technology,* 34–35.

Ivergard, T. (1989). *Handbook of control room design and ergonomics.* London: Taylor & Francis.

Jones, P. M., Wickens, C. D., & Deutsch, S. J. (1990). The display of multivariate information: An experimental study of an information integration task. *Human Performance, 3*(1), 1–17.

Kessel, C. J., & Wickens, C. D. (1982). The transfer of failure-detection skills between monitoring and controlling dynamic systems. *Human Factors, 24,* 49–60.

Kleinmuntz, B. (1990). Why we still use our heads instead of formulas: Toward an integrative approach. *Psychological Bulletin, 107*(3), 296–310.

Kragt, H., & Landeweerd, J. A. (1974). Mental skills in process control. In E. Edwards & F. P. Lees (eds.), *The human operator in process control.* London: Taylor & Francis.

Laios, L. (1978). Predictive aids for discrete decision tasks with input uncertainty. *IEEE Transactions on Systems, Man, and Cybernetics, SMC-8,* 19–29.

Landeweerd, J. A. (1979). Internal representation of a process fault diagnosis and fault correction. *Ergonomics, 22,* 1343–1351.

Landeweerd, J. A., Seegers, J. J., & Praageman, J. (1981). Effects of instruction, visual imagery, and educational background on process control performance. *Ergonomics, 24,* 133–141.

Laughery, K. R., Jr., & Drury, C. J. (1979). Human performance and strategy in a two-variable optimization task. *Ergonomics, 22,* 1325–1336.

Lee, A. T. (1989). Data link communications in the National Airspace System. *Proceedings of the 33rd annual meeting of the Human Factors Society* (pp. 57–60). Santa Monica, CA: Human Factors Society.

Lees, F. P. (1981). Computer support for diagnostic tasks in the process industries. In J. Rasmussen & W. B. Rouse (eds.), *Human detection and diagnosis of system failures.* New York: Plenum Press.

Lenorovitz, D. R., & Phillips, M. D. (1987). Human factors requirements engineering for air traffic control systems. In G. Salvendy (ed.), *Handbook of human factors* (pp. 1771–1789). New York: Wiley.

Lerner, E. J. (1983). The automated cockpit. *IEEE Spectrum, 20,* 57–62.

McCormick, E. J., & Sanders, M. S. (1987). *Human factors engineering and design.* New York: McGraw-Hill.

McLeod, P. (1976). Control strategies of novice and experienced controllers with a slow response system (a zero-energy nuclear reactor). In T. B. Sheridan & G. Johannsen (eds.), *Monitoring behavior and supervisory control.* New York: Plenum Press.

Madni, A. M. (1988). The role of human factors in expert systems design and acceptance. *Human Factors, 30,* 395–414.

Moray, N. (1981). The role of attention in the detection of errors and the diagnosis of errors in man-machine systems. In J. Rasmussen & W. Rouse (eds.), *Human detection and diagnosis of system failures.* New York: Plenum Press.

Moray, N. P., & Huey, B. M. (eds.). *Human factors research and nuclear safety.* Washington, DC: National Academy Press.

Moray, N., & Lee, J. (1990). *Trust and allocation of function in the control of automatic systems* (EPRL-90-05). Urbana: University of Illinois, Engineering Psychology Research Laboratory.

Moray, N., Lootsteen, P., & Pajak, J. (1986). Acquisition of process control skills. *IEEE Transactions on Systems, Man, and Cybernetics, SMC-16,* 497–504.

Moray, N., & Rotenberg, I. (1988). *Fault management in process control: Eye movement and action* (Technical Report EPRL–88-16). Urbana: University of Illinois, Engineering Psychology Research Laboratory.

Morris, N. M., & Rouse, W. B. (1986). *Adaptive aiding for human-computer control: Experimental studies of dynamic task allocation* (Technical Report AAMRL-TR-86-005). Wright-Patterson AFB, OH: Armstrong Aerospace Medical Research Laboratory.

Muir, B. M. (1988). Trust between humans and machines, and the design of decision aids. In E. Hollnagel, G. Mancini, & D. D. Woods (eds.), *Cognitive engineering in complex dynamic worlds* (pp. 71–83). London: Academic Press.

National Transportation Safety Board. (1984). *Scandinavian airlines system, DC-10-30, John F. Kennedy International Airport, Jamaica, New York, February 28, 1984* (Report NTSB-AAR-84/15). Washington, DC: Author.

National Transportation Safety Board. (1988). *Northwest Airlines, Inc. McDonnell Douglas DC-9-82 N312RC, Detroit Metropolitan Wayne County Airport, Romulus, Michigan, August 16, 1987* (Report No. NTSB-AAR-88-05). Washington, DC: Author.

Osborne, P. D., Barsam, H. F., & Burgy, D. C. (1981). Human factors considerations for implementation of a "green board" concept in an existing "red/green" power plant control room. In R. C. Sugarman (ed.), *Proceedings of the 25th annual meeting of the Human Factors Society.* Santa Monica, CA: Human Factors Society.

Patrick, J., Haines, B., Munley, G., & Wallace, A. (1989). Transfer of fault-finding between simulated chemical plants. *Human Factors, 31,* 503–518.

Petersen, R. J., Banks, W. W., & Gertman, D. I. (1981). Performance based evaluation of graphic displays for nuclear power plant control rooms. *Proceedings of the Conference on Human Factors in Computer Systems.* Gaithersburg, MD.

Queinnec, Y., DeTerssac, G., & Thon, P. (1981). Field study of the activities of process controllers. In H. G. Stassen (ed.), *First European Annual Conference on Human Decision Making and Manual Control*. New York: Plenum Press.

Rasmussen, J. (1983). Skills, rules, and knowledge: Signals, signs and symbols, and other distinctions in human performance models. *IEEE Transactions on Systems, Man, and Cybernetics, SMC-13*, 257–266.

Rasmussen, J. (1986). *Information processing and human-machine interaction: An approach to cognitive engineering*. New York: North Holland.

Rasmussen, J., & Rouse, W. B. (1981). *Human detection and diagnosis of system failures*. New York: Plenum Press.

Reason, J. (1990). *Human error*. Cambridge University Press.

Roth, E. M., & Woods, D. D. (1988). Aiding human performance: I. Cognitive analysis. *Le Travail Humain, 51*, 39–64.

Rouse, W. B. (1988). Adaptive aiding for human/computer control. *Human Factors, 30*(4), 431–443.

Rouse, W. B., Geddes, N. D., & Curry, R. E. (1987). An architecture for intelligent interfaces: Outline of an approach to supporting operators of complex systems. *Human-Computer Interaction, 3*, 87–122.

Rouse, W. B., Geddes, N. D., & Hammer, J. M. (1990, March). Computer-aided fighter pilots. *IEEE Spectrum*, pp. 38–40.

Rouse, W. B., & Morris, N. M. (1986). Understanding and enhancing user acceptance of computer technology. *IEEE Transactions on Systems, Man, and Cybernetics, SMC-16*, 539–549.

Rumelhart, D. E., McClelland, J. L., & the PDP Group. (1986). *Parallel distributed processing, Vol. 1: Foundations*. Cambridge, MA: MIT Press.

Salvendy, G. (ed.). (1987). *Handbook of human factors*. New York: Wiley.

Sanderson, P. M. (1989). The human planning and scheduling role in advanced manufacturing systems: An emerging human factors domain. *Human Factors, 31*, 635–666.

Seminara, J. L., Pack, R. W., Seidenstein, S., & Eckert, S. K. (1980). Human factors engineering enhancement of nuclear power-plant control rooms. *Nuclear Safety, 21*(3), 351–363.

Sharit, J. (1985). Supervisory control of a flexible manufacturing system. *Human Factors, 27*, 47–59.

Sheridan, T. B. (1981). Understanding human error and aiding human diagnostic behavior in nuclear power plants. In J. Rasmussen & W. B. Rouse (eds.), *Human detection and diagnosis of system failures*. New York: Plenum Press.

Sheridan, T. B. (1987). Supervisory control. In G. Salvendy (ed.), *Handbook of human factors* (pp. 1243–1268). New York: Wiley.

Sheridan, T. B., & Hennessy, R. T. (eds.) (1984). *Research and modeling of supervisory control*. New York: Plenum Press.

Sheridan, T. B., & Johannsen, G. (eds.). (1976). *Monitoring behavior and supervisory control*. New York: Plenum Press.

Shortliffe, E. H. (1983). Medical consultation systems: Designing for doctors. In M. E. Sime & M. J. Coombs (eds.), *Designing for human-computer communication* (pp. 209–238). New York: Academic Press.

Smith, H., & Crabtree, R. G. (1975). Interactive planning: A study of computer aiding in the execution of a simulated scheduling task. *International Journal of Man-Machine Studies, 7,* 213–231.

Sorkin, R. D., Kantowitz, B. H., & Kantowitz, S. C. (1988). Likelihood alarm displays. *Human Factors, 30,* 445–460.

Sorkin, R. D., & Woods, D. D. (1985). Systems with human monitors: A signal detection analysis. *Human-Computer Interaction, 1,* 49–75.

Stokes, A. F., Wickens, C. D., & Kite, K. (1990). *Display technology: Human factors concepts.* Warrendale, PA: Society of Automotive Engineers, Inc.

Thompson, D. A. (1981). Commercial aircrew detection of system failures: State of the art and future trends. In J. Rasmussen & W. B. Rouse (eds.), *Human detection and diagnosis of system failures.* New York: Plenum Press.

Umbers, I. G. (1976). *A study of cognitive skills in complex systems.* Unpublished Ph.D. dissertation, Aston, England: University of Aston.

Umbers, I. G. (1979). Models of the process operator. *International Journal of Man-Machine Studies, 11,* 263–284.

U.S. News & World Report. (1989, January). High tech and human error above the clouds.

Vicente, K. J., & Rasmussen, J. (1990). The ecology of human-machine systems II: Mediating "direct perception" in complex work domains. *Ecological Psychology, 2*(3), 207–249.

Wagenaar, W. A., & Groeneweg, J. (1988). Accidents at sea: Multiple causes and impossible consequences. In E. Hollnagel, G. Mancini, & D. D. Woods (eds.), *Cognitive engineering in complex dynamic worlds* (pp. 133–144). London: Academic Press.

West, B., & Clark, J. A. (1974). Operator interaction with a computer controlled distillation column. In E. Edwards & F. P. Lees (eds.), *The human operator in process control.* London: Taylor & Francis.

Whitefield, D., Ball, R., & Ord, G. (1980). Some human factors aspects of computer-aiding concepts for air traffic controllers. *Human Factors, 22,* 569–580.

Wickens, C. D., Isreal, J., McCarthy, G., Gopher, D., & Donchin, E. (1976). The use of event-related-potentials in the enhancement of system performance. *12th Annual Conference on Manual Control.* NASA Technical Memorandum, NASA TM X-73, 170.

Wickens, C. D., & Kessel, C. (1979). The effect of participatory mode and task workload on the detection of dynamic system failures. *IEEE Transactions on Systems, Man, and Cybernetics, 13,* 21–31.

Wickens, C. D., & Kessel, C. (1980). The processing resource demands of failure detection in dynamic systems. *Journal of Experimental Psychology: Human Perception and Performance, 6,* 564–577.

Wickens, C. D., & Kramer, A. (1985). Engineering psychology. *Annual Review of Psychology, 36,* 307–348.

Wiener, E. L. (1988). Cockpit automation. In E. L. Wiener & D. C. Nagel (eds.), *Human factors in aviation* (pp. 433–461). San Diego, CA: Academic Press.

Wiener, E. L. (1977). Controlled flight into terrain accidents: System-induced errors. *Human Factors, 19,* 171.

Wiener, E. L., & Curry, R. E. (1980). Flight deck automation: Promises and problems. *Ergonomics, 23,* 995–1012.

Williams, A. R., Seidenstein, S., & Goddard, C. J. (1980). Human factors survey of electrical power control centers. In G. Corrick, E. Haseltine, & R. Durst (eds.), *Proceedings of the 24th annual meeting of the Human Factors Society.* Santa Monica, CA: Human Factors Society.

Wohl, J. (1983). Cognitive capability versus system complexity in electronic maintenance. *IEEE Transactions on Systems, Man, and Cybernetics, 13,* 624–626.

Woods, D. D. (1984). Visual momentum: A concept to improve the coupling of person and computer. *International Journal of Man-Machine Studies, 21,* 229–244.

Woods, D. D. (1988). Commentary: Cognitive engineering in complex and dynamic worlds. In E. Hollnagel, G. Mancini, & D. D. Woods (eds.), *Cognitive engineering in complex dynamic worlds* (pp. 115–129). London: Academic Press.

Woods, D. D., O'Brien, J. F., & Hanes, L. F. (1987). Human factors challenges in process control: The case of nuclear power plants. In G. Salvendy (ed.), *Handbook of human factors* (pp. 1724–1770). New York: Wiley.

Woods, D. D., & Roth, E. (1988). Aiding human performance: II. From cognitive analysis to support systems. *Le Travail Humain, 51,* 139–172.

Woods, D., Wise, J., & Hanes, L. (1981). An evaluation of nuclear power plant safety parameter display systems. In R. C. Sugarman (ed.), *Proceedings of the 25th annual meeting of the Human Factors Society.* Santa Monica, CA: Human Factors Society.

Young, L. R. A. (1969). On adaptive manual control. *Ergonomics, 12,* 635–657.

Zach, S. E. (1980). Control room operating procedures: Content and format. In G. E. Corrick, E. C. Haseltine, & R. T. Durst, Jr. (eds.), *Proceedings of the 24th annual meeting of the Human Factors Society.* Santa Monica, CA: Human Factors Society.

Index

ISBN 0-673-46161-0